ESSENTIALS
OF APPLIED PHYSICS

ESSENTIALS
OF APPLIED PHYSICS

John E. Betts

A RESTON BOOK
PRENTICE HALL
Upper Saddle River, New Jersey 07458

Library of Congress Cataloging-in-Publication Data

Betts, John E.
    Essentials of applied physics / John Betts.
      p. cm.
    "A Reston book."
    Includes index.
    ISBN 0-13-284407-9
    1. Physics. I. Title.
QC23.B563 1989
530--dc19
                                                88-15213
                                                    CIP

Editorial/production supervision and
   interior design: *Ellen Denning*
Cover design: *Wanda Lubelska*
Manufacturing buyer: *Robert Anderson*

Portions of this book are taken from *Elements of Applied Physics* by John E. Betts, © 1983 by Prentice Hall, Englewood Cliffs, NJ and *Physics for Technology*, 2nd ed., © 1981 by Prentice Hall, Englewood Cliffs, NJ.

© 1989 by Prentice-Hall, Inc.
A Pearson Education Company
Upper Saddle River, NJ 07458

All rights reserved. No part of this book may be
reproduced, in any form or by any means,
without permission in writing from the publisher.

Printed in the United States of America

10 9 8 7 6 5 4 3 2

ISBN 0-13-284407-9

Prentice-Hall International (UK) Limited, London
Prentice-Hall of Australia Pty. Limited, Sydney
Prentice-Hall Canada Inc., Toronto
Prentice-Hall Hispanoamericana, S.A., Mexico
Prentice-Hall of India Private Limited, New Delhi
Prentice-Hall of Japan, Inc., Tokyo
Pearson Education Asia Pte. Ltd., Singapore
Editora Prentice-Hall do Brasil, Ltda., Rio de Janeiro

CONTENTS

	PREFACE	xiii
CHAPTER 1	**MEASUREMENT AND UNITS**	1

    1-1   Matter 1
    1-2   Scientific Notation 3
    1-3   Measurement and Standards 5
    1-4   International System of Units 7
    1-5   United States Customary System of Units (Engineering Units) 15
           Summary 16
           Questions 17
           Review Problems 17

CHAPTER 2	**ANALYSIS OF EXPERIMENTAL DATA**	19

    2-1   Formulas and Problem Solving 19
    2-2   Ratio and Proportion 28
    2-3   Graphs 31
    2-4   Density 37
    2-5   Trigonometry 41
           Summary 45
           Questions 46
           Review Problems 46

CHAPTER 3 INTRODUCTION TO VECTOR QUANTITIES 50

- 3-1 Vector and Scalar Quantities 50
- 3-2 Distance and Displacement 53
- 3-3 Speed and Velocity 55
- 3-4 Acceleration 58
- 3-5 Collinear Vectors 59
- Summary 61
- Questions 62
- Review Problems 62

CHAPTER 4 STRAIGHT-LINE MOTION 64

- 4-1 Uniformly Accelerated Motion 64
- 4-2 Freely Falling Bodies 70
- Summary 73
- Questions 73
- Review Problems 74

CHAPTER 5 FORCES 76

- 5-1 Newton's Laws of Motion 76
- 5-2 Friction 86
- 5-3 Momentum 89
- Summary 92
- Questions 93
- Review Problems 93

CHAPTER 6 ADDITION OF VECTORS 95

- 6-1 Vector Addition by Scale Diagrams 95
- 6-2 Components of Vectors 101
- 6-3 Vector Addition by Components 105
- 6-4 Vector Subtraction 108
- Summary 109
- Questions 110
- Review Problems 110

CHAPTER 7 EQUILIBRIUM 112

- 7-1 First Equilibrium Condition 112
- 7-2 Concurrent Forces in Equilibrium 114
- 7-3 Moment of Force and Torque 124
- 7-4 Nonconcurrent Forces in Equilibrium 127
- Summary 132
- Questions 132
- Review Problems 133

Contents

CHAPTER 8   **WORK, ENERGY, AND POWER**   135

     8-1   Work   135
     8-2   Energy   139
     8-3   Conservation of Mechanical Energy   144
     8-4   Power   146
           Summary   151
           Questions   152
           Review Problems   152

CHAPTER 9   **CIRCULAR AND ROTARY MOTION**   154

     9-1   Circular Motion   154
     9-2   Measurement of Angles   159
     9-3   Angular Motion   162
     9-4   Rotational Energy and Power   166
           Summary   169
           Questions   170
           Review Problems   170

CHAPTER 10   **SIMPLE MACHINES**   172

     10-1   Mechanical Advantage and Efficiency   173
     10-2   Levers   176
     10-3   Pulleys   180
     10-4   Gears   186
     10-5   Inclined Planes   191
     10-6   Compound Machines   194
           Summary   194
           Questions   195
           Review Problems   195

CHAPTER 11   **FLUIDS**   197

     11-1   Pressure   197
     11-2   Measurement of Pressures in Fluids   203
     11-3   Pascal's Principle   206
     11-4   Buoyancy and Archimedes' Principle   209
     11-5   Fluid Flow   211
           Summary   212
           Questions   213
           Review Problems   213

CHAPTER 12   **HEAT**   215

     12-1   Temperature and Heat   215
     12-2   Heat Transfer   219

12-3	Specific Heat Capacity	224
12-4	Phase Changes	228
12-5	Heat of Combustion	231
	Summary	233
	Questions	234
	Review Problems	234

CHAPTER 13 THERMAL EXPANSION 236

13-1	Linear Expansion	236
13-2	Area Expansion	239
13-3	Volume Expansion	240
13-4	Gas Laws	242
	Summary	247
	Questions	248
	Review Problems	248

CHAPTER 14 VIBRATIONS AND WAVES 250

14-1	Vibrations	250
14-2	Waves	252
14-3	Wavefronts	257
14-4	Interference	261
14-5	Standing Waves	264
	Summary	265
	Questions	266
	Review Problems	266

CHAPTER 15 SOUND 267

15-1	Sources of Sound	267
15-2	Material Medium	272
15-3	Doppler Effect	274
15-4	Human Ear	275
15-5	Reverberation	276
15-6	Reproduction of Sound	277
	Summary	277
	Questions	278
	Review Problems	278

CHAPTER 16 LIGHT 280

16-1	Nature of Light	281
16-2	Light Sources	285
16-3	Illumination of a Surface	289
	Summary	291
	Questions	292
	Review Problems	292

Contents

CHAPTER 17	**REFLECTION AND MIRRORS**		294
	17-1 Laws of Reflection 295		
	17-2 Spherical Mirrors 296		
	17-3 Spherical Aberration and Parabolic Reflectors 304		
	Summary 306		
	Questions 306		
	Review Problems 306		
CHAPTER 18	**REFRACTION AND LENSES**		308
	18-1 Refraction 308		
	18-2 Total Internal Reflection 312		
	18-3 Lenses 315		
	18-4 Simple Optical Instruments 323		
	18-5 Lens Defects 326		
	Summary 328		
	Questions 328		
	Review Problems 329		
CHAPTER 19	**ELECTROSTATICS**		330
	19-1 Electric Charge 330		
	19-2 Charging Processes and Ground 334		
	19-3 Coulomb's Law 338		
	19-4 Potential Difference 341		
	Summary 344		
	Questions 344		
	Review Problems 345		
CHAPTER 20	**DIRECT-CURRENT CIRCUITS**		346
	20-1 Charge and Electric Current 346		
	20-2 Current, Voltage, and Resistance 348		
	20-3 Ohm's Law 355		
	20-4 Resistivity 356		
	20-5 Combinations of Resistances 357		
	Summary 365		
	Questions 365		
	Review Problems 366		
CHAPTER 21	**ELECTRIC ENERGY AND POWER**		367
	21-1 Electric Energy and Power 367		
	21-2 Sources of Electric Energy 370		
	Summary 378		
	Questions 378		
	Review Problems 379		

CHAPTER 22 MAGNETISM 380

- 22-1 Magnetic Properties 380
- 22-2 Magnetic Fields 382
- 22-3 Ferromagnetism 384
- 22-4 Magnetic Forces 385
- 22-5 Magnetic Fields of Currents 386
- 22-6 Magnetic Forces on Currents: Motor Principle 392
- Summary 396
- Questions 396
- Review Problems 397

CHAPTER 23 ELECTROMAGNETIC INDUCTION 398

- 23-1 Electromagnetic Induction 398
- 23-2 Transformers 402
- 23-3 Generators 405
- Summary 409
- Questions 409
- Review Problems 410

CHAPTER 24 ALTERNATING CURRENT 411

- 24-1 Alternating Current 411
- 24-2 Inductive Reactance 413
- 24-3 Capacitive Circuits 414
- 24-4 AC Series Circuits 417
- 24-5 Resonance 423
- 24-6 AC Power Transmission 424
- Summary 430
- Questions 430
- Review Problems 430

APPENDIX A TECHNICAL MATHEMATICS 432

- A-1 Signed Numbers 432
- A-2 Exponents 434
- A-3 Rounding Off and Significant Figures 436
- A-4 Formula Manipulation 438

APPENDIX B CONVERSION FACTORS 441

APPENDIX C DEFINITIONS OF SI BASE UNITS 444

APPENDIX D SI UNITS AND PREFIXES 445

- D-1 Rules for Units 445
- D-2 Rules for Prefixes 446

APPENDIX E	PHYSICAL CONSTANTS	447
APPENDIX F	ANSWERS TO ODD-NUMBERED PROBLEMS	448
	INDEX	457

PREFACE

Physics is the applied science from which most of the engineering technologies have derived. A thorough knowledge of the basic principles of physics will help students to understand and apply many aspects of technology.

This book is intended for use in a one- or two-semester course of applied physics. No previous knowledge of physics is assumed, and the required mathematics is available for review, if necessary. Students will not normally be expected to study all topics in the book. However, a sufficient number of examples and applications in each of the engineering technologies are included to make physics interesting and relevant.

The analysis of data and the relationships between the theoretical and practical applications of physics are stressed. Many practical examples are included.

Since problem solving and the analysis of data are fundamental to science and engineering, a consistent approach is used. It is hoped that this approach will assist students in acquiring skills in these important areas.

The International System of Units (SI) and the United States Customary System of Units (USCS) are introduced and used throughout the text. Many problems are solved using both systems of units simultaneously so that the student can see the differences and relationships between them.

IMPORTANT FEATURES

1. Thorough introduction and proper use of SI and USCS units are given.
2. Numerous worked examples and student exercises are provided.
3. Emphasis is placed on the analysis of data and problem solving. A consistent approach to each is used.
4. Practical applications of physics in the engineering sciences are stressed, making the text relevant to the areas of study.

5. No previous knowledge of physics is assumed. A mathematics review is available if required.
6. Clear and concise descriptions of the physical principals are given, making the book easy to read.
7. Each chapter ends with:
   a. A summary to review the important concepts
   b. Questions to test the theoretical aspects
   c. Numerous students exercises for problem-solving practice
8. Appendices include:
   a. A mathematics review
   b. Conversion tables
   c. Definitions of SI base units
   d. Rules for the use of systems of units
   e. Table of constants
9. Where possible, each chapter is independent, allowing the reader to consider the material in different sequences.
10. Many line diagrams are included.
11. A complete solutions manual is available for instructors.

I would like to express my appreciation to my colleagues Eugene Tong, Russ Pierce, Malcolm Bancroft, Norman Preston, David Knapton, and particularly Robert Chapman for the considerable amount of time that they spent in assisting me in the preparation of the manuscript. My thanks also to the reviewers of the manuscript, Charles Korn of Middlesex County College, Jesse Waters of Fayetteville Technical Institute, and Greg Halfar of Ranken Technical Institute for their helpful suggestions and to Ellen Denning and the production staff at Prentice Hall for their help and cooperation. Finally, I would like to thank my wife, Ursule, and my daughter, Teresa, for their contributions to both the typing and the proofreading.

To the reader: I sincerely hope that you enjoy this book and find it useful. I would be very pleased to receive any comments.

John E. Betts

1
MEASUREMENT AND UNITS

The term **natural science** is used to denote a knowledge of facts and laws that have been obtained systematically by the scientific method. **Physics** is a branch of natural science that is concerned with the study of energy, forces, motion, and matter. The extremely large scope of physics includes mechanics, fluids, waves, sound, light, heat, electricity, magnetism, solids, the atom, and the atomic nucleus.

Before you begin the study of physics it is important to review the necessary mathematical skills outlined in the appendices of this book. Check with your instructor to determine the mathematical topics that you should study.

1-1 MATTER

All matter is composed of one or more pure basic substances called **elements.** Although there are more than 100 known elements (some of which no longer exist naturally), they rarely occur in the pure state. They are usually mixed or chemically combined with other elements.

Atoms are the smallest complete particles of an element. They are very small, not even visible under the most powerful microscope. Each atom has a very small core called the **nucleus,** which is extremely dense. One cubic meter of this material would have a mass of about 200 trillion metric tons (2×10^{17} kilograms). This dense nucleus is surrounded by a cloud of one or more even smaller particles called **electrons,** which orbit at very high speeds. In some respects, an atom resembles a minute solar system (Fig. 1-1). Electrons of the same atom perform their orbits at different distances from the nucleus in much the same way that the planets have different orbits about the sun. Like the solar system, most of an atom is actually empty space. However, whereas the orbits of the planets are relatively fixed, electron orbits are not always the same, and we can only give probabilities that electrons are at given locations. On a relative scale, if the nucleus were the size of a tennis ball, the electrons would be like peas in orbits several kilometers (or miles) wide.

Figure 1-1 Often used as a symbol of the "atomic age," this diagram does *not* accurately represent our present atomic model because of the electron orbits.

In a **mixture** of two or more elements, each element retains its own chemical and physical properties. For example, air is mainly a mixture of the elements oxygen and nitrogen, although other gases are also present.

Atoms frequently combine chemically with other atoms to form **molecules** in order to reduce their total energy. A chemical combination of two or more elements results in a substance called a **compound** which behaves quite differently from its component elements. For example, table salt is a chemical combination of the very reactive metal sodium and the poisonous gas chlorine, yet table salt is a very important, edible compound. A molecule is the smallest particle of a compound.

Matter may exist as a solid, liquid, or gas, depending on its energy. **Gases** and **vapors** are higher-energy states of matter; they flow to take the shape and to occupy the total volume of any container. When a gas loses sufficient energy (in the form of heat or by doing work), it condenses to a liquid. **Liquids** are also able to flow, but they are virtually incompressible. Matter that flows (i.e., a gas or liquid) is called a **fluid.** Liquid molecules are able to move and they flow past each other, but their motion is much slower than that of gases.

Solids are the lowest-energy states of matter. They usually possess rigid structures, they do not flow, and they are relatively difficult to compress. The atoms in a solid cannot move from one place to another, but they do vibrate about fixed positions in the structure. These vibrations are due to thermal energy.

When a solid is heated, the magnitude of these thermal vibrations increases until they "shake apart" the solid structure and the solid melts, becoming a liquid. If heat is applied to the liquid, the molecules increase their energy, and eventually they have sufficient energy to escape through the liquid surface, becoming a vapor or gas. When a liquid boils, vapor molecules form inside the body of the liquid and then bubble to the surface, where they escape. Some molecules escape from a liquid even at temperatures below the boiling point, because one molecule may obtain energy from several others, so that its energy is sufficient to enable it to escape through the surface. This process is called **evaporation,** and since it leaves the other molecules with less average energy, the remaining liquid is cooled.

Properties of Matter

All matter possesses the basic property of **inertia,** which is an opposition to any change in motion; this property is described by the quantity "mass." The **mass** of an object is an indication of the quantity of matter that it possesses; the more massive an object, the greater its inertia. An object has the same mass at any location; for example, its mass is the same on the earth, on the moon, and in free space.

A **force** is an action (such as a push or a pull) that tends to make a stationary object move, or changes the speed or direction of motion of a moving object. One type of force is due to the natural physical phenomenon that every particle of matter attracts all other particles of matter in the universe. This attraction between masses is called the **force of gravity.** The magnitude of the force of gravity between any two objects depends on their separation and their masses. The **weight** of an object is actually due mainly to its gravitational attraction toward a celestial body (such as the earth). Therefore, the weight of an object depends on its location. For example, the weight of an object at the earth's surface is not the same as its weight on the moon because the masses and the diameters of the earth and moon are quite different. Objects actually weigh less on the moon than on the earth. This is why astronauts can carry such large masses on the moon's surface.

In mechanics, some objects are considered as rigid bodies, but in reality, rigid bodies do not exist. All known materials are deformed to some extent by the application of a force. To avoid structural failures, engineers must know the limitations of the materials they use, and they must be able to calculate the magnitudes of any deformations.

A deformation depends on the type of material used and on the nature (magnitude and orientation) of the applied loads. Materials are usually utilized according to their properties and cost. For example, steel girders are relatively strong; therefore, they are often used as the main supports in large structures. Wooden beams are used in smaller structures because they have sufficient strength and are lighter and less expensive than steel girders.

When a material is subjected to repeated varying loads over a long period, it gradually loses its strength. This is known as **fatigue**. It occurs more rapidly if the material has a flaw. Fatigue is a common cause of failure in machinery.

The **hardness** of a material is defined in terms of its ability to resist scratching or penetration by other materials. Most tests for hardness consist of applying a standard load to force a standard object into the material. In the **Brinell hardness test**, standard loads are used to press a hardened steel ball into the material. The area of the resulting indentation is related to the hardness of the material. The **Rockwell hardness test** involves measuring the depth of the penetration when a standard indenter is forced into the material by a standard load.

In some cases, when a material is deformed by the application of a load, it will return to its original shape and volume when the load is removed. This property is called **elasticity**. For example, if we squeeze a tennis ball, we distort the shape, but the ball returns to its round shape when we stop squeezing it. The tennis ball therefore behaves elastically. Many metals behave elastically up to certain limits. This is an important property that allows us to use them in structures and machinery.

Most metals can actually be deformed permanently by the applications of various loads. Metal rods can be drawn into wires by pulling them through a die. This property is called **ductility**. Heating the metal will often improve its ductility. Many metals are also **malleable**; that is, their shapes can be changed by hammering and rolling them. This property allows us to produce sheets of such metals as tin and steel.

In some uses, materials are required to withstand excessive temperatures or temperature variations, resist corrosion, conduct heat or resist heat flow, and to conduct or resist electric currents. The choice of a suitable material is often quite critical.

1-2 SCIENTIFIC NOTATION

Before beginning the study of scientific notation, you must be familiar with the operations of signed numbers and exponents. These topics are reviewed in Appendix A. Check with your instructor to see what is required.

In physics and technology we must frequently deal with both very large and very small numbers. Writing these numbers in the conventional way would be tedious and it would take up a lot of space; therefore, we use a shorter form called **scientific notation**.

Rule: *To express a number in scientific notation it is written as a number between 1 and 10 multiplied by a power of 10.*

That is,

$$\text{original number} = N \times 10^n$$

where N is a number between 1 and 10 and n is the number of positions that the decimal point was moved; n is positive-valued if N was obtained by moving the decimal place to the left, and negative-valued if N is obtained by moving the decimal place to the right. When the decimal place is moved one position to the left we are really dividing by 10, and when it is moved one position to the right we are multiplying by 10.

On many calculators numbers can be expressed in scientific notation by the use of an EE (or EXP) button. For example, to place the number 7.40×10^8 into a calculator, key as follows:*

$$7.4 \quad EE \quad 8$$

It is not necessary to key the zero. This should appear as

$$7.40 \qquad 08$$

on your calculator; there may even be more zeros, depending on the number of digits that are being kept by the calculator.

For negative exponents use the $\pm$ key. For example, to place the number 4.50×10^{-3} in the calculator, key as follows:

$$4.5 \quad EE \quad 3 \pm$$

This should appear as

$$4.50 \qquad -03$$

on the calculator display.

EXAMPLE 1-1

Express the following in correct scientific notation:[†] (a) 5 800 000; (b) 0.000 042; (c) 325×10^7; (d) 82×10^{-4}; (e) 0.0056×10^6; (f) 0.081×10^{-5}.

Solution: To obtain a number N between 1 and 10, the decimal places must be moved as shown by a number of places producing the power of 10. The rules of exponents are also used to simplify the powers of 10 where necessary.
  (a) $5\ 800\ 000 = 5.8 \times 10^6$ In this case the decimal place was moved six positions to the left and the power of 10 is 6.
  (b) $0.000\ 042 = 4.2 \times 10^{-5}$ Decimal place was moved five positions to the right; the power of 10 is -5.
  (c) $325 \times 10^7 = 3.25 \times 10^2 \times 10^7 = 3.25 \times 10^{(2+7)} = 3.25 \times 10^9$
  (d) $82 \times 10^{-4} = 8.2 \times 10^1 \times 10^{-4} = 8.2 \times 10^{-3}$
  (e) $0.0056 \times 10^6 = 5.6 \times 10^{-3} \times 10^6 = 5.6 \times 10^{(-3+6)} = 5.6 \times 10^3$
  (f) $0.081 \times 10^{-5} = 8.1 \times 10^{-2} \times 10^{-5} = 8.1 \times 10^{(-2-5)} = 8.1 \times 10^{-7}$

To convert a number from scientific notation to decimal form, we move the decimal place the number of positions indicated by the power of 10. Zeros must be inserted to maintain the numerical value. Remember, a negative power represents division; therefore, the decimal place is moved to the left if the power is negative valued. The decimal place is moved to the right if the power is positive valued.

EXAMPLE 1-2

Write the following numbers in decimal form: (a) 3.6×10^3; (b) 8.4×10^{-5}; (c) 320×10^{-4}; (d) 0.0082×10^{-5}; (e) 0.041×10^3.

---

*This notation may be different for some calculators. Check the instruction manual.
[†]To simplify the reading of very large and very small numbers, a space is left after every third digit counting from the decimal place, but space is not required when only four digits precede or follow the decimal point.

Solution: (a) $3.6 \times 10^3 = 3600$ The decimal place is moved three places to the right because the power of 10 is +3.
(b) $8.4 \times 10^{-5} = 0.000\,084$ The power is −5; therefore, the decimal place is moved five positions to the left.
(c) $320 \times 10^{-4} = 0.0320$ or 0.032 The decimal place is moved four positions to the left. The zero in front of the 3 is required for this purpose; otherwise, we would be altering the value. The number must then be written to two digits as in the original number.
(d) $0.0082 \times 10^{-5} = 0.000\,000\,082$ Five positions left.
(e) $0.041 \times 10^3 = 41$ Three positions right.

PROBLEMS

1. Write the following in correct scientific notation, keeping the proper number of digits: (a) 93 000 000; (b) 0.000 000 082; (c) 526 000; (d) 4 790 000; (e) 0.0042; (f) 0.000 058 0; (g) 0.003 000; (h) 0.000 000 012; (i) 340×10^3; (j) $736\,000 \times 10^{-8}$; (k) $0.000\,051 \times 10^{-5}$; (l) 1200×10^6; (m) $0.000\,000\,12 \times 10^4$.

2. Write the following numbers in decimal form, keeping the correct number of digits: (a) 2.3×10^3; (b) 5.60×10^5; (c) 4.2×10^{-3}; (d) 8.10×10^{-5}; (e) 45.0×10^4; (f) 360×10^{-2}; (g) 8.9×10^7; (h) 4.20×10^{-7}.

1-3 MEASUREMENT AND STANDARDS

In science, our knowledge is usually acquired by systematic observation and measurement. Theories are then developed in attempts to describe these observations and to predict new results. It is important that we have a set of precisely defined standards for measured values, so that we can transmit our knowledge to others.

Any scientific observation, and the use of the data taken, depends on accurate measurement and on a statement of the uncertainties involved in taking that measurement. The statement of precision is extremely important because it gives the limitations of the data. Scientists must never underestimate the uncertainties in their measurements.

Measurement is the comparison of an unknown quantity with some well-defined *standard* quantity. The measured value is expressed in terms of that standard. For example, before we can measure a distance, we must define a standard of length, such as a meter; this is called a **unit**. We then measure how many times (or what fraction of) that standard will equal the distance. The result is expressed as a multiple or fraction of the meter. A measured value of 5.3 meters implies that the distance is 5.3 times the standard length, the meter.

It is important to remember that *the units must be considered as a part of any statement of a measured value for a physical quantity*. If the units are omitted, the statement is useless. For example, if a length is expressed only as 5.0, it could mean 5.0 meters, 5.0 kilometers, and so on. The statement is not complete without the units!

To make a measurement, we must first choose a suitable instrument, such as a meter stick, thermometer, or chemical balance. These instruments are normally calibrated so that we can read the measured value directly. The accuracy of any measurement is limited by the precision of the instrument used and by our ability to take the readings. As the precision of the measuring instrument increases, more digits are retained in the stated measured value. In practice, we usually retain only one doubtful digit (the last). The stated digits are then called **significant figures.** Thus in the measured value 1.8625 meters there are five significant figures, and the last digit, 5, is uncertain since it was only estimated.

We may actually indicate the precision in terms of the number of digits used. For example, if we wish to measure the length of a tabletop, we could visually estimate it to be 2 meters. Using a stick graduated in tenths of a meter, we might estimate the length to be 1.87 meters (i.e., between 1.8 and 1.9 meters). Finally, using a meter stick graduated in millimeters (one-thousandth of a meter), we may measure the length to be 1.862 meters (i.e., between 1.861 and 1.863 meters). This can also be written as 1.862 ± 0.001 meters since $1.862 + 0.001 = 1.863$ and $1.862 - 0.001 = 1.861$.

Vernier and micrometer calipers are often used to make measurements of small lengths (Fig. 1-2).

In general, a digit is significant when it is a part of a statement indicating precision.

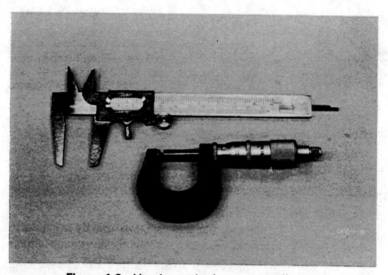

Figure 1-2 Vernier and micrometer calipers.

Rule: *Zeros are not significant when they can be replaced by a power of 10 using scientific notation, since they then indicate "size" rather than precision.*

For example, in 93 000 meters and 0.000 583 meter, the zeros are *not* significant since the values can be written as 9.3×10^4 meters and 5.83×10^{-4} meter. However, the zeros are significant in 4.5080 meters and 9.20×10^8 meters, since these zeros cannot be replaced by a power of 10.

EXAMPLE 1-3

Determine how many significant figures there are in the following measured values: **(a)** 5 609 000 meters; **(b)** 0.000 082 second; **(c)** 0.000 120 meter; **(d)** 32.00 feet; **(e)** 4.500×10^{-4} second; **(f)** 0.50×10^3 seconds.

Solution: **(a)** This value can be written as 5.609×10^6 meters; therefore, there are four significant figures. All other zeros merely indicate size and are represented in the power of 10.

(b) This can be written as 8.2×10^{-5} second; therefore, there are two significant figures.

(c) This is equivalent to 1.20×10^{-4} meter, which contains three significant figures. Note that the zero at the end of the number cannot be written as a part of the power of 10; therefore, it is significant.

(d) In this case there are four significant figures because the value can be written as 3.200×10 feet; both of the zeros are significant because they indicate precision rather than a size.

(e) Four significant figures. The zeros are significant because they cannot be replaced by a power of 10.

(f) This can be written as 5.0×10^2 seconds; the zero after the decimal place is significant because it cannot be replaced by a power of 10. There are therefore two significant figures.

Rounding Off

The use of calculators in science has enabled us to spend more time understanding concepts rather than completing tedious calculations. However, in many cases the solutions produced by calculators contain far too many digits and they must be modified to maintain the correct number of significant figures. This is accomplished by a procedure called **rounding off** and by the use of certain rules for the use of significant figures. These procedures are outlined in Appendix A.

It should be noted that in many common formulas the constants are usually assumed to have as many significant figures as necessary. For example, the volume of a sphere of radius r is given by the formula

$$V = \tfrac{4}{3}\pi r^3$$

The constants 4, 3, and π are assumed to be exact,* and only the precision of the measurement of the radius r limits the number of significant figures.

Finally, there is no reason why a zero should not be included in the readings for the precision. It is often necessary to express the zeros at the end of large numbers as being significant. This is accomplished by the use of scientific notation. We retain zeros at the end of the number after the decimal place if they are significant.

EXAMPLE 1-4

If the measurement 93 000 000 meters is correct to four significant figures, write it in the proper form.

Solution: We must round off the number to four significant figures; this digit is indicated below by the caret:

$$93\ 0\overset{\wedge}{0}0\ 000 \text{ becomes } 9.300 \times 10^7$$

The zeros at the end of the number now cannot be written as a power of 10, and they are therefore significant.

PROBLEMS

3 How many significant figures are there in (a) 320 meters; (b) 0.003 650 second; (c) 2.650×10^{-3} meter; (d) 8.006×10^4 meters; (e) 40.05 meters; (f) 420×10^7 meters?

4 Round off the following to three significant figures: (a) 7 688 000 meters; (b) 0.000 832 6 second; (c) 1.9834×10^4 meters; (d) 4.9973×10^{-6} second; (e) 17.159 meters; (f) 1083 seconds.

1-4 INTERNATIONAL SYSTEM OF UNITS

Physical quantities that belong to the same category (e.g., length) can be compared (e.g., your height and the width of this book). This is accomplished by the measurement process; the result gives a compared value for the physical quantity. It is,

---

*π is the Greek letter pi. In this context it has a constant value of 3.1416 . . .

of course, convenient to define standards for physical quantities that are generally accepted by others. Scientists have found that we can measure and describe all physical quantities using a few basic standards to define units. The result is called a **system of units.**

To define a complete system, we must specify a complete set of **base units** for some chosen fundamental physical quantities. These fundamental physical quantities must be independent of each other. That is, one base unit cannot be expressed in terms of the other base units. However, there must be a sufficient number of base units so that they can describe measured values for all physical quantities. All other physical quantities can therefore be written in terms of a combination of base units; these combinations are called **derived units.** For convenience, some derived units may be given special names and symbols.

The **International System of Units** (abbreviated **SI**) was adopted at the General Conference of Weights and Measures in 1960 and it is recommended for use in all areas of science and technology. Although the United States Customary Units are still used in North America, they have been partially replaced by SI units, and eventually they will be totally replaced by SI. Of course, the conversion is rather expensive and the public must be reeducated in the use of the new units. This will take time.

There are three classes of SI units:

1. **Seven base units** represent seven "fundamental" physical quantities that were chosen because they are independent of each other (one cannot be expressed in terms of the other) and they can be measured very accurately (see Figs. 1-3 and 1-4). Some of the fundamental physical quantities and the corresponding base units are listed in Table 1-1. The definitions of these units are given in Appendix C.

Figure 1-3 Standard kilogram mass. (Courtesy of the National Research Council of Canada.)

Sec. 1-4 / International System of Units

Figure 1-4 Atomic clock. (Courtesy of the National Research Council of Canada.)

2. **Two supplementary units** are used to represent angles and solid angles. These units are also listed in Table 1-1.
3. **Derived units** are then obtained by combining (by multiplication or division) base and supplementary units. There are three types of derived units. Some are written in terms of base units and supplementary units (Table 1-2). Others are given special names for convenience (Table 1-3). Some derived units are also written in terms of the derived units with special names (Table 1-4).

The use of a few other special units is also acceptable even though they are not really SI units; these units, and their symbols and definitions, are listed in Table 1-5. A useful fact to remember is that a 1.0 kg mass of pure liquid water occupies a volume of approximately 1.0 L. A number of rules have been adopted for the general use of SI units. These rules are outlined in Appendix D. They should be studied and used consistently.

TABLE 1-1
SI Base Units

Physical quantity	Quantity symbol	Unit	Unit symbol
Length	l	meter[a]	m
Mass	m	kilogram	kg
Time	t	second	s
Electric current	I	ampere	A
Quantity of a substance	n	mole	mol
Temperature	T	kelvin	K
Luminous intensity	I_L	candela	cd
Supplementary Units			
Angle	θ	radian	rad
Solid angle	Ω	steradian	sr

[a]Also written as metre.

TABLE 1-2
Examples of SI Derived Units Expressed in Terms of Base Units

Physical quantity	Quantity symbol	SI unit description	Unit symbol
Area	A	square meter	m^2
Volume	V	cubic meter	m^3
Speed, velocity	v, **v**	meter per second	m/s
Acceleration	**a**	meter per second squared	m/s^2
Density	ρ	kilogram per cubic meter	kg/m^3
Luminance	L	candela per square meter	cd/m^2
Angular velocity	ω	radian per second	rad/s
Radiant intensity	I	watt per steradian	W/sr

TABLE 1-3
Derived SI Units with Special Names

Physical quantity	Quantity symbol	Unit	Unit symbol	Definition in base units
Force (weight)	**F(w)**	newton	N	$kg \cdot m/s^2$
Pressure	p	pascal	Pa	$kg/(m \cdot s^2)$
Work (energy)	$W(E)$	joule	J	$kg \cdot m^2/s^2$
Power	P	watt	W	$kg \cdot m^2/s^3$
Frequency	f	hertz	Hz	1/s
Electric charge	Q	coulomb	C	$A \cdot s$
Electric potential difference	$V(U)$	volt	V	$kg \cdot m^2/(A \cdot s^3)$
Electric resistance	R	ohm	Ω	$kg \cdot m^2/(A^2 \cdot s^3)$
Electric conductance	G	siemen	S	$A^2 \cdot s^3/(kg \cdot m^2)$
Electric capacitance	C	farad	F	$A^2 \cdot s^4/(kg \cdot m^2)$
Magnetic flux	ϕ_m	weber	Wb	$kg \cdot m^2/(s^2 \cdot A)$
Magnetic flux density (magnetic induction)	**B**	tesla	T	$kg/(A \cdot s^2)$
Inductance	L	henry	H	$kg \cdot m^2/(s^2 \cdot A^2)$
Luminous flux	F_L	lumen	lm	cd/sr
Illuminance	E	lux	lx	$cd/(sr \cdot m^2)$

TABLE 1-4
Examples of Derived SI Units Expressed in Terms of Units with Special Names

Physical quantity	Quantity symbol	Unit	Unit symbol	Expression in base units
Moment of force (torque)	$M_0(\tau)$	newton meter	$N \cdot m$	$m^2 \cdot kg/s^2$
Specific heat capacity	c	joule per kilogram kelvin	$J/(kg \cdot K)$	$m^2/(s^2 \cdot K)$
Thermal conductivity	K	watt per meter kelvin	$W/(m \cdot K)$	$m \cdot kg/(s^3 \cdot K)$
Electric field strength (intensity)	**E**	volt per meter	V/m	$m \cdot kg/(s^3 \cdot A)$
Magnetic field strength (intensity)	**H**	ampere per meter	A/m	A/m

Sec. 1-4 / International System of Units

TABLE 1-5
Units Accepted for Use with SI

Physical quantity	Unit	Unit symbol	Definition
Mass	metric ton	t	1 t = 1000 kg
Time	minute	min	1 min = 60 s
	hour	h	1 h = 3600 s
	day	d	1 d = 86 400 s
	year	a	1 a ≈ 31.56 × 10⁶ s*
Area	hectare	ha	1 ha = 10⁴ m²
Volume	liter	L	1 L = 10⁻³ m³
Temperature	degree Celsius	°C	0°C = 273.15 K: in terms of intervals (or changes) in temperature 1°C = 1 K
Plane angle	degree	°	1° = (π/180) rad
	minute	′	1′ = (π/10 800) rad
	second	″	1″ = (π/648 000) rad

*This depends on the type of year (i.e., calendar, sidereal, or tropical).

Since we frequently multiply or divide physical quantities, we must also multiply and divide the units. Central dots between the symbols are used to indicate the multiplication of different units. Like units may also be squared, cubed, and so on. For example, we write (2 kg)(5 m) = 10 kg·m; also, (2 m)(3 m) = 6 m².

A stroke or negative exponent is used to indicate division of units. For example,

$$\frac{m}{s^2} = m/s^2 \quad \text{or} \quad m \cdot s^{-2}$$

The negative power represents division in exactly the same way, that is,

$$10^{-2} = \frac{1}{10^2}$$

The word "per" also means division.

Prefixes

Certain prefixes are used to designate multiples and submultiples of SI units. The prefix itself is equivalent to some power of 10 (Table 1-6).

Prefixes are written without a space between the prefix symbol and the unit symbol, and the prefix name is combined with the unit name. For example, one kilometer (1 km) is equal to one thousand meters (1000 m or 10^3 m), and one milligram (1 mg) is equivalent to one one-thousandth of a gram (0.001 g or 10^{-3} g). It should be noted that prefixes are considered to be combined directly with the unit for all algebraic manipulations. This is somewhat different from the rules for algebra! For example, in algebraic form,

$$x \cdot y^2 = x \cdot y \cdot y$$

But with SI prefixes and units,

$$1 \text{ mm}^3 = (1 \text{ mm})(1 \text{ mm})(1 \text{ mm}) = (10^{-3} \text{ m})^3 = 10^{-9} \text{ m}^3$$

The prefix "milli" is assumed to be a part of the total unit that is cubed. *You should become very familiar with this very important rule!*

TABLE 1-6

Numerical Prefixes for SI Units

Prefix	Multiplying factor	Symbol
exa	10^{18}	E
peta	10^{15}	P
tera	10^{12}	T
giga	10^{9}	G
mega	10^{6}	M
kilo	10^{3}	k
hecto	10^{2}	h
deka	10^{1}	da
deci	10^{-1}	d
centi	10^{-2}	c
milli	10^{-3}	m
micro	10^{-6}	μ
nano	10^{-9}	n
pico	10^{-12}	p
femto	10^{-15}	f
atto	10^{-18}	a

When a prefix is used to replace scientific notation, the number in front of the unit should be between 0.1 and 1000. For example, 2.5×10^{-4} s can be written as 0.25 ms or 250 μs. However, 1500 km and 0.02 ns should be written as 1.5 Mm and 20 ps, respectively.

A complete set of rules for the use of SI prefixes is given in Appendix D. To convert a number from prefix form into scientific notation, merely replace the prefix by the corresponding power of 10.

EXAMPLE 1-5

Write the following in terms of base units using proper scientific notation: **(a)** 2.3 Ms; **(b)** 45 Mm; **(c)** 560 ns.

Solution: Remember, for scientific notation we increase the power of 10 if the decimal point is moved to the left, and decrease the power if the decimal point is moved to the right.

(a) 2.3 Ms = 2.3×10^6 s, since M represents the power 10^6.
(b) 45 Mm = 45×10^6 m = 4.5×10^7 m.
(c) 560 ns = 560×10^{-9} s = 5.6×10^{-7} s.

To convert a number from scientific notation into proper prefix form, change the power of 10 so that it is exactly divisible by 3 and then use the corresponding prefix.

EXAMPLE 1-6

Write the following in proper prefix form: **(a)** 1.4×10^6 m; **(b)** 2.5×10^4 m; **(c)** 3.1×10^5 s; **(d)** 6.2×10^{-3} g; **(e)** 9.1×10^{-4} s; **(f)** 4.3×10^4 kg.

Solution: **(a)** 10^6 is equivalent to the prefix mega, symbol M; therefore, replace the power of 10 with the prefix directly.

$$1.4 \times 10^6 \text{ m} = 1.4 \text{ Mm}$$

(b) We must rewrite this number so that the power of 10 is divisible by 3. Thus

$$2.5 \times 10^4 \text{ m} = 25 \times 10^3 \text{ m} = 25 \text{ km}$$

Remember, we subtract from the power of 10 when the decimal place is moved to the right, and add when the decimal place is moved to the left.

Sec.1-4 / International System of Units

(c) 3.1×10^5 s = 310×10^3 s = 310 ks,
or 3.1×10^5 s = 0.31×10^6 s = 0.31 Ms. Both answers are acceptable.
(d) 6.2×10^{-3} g = 6.2 mg.
(e) 9.1×10^{-4} s = 0.91×10^{-3} s = 0.91 ms,
or 9.1×10^{-4} s = 910×10^{-6} s = 910 μs.
(f) In this case we must first remove the prefix kilo because combined prefixes are incorrect. 4.3×10^4 kg = $4.3 \times 10^4 \times 10^3$ g = 4.3×10^7 g = 43×10^6 g = 43 Mg.

Conversion of Units

In terms of SI prefixes, the conversion from one unit to another is relatively simple because prefixes represent powers of 10.

Rule: *The conversion factor is merely 10 raised to an exponent equal to the original exponent minus the required exponent.*

We can use direct substitution by applying this conversion factor.

EXAMPLE 1-7

How many nanometers are equivalent to 78 km?

Solution: The prefixes "kilo" and "nano" represent 10^3 and 10^{-9}, respectively. Since we are converting from kilo to nano, we subtract the exponent of nano from that of kilo in order to get the power of 10 for the conversion factor: $3 - (-9) = 12$. This can be seen if the powers are indicated on a number line (Fig. 1-5). Thus

$$1 \text{ km} = 10^{12} \text{ nm}$$

Therefore, 78 km = 78×10^{12} nm = 7.8×10^{13} nm.

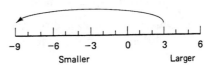

Figure 1-5 Number line.

EXAMPLE 1-8

Convert 7.8×10^8 μs into kiloseconds.

Solution: The prefixes must change from "micro" (10^{-6}) to "kilo" (10^3), and the power of 10 for the conversion factor is $-6 - 3 = -9$. That is,

$$1 \text{ } \mu\text{s} = 10^{-6-3} \text{ ks} = 10^{-9} \text{ ks}$$

Therefore, 7.8×10^8 μs = $7.8 \times 10^8 \times (10^{-9}$ ks$) = 7.8 \times 10^{-1}$ ks or 0.78 ks.

If the units are raised to a power, we must also raise the conversion factor to that same power since the prefix is considered to be a part of the total unit.

EXAMPLE 1-9

How many cubic millimeters are there in 0.8 km^3?

Solution: Milli represents 10^{-3}, and kilo represents 10^3. Therefore, since the power of 10 of the conversion factor between milli and kilo is $3 - (-3) = 6$,

1 km = 10^6 mm. But in this case the units are cubed; therefore, we must cube the conversion factor, numbers, and units.

$$(1 \text{ km})^3 = (10^6 \text{ mm})^3 = 10^{18} \text{ mm}^3$$

Thus 0.8 km^3 = 0.8 × 10^{18} mm^3 = 8 × 10^{17} mm^3.

EXAMPLE 1-10

Convert 2.8 × 10^6 cm^2 into square kilometers.

Solution: The prefixes change from "centi" (10^{-2}) to "kilo" (10^3), and thus the general conversion factor is

$$1 \text{ cm} = 10^{-2-3} \text{ km} = 10^{-5} \text{ km}$$

But in this case the units are squared and we must square the conversion factor before it is applied.

$$1 \text{ cm}^2 = (10^{-5} \text{ km})^2 = 10^{-10} \text{ km}^2$$

Then, by direct substitution,

$$2.8 \times 10^6 \text{ cm}^2 = 2.8 \times 10^6 \times (10^{-10} \text{ km}^2)$$
$$= 2.8 \times 10^{-4} \text{ km}^2$$

Remember, *if prefixes are used, they are raised to the same power as the unit.* It is often necessary to convert units that are not related by prefixes. In such a case the conversion factor is not merely a power of 10 and must be found in a table of conversions. The conversion is again accomplished by substituting the equivalent term in the expression.

EXAMPLE 1-11

Convert 54 km/h into meters per second.

Solution: In this case there are two conversion factors: in the numerator, 1 km = 10^3 m, and in the denominator, 1 h = 3600 s. Thus by direct substitution,

$$54 \frac{\text{km}}{\text{h}} = \frac{54 \times 10^3 \text{ m}}{3600 \text{ s}} = 15 \text{ m/s}$$

EXAMPLE 1-12

Convert 500 m/s into kilometers per hour.

Solution: The conversion factors are: 1 km = 10^3 m, therefore 1 m = 10^{-3} km (numerator), and 1 h = 3600 s, thus 1 s = (1/3600) h (denominator). Thus

$$500 \frac{\text{m}}{\text{s}} = \frac{500 \times 10^{-3} \text{ km}}{(1/3600) \text{ h}} = 1800 \frac{\text{km}}{\text{h}}$$

PROBLEMS

5. What is the area of a rectangular field with a length of 2.0 × 10^2 m and a width of 120 m?
6. What is the volume of a room 3.00 m high, 8.00 m wide, and 12.0 m long?
7. A circular swimming pool has a diameter of 5.50 m and is filled to an average depth of 1.50 m. Determine **(a)** the volume of water; **(b)** the mass of water in the pool.
8. Write the following in terms of SI base units using scientific notation: **(a)** 36 cm; **(b)**

930 km; **(c)** 260 μs; **(d)** 130 ps; **(e)** 120 Gm; **(f)** 86 ms; **(g)** 300 μA; **(h)** 32 Mg; **(i)** 8.0 L; **(j)** 760 mm²; **(k)** 0.94 cm³.

9. Write the following in terms of a unit with a prefix: **(a)** 4.6×10^3 m; **(b)** 2×10^{-6} g; **(c)** 8.3×10^{10} Hz; **(d)** 5.2×10^{-7} s; **(e)** 1.2×10^{-11} F; **(f)** 3.2×10^3 kg; **(g)** 4.2×10^7 N/C; **(h)** 8.6×10^2 V/mm; **(i)** 7.8 g/cm³.
10. How many kilograms are equal to **(a)** 3 Gg; **(b)** 5.0 cg; **(c)** 420 μg?
11. How many centimeters are equal to **(a)** 8.0 μm; **(b)** 5 mm; **(c)** 4 km?
12. Convert the following to prefix form: **(a)** 8.1×10^9 m³; **(b)** 4.0×10^6 m²; **(c)** 2.50×10^{-4} m²; **(d)** 1.96×10^{-10} m³.
13. Express the density 6.8 g/cm³ in SI base units.
14. Express 90 km/h in meters per second.
15. Express 20 m/s in kilometers per hour.
16. A car uses 54.0 L of gasoline to travel 351 km. Determine the gasoline consumption **(a)** in kilometers per liter; **(b)** in liters per 100 km.

1-5 UNITED STATES CUSTOMARY SYSTEM OF UNITS (ENGINEERING UNITS)

The **United States Customary System of Units** (USCS) are still widely used in North America. It will take time and considerable expense to complete the conversion to SI, and the conversion is still widely opposed by many. Some common USCS units are listed in Table 1-7. Many of these units are now defined in terms of SI units. For example,

1. **Length.** The **inch** (in.) is defined as 2.54×10^{-2} m, and a **foot** (ft) is equal to 12 in. or 0.3048 m. 1 **mile** (mi) = 1.609 km.
2. **Weight.** The **pound** (lb) is the weight of a 0.453 592 37 kg mass at sea level and a latitude of 45° on the earth's surface. Remember, weight (or force) units are not equivalent to mass units; a stationary object always has the same mass, but it may have different weights at different locations. **Ounces** (oz) are also frequently used for weights.

$$16 \text{ oz} = 1 \text{ lb}$$

TABLE 1-7
United States Customary Units (USCS) or Engineering Units

Physical quantity	Quantity symbol	Unit	Unit symbol (or definition)
Base Units			
Length	L	foot	ft
Weight (force)	$w(F)$	pound	lb
Time	t	second	s
Derived Units			
Mass	m	slug	lb·s²/ft
Pressure	p	pound per square foot	lb/ft²
Work (energy)	$W(E)$	foot pound	ft·lb
Power	P	horsepower	$1 \text{ hp} = 550 \frac{\text{ft} \cdot \text{lb}}{\text{s}}$

Conversion between SI and USCS Units

Some of the conversion factors between SI and USCS units are listed in Appendix B. The process of direct substitution can also be used to convert between systems of units.

EXAMPLE 1-13

What are the corresponding lengths in SI units of a lumber stud 2.00 in. × 4.00 in. × 8.00 ft?

Solution: The conversion factors are 1 in. = 2.54 cm and 1 ft = 0.3048 m (see Appendix B). Therefore, by direct substitution,

$$2.00 \text{ in.} = 2.00 \times (2.54 \text{ cm}) = 5.08 \text{ cm}$$
$$4.00 \text{ in.} = 4.00 \times (2.54 \text{ cm}) = 10.2 \text{ cm}$$
$$8.00 \text{ ft} = 8.00 \times (0.3048 \text{ m}) = 2.44 \text{ m}$$

EXAMPLE 1-14

Convert 48.0 ft^3 into cubic meters.

Solution: 1 ft = 0.305 m. But in this case the unit is cubed; therefore, we must also cube the *number* and the units for the conversion factor.

$$1 \text{ ft}^3 = (0.305 \text{ m})^3$$

Thus 48.0 ft^3 = 48.0 × (0.305)^3 m^3 = 1.36 m^3.

EXAMPLE 1-15

Convert 2700 kg/m^3 into slug/ft^3.

Solution: The conversion factors are 1 kg = 0.0685 slug and 1 m = 3.281 ft. The last expression must be cubed (both the number and the units) because the original statement involves cubic meters. Thus

$$1 \text{ m}^3 = (3.281 \text{ ft})^3$$

and

$$2700 \text{ kg} = \frac{2700(0.0685 \text{ slug})}{(3.281 \text{ ft})^3} = 5.24 \text{ slug/ft}^3$$

PROBLEMS

17. How many meters are there in 1.00 mi (5280 ft)?
18. How many feet are there in 1.00 km?
19. Convert (a) 1.000 mi (5280 ft) into kilometers; (b) 35 kg into slugs; (c) 120 lb into newtons; (d) 30 ft/s into m/s; (e) 25 m/s^2 into ft/s^2; (f) 4.5 slug/ft^3 into kg/m^3; (g) 120 ft^2 into m^2; (h) 1.2 m^3 into ft^3; (i) 101 Pa (N/m^2) into lb/ft^2 and lb/in^2; (j) 10.0 hp into watts (N · m/s).

SUMMARY

Inertia (opposition to changes in motion) is a fundamental property described by the term *mass*. A *force* is an action such as a push or pull.

To express a number in scientific notation, it is written as a number N between 1 and 10, multiplied by some power n of 10;

$$\text{original number} = N \times 10^n$$

A *measurement* is a process of comparison with a standard. Accepted standards are called *units*. The *International System of Units* (*SI*) has seven base units. *Derived units* are combinations of the base units. *Prefixes* are substitutes for powers of 10.

To convert between prefix values, use direct substitution with the *conversion factor* as the power of 10 raised to an exponent equal to the original exponent minus the required exponent.

United States Customary Units (*USCS*) are still widely used in North America. Direct substitution using the conversion factor allows us to convert between systems of units.

QUESTIONS

1. Define the following terms: (a) inertia; (b) ductile; (c) malleable; (d) elastic material; (e) measurement process; (f) unit; (g) fluid.
2. Describe what happens when a material is heated to the melting and then the boiling points.
3. Describe the structure of atoms.
4. Describe three objects that appear to behave elastically.
5. What are the seven SI base units?
6. Can a mass ever have no weight? Explain.
7. Describe a process in which sheet metal is produced.
8. Explain how metal wires are made.
9. What is meant by the term *hardness*?
10. Can an object ever have no mass or no weight?
11. Discuss the formation of clouds in terms of evaporation.
12. It is very dangerous to open the radiator cap of a car when the motor is very hot. Explain why.
13. We should choose suitable standards for the base units which are universally reproducible, accurately determined, and do not vary. Explain why.
14. The SI unit for mass (the kilogram) uses a prefix and it is based on an artifact. What are the disadvantages of this standard?
15. Why is a mass unit (the kilogram) rather than a force or weight unit chosen as a base unit in SI?

REVIEW PROBLEMS

1. Write the following in correct scientific notation keeping the correct number of significant figures: (a) 46 000 m; (b) 0.000 001 90 s; (c) 5200 m; (d) 0.000 000 420 000 s; (e) 186 000 m; (f) 426×10^6 ft; (g) 523×10^{-8} s; (h) 0.0052×10^{-5} s; (i) $0.008\,20 \times 10^7$ m.
2. Write the following in decimal form: (a) 1.9×10^5 m; (b) 4.20×10^3 m; (c) 9.1×10^{-4} s; (d) 72.0×10^4 s; (e) 420×10^{-7} s; (f) 8.9×10^6 in.
3. Round off the following to three significant figures: (a) 9200 m; (b) 0.000 493 567 s; (c) 4.190×10^7 g; (d) 3.9184×10^{-3} s; (e) 4.399×10^3 s; (b) $2.498\,000 \times 10^{-5}$ s; (g) 0.003 999 s; (h) 999 999 m; (i) 359 900 145 m; (j) 186 000 mi.

4   (a) Determine the volume of a room that has dimensions of 2.4 m × 5.0 m × 8.0 m.
    (b) What is the volume in liters?

5   A rectangular swimming pool is 100 m long, 30 m wide, and is filled to an average depth of 1.5 m. Determine (a) the volume of water in cubic meters and liters; (b) the mass of water in the pool. (*Hint:* One liter of water has a mass of 1 kg)

6   The gasoline tank of a car has a capacity of 55 L. How far can the car travel on a full tank if the average gasoline consumption is (a) 9.2 km/L; (b) 11 L/100 km?

7   Complete the following: (a) 36 m = _____ cm = _____ nm = _____ km; (b) 188 mL = _____ L = _____ m³ = _____ cm³; (c) 120 kg = _____ mg = _____ Mg = _____ g; (d) 30 m³ = _____ mm³ = _____ m³; (e) 9 × 10⁴ mm³ = _____ cm³ = _____ m³.

8   Express the following in SI base units, using scientific notation where appropriate: (a) 36 μm; (b) 240 ns; (c) 800 μA; (d) 18 Gg; (e) 28 mL; (f) 0.560 Gm; (g) 0.38 Mg; (h) 36 mm²; (i) 14.0 ps; (j) 9.3 cm³; (k) 14.0 μm²; (l) 0.250 μm³.

9   Express the following in prefix form: (a) 5.7×10^{-8} s; (b) 1.2×10^4 kg; (c) 3.4×10^{-4} m; (d) 7.2×10^{-2} L; (e) 9×10^{10} Hz; (f) 1.25×10^{-10} F; (g) 3.80×10^{-11} s; (h) 4.9×10^4 cm; (i) 7.8×10^{-5} m²; (j) 3.9×10^8 m³; (k) 7.00×10^{-10} m³.

10 How many 250 mL flasks of water are needed to fill completely an aquarium with inside dimensions of 1.2 m × 15 cm × 40 cm?

11 Convert 57.6 km/h into meters per second.

12 Convert 18.0 m/s into kilometers per hour.

13 Express the following in SI units in the correct form: (a) 2.4 g/cm³; (b) 3.6 μW/mm²; (c) 5×10^4 kV/cm; (d) 26 mJ/g; (e) 36 μA/mm²; (f) 15 mA/mm².

14 Determine the volume of a room that has dimensions of 12.0 ft × 22.0 ft × 8.00 ft.

15 How many cartons 2.0 in. × 4.0 in. × 8.0 in. can be stored in a box with inside dimensions of 2.0 ft × 3.0 ft × 4.0 ft?

16 Convert 60.0 mi/h into feet per second.

17 Convert (a) 1.000 mi into centimeters; (b) 12.0 slugs into kilograms; (c) 8.0 m² into square feet; (d) 25.0 ft³ into cubic meters; (e) 5.00 mi² into square meters; (f) 7880 kg/m³ into slug/ft³.

2
ANALYSIS OF EXPERIMENTAL DATA

Formulas and graphs are frequently used to illustrate relationships between physical quantities and to analyze experimental data. We must therefore be very familiar with formula manipulation and graphical analysis.

2-1 FORMULAS AND PROBLEM SOLVING

An **equation** is a mathematical relationship between two equal quantities; for example, $3 + 2 = 5$ and $2x - 5 = 3$ are both equations.

For convenience we often relate one physical quantity to another by means of an expression called a **formula** in which the physical quantities are usually represented by symbols. We may then determine values for one physical quantity by substituting corresponding values for the other quantities in the formula. For example, the area (quantity symbol A) of a square in terms of the lengths (quantity symbol L) of its sides is determined from the formula

$$A = L \times L = L^2 \tag{2-1}$$

Using this formula, if we are given the length of a side of any square, we can determine its area. For each value of length L there is a single corresponding value for the area A and we say that A is a function of L. Also, since both A and L may assume a number of different values, they are called **variables.** In the formula the value for A depends on the value for L; therefore, we call A the **dependent variable** and L the **independent variable.** However, if the question is rearranged to give

$$L = \sqrt{A}$$

the roles are reversed and L becomes the dependent variable.

Formulas are usually obtained from the analysis of observations of natural phenomena by making a series of experiments under controlled conditions. Some useful formulas involving geometric shapes are listed in Fig. 2-1.

In practice, formulas and equations must often be rearranged in order to solve for some other variable. This process involves the manipulation of algebraic quantities. The basic rules for algebraic manipulation are outlined in Appendix A.

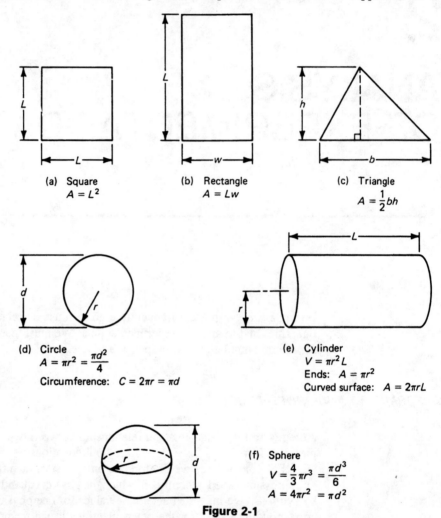

(a) Square
$A = L^2$

(b) Rectangle
$A = Lw$

(c) Triangle
$A = \frac{1}{2}bh$

(d) Circle
$A = \pi r^2 = \frac{\pi d^2}{4}$
Circumference: $C = 2\pi r = \pi d$

(e) Cylinder
$V = \pi r^2 L$
Ends: $A = \pi r^2$
Curved surface: $A = 2\pi r L$

(f) Sphere
$V = \frac{4}{3}\pi r^3 = \frac{\pi d^3}{6}$
$A = 4\pi r^2 = \pi d^2$

Figure 2-1

Problem-Solving Techniques

The solution of problems can often be simplified if a logical approach is used. Occasionally, it is possible to omit some of the steps, but the following procedure may assist in problem solving:

1. **Read.** Read the question carefully.
2. **Sketch.** A very simple sketch of the "system" often aids in the solution. In many cases it may be convenient to omit details and draw only the forces present (see free-body diagrams).
3. **Data.** List all of the given and known quantities in terms of their symbols and values. Include the symbols of the unknown quantities. This information is obtained by carefully examining the question.
4. **Equation(s).** State the equation(s) that will be used. The equation should

Sec. 2-1 / Formulas and Problem Solving

only have one unknown quantity (symbol); all other quantities should be listed in step 3.

5. **Rearrange.** Solve the equation for the symbol of the unknown quantity. The techniques are reviewed in Appendix A.
6. **Substitute.** Insert the values and *units* into the equation and complete the mathematical calculations for both the numbers and the units to find the unknown quantity. Remember, we must always round off the answer to the correct number of significant figures.
7. **Check.** See if the answer is reasonable and that the units appear to be correct.

EXAMPLE 2-1

Find the length of a cylinder that has **(a)** a radius of 1.50 cm if its volume is 65.0 cm³; **(b)** a radius of 1.20 in. if its volume is 55.0 in³.

Solution

Sketch: See Fig. 2-2. (A sketch is really not necessary in this case.)

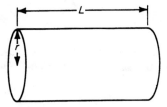

Figure 2-2 Cylinder.

Data:
(a) $V = 65.0$ cm³, $r = 1.50$ cm, $L = ?$ **(b)** $V = 55.0$ in³, $r = 1.20$ in., $L = ?$

Equation: $V = \pi r^2 L$ (see Fig. 2-1)

Rearrange: $L = \dfrac{V}{\pi r^2}$

Substitute:

(a) $L = \dfrac{65.0 \text{ cm}^3}{\pi (1.50 \text{ cm})^2} = 9.20$ cm   **(b)** $L = \dfrac{55.0 \text{ in}^3}{\pi (1.20 \text{ in.})^2} = 12.2$ in.

Check: The final units are both length units, and the answers seem reasonable. The answers must be written with three significant figures.

EXAMPLE 2-2

Determine the area of a piston head which has a diameter of **(a)** 8.6 cm; **(b)** 3.40 in.

Solution

Shape: Circle; a sketch is not required.

Data:
(a) $d = 8.6$ cm, $A = ?$ **(b)** $d = 3.40$ in., $A = ?$

Equation: $A = \dfrac{\pi d^2}{4}$

Substitute:

(a) $A = \dfrac{\pi (8.6 \text{ cm})^2}{4}$ **(b)** $A = \dfrac{\pi (3.40 \text{ in.})^2}{4}$

$= 58$ cm² $= 9.08$ in²

EXAMPLE 2-3

Find the volume of copper in a pipe **(a)** that is 2.00 m long if its inner and outer diameters are 2.500 cm and 2.650 cm, respectively; **(b)** that is 8.00 ft long if its inner and outer diameters are 0.500 in. and 0.550 in., respectively (see Fig. 2-3).

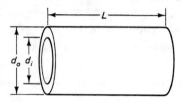

Figure 2-3

Solution

Shapes: Cylinders.

Data:

(a) $d_i = 2.500$ cm, $d_o = 2.650$ cm, $L = 2.00$ m $= 200$ cm, $V = ?$

(b) $d_i = 0.500$ in., $d_o = 0.550$ in., $L = 8.00$ ft $= 96.0$ in., $V = ?$

Equation: $V = \dfrac{\pi d^2 L}{4}$

Rearrange: The volume of copper V is the difference between the volume V_s of a solid rod of the same outer diameter and the volume V_c of the cavity (hole):

$$V = V_s - V_c = \frac{\pi d_o^2 L}{4} - \frac{\pi d_i^2 L}{4}$$

Substitute:

(a) $V = \dfrac{\pi (2.650 \text{ cm})^2 (200 \text{ cm})}{4} - \dfrac{\pi (2.500 \text{ cm})^2 (200 \text{ cm})}{4}$

$= 1103.09 \text{ cm}^3 - 981.75 \text{ cm}^3$

$= 120 \text{ cm}^3$ (note the application of the rules for significant figures)

(b) $V = \dfrac{\pi (0.550 \text{ in.})^2 (96.0 \text{ in.})}{4} - \dfrac{\pi (0.500 \text{ in.})^2 (96.0 \text{ in.})}{4}$

$= 22.81 \text{ in}^3 - 18.85 \text{ in}^3 = 4.0 \text{ in}^3$

The **cross-sectional area** of an object is the area of the surface that would be obtained when it is cut perpendicular to the axis. For example, the cross-sectional area of a pipe that is perpendicular to its length is the area of the surface shown in Fig. 2-4.

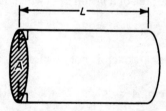

Figure 2-4 The cross-sectional area is perpendicular to the length and $V = AL$.

In general, if the dimensions are uniform, the volume V of any right solid (area perpendicular to the length) is the product of its cross-sectional area A and its length L along the perpendicular axis.

$$V = A \times L \tag{2-2}$$

Sec. 2-1 / Formulas and Problem Solving

The determination of cross-sectional areas is often needed in many branches of technology and science.

EXAMPLE 2-4

Determine the cross-sectional area of a solid cylindrical drive shaft perpendicular to its length if its diameter is (a) 2.5 cm; (b) 1.2 in.

Solution

Shape: Circle.

Data:

(a) $d = 2.5$ cm, $A = ?$ (b) $d = 1.2$ in., $A = ?$

Equation: $A = \dfrac{\pi d^2}{4}$

Substitute:

(a) $A = \dfrac{\pi (2.5 \text{ cm})^2}{4}$ (b) $A = \dfrac{\pi (1.2 \text{ in.})^2}{4}$

    $= 4.91$ cm² or 4.9 cm² $= 1.13$ in² or 1.1 in²

EXAMPLE 2-5

Calculate the cross-sectional area of the steel I-beam shown in Fig. 2-5, using (a) SI units; (b) USCS units.

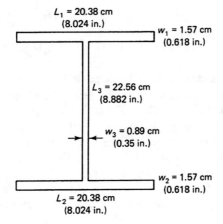

Figure 2-5 Cross section of an I-beam.

Solution

Shapes: Rectangles.

Data:

(a) $L_1 = L_2 = 20.38$ cm,   (b) $L_1 = L_2 = 8.024$ in.,
    $L_3 = 22.56$ cm, $L_3 = 8.882$ in.,
    $w_1 = w_2 = 1.57$ cm, $w_1 = w_2 = 0.618$ in.,
    $w_3 = 0.89$ cm, $A = ?$ $w_3 = 0.35$ in., $A = ?$

Equation: $A = Lw$

Rearrange: Total area $= A_1 + A_2 + A_3 = L_1 w_1 + L_2 w_2 + L_3 w_3$

Substitute:

(a) Area $= (20.38 \text{ cm})(1.57 \text{ cm}) + (20.38 \text{ cm})(1.57 \text{ cm})$
                $+ (22.56 \text{ cm})(0.89 \text{ cm})$
           $= 32.0$ cm² $+ 32.0$ cm² $+ 20$ cm²

Note that the last term has only two significant figures (the zero is significant). Thus

                area $= 84$ cm²   (two significant figures)

(b) Area = (8.024 in.)(0.618 in.) + (8.024 in.)(0.618 in.)
+ (8.882 in.)(0.35 in.)
= 4.96 in² + 4.96 in² + 3.1 in² = 13.0 in²

The area of the exterior surface of an object is called its **surface area.** We live on the surface area of the earth, which is approximately a sphere.

EXAMPLE 2-6

Determine the surface area of the curved surface of a cylinder which has **(a)** a radius of 3.2 cm and a height of 8.0 cm; **(b)** a radius of 1.2 in. and a height of 3.2 in.

Solution

Shape: Cylinder.

Data:
(a) $r = 3.2$ cm, $h = 8.0$ cm, $A = ?$ (b) $r = 1.2$ in., $h = 3.2$ in., $A = ?$

Equation: $A = 2\pi rh$

Substitute:
(a) $A = 2\pi(3.2 \text{ cm})(8.0 \text{ cm})$
    $= 161 \text{ cm}^2$ or 160 cm^2

(b) $A = 2\pi(1.2 \text{ in.})(3.2 \text{ in.})$
    $= 24.1 \text{ in}^2$ or 24 in^2

PROBLEMS

1. Find the area of a rectangle which has sides **(a)** 11.0 cm by 15.0 cm; **(b)** 4.20 in. by 8.90 in.
2. Determine the areas of each of the shapes in Fig. 2-6 **(a)** in SI units; **(b)** in USCS units.
3. Determine the cross-sectional area perpendicular to the length of a rivet which has a diameter of **(a)** 0.675 cm; **(b)** 0.375 in.
4. Determine **(i)** the volume of the cylinder, **(ii)** the cross-sectional area perpendicular to the length, and **(iii)** the total surface area of an automobile engine cylinder which is **(a)** 8.00 cm in diameter and 10.0 cm long; **(b)** 3.20 in. in diameter and 4.00 in. long.
5. Determine the volume of water that can be stored in a circular barrel that is **(a)** 1.20 m high and has a diameter of 0.750 m; **(b)** 3.60 ft high and has a diameter of 2.25 ft.
6. A rectangular city lot measures 25.0 m by 52.0 m. What is the area in **(a)** square meters; **(b)** hectares?
7. Determine the area **(a)** in square feet and **(b)** in acres of a rectangular lot which measures 225 ft by 825 ft.
8. Determine the length of a rectangular city lot that has **(a)** an area of 0.125 ha if its width is 28.0 m; **(b)** an area of 0.250 acres if its width is 75.0 ft.
9. Calculate the cross-sectional area perpendicular to the length of the solid part of a hollow pipe with inner and outer diameters of **(a)** 1.500 cm and 1.575 cm, respectively; **(b)** 0.500 in. and 0.530 in., respectively.
10. What is the cross-sectional area perpendicular to the length of a piston which has a diameter of **(a)** 9.00 cm; **(b)** 3.85 in.?
11. Find the cross-sectional areas of the steel beam sections shown in Fig. 2-7 using **(a)** SI units; **(b)** USCS units.
12. A large plot of land ABCD is divided into four lots (see Fig. 2-8). Determine the area of each lot in **(a)** square meters and **(b)** hectares if sides *AB* and *CD* are parallel.
13. A plot of land *ABCD* is divided into four lots (see Fig. 2-9). Find the area of each lot in **(a)** square feet and **(b)** acres if sides *AB* and *CD* are parallel.
14. How many tiles 15.0 cm by 15.0 cm are required to completely cover a wall 2.50 m by 1.55 m? Assume no spaces between the tiles.

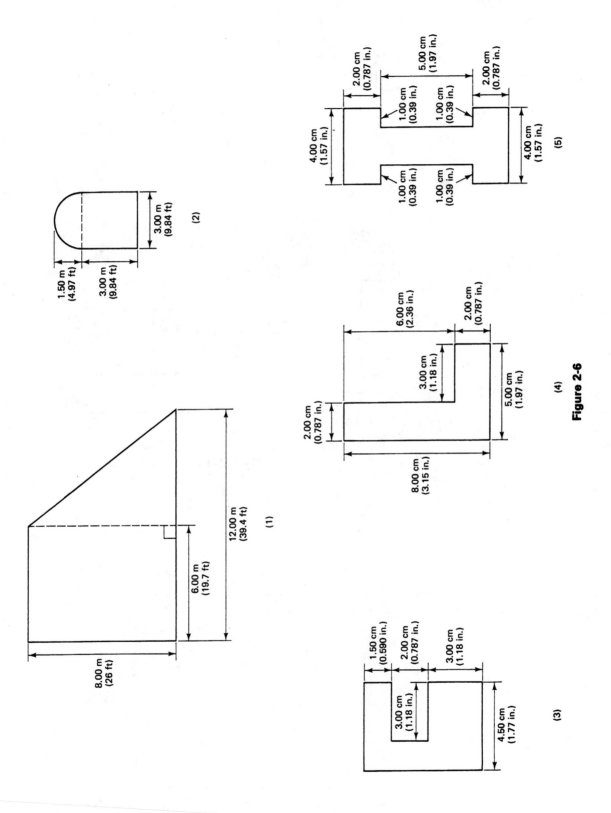

Figure 2-6

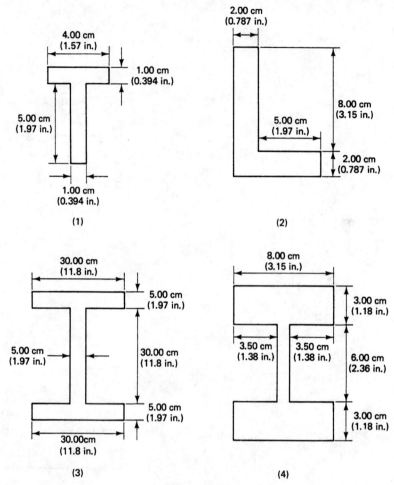

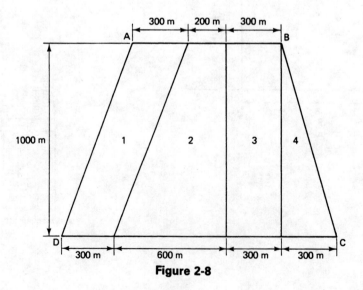

Figure 2-7

Figure 2-8

Sec. 2-1 / Formulas and Problem Solving

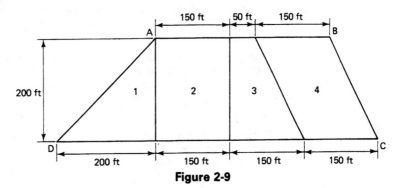

Figure 2-9

15. How many liters of water are required to fill a 5.50 m diameter circular pool to a depth of 1.25 m? (*Hint:* 1 L = 10^{-3} m^3.)
16. What surface area of a foundation could be poured to a depth of (a) 15.0 cm with 3.00 m^3 of concrete; (b) 6.00 in. with 27.0 ft^3 of concrete?
17. How many liters of water are required to fill a cylindrical tank that has an internal diameter of 40.0 cm and an internal height of 1.85 m?
18. Determine the number of sheets of plywood 1.25 m by 2.50 m that are required to cover an area that is 10.0 m by 8.75 m.
19. Determine the number of sheets of plywood 4.00 ft by 8.00 ft that are required to cover an area that measures 12.0 ft by 24.0 ft.
20. Determine the cross-sectional area of a wire that has a diameter of (a) 0.750 mm; (b) 1.15 cm; (c) 0.0820 in.
21. Calculate the volume of concrete required to construct a cylindrical support (a) which is 3.15 m long and has a diameter of 37.5 cm; (b) which is 10 ft 8 in. long and has a diameter of 15.0 in.
22. What is the height of a room if (a) its volume is 285 m^3 and its floor area is 115 m^2; (b) its volume is 875 ft^3 and its area is 108 ft^2?
23. Determine the total surface area and the volume of a brick which has sides (a) 10.0 cm by 6.25 cm by 20.0 cm; (b) 4.00 in. by 2.50 in. by 7.75 in.
24. Calculate the surface area and the volume of a ball bearing that has a diameter of (a) 2.20 cm; (b) 1.25 cm; (c) 0.750 in.

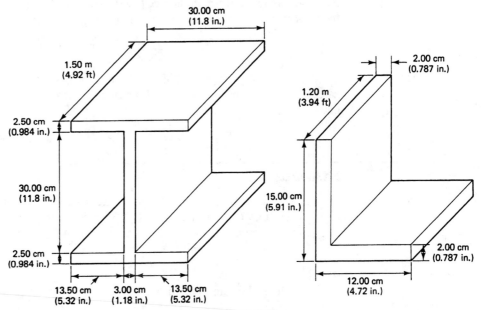

Figure 2-10

25 Find the cross-sectional area, the volume, and the surface area of a car cylinder which is **(a)** 12.0 cm long and has a diameter of 9.80 cm; **(b)** 4.15 in. long and has a diameter of 3.85 in.

26 Determine the cross-sectional area perpendicular to the length, the volume, and the total surface area of a drive shaft that is **(a)** 65.0 cm long and has a diameter of 7.15 cm; **(b)** 25.5 in. long and has a diameter of 2.75 in.

27 Calculate the volume of a cylinder that is **(a)** 5.2 cm high and has a base radius of 3.8 cm; **(b)** 2.00 in. high and has a base of radius 1.50 in.

28 Determine the cross-sectional areas and the volumes of steel in the beams illustrated in Fig. 2-10 **(a)** in SI units; **(b)** USCS units.

2-2 RATIO AND PROPORTION

Formulas are often established from experimental data either by inspection or by graphical analysis. In each case the objective is to determine relationships between the variables.

A **ratio** is a comparison of two quantities by a process of division. Many physical quantities are actually defined in terms of a ratio. For example, the density ρ* of a material is defined as the ratio of its mass m to its volume V.

$$\rho = \frac{m}{V} \tag{2-3}$$

A **proportion** is an equality between two or more ratios. In this case the ratio of the variables is equal to a constant. The ratio of the circumference C of any circle to its diameter d is a constant value called $\pi = 3.1416$†, a value that is the same for all circles.

$$\frac{C}{d} = \pi = 3.1416... \tag{2-4}$$

Since this ratio is constant, a circle with twice the diameter has twice the circumference, if we triple the diameter we triple the circumference, and so on, and we therefore say that the circumference C varies directly as (or is directly proportional to) the diameter d. Symbolically, this is written as

$$C \propto d$$

To change the proportionality sign into an equality, we merely insert a constant (in this case π). These constants may even have units which must then be included in the equation. Thus

$$C = \pi d$$

EXAMPLE 2-7

If a car travels at a constant speed, the distance d that it travels varies directly as the time t taken. Determine **(i)** the relationship between distance d and the time t taken and **(ii)** the distance traveled in 4.00 h, if a car travels **(a)** 75.0 km in 1.50 h; **(b)** 48.0 mi in 1.50 h.

Solution

    (i) *Data:* We are given $d \propto t$.

---
* The ρ is the Greek letter rho.
† The π symbol is Greek and is pronounced as "pie."

Sec. 2-2 / Ratio and Proportion

(a) $d = 75.0$ km, $t = 1.50$ h (b) $d = 48.0$ mi, $t = 1.50$ h
Equation: $d = vt$, where v is a constant.
Rearrange: Solving for v yields

$$v = \frac{d}{t}$$

Substitute:
(a) $v = \dfrac{75.0 \text{ km}}{1.50 \text{ h}} = 50.0$ km/h   (b) $v = \dfrac{48.0 \text{ mi}}{1.50 \text{ h}} = 32.0$ mi/h

Therefore, the final formulas are
(a) $d = (50.0 \text{ km/h})t$ (b) $d = (32.0 \text{ mi/h})t$

(ii) Substituting the value $t = 4.0$ h in these formulas gives us
(a) $d = (50.0 \text{ km/h})(4.00 \text{ h})$ (b) $d = (32.0 \text{ mi/h})(4.00 \text{ h})$
 $= 2.00 \times 10^2$ km $= 128$ mi

Note that the unit for time (h) cancels.

In many cases one physical quantity may vary as some power of another quantity. If this exponent is negative-valued (indicating a reciprocal), the variation is called *inverse*.

EXAMPLE 2-8

The mass m of a steel ball bearing varies as the cube of its diameter d. If a bearing with a diameter of 1.50 cm has a mass of 13.8 g, (a) find the formula; (b) determine the mass of a bearing with a 0.75 cm diameter.

Solution

(a) Data: $m \propto d^3$; $m = 13.8$ g when $d = 1.50$ cm
Equation: $m = kd^3$, where k is the proportionality constant
Rearrange: $k = \dfrac{m}{d^3}$

Substitute: $k = \dfrac{13.8 \text{ g}}{(1.50 \text{ cm})^3} = 4.09$ g/cm^3

Thus $m = (4.09 \text{ g/cm}^3)d^3$ is the formula.

(b) Data: $d = 0.75$ cm, $m = ?$
Equation: In part (a) we derived a formula that can now be used for other values.

$$m = (4.09 \text{ g/cm}^3)d^3$$

Substitute: $m = (4.09 \text{ g/cm}^3)(0.75 \text{ cm})^3 = 1.73$ g.

EXAMPLE 2-9

The illumination I of a surface varies inversely as the square of the distance d to the light source. If a source produces an illumination of 188 lm/m² (lumens per square meter) at 3.00 m, find (a) the formula; (b) the illumination that the same source would produce at 4.00 m.

Solution

(a) Data: $I \propto \dfrac{1}{d^2}$, $I = 188$ lm/m² when $d = 3.00$ m

Equation: $I = \dfrac{k}{d^2}$

Rearrange: $k = Id^2$

Substitute: $k = (188 \text{ lm/m}^2)(3.00 \text{ m})^2 = 1690 \text{ lm}$
Thus the formula is

$$I = \frac{1690 \text{ lm}}{d^2}$$

(b) *Data:* $d = 4.00$ m, $I = ?$

Equation: $I = \dfrac{1690 \text{ lm}}{d^2}$

Substitute: $I = \dfrac{1690 \text{ lm}}{(4.00 \text{ m})^2} = 106 \text{ lm/m}^2$

PROBLEMS

29 The compression ratio of an engine is defined simply as the ratio of the maximum to the minimum volume of the cylinder above the piston. What is the compression ratio **(a)** if the maximum volume is 1800 cm³ and the minimum volume is 160 cm³; **(b)** if the maximum volume is 1750 cm³ and the minimum volume is 175 cm³; **(c)** if the maximum volume is 110 in³ and the minimum volume is 9.80 in³?

30 The maximum weight w that a rope can support varies directly as its cross-sectional area A. If a rope with a cross-sectional area of 3.00 mm² can support 60.0 N (newtons), find **(a)** the formula; **(b)** the maximum weight that a rope with an 8.00 mm² area can support.

31 The weight w of a ball bearing varies as the cube of its diameter d. If a bearing with a diameter of 0.591 in. weighs 0.485 oz, find **(a)** the formula; **(b)** the weight of a ball bearing which has a diameter of 0.295 in.

32 The electrical resistance R of a wire varies directly as its length L. If a wire 25.0 m long has a resistance of 2.00 Ω* (ohms), find **(a)** the formula; **(b)** the resistance of a similar wire 75.0 m long.

33 The electric current I in an electric resistance varies directly as the potential difference V across it. If a potential difference of 25.0 V (volts) produces a current of 5.00 A (amperes), find **(a)** the formula; **(b)** the current produced by a 31.0 V potential difference; **(c)** the potential difference required to produce a 2.00 A current.

34 When a car travels at a constant speed, the distance d that it travels is directly proportional to the elapsed time t. If the car travels 525 km in 6.00 h, find **(a)** the formula; **(b)** the distance that it travels in 7.50 h if the speed is constant; **(c)** the time that it takes to travel 245 km.

35 The distance d that a car travels is directly proportional to the elapsed time t. If the car travels 330 mi in 6.00 h, find **(a)** the formula; **(b)** the distance that it travels in 7.50 h; **(c)** the time it takes to travel 150 mi at the same speed.

36 Ignoring friction, the distance d that an object falls from rest varies as the square of the elapsed time t. If an object falls 491 m in 10.0 s, find **(a)** the formula; **(b)** the distance it falls in 5.00 s; **(c)** the time required for it to fall 325 m.

37 Ignoring friction, the distance d that an object falls from rest varies as the square of the elapsed time t. If an object falls 1601 ft in 10.0 s, find **(a)** the formula; **(b)** the distance it falls in 5.00 s; **(c)** the time required for it to fall 1200 ft.

38 The elongation e of a wire varies directly as the applied force F. If an applied force of 158 N produces an elongation of 1.90 cm, find **(a)** the formula; **(b)** the elongation produced by a 90.0 N force; **(c)** the force required to produce a 3.00 cm elongation.

39 With a constant load, the deflection D of a (cantilever) beam varies as the cube of the length L projecting beyond the support. If the projecting length is 8.0 m, the deflection of a beam is 2.4 cm. Determine **(a)** the formula; **(b)** the deflection when the projecting length is 6.0 m; **(c)** the projecting length when the deflection is 3.2 cm.

40 The force F required to keep an object moving in a circle varies directly as the square of the speed v. If a 1460 N force is required to keep the object moving in a circle at 50.0 m/s,

---

*Represented by Ω, the capital Greek letter omega

find (a) the formula; (b) the force required when the object's speed is 35.0 m/s; (c) the speed if the force is 1250 N.

41. The electric power P dissipated across a resistance varies directly as the square of the current I through it. If the power dissipated is 8.00 W (watts) when the current is 0.220 A (amperes), find (a) the formula; (b) the power dissipated when the current is 0.350 A; (c) the current when the power is 12.0 W.

42. The volume V of a fixed mass of gas varies inversely as the total pressure p. If a mass of gas occupies 350 L at a total pressure of 105 kPa (kilopascal), determine (a) the formula; (b) the volume when the total pressure is 315 kPa; (c) the total pressure when the volume is 525 L.

43. The volume V of a fixed mass of gas varies inversely as the total pressure p. If a mass of gas occupies 12.0 ft^3 at a total pressure of 15.0 lb/in^2, find (a) the formula; (b) the volume when the total pressure is 45.0 lb/in^2; (c) the total pressure when the volume is 18.0 ft^3.

44. The frequency f of radio waves varies inversely as their wavelength λ.* If the frequency is 306 MHz (megahertz) when the wavelength is 98.0 cm, find (a) the formula; (b) the frequency when the wavelength is 75.0 cm; (c) the wavelength when the frequency is 108 MHz.

45. If the intensity I of a wave varies inversely as the square of the distance d from the source, and a wave has an intensity of 3.60 μW/m^2 (microwatts per square meter) at 2.00 m from the source, determine (a) the formula; (b) the intensity of the wave at 6.00 m from the source; (c) the distance at which the intensity is 1.20 μW/m^2.

46. The speed v of a wave in a vibrating string varies as the square root of the tension T applied. If a tension of 25.0 N produces a wavespeed of 41.0 m/s, find (a) the formula; (b) the wavespeed when the tension is 35.0 N; (c) the tension required to produce a wavespeed of 35.0 m/s.

2-3 GRAPHS

A **graph** is a diagram that shows the relationships between variables. In addition to giving a "picture" of experimental results, they may be used to establish formulas and predict new results either directly from the graph or by extrapolation.

We shall only consider graphs drawn on a Cartesian coordinate system in which two perpendicular lines form **axes**. Each axis is then graduated into increments representing changes in one of the variables. A "data point" is "plotted" at the intersection of perpendicular lines from the corresponding values of the variables along the axes (see point 122.6 km, 5 s). A typical graph corresponding to the data in Table 2-1 is shown in Fig. 2-11.

TABLE 2-1
Distance That an Object Falls in Certain Elapsed Times

Distance/m	0	4.9	19.6	44.2	78.5	122.6
Elapsed time/s	0	1	2	3	4	5

Note that the following information should be included with a graph:

1. There should be a title indicating what the graph represents.
2. The axes should be fully labeled with numerical values and the corresponding units where applicable. The units need not be written with each numerical value provided that they are indicated under a slash showing that each

---

*λ is the Greek lowercase letter lambda. It is commonly used to represent the physical quantity wavelength.

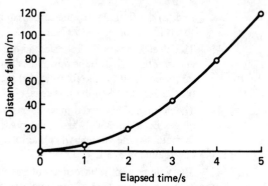

Figure 2-11 Graph of the distance versus the elapsed time that an object falls.

value has been divided by the unit, or in parentheses to show that the values must be multiplied by the unit. However, when values are obtained from the graph, these units must be reinserted. The independent variable is usually assigned the horizontal axis.

3. The scales used on the axes should be arranged so that the graph takes up as much of the page as practical. It is not necessary to choose awkward scales merely to use the whole page; some discretion should be used. In many cases the values on the axes need not begin at zero.

4. Normally, data points should be joined by a smooth, "best-fit" line using a French curve or a universal curve. However, on occasion one or more data points may not fall on the smooth line. Check to see if these points have been plotted correctly. If they are correct, the data may be in error, in which case these points may be noted but ignored when drawing the smooth line. Some caution is advised whenever data points are ignored.

5. If possible, a formula should be determined from the graph rather than from the numbers obtained during the experiment because the graph effectively averages the results, whereas pairs of numbers may be in error.

If a straight-line graph is obtained, we may determine the formula from its slope m and intercept b with the vertical axis (see Fig. 2-12). The **slope** m of a graph is defined as the ratio of the vertical change (the rise) to the corresponding horizontal change (the run), that is,

$$m = \frac{y_2 - y_1}{x_2 - x_1} = \frac{\Delta y}{\Delta x} = \frac{\text{rise}}{\text{run}} \qquad (2\text{-}5)$$

(the Greek capital letter delta, Δ, is often used to represent "a change in").

To minimize errors, use large values for the rise and run when the slope is found. The general formula is

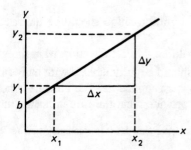

Figure 2-12 Graph of $y = mx \times b$, where $m = \Delta y/\Delta x = $ slope.

Sec. 2-3 / Graphs

$$y = mx + b \qquad (2\text{-}6)$$

For example, let us analyze a graph of the equation

$$d = (50 \text{ km/h})t + 20 \text{ km}$$

which represents the distance d of a car from some reference position as a function of the elapsed times t if it starts (at $t = 0$ h) at a point 20 km from some reference position. We shall draw the graph for a total elapsed time of 4.0 h. By substituting values for t in the equation, we may tabulate the distances and the corresponding elapsed times (Table 2-2) and the graph is illustrated in Fig. 2-13.

TABLE 2-2

Position/km	20	70	120	170	220
Elapsed time/h	0	1	2	3	4

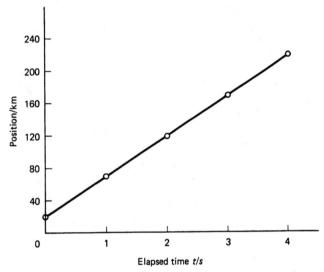

Figure 2-13 Position of the car from some reference as a function of the elapsed time.

Since the graph is a straight line, the slope is constant and it may be found between any two points on the line:

$$\text{slope } m = \frac{\text{rise}}{\text{run}} = \frac{220 \text{ km} - 70 \text{ km}}{3.0 \text{ h}} = 50 \frac{\text{km}}{\text{h}}$$

The intercept $b = 20$ km; therefore, the equation is

$$d = mt + b = (50 \text{ km/h})t + 20 \text{ km}$$

which is the same as the original.

Note that other values can be read directly from the graph. For example, an elapsed time of 1.5 h corresponds to a distance of 95 km. Similarly, a distance of 145 km corresponds to an elapsed time of 2.5 h.

We can also extrapolate data by extending the line beyond the limits given in the table by assuming that the equation is still valid. Therefore, for an elapsed time of 5.0 h, we would expect a distance of 270 km, and so on. Again, some caution is advised when extrapolating data, however, because the graph may not be valid for values not indicated.

If a graph of two variables (x, y) is a straight line and *it passes through the origin* (point $x = 0$, $y = 0$), the intercept $b = 0$ and $y = mx$. Since the slope m is constant, y varies directly as x, $(y \propto x)$. Whenever we manage to produce a straight-line graph *that passes through the origin*, there is a direct proportion between that which is plotted vertically and that plotted horizontally. The slope of the graph is the proportionality constant.

EXAMPLE 2-10

Determine the equations from the graphs in Fig. 2-14.

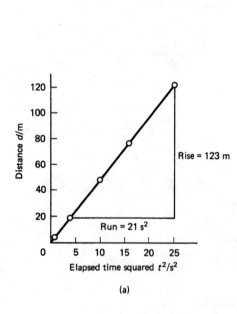
(a)

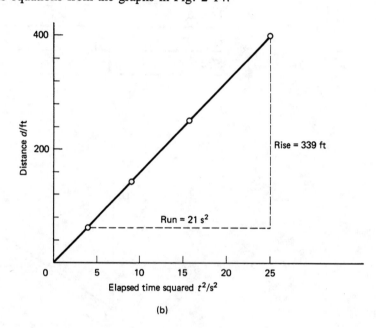
(b)

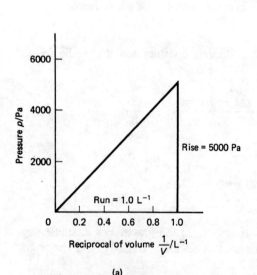
(a)

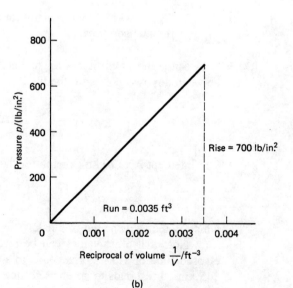
(b)

Figure 2-14

Solution

(i) *Equation:* Slope $m = \dfrac{\text{rise}}{\text{run}}$

Substitute:

(a) $m = \dfrac{123 \text{ m} - 20 \text{ m}}{25 \text{ s}^2 - 4 \text{ s}^2}$
    $= 4.9 \text{ m/s}^2$

(b) $m = \dfrac{403 \text{ ft} - 64 \text{ ft}}{25 \text{ s}^2 - 4 \text{ s}^2}$
    $= 16 \text{ ft/s}^2$

Since the graphs are straight lines that pass through the origin, $d \propto t^2$, and therefore $d = mt^2$. Thus

(a) $d = (4.9 \text{ m/s}^2)t^2$

(b) $d = (16 \text{ ft/s}^2)t^2$

(ii) *Equation:* Slope $m = \dfrac{\text{rise}}{\text{run}}$

Substitute:

(a) $m = \dfrac{5000 \text{ Pa}}{1.0/\text{L}}$
    $= 5000 \text{ Pa} \cdot \text{L}$

(b) $m = \dfrac{700 \text{ lb/in}^2}{0.0035 \text{ ft}^{-3}}$
    $= 2.0 \times 10^5 \text{ lb} \cdot \text{ft}^3/\text{in}^2$

Again, since the straight lines pass through the origin,

$$p \propto \dfrac{1}{V} \quad \text{or} \quad p = \dfrac{m}{V}$$

Thus

(a) $p = \dfrac{5000 \text{ Pa} \cdot \text{L}}{V}$

(b) $p = \dfrac{2.0 \times 10^5 \text{ lb} \cdot \text{ft}^3/\text{in}^2}{V}$

PROBLEMS

47 Determine the equations from the graphs in Fig. 2-15.

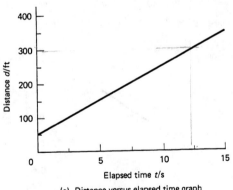
(a) Distance versus elapsed time graph

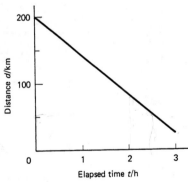
(b) Distance versus elapsed time graph

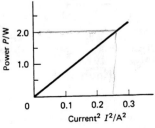
(c) Power versus square of current

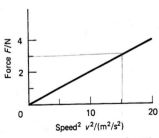
(d) Force versus square of speed

Figure 2-15

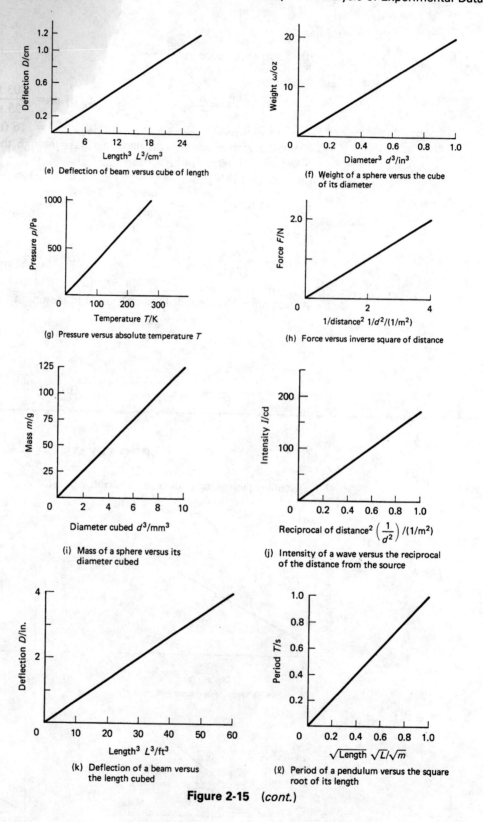

(e) Deflection of beam versus cube of length

(f) Weight of a sphere versus the cube of its diameter

(g) Pressure versus absolute temperature T

(h) Force versus inverse square of distance

(i) Mass of a sphere versus its diameter cubed

(j) Intensity of a wave versus the reciprocal of the distance from the source

(k) Deflection of a beam versus the length cubed

(ℓ) Period of a pendulum versus the square root of its length

Figure 2-15 (*cont.*)

48 Draw a graph of d versus t for the relationships (a) $d = (2.5 \text{ m/s})t + 12 \text{ m}$ and (b) $d = (8.00 \text{ ft/s})t + 25 \text{ ft}$ for values of t from 0 s to 10 s.

49 (a) Draw a graph of force versus elongation for the data in Table 2-3; (b) determine the equation.

Sec. 2-4 / Density

TABLE 2-3

Force F/N	0	10	20	30	40	50	60
Elongation e/mm	0	2	4	6	8	10	12

50 Using the information in Table 2-4, draw graphs of (i) distance versus elapsed time; (ii) distance versus the square of the elapsed time. (iii) Determine the formula in (a) SI units; (b) USCS units.

TABLE 2-4

Distance d/m	0	4.9	19.6	44.1	78.4	123	176
Distance d/ft	0	16.1	64.3	145	257	404	577
Elapsed time t/s	0	1	2	3	4	5	6

51 Using the data in Table 2-5, draw a graph of (a) mass versus the diameter and (b) mass versus the cube of the diameter of a steel ball bearing. (c) Find the equation.

TABLE 2-5

Mass m/g	0.033	0.26	0.88	2.10	4.08	7.06
Diameter d/cm	0.20	0.40	0.60	0.80	1.00	1.20

52 (a) Draw a graph of pressure p versus volume V using the data in Table 2-6. (b) Draw a graph of pressure versus the reciprocal of the volume $(1/V)$. (c) Find the formula.

TABLE 2-6

Pressure p/kPa	100	200	300	400	500
Volume V/cm³	275	138	91.7	68.8	55.0

53 (a) Draw a graph of force F versus distance using the data in Table 2-7. (b) Plot a graph of the force F versus the reciprocal of the distance squared $(1/d^2)$. (c) From the graph, find the formula.

TABLE 2-7

Force F/lb	10	2.5	1.1	0.63	0.40	0.28
Distance d/ft	1	2	3	4	5	6

2-4 DENSITY

Even if we consider equal masses of two materials, such as lead and cork, we usually assume that one material (lead) is "heavier" than the other. Of course, the volumes of the equal masses of the two materials would be quite different. To indicate this characteristic property of "heaviness," we must include the volumes.

We define the **density** ρ (or **mass density**) of a substance as its mass m per unit volume V:

$$\rho = \frac{\text{mass}}{\text{volume}} = \frac{m}{V} \tag{2-7}$$

This is an important example of a physical quantity that is defined in terms of a ratio.

If we substitute the units for mass and volume in this formula, we see that density units are kilograms per cubic meter (kg/m³) in SI and slugs per cubic foot (slug/ft³) in USCS. Density is the constant of proportionality between the mass m and the volume V of the material, and it is therefore independent of the length, mass, area, and so on. For example, if we double the volume, we double the mass, but the ratio (the density) remains constant for the material under fixed conditions.

In USCS, since weight is used more frequently than mass, it is convenient to define a quantity called **weight density** D as the ratio of the weight w to the volume V of the substance.

$$D = \frac{w}{V} \tag{2-8}$$

The corresponding units are pounds per cubic foot (lb/ft³) in USCS, and newtons per cubic meter (N/m³) in SI.

Some typical values for density and weight density are listed in Table 2-8. Density is characteristic of all states of matter: solids, liquids, *and* gases.

TABLE 2-8

Substance	Relative density (specific gravity) ρ_r	Density ρ kg/m³	Density ρ slug/ft³	Weight density D lb/ft³
Solids	*(relative to water)*			
Aluminum	2.7	2700	5.25	169
Brass	8.6	8600	16.8	537
Concrete	2.3	2300	4.5	144
Copper	8.89	8890	17.3	555
Gold	19.3	19 300	37.7	1204
Ice	0.92	920	1.8	57.5
Iron	7.85	7850	15.3	490
Lead	11.3	11 300	22.0	705
Steel	7.8	7800	15.2	486
Liquids at 20°C				
Alcohol, ethyl	0.79	790	1.53	49.3
Benzene	0.88	880	1.71	54.7
Gasoline	0.68	680	1.32	42.4
Mercury	13.6	13 600	26.5	850
Water	1.0	1000	1.95	62.4
Gases at 0°C and 760 mm Hg	*(relative to air)*			
Air	1.0	1.29	2.52×10^{-3}	8.07×10^{-2}
Carbon dioxide	1.53	1.98	3.85×10^{-3}	1.24×10^{-1}
Helium	1.40	0.18	3.5×10^{-4}	1.12×10^{-2}
Hydrogen	0.07	0.09	1.75×10^{-4}	5.6×10^{-3}
Nitrogen	0.97	1.25	2.44×10^{-3}	7.8×10^{-2}
Oxygen	1.11	1.43	2.78×10^{-3}	8.9×10^{-2}
Propane	1.56	2.01	3.93×10^{-3}	1.26×10^{-1}

EXAMPLE 2-11

Determine **(a)** the density of a 755 kg rectangular concrete block which has dimensions of 35.0 cm × 75.0 cm × 1.25 m; **(b)** the weight density of a 1660 lb block which has dimensions of 1.15 ft × 2.46 ft × 4.10 ft.

Solution

(a) Data: $L = 1.25$ m, $b = 0.350$ m, $h = 0.750$ m, $m = 755$ kg, $\rho = ?$

(b) Data: $L = 4.10$ ft, $b = 1.15$ ft, $h = 2.46$ ft, $w = 1660$ lb, $D = ?$

Sec. 2-4 / Density

Equations: $V = Lbh$, $\rho = \dfrac{m}{V}$ 　　　　Equations: $V = Lbh$, $D = \dfrac{w}{V}$

Rearrange: $\rho = \dfrac{m}{Lbh}$ 　　　　　　　　Rearrange: $D = \dfrac{w}{Lbh}$

Substitute: 　　　　　　　　　　　　　　　　Substitute:

$\rho = \dfrac{755 \text{ kg}}{(1.25 \text{ m})(0.350 \text{ m})(0.750 \text{ m})}$ 　　$D = \dfrac{1660 \text{ lb}}{(4.10 \text{ ft})(1.15 \text{ ft})(2.46 \text{ ft})}$

$= 2.30 \times 10^3 \text{ kg/m}^3$ 　　　　　　　　$= 143 \text{ lb/ft}^3$

EXAMPLE 2-12

Determine the mass of a steel ball bearing that has a diameter of (a) 3.00 cm; (b) 1.18 in.

Solution

Data:

(a) $\rho = 7800 \text{ kg/m}^3$ (Table 2-8), $d = 3.00$ cm, therefore, the radius $r = 1.50$ cm or $r = 1.50 \times 10^{-2}$ m, $m = ?$

(b) $\rho = 15.2 \text{ slug/ft}^3$ (Table 2-8), $d = 1.18$ in., $r = 0.59$ in., or $r = \dfrac{0.59}{12}$ ft, $m = ?$

Equations: $V = \dfrac{4\pi r^3}{3}$, $\rho = \dfrac{m}{V}$

Rearrange: $m = \rho V = \dfrac{\rho 4\pi r^3}{3}$

Substitute:

(a) $m = (7800 \text{ kg/m}^3)\dfrac{4}{3}\pi(1.50 \times 10^{-2} \text{ m})^3 = 0.11$ kg or 110 g

(b) $m = (15.2 \text{ slug/ft}^3)\dfrac{4}{3}\pi\left\{\dfrac{0.59}{12}\text{ ft}\right\}^3 = 7.57 \times 10^{-3}$ slug

EXAMPLE 2-13

Determine the volume of (a) 10.0 kg and (b) 22.0 lb of gasoline at 20°C.

Solution

Data:

(a) $\rho = 680 \text{ kg/m}^3$ (Table 2-8), $m = 10.0$ kg, $V = ?$

(b) $D = 42.4 \text{ lb/ft}^3$ (Table 2-8), $w = 22.0$ lb, $V = ?$

Equation: $\rho = \dfrac{m}{V}$ 　　　　　　　　$D = \dfrac{w}{V}$

Rearrange: $V = \dfrac{m}{\rho}$ 　　　　　　　　$V = \dfrac{w}{D}$

Substitute: $V = \dfrac{10.0 \text{ kg}}{680 \text{ kg/m}^3}$ 　　　$V = \dfrac{22.0 \text{ lb}}{42.4 \text{ lb/ft}^3}$

$= 1.47 \times 10^{-2} \text{ m}^3$ 　　　　　　　$= 0.519 \text{ ft}^3$

Note the analysis of the units:

$$\dfrac{\text{kg}}{\text{kg/m}^3} = \dfrac{\text{kg} \cdot \text{m}^3}{\text{kg}} = \text{m}^3$$

Since solids and liquids are only slightly compressible, their densities are almost constant. Gases and vapors are relatively easy to compress; therefore, their densities will vary considerably with the conditions under which they are measured.

The **relative density (specific gravity)** of a substance is the ratio of its density to that of some reference material at a certain temperature. Pure water at 4°C is usually taken as the reference for solids and liquids; air is usually used for gases unless otherwise stated. Thus the relative density of some substance, which has a density ρ is given by

$$\rho_r = \frac{\rho}{\rho_{\text{reference}}} \quad (2\text{-}9)$$

where $\rho_{\text{reference}}$ is the density of the reference material. Also, since $\rho V = m$, the relative density or specific gravity at some reference temperature

$$\rho_r = \frac{\text{mass of material}}{\text{mass of equal volume of reference material}} \quad (2\text{-}10\text{a})$$

or

$$\rho_r = \frac{\text{weight of material}}{\text{weight of equal volume of reference material}} \quad (2\text{-}10\text{b})$$

Note that the numerical value of density will vary with the system of units and the temperature, but relative density is a dimensionless quantity (i.e., it has no units) that has the same value for a particular reference temperature. For simplicity, we will assume the same reference temperature.

EXAMPLE 2-14

An irregularly shaped block of metal has a mass of 12.0 kg. When it is lowered into a container, which is completely full of water, 1.40 L of water overflows. Determine the density of the metal.

Solution

Data: The mass $m = 12.0$ kg and the volume of the block must be equal to the volume of water displaced; thus

$$V = 1.40 \text{ L} = 1.40 \times 10^{-3} \text{ m}^3$$

Equation: $\rho = \dfrac{m}{V}$

Substitute: $\rho = \dfrac{12.0 \text{ kg}}{1.40 \times 10^{-3} \text{ m}^3} = 8570 \text{ kg/m}^3$

PROBLEMS

54 What is the average density of the earth (a) if its mass is 5.98×10^{24} kg and its radius is 6.38×10^6 m; (b) if its mass is 4.093×10^{23} slug and its radius is 2.07×10^7 ft? Consider the earth to be a sphere.

55 Determine the mass of the sun (a) if it has an average density of 1.4×10^3 kg/m³ and a radius of 6.97×10^8 m; (b) if its average density is 2.73 slug/ft³ and its radius is 2.29×10^9 ft.

56 What is the volume of a pure copper block which has (a) a mass of 2.50 kg; (b) a weight of 8.75 lb?

57 Determine the density of a rectangular block 30.0 cm × 20.0 cm × 25.0 cm which has a mass of 135 kg.

58 What is the mass of a rectangular concrete block that has dimensions of (a) 25.0 cm × 35.0 cm × 32.0 cm; (b) 10.0 in. × 14.0 in. × 12.5 in.?

Sec. 2-5 / Trigonometry

59. Determine the mass of a steel ball bearing that has a diameter of (a) 1.25 cm; (b) 3.00 mm; (c) 7.50 mm; (d) 0.75 in.
60. What volume of aluminum has a mass of (a) 7.50 kg; (b) 25.0 kg; (c) 865 g; (d) 0.163 slug; (e) 0.0565 slug?
61. What is the weight of a concrete foundation that has dimensions of (a) 15.0 cm × 8.50 m × 11.2 m; (b) 6.00 in. × 28.0 ft × 36.0 ft?
62. What is the mass of (a) 35.0 L of gasoline; (b) 2.00 ft³ of benzene?
63. What volume of mercury has a mass of (a) 1.25 kg; (b) 650 g; (c) 0.25 slug?
64. What volume is occupied by (a) 8.00 kg and (b) 10.0 lb of propane at 0°C and 1 atm pressure?
65. Determine the diameter of a steel ball bearing that has a mass of (a) 25.0 g; (b) 58.0 g; (c) 125 g; (d) 0.250 slug.
66. Find the mass of a cylinder of aluminum that has (a) a length of 50.0 cm and a diameter of 12.5 cm; (b) a length of 1.25 ft and a diameter of 8.00 in.
67. What is the mass of an iron girder 25.0 cm × 8.00 cm × 12.0 m?
68. What volume of gold could you buy for $680.00 if gold costs (a) $2.54 per gram; (b) $325.00 per ounce?
69. A man attempts to sell two 150 g nuggets, which he claims are pure gold. When the nuggets are lowered into a container which is completely full of water, the first nugget displaces 13.3 mL of water and the second displaces 7.77 g of water. Check the densities of the nuggets to see if they are really gold.
70. An irregularly shaped metal block has a mass of 11.0 kg. When it is totally immersed in a full container of water, 1.25 kg of water overflows. What is the density of the block?
71. What is the mass of steel pipe (a) that is 12.0 m long if its inner and outer diameters are 1.90 cm and 2.25 cm, respectively; (b) if it is 15.0 ft long and its inner and outer diameters are 2.00 in. and 2.15 in., respectively?
72. Find the mass of a lead pipe that is 3.00 m long if it has inner and outer diameters of 1.25 cm and 1.40 cm.

2-5 TRIGONOMETRY

The analysis and solution of many problems may be simplified by the use of trigonometry. In many cases it is only necessary to consider a right-angled triangle (a triangle with one angle of 90°). The longest side (opposite to the right angle) is called the **hypotenuse** (Fig. 2-16). The sum of the angles in any triangle is 180°. Also, in

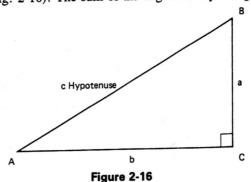

Figure 2-16

any right-angled triangle the square of the hypotenuse is equal to the sum of the squares of the other two sides. This is known as the **Pythagorean theorem**. Thus

$$c^2 = a^2 + b^2 \tag{2-11}$$

Consider the right-angled triangles *ABC*, *ADE*, and *AFG* in Fig. 2-17. These triangles are similar because they are all right angled and the angle *A* is common; therefore, the ratios of their sides are equal.

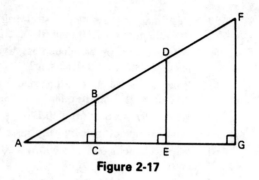

Figure 2-17

The ratio of the side opposite to angle A to the hypotenuse is a constant known as the **sine** of angle A (abbreviated sin A)

$$\sin A = \frac{BC}{AB} = \frac{DE}{AD} = \frac{FG}{AF} = \text{constant} = \frac{\text{side opposite}}{\text{hypotenuse}} \qquad (2\text{-}12)$$

The ratio of the side adjacent (next to) A and the hypotenuse is known as the **cosine** of A (abbreviated cos A).

$$\cos A = \frac{AC}{AB} = \frac{AE}{AD} = \frac{AG}{AF} = \text{constant} = \frac{\text{adjacent}}{\text{hypotenuse}} \qquad (2\text{-}13)$$

Finally, the ratio of the side opposite to A and the side adjacent to A is known as the **tangent** of A (abbreviated tan A).

$$\tan A = \frac{BC}{AC} = \frac{DE}{AE} = \frac{FG}{AG} = \text{constant} = \frac{\text{opposite}}{\text{adjacent}} \qquad (2\text{-}14)$$

In each case the value of the sine, cosine, and tangent depends only on the magnitude of the angle A. These relationships may be summarized for the general right-angled triangle in Fig. 2-16 as

$$\sin A = \frac{\text{opposite}}{\text{hypotenuse}} = \frac{a}{c}$$

$$\cos A = \frac{\text{adjacent}}{\text{hypotenuse}} = \frac{b}{c} \qquad (2\text{-}15)$$

$$\tan A = \frac{\text{opposite}}{\text{adjacent}} = \frac{a}{b}$$

EXAMPLE 2-15

Find the sine, cosine, and tangents of angles A and B in the right-angled triangle in Fig. 2-18.

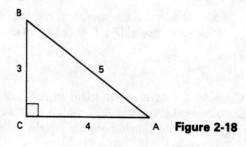

Figure 2-18

Sec. 2-5 / Trigonometry

Solution

$$\sin A = \frac{a}{c} = \frac{3}{5} = 0.6 \qquad \sin B = \frac{b}{c} = \frac{4}{5} = 0.8$$

$$\cos A = \frac{b}{c} = \frac{4}{5} = 0.8 \qquad \cos B = \frac{a}{c} = \frac{3}{5} = 0.6$$

$$\tan A = \frac{a}{b} = \frac{3}{4} = 0.75 \qquad \tan B = \frac{b}{a} = \frac{4}{3} = 1.3$$

In Example 2-15 it can be seen that since the sum of the angles in any triangle is 180°, in a right-angled triangle where one angle C is 90°,

$$A = 90° - B \quad \text{and} \quad B = 90° - A$$

Also,

$$\sin A = \cos(90° - A) = \cos B$$

$$\cos A = \sin(90° - A) = \sin B \qquad (2\text{-}16)$$

The values for the sine, cosine, and tangents of angles may be found in tables or from a calculator. To find the values on the calculator, first make sure that it is in the correct mode (i.e., degrees). Key the angle first and then the function. For example, to find $\cos 28°$, key as follows:

$$28 \cos$$

You should see 0.883 on the display. Similarly, $\sin 30° = 0.5$, $\cos 30° = 0.866$, and $\tan 30° = 0.577$.

EXAMPLE 2-16

Determine the length of the sides a and b in the right-angled triangle in Fig. 2-19.

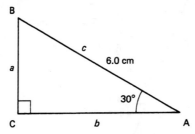

Figure 2-19

Solution: $\sin A = \dfrac{a}{c}$; therefore,

$$a = c \sin A = (6.0 \text{ cm}) \sin 30° = 3.0 \text{ cm}$$

$\cos A = \dfrac{b}{c}$; thus

$$b = c \cos A = (6.0 \text{ cm}) \cos 30° = 5.2 \text{ cm}$$

These results may be checked with the Pythagorean theorem:

$$c^2 = a^2 + b^2 \quad \text{or} \quad (6.0 \text{ cm})^2 = (3.0 \text{ cm})^2 + (5.2 \text{ cm})^2$$

EXAMPLE 2-17

To measure the height of a vertical building, a man measured a distance of 120 m in a straight line from its base and then from that location determined that the angle from the horizontal to the top of the building was 42° (Fig. 2-20). How high is the building?

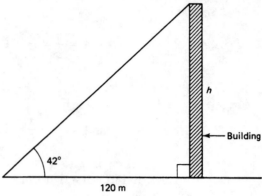

Figure 2-20

Solution: In the right-angled triangle, $\tan 42° = \dfrac{h}{120 \text{ m}}$; therefore,

$$h = (120 \text{ m}) \tan 42° = 108 \text{ m}$$

EXAMPLE 2-18

To measure the distance AB between two points across a river, a surveyor sights one point B on the opposite side of the river from the other location A. He then moves perpendicular to this line AB a distance of 70.0 m to a point C and measures the angle C to be 68° (see Fig. 2-21). What is the distance between A and B?

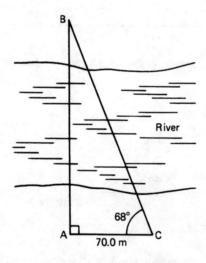

Figure 2-21

Solution: $\tan 68° = \dfrac{AB}{AC}$; therefore,

$$AB = AC \tan 68° = (70.0 \text{ m}) \tan 68° = 173 \text{ m}$$

PROBLEMS

73 Find the sines, cosines, and tangents of the angles A and B in Fig. 2-22.
74 Find the lengths of the sides a and b in the right-angled triangle in Fig. 2-23.
75 To see if his forms are square, a builder measures the lengths of the diagonals to the

Sec. 2-5 / Trigonometry

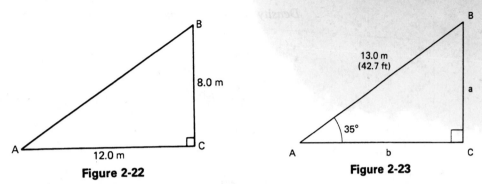

Figure 2-22

Figure 2-23

corners and finds that they are both 5.25 m long. If the length of one side of the forms is 3.25 m, (a) how long is the other side; (b) what are the angles between the sides and the diagonals?

76 The members of symmetric roof trusses are illustrated in Fig. 2-24. Find the angles and the lengths of all the sides.

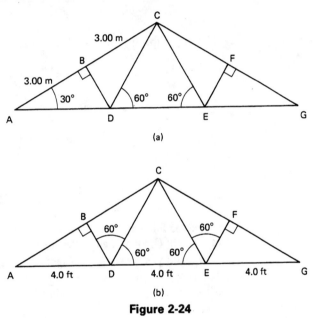

Figure 2-24

77 To measure the height of a tree, a person measures 80.0 m in a straight line from its base and from that location finds that the angle from the horizontal to the tree top is 36°. How high is the tree?

SUMMARY

Formulas relate physical quantities. A *ratio* is a comparison by division and a *proportion* is an equality of ratios.

Graphs are diagrams illustrating relationships between variables. The slope of a graph is the ratio of the rise to the run:

$$\text{slope} = \frac{\text{rise}}{\text{run}}$$

If the graph is a straight line that passes through the origin, the quantities are in proportion and the proportionality constant is equal to the slope of the line.

Density is the ratio of the mass m to the volume V:

$$\rho = \frac{m}{V}$$

Weight density is the ratio of the weight to the volume:

$$D = \frac{w}{V}$$

Relative density

$$\rho = \frac{\text{mass of material}}{\text{mass of equal volume of the reference material}}$$

$$= \frac{\text{weight of material}}{\text{weight of equal volume of the reference material}}$$

For any right-angled triangle that has sides a and b and hypotenuse c (Fig. 2-16):

$$c^2 = a^2 + b^2 \quad \text{Pythagorean theorem}$$

$$\sin A = \frac{a}{c} \quad \cos A = \frac{b}{c} \quad \tan A = \frac{a}{b}$$

$$\sin(90° - A) = \cos A \quad \cos(90° - A) = \sin A$$

QUESTIONS

1. Define the following terms: (a) formula; (b) ratio; (c) proportion; (d) density; (e) slope of a graph; (f) sine of an angle; (g) cosine of an angle.
2. How could you measure the height of a tree from the ground?
3. To check to see if the forms are "square" (at right angles 90°), we need only check to see if the diagonals are equal. Explain why.
4. What information could you obtain from a graph that was a straight line passing through the origin?
5. What is meant by the term "cross-sectional area"?
6. How could you check to see if an ingot is pure gold?

REVIEW PROBLEMS

1. Find the areas of the shapes in Fig. 2-25.
2. Determine the volume of a sphere of radius (a) 3.50 cm; (b) 4.80 in.
3. Find the total surface areas of cement blocks that have sides with lengths (a) 35 cm × 25 cm × 8.0 cm; (b) 12.0 in. × 10.0 in. × 3.00 in.
4. What surface area of a foundation could be poured (a) to a depth of 25.0 cm with 5.00 m³ of concrete; (b) to a depth of 6.00 in. with 54.0 ft³ of concrete?
5. How many liters of water are required to fill completely a barrel that has an inside diameter of 45.0 cm and a depth of 1.15 m?
6. How many liters of concrete would be required to pour to a depth of 18.0 cm, a foundation 15.0 m × 25.0 m?

Review Problems

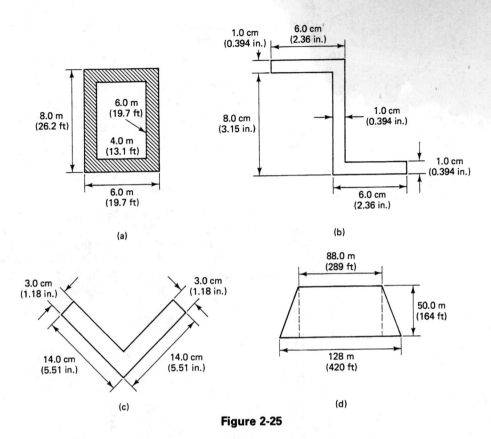

Figure 2-25

7 Determine the cross-sectional area of the steel beam illustrated in Fig. 2-26.

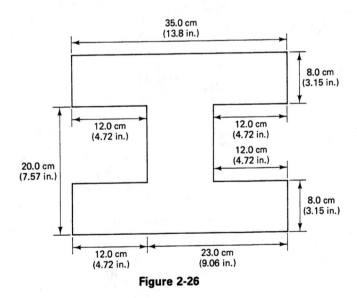

Figure 2-26

8 The cost C of gold varies directly as its mass m. If 28.5 g of gold costs $350, find (a) the formula; (b) the cost of 165 g of gold; (c) the mass of gold that can be purchased for $850.

9 The kinetic energy E of an object varies directly as the square of its speed v. If an object moving at 8.20 ft/s has a kinetic energy of 11.4 ft · lb (foot pounds), find (a) the formula;

(b) the kinetic energy when the mass moves at 4.00 ft/s; (c) the speed of the mass when its kinetic energy is 600 ft·lb.

10 The mass m of a sphere of a certain material varies as the cube of its diameter d, and a sphere of radius 5.00 cm has a mass of 205 g. Determine (a) the formula; (b) the mass of a sphere of the same material that has a diameter of 12.5 cm; (c) the diameter of the sphere of the same material that has a mass of 545 g.

11 For a constant force, the acceleration a of an object varies inversely as its mass m, and a mass of 5.0 kg accelerates at 3.0 m/s². Find (a) the formula; (b) the acceleration if the mass is 15.0 kg; (c) the mass that the force would accelerate at 1.50 m/s².

12 (a) Draw a graph of the relationship

$$s = (2.50 \text{ m/s})t + 8.00 \text{ m}$$

for values of t from 2.0 s to 12.0 s. (b) From the graph determine the values of s for the following elapsed times: 4.50 s, 5.00 s, and 9.50 s. (c) Check the results of part (b) using the formula. (d) From the graph, determine the values of t for the following values s: 15.5 m, 20.5 m, and 35.5 m. (e) Check the answers to part (d) using the formula.

13 Determine the equations for the graphs in Fig. 2-27.

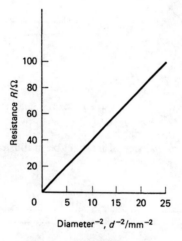

(a) Electric resistance of a wire versus the inverse square of its diameter

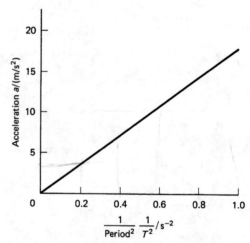

(b) Acceleration versus the inverse square of the period for an object moving in a circle

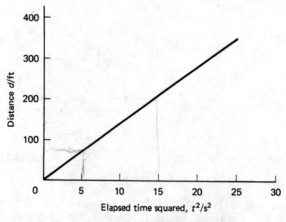

(c) Distance versus elapsed time squared

Figure 2-27

Review Problems

14 (a) Draw a graph of the force F between two objects versus the distance d between their centers using the data in Table 2-9. (b) Draw a graph of the force F versus the inverse square of their separation ($1/d^2$). (c) Find the formula.

TABLE 2-9

Force F/N	540	135	60	34	22
Distance d/cm	1.0	2.0	3.0	4.0	5.0

15 What is the density of an object that has (a) a mass of 68.0 g and a volume of 8.20 cm³; (b) a mass of 7.76×10^{-3} slug and a volume of 1.20 in³?

16 What is the volume of (a) a 25.0 kg block of steel; (b) a 12.5 lb block of steel?

17 An irregularly shaped block of metal has a mass of 15.0 kg. When it is lowered into a container that is completely full of water, 1.80 L overflows. Determine the density of the metal.

18 What volume is occupied by (a) 5.75 kg of propane gas and (b) 20.0 lb of propane gas at 1°C and 1 atm pressure?

19 Determine the diameter of a steel ball bearing that has a mass of (a) 350 g; (b) 225 g; (c) 0.0150 slug.

20 If 1.00 oz of gold costs $330.00, what volume could you buy for (a) $3000.00; and (b) $750.00?

21 Find the mass and weight of a 5.00 cm long copper pipe that has inner and outer diameters of 1.25 cm and 1.45 cm.

22 Find the sines, cosines, and tangents of the angles A and B in Fig. 2-28.

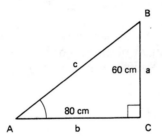

Figure 2-28

23 Find the lengths of sides a and b in Fig. 2-29.

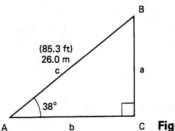

Figure 2-29

24 To measure the distance AB between two points across a river, a surveyor sights one point B on the opposite side from the other location A. She then moves perpendicular to line AB a distance of 56.0 m to point C and measures the angle C to be 54°. What is the distance between A and B?

3
INTRODUCTION TO VECTOR QUANTITIES

Vector quantities are frequently used when we analyze motion, forces, electricity, magnetism, and many other areas of science and engineering technology. Therefore, a knowledge of vectors is very important for science and engineering.

3-1 VECTOR AND SCALAR QUANTITIES

We can completely describe many physical quantities by stating their **magnitude** (size and units) only: these are called **scalar quantities.** They are usually measured directly by means of a scale such as that on a ruler, thermometer, or balance. Some typical categories of scalar quantities are: mass, temperature, length, work, and energy.

For some other physical quantities a direction may be as important as a magnitude. These are known as **vector quantities.** Some common categories of vector quantities are: displacement, velocity, acceleration, and force.

Representation of Vectors

In most textbooks the symbols representing vector quantities are printed in boldface (heavy) type and scalar quantities are shown in italic type. For example, **F** is the symbol used to represent a force vector that has a magnitude of F, while m is the symbol used to represent the scalar quantity mass. In the written form we represent a force vector by the symbol $\vec{F}$.

Vectors are usually represented by arrows in scale diagrams (Fig. 3-1). The "tail" of the arrow is called the **origin** of the vector, and the "head" of the arrow is called its **terminal point.** The length of the arrow to scale (in the correct units) is the magnitude, and the direction of the vector corresponds to the direction of the arrow. Usually, the direction is given with respect to some reference. The scale and the angles must normally be included in the scale diagrams.

Sec. 3-1 / Vector and Scalar Quantities

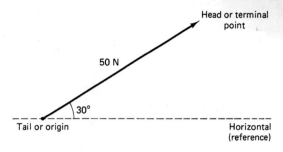

Note that the length of the arrow is 5.0 cm which corresponds to 50 N in the chosen scale

Scale 1 cm ≡ 10 N

Figure 3-1 Scale diagram of a 50 N force directed 30° above the horizontal.

Scale Factor

To determine the length of an arrow representing a vector in a scale diagram, we first determine a scale factor and then multiply the magnitude of the vector by that scale factor. The scale factor is determined from

$$\text{scale factor} = \frac{\text{scale length unit}}{\text{equivalent vector unit}}$$

For example, if we have a scale where 1.0 cm is equivalent to 5.0 km (written as 1.0 cm ≡ 5.0 km), then

$$\text{scale factor} = \frac{\text{scale length unit}}{\text{equivalent vector unit}} = \frac{1.0 \text{ cm}}{5.0 \text{ km}}$$

EXAMPLE 3-1

(a) Using a scale of 1.0 cm ≡ 25 km, determine the length that would represent the magnitude of a 150 km due east vector. (b) Determine the length that would represent the magnitude of an upward 75 lb force vector using a scale of 1.0 in. ≡ 50 lb.

Solution

(a) Scale factor $= \dfrac{1.0 \text{ cm}}{25 \text{ km}}$

length $= 150 \text{ km} \times$ scale factor

$= 150 \text{ km} \times \dfrac{1.0 \text{ cm}}{25 \text{ km}}$

$= 6.0 \text{ cm}$

(b) Scale factor $= \dfrac{1.0 \text{ in.}}{50 \text{ lb}}$

length $= 75 \text{ lb} \times$ scale factor

$= 75 \text{ lb} \times \dfrac{1.0 \text{ in.}}{50 \text{ lb}}$

$= 1.5 \text{ in.}$

Angles in Standard Position

When a vector is drawn with its origin at the origin of a coordinate system it is said to be in the **standard position** or in **polar form**. It is then expressed as a magnitude in a direction measured counterclockwise from the positive *x*-axis (see Fig. 3-2).

The following procedure may be used to draw a scale diagram of a vector:

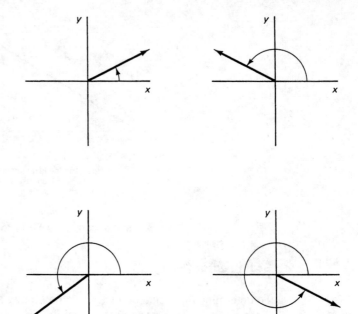

Figure 3-2 Angles are measured counterclockwise from the positive x-axis.

1. Choose a *convenient* scale to represent the vector so that the diagram will fit into the space to be used.
2. Draw the reference direction (such as the x-axis) on the paper and use a protractor to draw a line in the direction of the vector.
3. Use the scale factor to determine the length representing the magnitude of the vector.
4. Measure this length along the line in the direction of the vector and place the arrowhead at the terminal point.
5. Label the diagram with a title, scale, magnitudes, and angles of the vectors.

EXAMPLE 3-2

Using a scale of 1.0 cm ≡ 25 m, draw scale diagrams representing the following vectors: **(a)** 50 m at 30°; **(b)** 75 m at 120°; **(c)** 40 m at 250°.

Solution: In each case we draw the direction to represent the x-axis. The scale factor is $\frac{1.0 \text{ cm}}{25 \text{ m}}$.

(a) Draw a line at 30° counterclockwise from the x-axis (see Fig. 3-3a). This represents the direction of the vector. The length of this line (representing the magnitude of 50 m) is

$$50 \text{ m} \times \text{scale factor} = 50 \text{ m} \times \frac{1.0 \text{ cm}}{25 \text{ m}} = 2.0 \text{ cm}$$

(b) Draw a line at 120° counterclockwise from the x-axis (Fig. 3-3b). The length of the line should be

$$75 \text{ m} \times \frac{1.0 \text{ cm}}{25 \text{ m}} = 3.0 \text{ cm}$$

(c) Draw a line 250° counterclockwise from the x-axis (Fig. 3-3c) with a length

Sec. 3-2 / Distance and Displacement

$$40 \text{ m} \times \frac{1.0 \text{ cm}}{25 \text{ m}} = 1.6 \text{ cm}$$

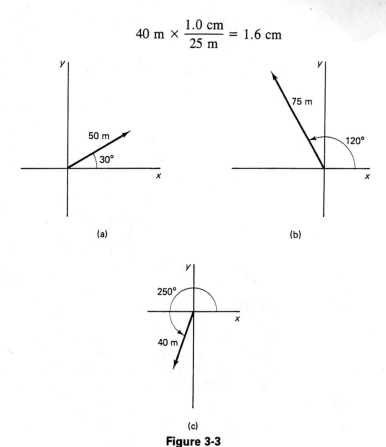

Figure 3-3

PROBLEMS

1. Using a scale of 2.0 cm ≡ 50 km, find the lengths that represent the following vectors: **(a)** 200 km east; **(b)** 75 km south; **(c)** 450 km north; **(d)** 325 km west; **(e)** 320 km north; **(f)** 35 km south.

2. Using a scale of 1.0 in. ≡ 50 mi, find the lengths that represent the following vectors: **(a)** 350 mi north; **(b)** 225 mi south; **(c)** 80 mi east; **(d)** 575 mi west.

3. Using a scale of 1.0 cm ≡ 25 N, find the lengths that represent the following force vectors: **(a)** 75 N east; **(b)** 420 N upward; **(c)** 720 N downward; **(d)** 35 N upward.

4. Draw the following vectors in standard position, using an appropriate scale: **(a)** 250 km at 45°; **(b)** 85 N at 160°; **(c)** 25 km at 120°; **(d)** 75 N at 58°; **(e)** 40 N at 235°; **(f)** 60 m at 280°; **(g)** 70 m/s at 310°; **(h)** 50 m/s at 225°; **(i)** 90 N at 145°; **(j)** 28 mi at 50°; **(k)** 80.0 lb at 120°; **(l)** 24 ft/s at 135°; **(m)** 142 lb at 315°.

3-2 DISTANCE AND DISPLACEMENT

Both scalar and vector quantities are used extensively in the analysis of motion. The term *position* is used to indicate a location relative to some reference point (e.g., 500 km north of New York) or in terms of some coordinate system (e.g., latitude and longitude).

A moving object continually changes its position. The **distance** that an object travels is the length measured along the path that it takes. It is a scalar quantity because no particular direction is given.

The change in position is called the **displacement s**. It is a vector quantity that represents the length in the direction of the straight line from the initial position to the final position (Fig. 3-4).

Since both distances and displacements are lengths, they have length units such as meters, kilometers, feet, and miles.

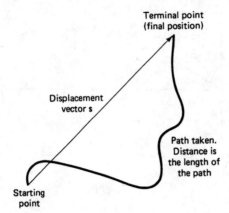

Figure 3-4 Difference between distance and displacement.

Directions

Directions are often given in terms of degrees measured from one direction toward some other direction. For example, 20° south of west is the direction measured from west and it is 20° toward the south (i.e., 20° anticlockwise from west). However, 20° west of south is the direction measured from the south and is 20° toward the west (Fig. 3-5).

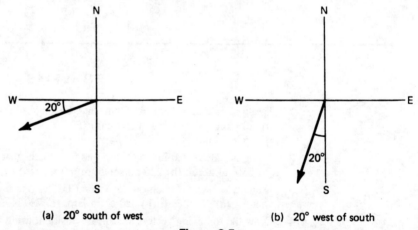

(a) 20° south of west   (b) 20° west of south

Figure 3-5

EXAMPLE 3-3

Draw the scale diagram for a 30 km displacement that is 35° south of east using a scale of 1 cm ≡ 10 km.

Solution: In the given scale 30 km is represented by a length of

$$30 \text{ km} \times (\text{scale factor}) = 30 \text{ km} \times \frac{1 \text{ cm}}{10 \text{ km}} = 3.0 \text{ cm}$$

Draw lines to indicate the north-south and east-west directions. The direction 35° south of east is measured 35° from east toward the south. An arrow 3.0 cm long is then drawn in that direction (Fig. 3-6).

Sec. 3-3 / Speed and Velocity

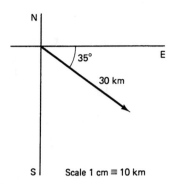

Figure 3-6 Displacement of 30 km in a direction 35° south of east.

PROBLEMS

5 Draw the scale diagrams of the following displacement vectors using a scale of 1 cm ≡ 20 km: **(a)** 50 km due west; **(b)** 80 km 25° south of west; **(c)** 130 km 35° east of north; **(d)** 35 km 22° west of south; **(e)** 79 km 48° north of east; **(f)** 123 km at 33° south of east; **(g)** 67 km at 38° south of west; **(h)** 28 km at 23° north of west; **(i)** 107 km at 46° east of north; **(j)** 95 km at 8° west of north; **(k)** 76 km at 52° east of north.

6 Using a scale of 1.0 in. ≡ 50 mi, draw the scale diagrams of the following displacement vectors: **(a)** 75 mi at 30° east of south; **(b)** 120 mi at 25° south of west; **(c)** 175 mi at 43° west of south; **(d)** 90 mi at 75° north of west; **(e)** 80 mi at 68° west of north; **(f)** 130 mi at 35° east of north; **(g)** 190 mi at 68° east of north; **(h)** 160 mi at 20° east of south; **(i)** 60 mi at 63° west of south; **(j)** 78 mi at 22° north of west.

3-3 SPEED AND VELOCITY

Speed is a scalar quantity that is used to describe how fast an object is moving; the direction of motion is not specified. The **average speed** v of an object is defined as the ratio of the distance d that it travels to the time t taken.

$$v = \frac{d}{t} \tag{3-1}$$

The corresponding units are meters per second (m/s), kilometers per hour (km/h), feet per second (ft/s), and miles per hour (mi/h).

If an object travels equal distances in equal time intervals, it is said to have a constant (or uniform) speed. For example, a car has a constant speed of 60 km/h when it travels 1 km in each minute.

EXAMPLE 3-4

Determine the average speed of an aircraft if it travels **(a)** 750 km in 3.0 h; **(b)** 828 mi in 4.00 h.

Solution

Data:
(a) $d = 750$ km, $t = 3.0$ h, $v = ?$ **(b)** $d = 828$ mi, $t = 4.00$ h, $v = ?$

Equation: $v = \dfrac{d}{t}$

Substitute:

(a) $v = \dfrac{750 \text{ km}}{3.0 \text{ h}} = 250$ km/h **(b)** $v = \dfrac{828 \text{ mi}}{4.00 \text{ h}} = 207$ mi/h

EXAMPLE 3-5

How far does a car travel in 2.5 h if its average speed is (a) 56 km/h; (b) 38 mi/h?

Solution

Data:
(a) $v = 56$ km/h, $t = 2.5$ h, $d = ?$ (b) $v = 38$ mi/h, $t = 2.5$ h, $d = ?$

Equation: $v = \dfrac{d}{t}$

Rearrange: $d = vt$

Substitute:
(a) $d = (56 \text{ km/h})(2.5 \text{ h})$
 $= 140$ km

(b) $d = (38 \text{ mi/h})(2.5 \text{ h})$
 $= 95$ mi

EXAMPLE 3-6

What is the average speed of an automobile if it travels at constant speeds of (a) 80.0 km/h for 1.50 h and then 50.0 km/h for an additional 30.0 min; (b) 50.0 mi/h for 1.50 h and then 30.0 mi/h for 30.0 min?

Solution

Data:
(a) $v_1 = 80.0$ km/h, $t_1 = 1.50$ h,
    $v_2 = 50.0$ km/h, $t_2 = 0.500$ h,
    average speed = ?

(b) $v_1 = 50.0$ mi/h, $t_1 = 1.59$ h,
    $v_2 = 30.0$ mi/h, $t_2 = 0.500$ h,
    average speed = ?

Equation: In each case we must first determine the total distance traveled and the total time taken using $v = d/t$ for each segment of the trip, and then for the totals.

Rearrange: $d = vt$, $d_{total} = v_1 t_1 + v_2 t_2$

Substitute:
(a) $d = (80.0 \text{ km/h})(1.50 \text{ h}) + (50.0 \text{ km/h})(0.500 \text{ h})$
  $= 145$ km
(b) $d = (50.0 \text{ mi/h})(1.50 \text{ h}) + (30.0 \text{ mi/h})(0.500 \text{ h})$
  $= 90.0$ mi

Equation: Average speed $= \dfrac{\text{total distance traveled}}{\text{total time taken}}$

Substitute:
(a) Average speed $= \dfrac{145 \text{ km}}{1.50 \text{ h} + 0.50 \text{ h}} = 72.5$ km/h

(b) Average speed $= \dfrac{90.0 \text{ mi}}{1.50 \text{ h} + 0.500 \text{ h}} = 45.0$ mi/h

---

The **velocity** of a moving object is a vector quantity defined as the time rate of change of position; it represents a speed in a given direction. We define the average velocity of an object as the ratio of its displacement **s** (or change in position $\Delta \mathbf{r}$)* to the time t taken.

$$\mathbf{v} = \frac{\Delta \mathbf{r}}{t} = \frac{\mathbf{s}}{t} \tag{3-2}$$

---

*As noted in Chapter 2, Δ is used to indicate "a change in"; thus $\Delta \mathbf{r}$ represents a change in position.

Sec. 3-3 / Speed and Velocity

If the object moves in a straight line at a constant speed, it is said to have a constant velocity. This is the simplest motion to describe because changes in direction are more difficult to describe.

Speed and velocity are expressed in units of length divided by time (e.g., m/s, km/h, ft/s, mi/h, etc.). Useful conversion factors are

$$1 \text{ m/s} = 3.6 \text{ km/h} \quad \text{and} \quad 88 \text{ ft/s} = 60 \text{ mi/h}$$

EXAMPLE 3-7

Determine (i) the average speed and (ii) the average velocity of an automobile (a) if it travels a total distance of 165 km in 3.00 h and its final position is 138 km due south of its starting point; (b) if it travels a total distance of 96.0 mi in 3.00 h and its final position is 66.0 mi north of its starting point.

Solution

Data:

(a) $d = 165$ km, $t = 3.00$ h,
 $\mathbf{s} = 138$ km south, $v = ?$,
 $\mathbf{v} = ?$

(b) $d = 96.0$ mi, $t = 3.00$ h,
 $\mathbf{s} = 66.0$ mi north, $v = ?$,
 $\mathbf{v} = ?$

(i) *Equation:* Average speed $v = \dfrac{d}{t}$

Substitute:

(a) $v = \dfrac{165 \text{ km}}{3.0 \text{ h}} = 55.0$ km/h

(b) $v = \dfrac{96.0 \text{ mi}}{3.00 \text{ h}} = 32.0$ mi/h

(ii) *Equation:* Average velocity $\mathbf{v} = \dfrac{\mathbf{s}}{t}$

Substitute:

(a) $\mathbf{v} = \dfrac{138 \text{ km}}{3.0 \text{ h}}$ south
  $= 46.0$ km/h south

(b) $\mathbf{v} = \dfrac{66.0 \text{ mi}}{3.00 \text{ h}}$ north
  $= 22.0$ mi/h north

PROBLEMS

7. What is the average speed of an aircraft if it travels (a) 520 km in 55 min; (b) 470 mi in 50.0 min?

8. What is the average speed of a car that travels 38.0 mi in 45.0 min (a) in feet per second; (b) in miles per hour?

9. A bicycle travels 8.20 km in 22.0 min. What is its average speed (a) in meters per second; (b) in kilometers per hour?

10. The star Sirius is 8.1×10^{13} km from the earth. If the speed of light is 3.0×10^8 m/s, how many years does it take the light emitted by Sirius to reach the earth?

11. If a car travels at an average speed of 65.0 km/h, (a) how long does it take to travel 315 km; (b) how far will it travel in 3.50 h?

12. If the average speed of a car is 92.5 km/h, (a) how far does it travel in 4.0 h; (b) how long does it take to travel 245 km?

13. If the average speed of a car is 55.0 mi/h, (a) how far does it travel in 4.00 h; (b) how long does it take to travel 245 mi?

14. A satellite in orbit about the earth travels at an average speed of 27 000 km/h. How far does it travel in (a) 3.50 h; (b) 2.00 days; (c) 20.0 min; (d) 15.0 s?

15. A satellite in orbit about the earth travels at an average speed of 17 000 mi/h. How far does it travel in (a) 3.50 h; (b) 2.00 days; (c) 20.0 min; (d) 15.0 s?

16. A car travels at 80.0 km/h for 2.50 h and then 55.0 km/h for 1.20 h. What was its average speed for the whole journey?
17. An aircraft travels at 350 km/h for 1.25 h and then 380 km/h for 1.80 h. What is its average speed for the complete journey?
18. What is the total average speed of a car if it travels 55.0 km in 1.50 h, then 85.0 km in the next 2.00 h, and finally 75.0 km in the next 1.80 h?
19. A car travels at 60.0 mi/h for 2.50 h and then at 35.0 mi/h for 1.20 h. What is its average speed?
20. A ship is initially 120 km north of a lighthouse. What is its average velocity if, after 6.0 h, it is 360 km north of the lighthouse?
21. Determine the average velocity of an aircraft if it flies in a straight line and travels 1250 km due north in 2.20 h.
22. Determine the average speed of a ship if it travels 120 km in 3.80 h (a) in kilometers per hour; (b) in meters per second.
23. If an aircraft flies with a velocity of 650 km/h due west, what is its displacement after an elapsed time of (a) 3.00 h; (b) 50.0 min; (c) 45.0 s?
24. Determine the average velocity of a ship in (a) miles per hour and (b) feet per second if it travels 75.0 mi south in 3.80 h.
25. A boy sees a flash of lightning and hears the thunder 8.0 s later. How far away is the storm (a) if the speed of sound in air is 330 m/s, and the speed of light is 3.0×10^8 m/s; (b) if the speed of sound in air is 1080 ft/s and the speed of light is 1.85×10^5 mi/s?

3-4 ACCELERATION

Objects rarely travel in straight lines at constant speeds. They may change their speed, direction of motion, or both speed and direction of motion. They are then said to be accelerating. The **acceleration a** of an object is defined as the time rate of change of velocity. Therefore, if the velocity of an object changes from $\mathbf{v}_0$ to $\mathbf{v}_1$ in a time interval t, the average acceleration is

$$\mathbf{a} = \frac{\mathbf{v}_1 - \mathbf{v}_0}{t} = \frac{\Delta \mathbf{v}}{t} \qquad (3\text{-}3)$$

The units for acceleration have dimensions of length divided by time squared: meters per second per second (m/s^2), and feet per second per second (ft/s^2).

EXAMPLE 3-8

What is the magnitude of the average acceleration of an aircraft if during takeoff it accelerates in a straight line from rest to (a) 44.0 m/s in 11.0 s; (b) 140 ft/s in 11.2 s?

Solution

Data:
(a) $v_0 = 0$, $v_1 = 44.0$ m/s, $t = 11.0$ s, $a = ?$

(b) $v_0 = 0$, $v_1 = 140$ ft/s, $t = 11.2$ s, $a = ?$

Equation: $a = \dfrac{v_1 - v_0}{t}$

Substitute:

(a) $a = \dfrac{44.0 \text{ m/s} - 0}{11.0 \text{ s}}$
$= 4.00$ m/s^2

$a = \dfrac{140 \text{ ft/s} - 0}{11.2 \text{ s}}$
$= 12.5$ ft/s^2

EXAMPLE 3-9

What is the final speed of a racing car that accelerates in a straight line from rest for 8.00 s (a) at 4.20 m/s²; (b) at 15.0 ft/s²?

Solution

Data:
(a) $v_0 = 0$, $a = 4.20$ m/s², $t = 8.00$ s, $v_1 = ?$

(b) $v_0 = 0$, $a = 15.0$ ft/s², $t = 8.00$ s, $v_1 = ?$

Equation: $a = \dfrac{v_1 - v_0}{t}$

Rearrange: $v_1 = v_0 + at$

Substitute:
(a) $v_1 = 0 + (4.20 \text{ m/s}^2)(8.00 \text{ s})$
$= 33.6$ m/s

(b) $v_1 = 0 + (15.0 \text{ ft/s}^2)(8.00 \text{ s})$
$= 120$ ft/s

PROBLEMS

26 Determine the acceleration of an automobile if it travels due east in a straight line and changes speed from (a) 5.80 m/s to 16.2 m/s in 2.00 s; (b) 3.00 m/s to 20.00 m/s in 1.80 s; (c) rest to 95.0 km/h in 12.0 s; (d) rest to 55.0 km/h in 15.0 s; (e) 75 km/h to 35 km/h in 15 s; (f) 15.0 ft/s to 40.0 ft/s in 2.00 s; (g) rest to 60.0 mi/h in 12.0 s.

27 What is the velocity of a car after 8.00 s if it accelerates in a straight line due east from rest at (a) 2.85 m/s²; (b) 8.00 ft/s²?

28 What is the average acceleration of an electron in an oscilloscope if it accelerates from rest to a final speed of 6.50×10^6 m/s in 12.0 ns?

29 What is the average acceleration of a rocket that moves in a straight line from rest to a speed of (a) 11 500 km/h in 25.0 min; (b) 4500 mi/h in 25.0 min?

30 An electron accelerates from rest in a straight line at 2.50×10^{14} m/s². Determine its final speed after (a) 36.0 ns; (b) 24.0 ns.

3-5 COLLINEAR VECTORS

Two or more vectors with the same units may be "added" to produce a single vector called the **resultant,** which represents their combined effect. However, the process of vector addition is different from the addition of scalar quantities because the directions of the vectors must be considered. The rules for the addition of vectors can be determined directly from our knowledge of displacements, forces, and geometry. We shall consider the more general cases for vector addition in Chapter 6. For the present we shall only use collinear vectors.

Vectors are said to be **collinear** when they are located on the same straight line; they are either parallel (in the same direction) or antiparallel (in opposite directions).

Suppose that a man walks four city blocks east and then an additional 3 blocks east. He has actually walked a total distance of 7 blocks and his final position is 7 blocks east of his starting position (Fig. 3-7). Therefore, the vector sum of the 4-block and 3-block displacements toward the east is 7 blocks east. In general, if two or more vectors are parallel, their vector sum is merely the sum of their magnitudes in the same direction. Also, the order of the vector addition does not matter; the same answer applies if we reverse the order of the addition. For any two vectors **A** and **B**,

$$\mathbf{A} + \mathbf{B} = \mathbf{B} + \mathbf{A}$$

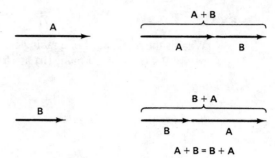

Figure 3-7

Negative Vectors

If two vectors are in opposite directions, the resultant representing their vector sum is equal to the difference in their magnitudes in the direction of the larger. For example, if a man walks 4 blocks east and then 3 blocks west, his final position is 1 block east of his starting position (Fig. 3-8).

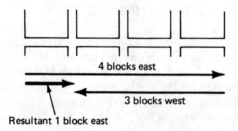

Figure 3-8 The resultant of displacements 4 blocks east and 3 blocks west is 1 block east.

To simplify the addition of vectors, we often designate one direction as the positive direction; vectors in the opposite direction are given a negative sign. Thus down is taken as the negative of up, south is the negative of north, left is the negative of right, and so on. Using this sign convention, we see that the vector sum of collinear vectors (with the same units) is the algebraic sum in the positive direction.

EXAMPLE 3-10

A car travels 120 km east and then turns around and goes 80 km west. If the total driving time is 4.0 h, determine (a) the total distance traveled; (b) the total displacement; (c) the average speed; (d) the average velocity.

Solution

Data: $s_1 = 120$ km east, $s_2 = 80$ km west, $t = 4.0$ h

(a) Equation: $d = 120$ km $+ 80$ km $= 200$ km (scalar quantities)

(b) Equation: Since west is the negative of east,

$$s = (120 \text{ km east}) + (-80 \text{ km east}) = 40 \text{ km east}$$

(c) Equation: $v = \dfrac{d}{t}$

Substitute: $v = \dfrac{200 \text{ km}}{4.0 \text{ h}} = 50$ km/h

(d) Equation: $\mathbf{v} = \dfrac{\mathbf{s}}{t}$

Substitute: $\mathbf{v} = \dfrac{40 \text{ km}}{4.0 \text{ h}}$ east $= 10$ km/h east

PROBLEMS

31. Determine the total displacement of an aircraft that flies (a) 45 km due west and then 65 km due east; (b) 320 mi north and then 110 mi south.

32. A woman drives a car 48 km due south and then 25 km due north. (a) What is the total distance that she travels? (b) What is the total displacement?

33. A car travels 85.0 km east and then turns around and goes 45.0 km west. If the total driving time is 2.50 h, determine (a) the total distance traveled; (b) the total displacement; (c) the average speed; (d) the average velocity.

34. A car travels 85.0 mi north and then 35.0 mi south in a total time of 2.00 h. Determine (a) the total distance traveled; (b) the total displacement; (c) the average speed; (d) the average velocity.

35. A ball is dropped a vertical height of 35.0 m to the ground and bounces back 12.5 m in a total time of 4.27 s. Find (a) the total distance traveled; (b) the total displacement; (c) the average speed; (d) the average velocity of the ball.

36. What is the average acceleration of a car if it changes its velocity from 15.0 m/s due west to (a) 25.0 m/s west in 12.0 s; (b) 7.25 m/s west in 8.00 s; (c) 12.5 m/s east in 25.0 s; (d) 32.0 m/s east in 2.50 min?

SUMMARY

Scalar quantities can be described by a magnitude only. *Vector quantities* have both magnitude and direction. Vector quantities can be represented by arrows in scale diagrams.

The *magnitude* of the vector quantity is represented by the length of the arrow to scale, and its direction is the direction of the arrow with respect to some reference. A *vector* is in *standard position* if its origin is at the origin of a coordinate system; the direction is then measured counterclockwise from the positive x-axis.

Distance is the length of the path traveled.

Displacement is a vector quantity equal to the change in position (a straight line from the origin to the terminal point of the path).

Average speed v is the ratio of the distance d traveled to the time t taken.

$$v = \frac{d}{t}$$

Velocity **v** is a vector quantity equal to the time rate of change of position. The average velocity is the ratio of the change in position $\Delta \mathbf{r}$ (or displacement **s**) to the time t taken.

$$\mathbf{v} = \frac{\Delta \mathbf{r}}{t} = \frac{\mathbf{s}}{t}$$

Acceleration is the time rate of change of velocity. Average acceleration **a** is the ratio of the change in velocity $\Delta \mathbf{v}$ to the time t taken.

$$\mathbf{a} = \frac{\Delta \mathbf{v}}{t} = \frac{\mathbf{v}_1 - \mathbf{v}_0}{t}$$

Collinear vectors are located on the same straight line. A *resultant vector* represents the combined effect (sum) of a number of vectors. *Negative vectors* are in the opposite direction to the designated positive direction.

QUESTIONS

1. Define the following terms: (a) vector quantity; (b) scalar quantity; (c) magnitude; (d) distance; (e) displacement; (f) average speed; (g) average velocity; (h) average acceleration; (i) collinear vectors; (j) negative vectors.
2. If a person travels around the world and returns to her starting position, what is her total displacement?
3. Is an object accelerating when it is traveling in a circle at a constant speed? Explain your answer.
4. Vector quantities are used in navigation. Explain how.
5. List several physical quantities that are vector quantities.

REVIEW PROBLEMS

1. Draw the scale diagrams of the following displacement vectors using a scale of 1 cm ≡ 50 km: (a) 450 km due east; (b) 575 km 38° west of north; (c) 375 km 26° east of south; (d) 250 km at 35° south of west; (e) 325 km at 67° east of north; (f) 275 km at 34° east of south.
2. Draw the scale diagrams of the following displacement vectors using a scale of 1 in. ≡ 35 mi: (a) 87.5 mi at 86° east of north; (b) 105 mi at 34° west of south; (c) 140 mi at 52° west of north; (d) 70 mi at 29° south of west.
3. What is the average speed of a car that travels 38.0 m in 2.20 s? Express the answer in kilometers per hour.
4. A ship transmits a radar pulse and receives the reflection from an aircraft after 250 μs. How far away is the aircraft if the speed of the radar pulse is (a) 3.00×10^8 m/s; (b) 1.86×10^5 mi/s?
5. How far does an aircraft travel in 45.0 s if it has a constant speed of (a) 535 km/h; (b) 485 mi/h?
6. Determine the average velocity of an aircraft if its position, measured from an airfield, changes (a) from 450 km due east to 225 km due east in 45 min; (b) from 835 mi due north to 285 mi due north in 55.0 min?
7. If a ship changes its position from 360 km due west of a lighthouse to 240 km due west of the same lighthouse in 16 h, what is its average velocity?
8. What is the average acceleration of a car that travels due east and changes its speed (a) from 15 m/s to 45 m/s in 8.0 s; (b) from 65.0 ft./s to 120 ft/s in 12.0 s?
9. An electron accelerates in a straight line at 4.5×10^{10} m/s² in a cathode-ray tube. If its initial speed was 2.5×10^4 m/s, what is its speed after 3.0×10^{-6} s?
10. What is the acceleration of an aircraft if it travels due north and changes its speed (a) from 128 m/s to 160 m/s in 15.0 s; (b) from 325 ft/s to 485 ft/s in 12.0 s?
11. Determine the acceleration of a car if it travels due south in a straight line and changes speed from (a) 7.70 m/s to 16.9 m/s in 8.00 s; (b) 4.50 m/s to 25.5 m/s in 7.00 s; (c) rest to 45.0 km/h in 15.0 s; (d) rest to 135 km/h in 25.0 s; (e) rest to 60.0 mi/h in 7.50 s; (f) 30.0 mi/h to 75.0 mi/h in 12.0 s.
12. How fast is a truck going after 12.0 s if it accelerates in a straight line from rest at (a) 2.50 m/s²; (b) 9.25 ft/s²?
13. A person walks 3.00 km west and then turns around and walks 5.00 km east. If the total time of the walk is 2.50 h, determine (a) the total distance traveled; (b) the total displacement; (c) the average speed; (d) the average velocity.
14. An aircraft flies 560 km due north, and then 285 km due south in a total time of 4.50 h. Determine (a) the total distance flown; (b) the total displacement; (c) the average speed; (d) the average velocity of the aircraft.

Review Problems

15. A ship travels 25.0 mi due north and then 65.0 mi due south in a total time of 5.00 h. Determine (a) the total distance traveled; (b) the total displacement; (c) the average speed; (d) the average velocity.

16. A boy fires a stone directly upward with a slingshot from the top of a bridge. The stone rises 45.0 m and then falls into the water 125 m below the bridge 8.92 s later. Determine (a) the total distance traveled; (b) the displacement; (c) the average speed; (d) the average velocity of the ball.

4
STRAIGHT-LINE MOTION

Even though the causes may differ, the motion of most objects, such as celestial bodies, trains, cars, golf balls, falling bricks, atoms, and electrons, may be described by similar equations. In this chapter we consider a special type of motion in which the acceleration is constant and the object travels in a straight line. Motion with constant acceleration is called **uniformly accelerated motion.**

4-1 UNIFORMLY ACCELERATED MOTION

The average velocity **v** of an object is the ratio of its displacement **s** (or change in position $\Delta \mathbf{r}$) to the corresponding elapsed time t:

$$\mathbf{v} = \frac{\mathbf{s}}{t} = \frac{\Delta \mathbf{r}}{t} \tag{4-1}$$

When an object is moving in a straight line and changes its position by equal amounts in equal time intervals, its velocity **v** is constant (i.e., it travels in a straight line at a constant speed). This is the simplest type of motion to analyze because the magnitude of the displacement **s** (or change in position) is directly proportional to the elapsed time t:

$$\mathbf{s} = \mathbf{v}t \quad \text{and} \quad s \propto t$$

since **v** is constant. Therefore, if we plot a graph with displacement **s** (or change in position $\Delta \mathbf{r}$) as the ordinate (y-axis), and elapsed time t as the abscissa (x-axis), we obtain a straight line passing through the origin. The slope of this line is equal to the velocity **v** (Fig. 4-1).

When an object has an **average velocity v** for an elapsed time t, its displacement is

Sec. 4-1 / Uniformly Accelerated Motion

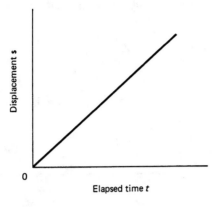

Figure 4-1 Displacement versus elapsed time graph for constant velocity.

$$s = \Delta r = vt \tag{4-2}$$

If its initial and final velocities are v_0 and v_1, its **average acceleration** is

$$a = \frac{v_1 - v_0}{t} = \frac{\Delta v}{t} \tag{4-3}$$

These two equations are valid for all types of motion.

Even though we are considering vector quantities, if an object travels in a straight line, the vectors are collinear. In this special case, with uniform (constant) acceleration, the magnitude of the acceleration is equal to the difference between the final speed v_1 and the initial speed v_0 divided by the time t taken.

$$a = \frac{v_1 - v_0}{t} \quad \text{or} \quad v_1 = v_0 + at \tag{4-4}$$

Remember, if one direction along the line of motion is taken as positive, then the opposite direction is taken as negative.

If we plot a graph of velocity versus the elapsed time for uniformly accelerated motion (Fig. 4-2), we obtain a straight line with a slope equal to the acceleration. Remember the units for these quantities:

Displacements and distances: meters (m), kilometers (km), feet (ft), or miles (mi)

Velocity and speed: meters per second (m/s), kilometers per hour (km/h), feet per second (ft/s), or miles per hour (mi/h)

Acceleration: meters per second squared (m/s^2) or feet per second squared (ft/s^2).

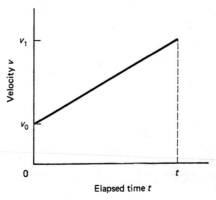

Figure 4-2 Velocity versus elapsed time graph for uniformly accelerated motion.

If v_0 and v_1 are the initial and final speeds, the average speed of an object in uniformly accelerated motion is given by

$$v = \frac{v_1 + v_0}{2} \qquad (4\text{-}5)$$

From Eq. 4-1, the magnitude of the displacement in an elapsed time t is

$$s = vt = \frac{(v_1 + v_0)}{2} t \qquad (4\text{-}6)$$

EXAMPLE 4-1

Determine **(i)** the distance traveled and **(ii)** the magnitude of the acceleration of a train if it accelerates **(a)** from 8.00 m/s to 28.0 m/s in 50.0 s; **(b)** from 22.0 ft/s to 88.0 ft/s in 50.0 s.

Solution

Data:
(a) $v_0 = 8.00$ m/s, $v_1 = 28.0$ m/s, $t = 50.0$ s, $s = ?$, $a = ?$

(b) $v_0 = 22.0$ ft/s, $v_1 = 88.0$ ft/s, $t = 50.0$ s, $s = ?$, $a = ?$

(i) Equation: $s = \left(\dfrac{v_1 + v_0}{2}\right) t$

Substitute:

(a) $s = \left(\dfrac{28.0 \text{ m/s} + 8.00 \text{ m/s}}{2}\right)(50.0 \text{ s}) = 900$ m

(b) $s = \left(\dfrac{88.0 \text{ ft/s} + 22.0 \text{ ft/s}}{2}\right)(50.0 \text{ s}) = 2750$ ft

(ii) Equation: $a = \dfrac{v_1 - v_0}{t}$

Substitute:

(a) $a = \dfrac{28.0 \text{ m/s} - 8.00 \text{ m/s}}{50.00 \text{ s}}$
$= 0.400$ m/s^2

(b) $a = \dfrac{88.0 \text{ ft/s} - 22.0 \text{ ft/s}}{50.0 \text{ s}}$
$= 1.32$ ft/s^2

EXAMPLE 4-2

Determine the time that an aircraft spends during its takeoff run if it accelerates from rest **(a)** at 3.00 m/s^2 to a final speed of 36.0 m/s; **(b)** at 8.00 ft/s^2 to a final speed of 120 ft/s.

Solution

Data:
(a) $a = 3.00$ m/s^2, $v_0 = 0$ m/s (rest), $v_1 = 36.0$ m/s, $t = ?$

(b) $a = 8.00$ ft/s^2, $v_0 = 0$ ft/s (rest), $v_1 = 120$ ft/s, $t = ?$

Equation: $a = \dfrac{v_1 - v_0}{t}$

Rearrange: $t = \dfrac{v_1 - v_0}{a}$

Substitute:

(a) $t = \dfrac{36.0 \text{ m/s} - 0 \text{ m/s}}{3.00 \text{ m/s}^2}$
$= 12.0$ s

(b) $t = \dfrac{120 \text{ ft/s} - 0 \text{ ft/s}}{8.00 \text{ ft/s}^2}$
$= 15.0$ s

Sec. 4-1 / Uniformly Accelerated Motion

We may also obtain two other useful equations from Eqs. 4-4 and 4-6. If we solve Eq. 4-6 for the final velocity v_1:

$$s = \frac{v_1 + v_0}{2} t \quad \text{and} \quad v_1 = \frac{2s}{t} - v_0$$

Also, in Eq. 4-4,

$$v_1 = v_0 + at$$

Equating these expressions,

$$\frac{2s}{t} - v_0 = v_0 + at$$

and rearranging gives us

$$s = v_0 t + \tfrac{1}{2} a t^2 \qquad (4\text{-}7)$$

Similarly, solving Eqs. 4-4 and 4-6 for t, we have

$$t = \frac{v_1 - v_0}{a} \quad \text{and} \quad t = \frac{2s}{v_1 + v_0}$$

Equating,

$$\frac{v_1 - v_0}{a} = \frac{2s}{v_1 + v_0}$$

and rearranging yields

$$v_1^2 = v_0^2 + 2as \qquad (4\text{-}8)$$

To summarize, the equations of **uniformly accelerated motion** are

$$a = \frac{v_1 - v_0}{t} \qquad (4\text{-}9)$$

$$s = vt \qquad (4\text{-}10)$$

$$s = vt = \left(\frac{v_1 + v_0}{2}\right) t \qquad (4\text{-}11)$$

$$s = v_0 t + \tfrac{1}{2} a t^2 \qquad (4\text{-}12)$$

$$v_1^2 = v_0^2 + 2as \qquad (4\text{-}13)$$

Remember, they apply only when the acceleration is constant!

If an object accelerates uniformly and any three of the quantities s, v_0, v_1, a, or t are known, the other quantities may be determined from these equations. Each of the four equations has a missing term. Therefore, when we are given three terms and are asked for a fourth term, pick the equation that does not contain the fifth term. Remember, these equations really involve vector quantities.

EXAMPLE 4-3

What is the magnitude of the acceleration of an automobile when the driver applies the brakes and slows uniformly (a) from 30.0 m/s to 2.0 m/s in 120 m; (b) from 90.0 ft/s to 10.0 ft/s in 340.0 ft?

Solution

Data:

(a) $v_0 = 30$ m/s, $v_1 = 2.0$ m/s, (b) $v_0 = 90.0$ ft/s, $v_1 = 10.0$ ft/s,
    $s = 120$ m, $a = ?$ $s = 340.0$ ft, $a = ?$

The missing term is t; therefore, pick the equation that does not have a t:

Equation: $v_1^2 = v_0^2 + 2as$

Rearrange: $a = \dfrac{v_1^2 - v_0^2}{2s}$

Substitute:

(a) $a = \dfrac{(2.00 \text{ m/s})^2 - (30.0 \text{ m/s})^2}{2(120 \text{ m})}$ (b) $a = \dfrac{(10.0 \text{ ft/s}) - (90.0 \text{ ft/s})^2}{2(340 \text{ ft})}$

$\quad = -3.7 \text{ m/s}^2$ $\quad = -11.8 \text{ ft/s}^2$

The negative sign indicates that the acceleration is in the direction opposite to the motion and slows down the automobile.

EXAMPLE 4-4

(i) What is the final speed of an object, and (ii) how far does it travel while accelerating (a) if its initial speed is 10.0 m/s and it accelerates at 8.00 m/s² for 5.00 s; (b) if its initial speed is 8.00 ft/s and it accelerates for 5.00 s at 4.00 ft/s²?

Solution

Data:

(a) $v_0 = 10$ m/s, $a = 8.0$ m/s²,   (b) $v_0 = 8.00$ ft/s, $t = 5.00$ s,
    $t = 5.0$ s, $v_1 = ?$ $s = ?$ $a = 4.00$ ft/s², $v_1 = ?$, $s = ?$

(i) *Equation:* The missing term is s; therefore, the equation is

$$v_1 = v_0 + at$$

Substitute:

(a) $v_1 = 10.0$ m/s $+ (8.00$ m/s²$)(5.00$ s$) = 50.0$ m/s
(b) $v_1 = 8.00$ ft/s $+ (4.00$ ft/s²$)(5.00$ s$) = 28.0$ ft

(ii) *Equation:* The missing term is v_1; therefore, the equation is

$$s = v_0 t + \tfrac{1}{2} at^2$$

Substitute:

(a) $s = (10.0$ m/s$)(5.0$ s$) + \tfrac{1}{2} (8.00$ m/s²$)(5.00$ s$)^2$
    $\quad = 150$ m
(b) $s = (8.00$ ft/s$)(5.00$ s$) + \tfrac{1}{2} (4.00$ ft/s²$)(5.00$ s$)^2$
    $\quad = 90.0$ ft

PROBLEMS

1. What is the average speed of a sprinter who runs (a) 100 m in 10.2 s; (b) 300 ft in 10.2 s?
2. What is the average speed of a car that travels (a) 45.0 km in 1.25 h; (b) 68.0 mi in 1.25 h?
3. A car accelerates from 50.0 km/h to 125 km/h in 20.0 s. (a) What is the magnitude of its acceleration in meters per second squared? (b) How far does the car travel while accelerating?

Sec. 4-1 / Uniformly Accelerated Motion

4. A car accelerates from 30.0 mi/h to 80.0 mi/h in 20.0 s. (a) What is the magnitude of its acceleration in feet per second squared? (b) How far does the car travel while accelerating?

5. A train starts from rest and accelerates uniformly to a speed of 20.0 mi/h in 35.0 s. (a) What is the magnitude of its acceleration in feet per second squared? (b) How far does it travel while accelerating?

6. During takeoff, an aircraft accelerates from rest to a final takeoff speed of 56.0 m/s in 18.0 s. (a) What is its acceleration? (b) How far does it travel along the runway before it takes off?

7. A rocket accelerates from rest in a straight line at 14.0 m/s^2 for 5.00 s. Determine (a) its final speed in meters per second and kilometers per hour; (b) the distance that it travels while accelerating.

8. What is the magnitude of the acceleration, and how far does a car travel while accelerating if it accelerates uniformly from rest to a speed of (a) 125 km/h in 13.0 s; (b) 75.0 mi/h in 12.0 s?

9. A train accelerates uniformly in a straight line from 20.0 mi/h to 81 mi/h in 2 min 25 s. (a) What is the magnitude of its average acceleration? (b) How far does it go while accelerating?

10. (a) How long does it take a car that accelerates from rest at 2.80 m/s^2 to reach 80.0 km/h? (b) How far does it travel while accelerating?

11. During takeoff, an aircraft accelerates from rest at 28.0 ft/s^2 to a final speed of 125 mi/h. (a) How long does it take? (b) How far does it travel down the runway before it reaches this speed?

12. An electron accelerates uniformly at 5.00×10^{11} m/s^2 for 2.00 μs. If the electron was initially at rest, calculate (a) its final speed; (b) the distance that it travels while accelerating.

13. An electron in a cathode-ray tube accelerates from rest to a speed of 415 km/s in an elapsed time of 1.50 μs. (a) What is its acceleration? (b) How far does it travel while accelerating?

14. An object accelerates uniformly at 10.0 m/s^2 for 10.0 s. If its initial speed was 30.0 m/s, (a) what is its final speed; (b) how far does it travel in the 10.0 s?

15. The driver of a car applies the brakes, bringing the car uniformly to rest in 4.00 s. (i) How far did the car travel after the brakes were applied, and (ii) what was the acceleration if the initial speed was (a) 80.0 km/h? (b) 62.5 mi/h?

16. A driver exceeds an 80.0 km/h speed limit by an average of 5.00 km/h on a 272 km journey. How much time does the driver save?

17. An object accelerates uniformly at 9.00 m/s^2 to a final speed of 90.0 m/s in a distance of 80.0 m. (a) What was the initial speed? (b) How long does it accelerate?

18. A car is started from rest and accelerates uniformly. If it travels 36.0 m during the fifth second, find (a) its acceleration; (b) the total distance that it travels from rest in 6.00 s. (*Hint:* Use Eq. 4-12 and find expressions for times of 4.00 s and 5.00 s. The difference in the displacements is 36.0 m.)

19. A car accelerates at 0.750 m/s^2 for 4.00 s to a final speed of 32.0 m/s. Determine (a) the initial speed; (b) the distance traveled during the 4.00 s.

20. A car slows uniformly to rest in an elapsed time of 2.50 s. Determine the magnitude of the average acceleration and the distance traveled while braking if the initial speed is (a) 18.0 m/s; (b) 7.50 m/s; (c) 50.0 km/h; (d) 30.0 mi/h; (e) 82.0 mi/h.

21. A car passes a sign post traveling at 50.0 km/h and immediately accelerates uniformly at 1.20 m/s^2 for 5.00 s. Determine (a) the final speed; (b) the distance that the car travels past the sign post in the 5.00 s.

22. A car approaching a speed zone slows uniformly from 60.0 mi/h to 30.0 mi/h in 5.00 s. Determine (a) the magnitude of the acceleration; (b) the distance that the car travels while decelerating.

23. A rocket sled accelerates uniformly from rest to 215 m/s in 4.20 s, then brakes to a stop with a uniform deceleration in 4.80 s. (a) How far did the sled travel during the first

4.20 s? **(b)** How far did it travel during the total 9.00 s? **(c)** Determine the accelerations during each part of the ride.

24 A racing car starts from rest and accelerates uniformly to a speed of 180 mi/h in a distance of 1200 ft. Determine **(a)** the magnitude of the acceleration; **(b)** the time it took to travel the 1200 ft.

25 A truck traveling at 50.0 km/h accelerates uniformly to a speed of 90.0 km/h in a distance of 475 m. Determine **(a)** the magnitude of the acceleration; **(b)** the time it took to travel the 475 m.

26 The engine of a 195 m long freight train passes a crossing at a speed of 16.0 km/h. If the train then accelerates uniformly at 0.460 m/s^2, **(a)** what is the speed of the train when the last car passes the crossing; **(b)** how long does it take the train to pass the crossing?

27 The engine of a 625 ft long freight train passes a crossing at a speed of 10.0 mi/h. If the train then accelerates uniformly at 1.50 ft/s^2, **(a)** what is the speed of the train when the last car passes the crossing; **(b)** how long does it take the entire train to pass?

28 A speeding car traveling at 125 km/h passes a stationary police car. At the instant that the speeding car passes, the police car accelerates uniformly at 6.20 m/s^2 from rest for 8.00 s, then travels at a constant speed until it overtakes the speeding car. **(b)** How long does it take the policeman to overtake the speeding car? **(b)** How far does the police car travel before overtaking the speeder?

29 A car traveling at 80.0 km/h decelerates uniformly to 50.0 km/h in 8.00 s to pass through a speed zone. It then travels through the speed zone at 50.0 km/h in 18.0 s and accelerates uniformly back to 80.0 km/h in an additional 7.50 s. Determine **(a)** the accelerations of the car at each stage of the time period; **(b)** the total distance that the car travels during this time.

4-2 FREELY FALLING BODIES

It is a well-known fact that when an object is dropped near the surface of the earth, it increases its speed as it falls. Therefore, freely falling objects must be accelerated toward the center of the earth. By rolling balls down inclined planes, Galileo discovered that this acceleration, which is called the **acceleration due to gravity,** is the same for all bodies, independent of their mass. This may be illustrated by simultaneously dropping a book of many pages and single sheet of paper, made into a compact ball, from the same height; they both hit the ground at the same instant. Of course, the speed, density, and shape of the object may affect this result because of air resistance and buoyancy. Because of this effect, objects falling freely may eventually reach a constant velocity called the **terminal velocity.** For example, a parachutist does not accelerate continually. Once the parachute is open the drag of the air eventually balances the force of gravity and the acceleration becomes zero, producing a constant terminal velocity.

The magnitude g of the acceleration due to gravity is approximately 9.81 m/s^2 or 32.2 ft/s^2 at the surface of the earth. Even though the acceleration due to gravity changes with the distance from the center of the earth, it is usually considered to be constant for small changes in height near the earth's surface. Therefore, the equations of uniformly accelerated motion may be used for objects falling through distances (or thrown vertically upward to heights) which are small compared with the radius of the earth. In these problems the displacement s corresponds to the change in height h measured from the origin to the terminal point of the path, and the acceleration **a** is the acceleration due to gravity **g**, which always acts downward. It is important to remember that these equations involve *vector* quantities but the vectors are collinear.

EXAMPLE 4-5

A stone is dropped from a cliff into the sea below. **(i)** How long does the fall last, and **(ii)** what is the speed of the stone as it strikes the water if the cliff is **(a)** 35.0 m high; **(b)** 101 ft high? Neglect air resistance.

Sec. 4-2 / Freely Falling Bodies

Solution

Data: If we take down as the positive direction:
(a) $s = h = 35.0$ m, $v_0 = 0$ m/s,   (b) $s = h = 101$ ft, $v_0 = 0$ ft/s
$a = g = 9.81$ m/s², $t = ?$, $a = g = 32.2$ ft/s², $t = ?$,
$v_1 = ?$ $v_1 = ?$

(i) Equation: $s = h = v_0 t + \frac{1}{2} a t^2 = v_0 t + \frac{1}{2} g t^2$

Our next step follows from the fact that $v_0 = 0$ m/s and 0 ft/s.

Rearrange: $h = \frac{1}{2} g t^2$ or $t = \sqrt{\dfrac{2h}{g}}$

Substitute:

(a) $t = \sqrt{\dfrac{2(35.0 \text{ m})}{9.81 \text{ m/s}^2}}$ (b) $t = \sqrt{\dfrac{2(101 \text{ ft})}{32.2 \text{ ft/s}^2}}$
    $= 2.67$ s $= 2.50$ s

(ii) Equation: $v_1^2 = v_0^2 + 2as$

Rearrange: $v_1^2 = 0 + 2gh$ or $v_1 = \sqrt{2gh}$

Substitute:

(a) $v_1 = \sqrt{2(9.81 \text{ m/s}^2)(35.0 \text{ m})}$ (b) $v_1 = \sqrt{2(32.2 \text{ ft/s}^2)(101 \text{ ft})}$
     $= 26.2$ m/s $= 80.6$ ft/s

Check: We can use another equation of uniformly accelerated motion, such as $v_1 = v_0 + at$, with our calculated values as a check. Since $v_0 = 0$ and $a = g$, then (a) $v_1 = gt = (9.81 \text{ m/s}^2)(2.67 \text{ s}) = 26.2$ m/s and (b) $v_1 = (32.2 \text{ ft/s}^2)(2.50 \text{ s}) = 80.6$ ft/s.

EXAMPLE 4-6

A ball is thrown vertically upward from the top of a bridge and falls into the water directly below the bridge 8.00 s later. **(i)** How high did the ball go above the bridge, and **(ii)** how high is the bridge if the initial speed of the ball is (a) 20.0 m/s; (b) 64.0 ft/s? Neglect air resistance.

Solution

(i) *Data:* We could take either up or down as the positive direction as long as we use the proper signs on the vector quantities. If up is taken as the positive direction,

(a) $v_0 = 20.0$ m/s   (b) $v_0 = 64.0$ ft/s
    $a = g = -9.81$ m/s² $a = -g = -32.2$ ft/s²
    (i.e., g acts downward)
    $v_1 = 0$ m/s $v_1 = 0$ ft/s

(i.e., the ball comes momentarily to rest at the top of its flight), and
    $s = h = ?$ $s = h = ?$

Equation: $v_1^2 = v_0^2 + 2as$ or $v_1^2 = v_0^2 + 2(-g)h$

Rearrange: $h = \dfrac{v_1^2 - v_0^2}{-2g}$

Substitute:

(a) $h = \dfrac{0 - (20.0 \text{ m/s})^2}{-2(9.81 \text{ m/s}^2)}$ (b) $h = \dfrac{0 - (64.0 \text{ ft/s})^2}{-2(32.2 \text{ ft/s}^2)}$
    $= 20.4$ m $= 63.6$ ft

(ii) *Data:*

(a) $v_0 = 20.0$ m/s, $t = 8.00$ s,   (b) $v_0 = 64.0$ ft/s, $t = 8.00$ s,
    $a = -g = -9.81$ m/s², $a = -g = -32.2$ ft/s²,
    $s = h = ?$ $s = h = ?$

Note that the *displacement* is a *vector quantity* measured from the origin to the terminal point of the path. In this case the origin of the motion was the top of the bridge and the terminal point was the water below the bridge. Therefore, the displacement magnitude is merely the height of the bridge, even though the ball was initially thrown upward to positions 20.4 m and 63.6 ft above the bridge (Fig. 4-3). We need not consider the distance that the stone travels because the equations of motion relate *vector quantities*.

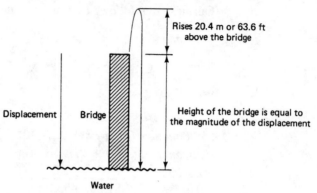

Figure 4-3

Equation: $s = v_0 t + \frac{1}{2} a t^2$
Rearrange: $h = v_0 t + \frac{1}{2}(-g) t^2$
Substitute:
(a) $h = (20.0 \text{ m/s})(8.00 \text{ s}) + \frac{1}{2}(-9.81 \text{ m/s}^2)(8.00 \text{ s})^2$
   $= -154 \text{ m}$
(b) $h = (64.0 \text{ ft/s})(8.00 \text{ s}) + \frac{1}{2}(-32.2 \text{ ft/s}^2)(8.00 \text{ s})^2$
   $= -518 \text{ ft}$

The minus signs indicate that the displacement is downward.

PROBLEMS

Neglect air resistance in each case.

30. A stone is dropped from an altitude of 275 m. (i) How long does it take a stone to reach the ground, and (ii) what is its speed as it strikes the ground if it is dropped from an altitude of (a) 275 m; (b) 500 ft?

31. A rock is dropped from the top of a 35.0 m building. (a) How long does it take to reach the ground? (b) What is the speed of the rock as it strikes the ground?

32. A stone dropped from a tall building takes 3.60 s to reach the ground. How high is the building?

33. While over enemy lines in his balloon, a soldier throws a bomb vertically downward with an initial speed of 12.0 m/s from a height of 1250 m. (a) What is the speed of the bomb as it strikes the ground? (b) How long does it take the bomb to reach the ground?

34. A stone is catapulted straight upward from a high bridge. On its return, it narrowly misses the bridge and plunges into the water below exactly 8.00 s after launch. How high is the bridge above the water if the initial upward speed of the stone is (a) 25.0 m/s; (b) 56.0 ft/s?

35. A ball thrown straight downward from a tower and hits the ground below exactly 2.50 s later. (i) How high is the tower, and (ii) what is the speed of the ball just before impact if its initial downward speed is (a) 8.00 m/s; (b) 25.0 ft/s?

36. A ball is thrown vertically upward with an initial speed of 32.0 m/s. (a) How high will it go? (b) How long will it take to return to the ground? (c) What is its speed as it strikes the ground?

37 A bomber in a vertical dive at 60.0 m/s releases a bomb that strikes the ground 18.0 s later. (a) What is the speed of the bomb just before it strikes the ground? (b) From what height was the bomb released?

38 A bullet is fired vertically upward from the earth's surface. (i) How high will it go, and (ii) how long will it take it to return to earth if its initial upward speed is (a) 225 m/s; (b) 600 ft/s? (*Hint:* The total displacement is zero. Why?)

39 A bullet fired vertically upward reaches a height of 2500 m. (a) What was its initial speed? (b) How long does it take to reach the height of 2500 m?

40 A ball is thrown vertically upward from the earth's surface with an initial speed of 75.0 mi/h. Determine (a) how high it goes and (b) the time it takes to return to the earth.

41 A rock is thrown vertically downward, into a lake, with an initial speed of 75.0 km/h from the top of a 135 m cliff. (a) How long does it take to reach the water? (b) What is the speed of the rock as it strikes the water?

42 (a) How long does it take an object to fall 355 m from rest? (b) What is its final speed?

SUMMARY

An object is in *uniformly accelerated motion* if its acceleration is constant and it travels in a straight line. The equations of uniformly accelerated motion are:

$$a = \frac{v_1 - v_0}{t} \tag{4-9}$$

$$s = vt \tag{4-10}$$

$$s = vt = \left(\frac{v_1 + v_0}{2}\right) t \tag{4-11}$$

$$s = v_0 t + \tfrac{1}{2} a t^2 \tag{4-12}$$

$$v_1^2 = v_0^2 + 2as \tag{4-13}$$

These equations relate *vector quantities*.

If air resistance is negligible, objects falling freely near the earth's surface have the same acceleration toward the earth. This acceleration is called the *acceleration due to gravity* **g**; its magnitude is approximately 9.81 m/s² or 32.2 ft/s² near the earth's surface. Since the acceleration is approximately constant for small changes in height near the earth, this type of motion is also uniformly accelerated.

QUESTIONS

1 Define the following terms: (a) displacement; (b) average velocity; (c) average acceleration; (d) uniformly accelerated motion; (e) terminal velocity; (f) deceleration.

2 Can an object ever have an acceleration at an instant when its velocity is zero? Explain your answer.

3 Can the displacement of an object ever exceed the distance that it travels?

4 If you are traveling at a constant speed in a car, could you also be accelerating?

5 If a car starts from rest and accelerates uniformly for a period of time and then decelerates uniformly to rest, you should not use the equations of uniformly accelerated motion for the entire journey. Why not? How could you use these equations to describe this type of motion?

REVIEW PROBLEMS

Neglect air resistance.

1. How long does it take an aircraft traveling at a constant speed of 975 km/h to fly 725 km?

2. If an aircraft travels at a constant speed of 625 mi/h, how long does it take to fly a distance of 450 mi?

3. If a car travels at a constant speed of 75.0 km/h, how long does it take to travel a distance of 68.0 km?

4. Determine the magnitude of the acceleration if an object is in uniformly accelerated motion and changes its speed from (a) 5.25 m/s to 12.0 m/s in 12.0 s; (b) 7.00 m/s to 35.0 m/s in 9.00 s; (c) 28.0 m/s to 12.5 m/s in 25.0 s; (d) 30.0 km/h to 60.0 km/h in 7.50 s; (e) 80.0 km/h to 35.0 km/h in 15.0 s; (f) 15.0 ft/s to 35.0 ft/s in 12.0 s; (g) 35.0 mi/h to 60.0 mi/h in 9.00 s.

5. An electron in a television tube accelerates uniformly from rest to a final speed of 6.00×10^6 m/s in a distance of 5.00 cm. Find (a) the acceleration; (b) the time it takes the electron to travel the 5.0 cm.

6. A racing car starts from rest and accelerates uniformly to 216 km/h in 12.0 s. (a) What is its acceleration? (b) How far does it travel in the 12.0 s?

7. A racehorse starts from rest with a uniform acceleration of 1.50 m/s^2 for 8.00 s. (a) What is its final speed? (b) How far does it travel in the 8.00 s?

8. A car slows uniformly to rest in 6.00 s. Determine the magnitude of the acceleration and the distance traveled while slowing down if the initial speed is (a) 45.0 m/s; (b) 28.0 m/s; (c) 85.0 km/h; (d) 54.0 km/h; (e) 75.0 ft/s; (f) 48.0 ft/s; (g) 75.0 mi/h; (h) 32.0 mi/h.

9. A rocket accelerates uniformly from 2.0 km/s to 8.0 km/s in a distance of 105 km. (a) What is its acceleration? (b) How long does it accelerate?

10. A 192 m freight train accelerates uniformly. If it travels at an initial speed of 8.00 km/h as it passes a crossing, and the entire train takes 52.0 s to pass the crossing, determine (a) the acceleration; (b) the final speed as it passes the crossing.

11. A car traveling at 85.0 mi/h decelerates uniformly at 9.50 ft/s^2 for 5.00 s. Determine (a) the final speed; (b) the distance traveled while decelerating.

12. A rocket sled accelerates uniformly from rest to 425 mi/h in 6.00 s and then brakes to rest uniformly in an additional 5.00 s. Determine (a) the acceleration and deceleration; (b) the total distance that the sled traveled.

13. A speeding motorist traveling at 115 km/h passes a stationary police patrol car. At the instant the patrol car is passed, it accelerates uniformly from rest to a speed of 135 km/h in 8.00 s, and then maintains that speed until the speeder is caught. Determine (a) the acceleration of the patrol car; (b) the total time required for the police to catch the speeder; (c) the total distance that the chase lasted.

14. How long does it take a rock to fall (a) 256 m; (b) 1250 m; (c) 235 ft; (d) 1450 ft?

15. A stone is dropped from a bridge and takes 6.20 s to reach the water below. How high is the bridge above the water?

16. A boy throws a stone vertically downward from a bridge at an initial speed of 65.0 ft/s. If the bridge is 185 ft above the water, (a) how long does it take the stone to reach the water; (b) what is the speed of the stone as it strikes?

17. A rock is thrown at an initial speed of 28.0 m/s vertically downward into a chasm and is heard to land at the bottom 12.5 s later. (a) How deep is the chasm? (b) What is the speed of the rock as it lands?

18. A man fires a bullet that strikes a duck 59.8 m directly above him. If the initial speed of the bullet is 355 m/s, (a) how long does it take the bullet to reach the duck; (b) with what speed does the bullet strike the duck?

19. A stone fired upward from the top of a 150 ft high bridge falls into the water below the bridge 12.0 s later. (a) What was the initial velocity of the stone? (b) How high above the bridge did it go? (c) What was the velocity of the stone just before it struck the water?

20. If an object is dropped from rest and allowed to fall freely toward the earth, the magnitude of its acceleration $g = 9.81$ m/s^2 or 32.2 ft/s^2. (a) Plot a graph of the distance that the

Review Problems

object falls versus the elapsed time for the first 8.00 s of the motion. **(b)** Plot a graph of the distance versus the square of the elapsed time. What can you say about the slope?

21 Plot a velocity versus elapsed time graph for the data in Table 4-1 using **(i)** SI units; **(ii)** USCS units. **(a)** Is the motion uniformly accelerated? **(b)** Determine the acceleration from the graph. (*Hint:* Consider the slope and relate it to Eq. 4-9.) **(c)** Find the displacement in the 5.0 s from the graph. (*Hint:* Find the area between the line and the time axis and relate the area to Eq. 4-11.)

TABLE 4-1

Velocity/(m/s)	5.0	7.0	9.0	11.0	13.0	15.0
Velocity/(ft/s)	16.4	23.0	29.5	36.1	42.7	49.2
Elapsed time/s	0	1.0	2.0	3.0	4.0	5.0

5
FORCES

Dynamics is the study of the relationships between changes in the motion of solid objects and the actions that caused these changes. These actions, called **forces,** tend to make stationary objects move or to change the motion of moving objects. Forces have many different origins; they may be mechanical, electrical, magnetic, gravitational, nuclear, and so on, but their results are similar.

5-1 NEWTON'S LAWS OF MOTION

The basic laws governing motion were first formulated by Sir Isaac Newton (1642–1727) and he presented them in his publication *Principia Mathematica* (1686). Newton's statement of the laws of motion was one of the most important developments in physics. These laws form the basis of what is now called **classical** (or **Newtonian**) **mechanics.** It was not until the twentieth century that Newton's laws proved to be inadequate because they could not fully describe the motion of extremely small atomic-sized particles, or particles that travel close to the speed of light (300 000 km/s or 186 000 mi/s). However, because of their mathematical simplicity and accuracy, the laws of classical mechanics are still used to predict and describe the motions of many objects, such as cars, aircraft, planets, and satellites.

Newton's First Law of Motion: The Law of Inertia

Newton's first law of motion merely describes a basic property of matter called **inertia.** The law states:

> *An object at rest will remain at rest, and an object in motion will continue that motion with a constant velocity (speed and direction) unless it is acted upon by some external unbalanced force.*

In other words, a net force such as a push or a pull is required to:

1. Make a stationary object move.
2. Stop a moving object.
3. Change the speed of a moving object.
4. Alter the direction of motion of a moving object.

In each case, since some kind of force is required to change the motion of a solid object, this law indicates that matter itself has inertia because it resists changes in motion.

We have all experienced this property of inertia. When a car (or aircraft) accelerates we feel pushed backward into the seat because the car must push us so that we accelerate with it. As a car turns a corner we tend to slide away from the direction of the turn because we would normally continue our motion in a straight line (Fig. 5-1).

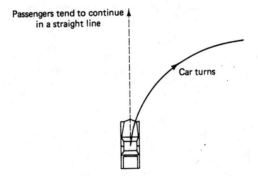

Figure 5-1 Passengers tend to continue in a straight line as the car turns. As a result, the passengers tend to "slide away" from the direction of the turn.

Mass is a numerical (quantitative) measure of inertia. Objects with large masses also have large inertias. For example, it is harder to accelerate a truck (more force is required) than a small car, because the truck has the greater mass.

Newton's Second Law of Motion: The Law of Acceleration

Newton's second law describes the changes in motion of an object when it is acted upon by some net external force. In a simple form the second law may be stated as follows:

The product of the mass m and acceleration **a** *of an object is equal to the net (vector sum) force acting on it.**

$$\text{net } \mathbf{F} = m\mathbf{a} \tag{5-1}$$

This is a very important relationship. It is used to define the SI derived unit of force called the *newton*. A *newton* (symbol N) is defined as the net force that accelerates a 1 kg mass at 1 m/s². Therefore, substituting in Eq. 5-1, we have

$$1 \text{ N} = (1 \text{ kg})(1 \text{ m/s}^2) = 1 \text{ kg} \cdot \text{m/s}^2$$

---

*Notice that this assumes that the mass is constant. Newton actually stated the law as follows: The net force varies directly as the time rate of change of momentum (the product of the mass and the velocity; see Section 5-3).

In USCS units, the mass m is described by a derived unit called a **slug,** which is defined in terms of the pound unit for force and the acceleration unit of feet per second per second:

$$m = \frac{\text{net } F}{a} \quad \text{or} \quad 1 \text{ slug} = \frac{1 \text{ lb}}{1 \text{ ft/s}^2}$$

Therefore, a slug is the mass that would be accelerated at 1 ft/s² by a 1 lb net force.

The direction and magnitude of any changes in motion depends on the direction and magnitude of the applied forces. However, if an external force results in a "push" on an object, the same effect could be obtained by "pulling" the object with an equal force. Consequently, the direction of the line extended both ahead and behind the actual force vector has a special significance; it is called the **line of action** of that force (Fig. 5-2). If the net external force on an object is equal to zero, there will be no change in its velocity, and the object is said to be in a **state of equilibrium.**

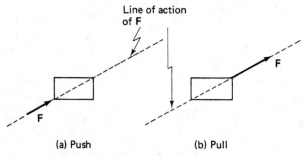

Figure 5-2 Lines of action.

EXAMPLE 5-1

Determine the magnitude of the net force that would cause **(a)** a 2.20×10^4 kg aircraft to accelerate at 2.45 m/s²; **(b)** a 50.0 slug car to accelerate at 2.50 ft/s².

Solution

Data:
(a) $m = 2.20 \times 10^4$ kg, $a = 2.45$ m/s², net $F = ?$

(b) $m = 50.0$ slug, $a = 2.50$ ft/s², net $F = ?$

Equation: net $F = ma$

Substitute:
(a) net $F = (2.20 \times 10^4 \text{ kg})(2.45 \text{ m/s}^2)$
$= 5.39 \times 10^4$ N

(b) net $F = (50.0 \text{ slug})(2.50 \text{ ft/s}^2)$
$= 125$ lb

EXAMPLE 5-2

Find the acceleration of **(a)** an 1800 kg car when it experiences a 4500 N net force; **(b)** a 60.0 slug car that experiences a 325 lb net force.

Solution

Data:
(a) $m = 1800$ kg, net $F = 4500$ N, $a = ?$

(b) $m = 60.0$ slug, net $F = 325$ lb, $a = ?$

Equation: Net $F = ma$

Rearrange: $a = \dfrac{\text{net } F}{m}$

Sec. 5-1 / Newton's Laws of Motion

Substitute:

(a) $a = \dfrac{4500 \text{ N}}{1800 \text{ kg}} = 2.5 \text{ m/s}^2$ \hspace{2em} (b) $a = \dfrac{325 \text{ lb}}{60.0 \text{ slug}} = 5.42 \text{ ft/s}^2$

Note: $1 \dfrac{\text{N}}{\text{kg}} = 1 \dfrac{\text{kg} \cdot \text{m/s}^2}{\text{kg}} = 1 \text{ m/s}^2$, and $1 \dfrac{\text{lb}}{\text{slug}} = 1 \dfrac{\text{slug} \cdot \text{ft/s}^2}{\text{slug}} = 1 \text{ ft/s}^2$.

EXAMPLE 5-3

If a net force gives an 18 kg mass an acceleration of 8.0 m/s², what acceleration would the same net force give a 24 kg mass?

Solution

Data: $m_1 = 18$ kg, $a_1 = 8.0$ m/s², same net F, $m_2 = 24$ kg, $a_2 = ?$
Equations: Net $F = m_1 a_1$, net $F = m_2 a_2$
Rearrange: Net $F = m_1 a_1 = m_2 a_2$; therefore,

$$a_2 = \frac{m_1 a_1}{m_2}$$

Substitute: $a_2 = \dfrac{(18 \text{ kg})(8.0 \text{ m/s}^2)}{24 \text{ kg}} = 6.0 \text{ m/s}^2$

Mass and Weight. When an object is allowed to fall freely to the earth from some height, it accelerates as it falls. This **acceleration due to gravity, g,** is directed toward the center of the earth, and its magnitude varies with the distance from the earth. However, if air resistance is neglected, it has the same magnitude for all objects at any location. Near the surface of the earth, the acceleration due to gravity has a magnitude of 9.81 m/s² or 32.2 ft/s².

Since all freely falling objects experience this acceleration, Newton's second law implies the presence of some net external force. This force is due to the attraction between the object and the earth. In fact, every object in the universe attracts every other object, with a force that depends on the amount of matter in each object, and the distance between their centers. This attractive force between objects is called the **force of gravity.**

Therefore, if g is the acceleration due to gravity, in free fall the net force on some mass m is the gravitational force F_g:

$$\text{net } F = ma$$

becomes

$$F_g = mg \tag{5-2}$$

However, objects do not have to be falling or accelerating in order to experience the force of gravity because it is due to the attraction between the masses. For a mass m at the earth's surface, this force of gravity is due to the attraction between the earth and the mass and it is also called the weight w (Fig. 5-3).* Therefore,

$$w = mg \tag{5-3}$$

This equation may be used to determine the weight of a particular mass, or vice versa, even if the mass is not falling.

---

*The measured value for weight is slightly different from the force of gravity, because of the effects of the earth's rotation, the buoyancy of air, and other local effects.

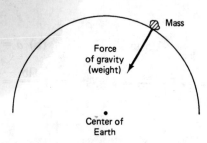

Figure 5-3 The weight of a mass m at the earth's surface is due to the attractive force of gravity (toward the center of the earth).

Mass is a fundamental quantity in SI; it is independent of the position of the object, but its measured value is affected by motion. This is due to relativity, which was proposed by Albert Einstein in the twentieth century.

The **force of gravity** exerted by the earth on an object (the weight of the object) varies with its distance from the center of the earth. The force of gravity (weight) decreases as the distance from the earth increases. In addition, the force of gravity depends on the magnitudes of the masses involved. For example, the earth is much more massive than our moon, and therefore the gravitational force on any mass at the surface of the earth is greater than the gravitational force on the same mass at the surface of the moon even though the radius of the moon is less than that of the earth. However, masses are the same at all locations.

EXAMPLE 5-4

Determine the weight of (a) a 75.0 kg man and (b) a 5.00 slug man (i) on the earth's surface, and (ii) on the moon, where the acceleration due to gravity is 1.63 m/s² or 5.35 ft/s².

Solution

Data:
(a) $m = 75.00$ kg, $w = ?$ (b) $m = 5.00$ slug, $w = ?$
(i) $g = 9.81$ m/s²   (i) $g = 32.2$ ft/s²
(ii) $g = 1.63$ m/s²   (ii) $g = 5.35$ ft/s²

Equation: $w = mg$

(i) Substitute:
(a) $w = (75.0$ kg$)(9.81$ m/s²$)$ (b) $w = (5.00$ slug$)(32.2$ ft/s²$)$
$= 736$ N $= 161$ lb

(ii) Substitute:
(a) $w = (75.0$ kg$)(1.63$ m/s²$)$ (b) $w = (5.00$ slug$)(5.35$ ft/s²$)$
$= 122$ N $= 26.8$ lb

EXAMPLE 5-5

What is the mass of a woman whose weight at the earth's surface is (a) 525 N; (b) 125 lb?

Solution

Data:
(a) $w = 525$ N, $g = 9.81$ m/s²,   (b) $w = 125$ lb, $g = 32.2$ ft/s²,
$m = ?$ $m = ?$

Equation: $w = mg$

Rearrange: $m = \dfrac{w}{g}$

Substitute:

(a) $m = \dfrac{525 \text{ N}}{9.81 \text{ m/s}^2} = 53.5$ kg   (b) $m = \dfrac{125 \text{ lb}}{32.2 \text{ ft/s}^2} = 3.88$ slug

Sec. 5-1 / Newton's Laws of Motion

To measure the mass of an object, we usually compare the force of gravity on it with the force of gravity on a calibrated standard mass (Fig. 5-4). We may use an equal-arm balance, adding calibrated masses to one side, until their total weight w_s balances the weight w of the mass on the other side. Since the acceleration due to gravity is the same for both sides, when the balance is achieved the masses are qual:

$$w_s = w$$

or

$$m_s g = mg \quad \text{and} \quad m_s = m$$

This type of balance can be used at any location where there is a net gravitational force.

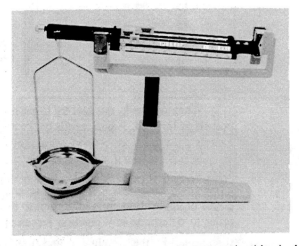

Figure 5-4 Balance used to measure masses. In this device, the torque due to the mass in the pan is balanced by the torque due to the movable constant mass along a calibrated arm.

Spring balances may also be used to determine masses, but they are calibrated by using standard masses at a certain location. They would be slightly inaccurate, to varying degrees, at other locations, because of the variations in the acceleration due to gravity. These devices operate on the principle that the elongation e of an elastic spring varies directly as the applied load F. That is,

$$F \propto e \quad \text{or} \quad F = ke \tag{5-4}$$

This relationship is known as **Hooke's law**; the proportionality constant k is called the **spring** or **elastic constant**.

Newton's Third Law of Motion: Action and Reaction

Newton's third law describes the effect of interactions between objects. The law states that:

Whenever one object exerts a force on a second object, the second object exerts a reactive force of equal magnitude and opposite direction on the first object.

For example, if a book is at rest on a flat horizontal table, the force of gravity (weight) on the book produces a force on the table, and the table exerts an equal but opposite reactive force on the book (Fig. 5-5). If the table did not exert this reactive force, the weight of the book would not be balanced, and according to Newton's second law, the book would accelerate and fall through the table. Reaction forces must always be considered in calculations involving forces.

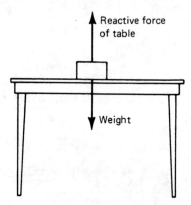

Figure 5-5 Reactive force of the table on the book balances the weight.

When we walk, our feet try to push the ground backward and the reaction force propels us forward. A gun recoils backward as it propels a bullet forward. Aircraft and rockets are propelled forward when they exhaust gases and air behind them.

In mechanics, it is usually convenient to draw a diagram of the system under study by considering all items only in terms of their effect *on* the system. This is called a **free-body diagram**. A free-body diagram is a sketch of the forces present and their points of application. It is important to note that Newton's third law must be used in order to determine any reaction forces in the system. Consider a load at rest on a horizontal surface (Fig. 5-6). The free-body diagram must include the force of gravity (represented by the weight **w**), and the reaction force of the table on the load (Fig. 5-6b). This reaction force is called a **normal force N**; it always acts perpendicularly to the surfaces in contact.

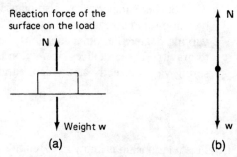

Figure 5-6 Load *w* on a table.

When a load **w** is suspended by a string, the string is in a state of **tension** because the load tends to elongate the string. If the string is in equilibrium and has negligible weight, the tensions at each end of the string must have equal magnitudes but opposite directions, otherwise the string would accelerate. Since a tension force **T** will oppose the force of gravity (weight **w**), the free-body diagram must contain the tension force as well as the weight (Fig. 5-7).

Sec. 5-1 / Newton's Laws of Motion

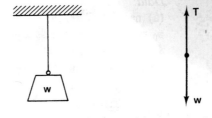

(a) Load w suspended from a fixed string

(b) Free body diagram

Figure 5-7

Newton's laws of motion may be used to solve many types of problems. The procedure is as follows:

1. Draw a free-body diagram, if necessary. All reaction forces must be included and the forces of gravity (weights) of the masses must be calculated using Eq. 5-3.
2. List the quantities given and those to be found in terms of their symbols. It may be necessary to choose a positive direction.
3. Use Newton's second law and kinematic equations to find the unknown quantities.

EXAMPLE 5-6

Determine the magnitude of a net force on **(a)** a 20.0 kg block if it accelerates uniformly from 2.50 m/s to 5.80 m/s in 3.00 s; **(b)** a 35.0 lb crate that accelerates uniformly from 1.00 ft/s to 6.00 ft/s in 3.00 s.

Solution

Sketch: Not required in this case.

Data:
(a) $v_0 = 2.50$ m/s, $v_1 = 5.80$ m/s, $t = 3.00$ s, $m = 20.00$ kg, net $F = ?$

(b) $v_0 = 1.00$ ft/s, $v_1 = 6.00$ ft/s, $t = 3.00$ s, $w = 35.0$ lb, $g = 32.2$ ft/s², net $F = ?$

Equations: net $F = ma$, $a = \dfrac{v_1 - v_0}{t}$, and $w = mg$, or $m = \dfrac{w}{g}$

Rearrange:
(a) net $F = ma = m\left(\dfrac{v_1 - v_0}{t}\right)$

(b) net $F = ma = \dfrac{w}{g}\left(\dfrac{v_1 - v_0}{t}\right)$

Substitute:
(a) net $F = (20.0 \text{ kg})\left(\dfrac{5.80 \text{ m/s} - 2.50 \text{ m/s}}{3.00 \text{ s}}\right) = 22.0$ N

(b) net $F = \dfrac{35.0 \text{ lb}}{32.2 \text{ ft/s}^2}\left(\dfrac{6.00 \text{ ft/s} - 1.00 \text{ ft/s}}{3.00 \text{ s}}\right) = 1.81$ lb

EXAMPLE 5-7

Determine **(i)** the acceleration and **(ii)** the distance traveled in 6.00 s **(a)** when a 7500 N net force accelerates a 1500 kg car uniformly from rest; **(b)** when a 1400 lb net force accelerates a 3200 lb car from rest.

Solution

Sketch: Not necessary

Data:
(a) net $F = 7500$ N, $m = 1500$ kg, $v_0 = 0$ m/s, $t = 6.0$ s, $a = ?$, $s = ?$

(b) net $F = 1400$ lb, $w = 3200$ lb, $v_0 = 0$ ft/s, $t = 6.00$ s, $g = 32.2$ ft/s^2, $a = ?$, $s = ?$

(i) *Equations:* net $F = ma$, $w = mg$
Rearrange:

(a) $a = \dfrac{\text{net } F}{m}$

(b) $m = \dfrac{w}{g}$, $a = \dfrac{\text{net } F}{w/g} = \dfrac{(\text{net } F)g}{w}$

Substitute:

(a) $a = \dfrac{7500 \text{ N}}{1500 \text{ kg}} = 5.0$ m/s^2

(b) $a = \dfrac{(1400 \text{ lb})(32.2 \text{ ft/s}^2)}{3200 \text{ lb}}$
$= 14.1$ ft/s^2

(ii) *Equation:* $s = v_0 t + \tfrac{1}{2} a t^2$
Substitute:
(a) $s = (0 \text{ m/s})(5.0 \text{ s}) + \tfrac{1}{2}(5.0 \text{ m/s}^2)(6.0 \text{ s})^2 = 90$ m
(b) $s = (0 \text{ ft/s})(6.00 \text{ s}) + \tfrac{1}{2}(14.1 \text{ ft/s}^2)(6.00 \text{ s})^2 = 253$ ft

EXAMPLE 5-8

At the instant of blast-off from the earth's surface, determine the acceleration of a Saturn rocket (a) if its mass is 2.82×10^6 kg and the total upward force (thrust) developed by its five engines is 3.50×10^7 N; (b) if its initial weight is 6.20×10^6 lb and the upward thrust is 7.87×10^6 lb.

Solution

Sketch: The free-body diagram should include the upward thrust F_{up} and the force of gravity (weight w), that act on the rocket (Fig. 5-8).

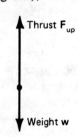

(a) Rocket   (b) Free-body diagram.   **Figure 5-8**

Data:
(a) $m = 2.82 \times 10^6$ kg,
$F = 3.50 \times 10^7$ N,
$g = 9.81$ m/s^2, $a = ?$

(b) $w = 6.20 \times 10^6$ lb,
$F = 7.87 \times 10^6$ lb,
$g = 32.2$ ft/s^2, $a = ?$

Equations: net $F_{up} = F - w = ma$, $w = mg$
Rearrange:

(a) $a = \dfrac{\text{net } F_{up}}{m} = \dfrac{F - mg}{m}$

(b) $a = \dfrac{F - w}{w/g}$

Substitute:
(a) $a = \dfrac{3.50 \times 10^7 \text{ N} - (2.82 \times 10^6 \text{ kg})(9.81 \text{ m/s}^2)}{2.82 \times 10^6 \text{ kg}}$ upward
$= 2.60$ m/s^2 upward

(b) $a = \dfrac{7.87 \times 10^6 \text{ lb} - 6.20 \times 10^6 \text{ lb}}{(6.20 \times 10^6 \text{ lb})/32.2 \text{ ft/s}^2} = 8.67$ ft/s^2

PROBLEMS

1. What net force is required to give (a) a 12.0 kg mass an acceleration of 5.0 m/s^2; (b) a 2.25 slug mass an acceleration of 4.15 ft/s^2?

2. Determine the net force required to accelerate the following masses as indicated: (a) 56.0 kg at 3.20 m/s^2; (b) 1.25 kg at 4.60 m/s^2; (c) 1500 kg at 0.755 m/s^2; (d) 1.20 slug at 8.50 ft/s^2; (e) 4.50 slug at 1.25 ft/s^2; (f) 0.525 slug at 12.5 ft/s^2; (g) 9.10×10^{-31} kg at 5.20×10^{12} m/s^2; (h) 2100 t at 0.450 m/s^2 [1 t (metric ton) = 1000 kg].

3. Determine the acceleration of the following masses when they experience the corresponding net forces: (a) 3.50 kg with 70.0 N; (b) 86.0 kg with 2.30 N; (c) 2500 kg with 850 N; (d) 3.20 slug with 58.0 lb; (e) 8.60 slug with 7.50 lb; (f) 25.0 slug with 1750 lb; (g) 56.0 t with 1250 MN; (h) 6.67×10^{-27} kg with 2.50×10^{-17} N.

4. Find the masses that would be accelerated at the following rates by the given net forces: (a) 8.00 m/s^2 with 2.40 N; (b) 1.50 m/s^2 with 26.0 N; (c) 0.0650 m/s^2 with 2250 N; (d) 3.45 ft/s^2 with 22.5 lb; (e) 3.25 ft/s^2 with 255 lb; (f) 5.60 ft/s^2 with 125 lb; (g) 3.20×10^{12} m/s^2 with 7.25×10^{-16} N.

5. Determine the acceleration of a 25.0 lb object if it experiences a net force of 70.0 lb.

6. What is the mass of an object that is accelerated at 2.45 m/s^2 by a 26.0 N net force?

7. If a 4.50 lb net force accelerates an object at 6.00 ft/s^2, what net force would accelerate that object at 8.00 ft/s^2?

8. An object is accelerated at 3.00 m/s^2 by some net force. What is the acceleration if the net force is tripled?

9. A net force **F** acts on a 5.00 kg mass and accelerates it at 5.00 m/s^2. What is the acceleration of a 25.0 kg mass when it is acted on by the same force?

10. A net force of 12.0 lb gives an object an acceleration of 2.00 ft/s^2. (a) What force will accelerate the same object at 8.00 ft/s^2? (b) If the same object is acted on by a 32.5 lb force, what is its acceleration?

11. If a force produces an acceleration of 35.0 m/s^2, what acceleration would the same force produce on half the mass?

12. Determine the forces of gravity (weights) of the following masses on the earth's surface: (a) 75.0 kg; (b) 4.25 kg; (c) 2500 kg; (d) 2.50 t; (e) 3.80 slug; (f) 25 slug; (g) 175 slug; (h) 9.1×10^{-31} kg; (i) 2.55 mg; (j) 3.50 Mg; (k) 2.75 mg; (l) 1.20 Mg; (m) 4.9 μg.

13. Determine the weights of the masses in Problem 12 if they were on the moon, where the acceleration due to gravity is 1.63 m/s^2 or 5.35 ft/s^2.

14. An electron, which as a mass of 9.10×10^{-31} kg, experiences an acceleration of 3.00×10^{8} m/s^2 when it passes through an electric field. What net force is acting on it?

15. Determine the mass of an object if it weighs 58.0 N on the earth's surface.

16. Determine the mass of an object whose weight is 52.8 lb on the earth's surface.

17. Determine the tension in a rope when it is used to pull 30.0 lb of bricks to the top of a house with (a) a constant speed of 3.00 ft/s; (b) a constant acceleration of 3.00 ft/s^2 upward.

18. What is the tension in the cable of a 1250 kg elevator (a) if it ascends with an acceleration of 5.00 m/s^2; (b) if it descends with an acceleration of 5.00 m/s^2; (c) if it descends with an acceleration of 9.81 m/s^2?

19. A loaded 2250 lb elevator is pulled upward by a 2600 lb force on the cable. What is the upward acceleration?

20. A 1.80×10^4 kg aircraft starts from rest and accelerates uniformly at 3.00 m/s^2 to its takeoff speed of 198 km/h. Calculate (a) the length of runway used; (b) the time required to attain takeoff speed; (c) the net force on the aircraft.

21. Determine the net force required to accelerate (a) a 1250 kg car from rest to 80.0 km/h in 12.5 s; (b) a 3200 lb car from rest to 65.0 mi/h in 9.20 s.

22. A 625 N net force acts on a 1850 kg truck, accelerating it uniformly from rest to a final speed of 21.0 m/s. (a) How long was the force applied? (b) How far did the truck travel while it experienced the force?

23 Determine the net force on a 7.50 × 10⁶ kg rocket that accelerates uniformly from 2.50 km/s to 8.25 km/s in a distance of 115 km.

24 A 1200 kg elevator is pulled upward from rest by a 11 900 N force in the cable until it reaches a final speed of 2.25 m/s. (a) How long does it take the elevator to reach the final speed? (b) How far does the elevator travel in that time?

5-2 FRICTION

There is no such thing as a perfectly smooth material; all known materials have some irregularities in their surfaces. Consequently, when two objects are in contact, these irregularities interlock, and the surfaces adhere to each other (Fig. 5-9). If a force is applied in such a way that these objects slide over each other, the adhesion between the surfaces results in a resistance to the relative motion of the objects. This resistance to the relative motion is called **friction**. The friction force always acts in the opposite direction to the motion, and it opposes any tendency of motion. Friction forces will slow moving objects.

Figure 5-9 Surface irregularities interlock to produce friction.

Friction is normally used to start fires, allows us to move and to stop, holds nails in wood, and can even be used to weld two surfaces together. However, friction is often undesirable, since it produces unwanted heat, causes wear, and reduces the efficiency of machines. Its effects can usually be minimized by lubrication, by smoothing the surfaces, and by the use of bearings and rollers. Since friction tends to oppose any motion, friction forces always act in the opposite direction to any tendency of motion.

Friction forces have the following properties:

1. They act parallel to the surfaces in contact, and in the opposite direction to any relative motion or tended motion.
2. Larger forces are required to start objects sliding than to keep them sliding at a constant speed. Therefore, starting friction forces are larger than sliding friction forces. However, once one surface is sliding over another, the friction forces do not depend on the sliding speed.
3. Friction forces are proportional to the forces pushing the surfaces together (the normal force **N**). The normal force always acts perpendicularly to the surfaces. Therefore,

$$F \propto N \quad \text{or} \quad F = \mu N \tag{5-5}$$

The constant μ is called the **coefficient of friction**; it depends on the nature of the two surfaces in contact. Some typical values are listed in Table 5-1. Note that these coefficients have no units since they are the ratio of two forces. That is,

$$\mu = \frac{F}{N}$$

The coefficient of friction corresponding to the **maximum** friction force before an object begins to slide is called the **coefficient of static friction**. The **coefficient of** sliding or kinetic friction is used when the object is sliding. Note that the coefficients of static friction are larger than those of sliding friction.

TABLE 5-1
Typical Values of Friction Coefficients between Two Surfaces

Surface	μ_s	μ_k
Glass on glass	0.95	0.80
Steel on steel	0.58	0.25
Wood on wood	0.35	0.30
Wood on metal	0.40	0.20
Wood on brick	0.60	0.25
Steel on wood	0.55	0.40
Steel on concrete	0.50	0.33
Rubber tire on dry concrete	0.95	0.71
Rubber tire on wet concrete	0.72	0.52
Wood on concrete	0.55	0.35

EXAMPLE 5-9

If a steel block has (a) a mass of 58.0 kg, or (b) a weight of 128 lb, (i) what minimum applied force is required to start it sliding over a horizontal concrete surface? (ii) What applied force will keep the block sliding at a constant velocity?

Solution

Sketch: See Fig. 5-10.

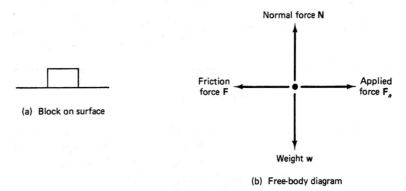

Figure 5-10

Data: In each case, the magnitude of the normal force N pushing the surfaces together is equal to the magnitude of the force of gravity (weight w) on the block. Also, to start the object sliding, the applied force must overcome the maximum friction force, and to keep it sliding at a constant speed the applied force must balance the friction force. (i) $\mu = 0.50$, (ii) $\mu = 0.33$ (Table 5-1)
(a) $m = 58.0$ kg, $g = 9.81$ m/s²,   (b) $w = 128$ lb, $g = 32.2$ ft/s²,
    $F_a = ?$ $F_a = ?$

Equations: $F = \mu N$, $w = mg$, and $N = mg$
Rearrange: $F = \mu mg$ and $F = \mu w$
(i) *Substitute:*
(a) $F = (0.50)(58.0 \text{ kg})(9.81 \text{ m/s}^2)$ (b) $F = (0.50)(128 \text{ lb})$
       $= 280$ N $= 64$ lb

(ii) *Substitute*:
(a) $F = (0.33)(58.0 \text{ kg})(9.81 \text{ m/s}^2)$
  $= 190 \text{ N}$
(b) $F = (0.33)(128 \text{ lb})$
  $= 42 \text{ lb}$

EXAMPLE 5-10

What mass of wood can a man start sliding over a horizontal concrete surface if he applies a horizontal force of (a) 725 N; (b) 75.0 lb?

Solution

Sketch: See Fig. 5-11.

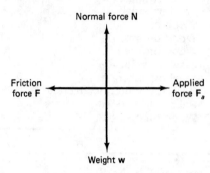

Figure 5-11 Free-body diagram.

Data: The normal force is equal in magnitude to the force of gravity (weight) $w = N = mg$; $\mu = 0.55$.

(a) $F_a = 725 \text{ N}$, $g = 9.81 \text{ m/s}^2$,
  $m = ?$
(b) $F_a = 75.0 \text{ lb}$, $g = 32.2 \text{ ft/s}^2$,
  $m = ?$

Equations: $w = mg$, $F = \mu N$

Rearrange: $F = \mu mg$; therefore, $m = \dfrac{F}{\mu g}$

Substitute:

(a) $m = \dfrac{725 \text{ N}}{(0.55)(9.81 \text{ m/s}^2)}$
  $= 130 \text{ kg}$
(b) $m = \dfrac{75.0 \text{ lb}}{(0.55)((32.2 \text{ ft/s}^2)}$
  $= 4.2 \text{ slug}$

Reduction of Friction

Friction reduces the efficiency of machines, and it can cause damage to moving parts. Lubricants and bearings are usually employed to reduce these effects. However, in spite of lubricants, about 20% of the power developed by an automobile engine is actually lost internally to friction. The control of dust and dirt and the use of new long-lasting, hermetically sealed greases are also used to control friction. Gases, such as air, are excellent lubricants. Hovercraft even float on a cushion of air, which reduces the friction between the craft and the surface.

The invention of the wheel enabled human beings to move heavy objects with much less effort. Friction effects between a rolling object (a wheel or sphere) and a flat surface are much less than the friction between two flat surfaces. Therefore, wheels and bearings allow us to achieve greater efficiency and are used extensively in machinery.

A **bearing** is a part of a machine that acts as a support for some other part, such as a rotating **shaft** or **journal**. The journal may turn or slide inside the bearing, or the bearing may rotate about a fixed journal. There are several types of bearings in common use (Fig. 5-12).

Recently, new tires for vehicles have been developed from special kinds of rubber. These tires are able to produce a high friction force on wet surfaces, and therefore they are safer.

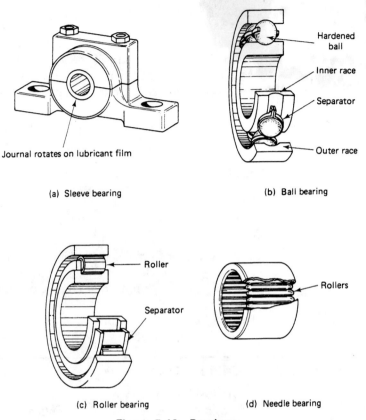

Figure 5-12 Bearings.

PROBLEMS

25. (a) What minimum force is required (a) to start a 29.5 kg steel block sliding over a horizontal wood surface; (b) to keep the block sliding at a constant velocity?

26. (a) What minimum force is required (a) to start a 6.5 lb wooden crate sliding over a horizontal brick surface; (b) to keep the crate sliding at a constant velocity?

27. A minimum force of 1560 N is required to start a 285 kg crate sliding over a horizontal surface. What is the coefficient of static friction?

28. Determine the coefficient of static friction if a 47.5 lb force is required to start a 102 lb block sliding over a horizontal surface.

29. Determine the mass of a steel crate that can be pushed at a constant velocity over a horizontal steel surface by an applied force of (a) 525 N; (b) 150 lb.

30. Determine the acceleration of a 65.0 kg wooden crate if it is pushed over a horizontal concrete surface by a 275 N force. (*Hint:* Find the net force.)

5-3 MOMENTUM

Momentum is another property that a moving object has because of its motion. It is a vector quantity that is used primarily in the solution of collision and recoil problems.

Impulse and Momentum

The **linear momentum p** of an object is a vector quantity defined as the product of its mass m and velocity **v**.

$$\mathbf{p} = m\mathbf{v} = \left(\frac{w}{g}\right)\mathbf{v} \tag{5-6}$$

where w is the weight and g is the acceleration due to gravity.

Newton's second law may be stated (more accurately) as: The net force **F** on an object is equal to its time rate of change of momentum.

$$\mathbf{F} = \frac{\Delta(m\mathbf{v})}{\Delta t} = \frac{\Delta \mathbf{p}}{\Delta t} \tag{5-7}$$

Therefore,

$$\mathbf{F}\Delta t = \Delta(m\mathbf{v}) = \Delta\left(\frac{w\mathbf{v}}{g}\right)$$

The product of the force **F** and the elapsed time Δt is known as the **impulse**. Therefore,

impulse = change in linear momentum

The units for impulse and linear momentum are therefore $N \cdot s$ or $kg \cdot m/s$ in SI, and $lb \cdot s$ or $slug \cdot ft/s$ in USCS.

EXAMPLE 5-11

Determine the momentum of **(a)** a 1250 kg car that travels due west at 22.5 m/s; **(b)** a 75.0 lb skater traveling south at 80.0 ft/s.

Solution

(a) Data: $m = 1250$ kg,
$\mathbf{v} = 22.5$ m/s west,
$\mathbf{p} = ?$

Equation: $\mathbf{p} = m\mathbf{v}$

Substitute:

$\mathbf{p} = (1250 \text{ kg})(22.5 \text{ m/s west})$
$= 2.81 \times 10^4 \text{ N} \cdot \text{s west}$

(b) Data: $w = 75.0$ lb,
$g = 32.2$ ft/s^2,
$\mathbf{v} = 80.0$ ft/s south,
$\mathbf{p} = ?$

Equation: $\mathbf{p} = \dfrac{w\mathbf{v}}{g}$

Substitute:

$\mathbf{p} = \dfrac{(75.0 \text{ lb})(80.0 \text{ ft/s south})}{32.2 \text{ ft/s}^2}$
$= 186 \text{ lb} \cdot \text{s south}$

EXAMPLE 5-12

A 0.250 kg ball is thrown with an initial speed of 20.0 m/s to a batter, who hits it back along its original path with a speed of 30.0 m/s. If the duration of the impact was 10.0 ms, what was the magnitude of the average force on the ball?

Solution

Data: The ball changes direction after impact; therefore, if the final direction of the ball's motion is taken as positive, the final velocity $v = 30.0$ m/s, and the initial velocity $u = -20.0$ m/s. $m = 0.250$ kg, $t = 10.0$ ms $= 0.0100$ s.

Equation: $F\Delta t = mv - mu = m(v - u)$

Rearrange: $F = \dfrac{m(v - u)}{\Delta t}$

Substitute: $F = \dfrac{(0.250 \text{ kg})[(30.0 \text{ m/s}) - (-20.0 \text{ m/s})]}{0.0100 \text{ s}} = 1250 \text{ N}$

Conservation of Linear Momentum

In any interaction, such as a collision or a recoil (Fig. 5-13), the total linear momentum remains constant. This statement is known as the **law of conservation of linear momentum.** It is really another statement of Newton's third law of motion:

$$\text{total linear momentum before} = \text{total linear momentum after}$$

For an interaction between two masses m_1 and m_2 that had initial velocities $\mathbf{u}_1$ and $\mathbf{u}_2$ and final velocities $\mathbf{v}_1$ and $\mathbf{v}_2$:

$$m_1\mathbf{u}_1 + m_2\mathbf{u}_2 = m_1\mathbf{v}_1 + m_2\mathbf{v}_2 \qquad (5\text{-}8)$$
$$\text{initial momentum} \quad \text{final momentum}$$

or, since the weights of the objects are $w_1 = m_1 g$ and $w_2 = m_2 g$, this relationship may be written as

$$w_1\mathbf{u}_1 + w_2\mathbf{u}_2 = w_1\mathbf{v}_1 + w_2\mathbf{v}_2 \qquad (5\text{-}9)$$

because the acceleration due to gravity g cancels. Note that we must consider vector quantities in this relationship.

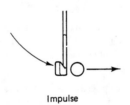

Figure 5-13

EXAMPLE 5-13

If a rifle is initially at rest, what is its recoil velocity if **(a)** its mass is 3.65 kg and it fires a 25.0 g bullet with a muzzle velocity of 788 m/s; **(b)** its weight is 8.00 lb and it fires a 0.880 oz bullet with a muzzle velocity of 2560 ft/s?

Solution

Data: The initial velocities of bullet u_b and gun u_g were zero (stationary). If the direction of the motion of the bullet is taken as the positive direction:

(a) $v_b = 788$ m/s, $m_g = 3.65$ kg, $m_b = 25.0$ g $= 25.0 \times 10^{-3}$ kg, $v_g = ?$

(b) $v_b = 2560$ ft/s, $w_g = 8.00$ lb, $w_b = 0.880$ oz $= \dfrac{0.880}{16}$ lb $= 0.0550$ lb, $v_g = ?$

Equations:
(a) $m_b u_b + m_g u_g = m_b v_b + m_g v_g$
(b) $w_b u_b + w_g u_g = w_b v_b + w_g v_g$

Rearrange: Since u_b and u_g are zero:

(a) $v_g = \dfrac{-m_b v_b}{m_g}$

(b) $v_g = \dfrac{-w_b v_b}{w_g}$

Substitute:

(a) $v_g = \dfrac{-(25.0 \times 10^{-3} \text{ kg})(788 \text{ m/s})}{3.65 \text{ kg}} = -5.40 \text{ m/s}$

(b) $v_g = \dfrac{-(0.0550 \text{ lb})(2650 \text{ ft/s})}{8.00 \text{ lb}} = -18.2 \text{ ft/s}$

Rockets and jet propulsion involves the ejection of hot gases at high speed from the rear of the craft, causing it to recoil in the forward direction. Air resistance, variations in the acceleration due to gravity, and changes in the total mass of the craft as it ejects fuel must be considered.

While momentum and energy are always conserved in a closed system, some mechanical kinetic energy is often converted into other energy forms during an interaction. In some cases, the kinetic energy of the system actually increases as particles are projected at high speed from their original positions as a result of explosions or radioactive decay. However, in collisions, the total kinetic energy usually is either constant or reduced.

PROBLEMS

31 What is the linear momentum of (a) a 1250 kg car traveling due east at 14.0 m/s; (b) a 5800 lb truck traveling south at 50.0 mi/h?

32 Determine the linear momentum of a 2500 kg truck traveling north at 75.0 km/h.

33 What is the magnitude of the linear momentum of a 6.25×10^6 kg rocket traveling at 23 000 km/h?

34 What is the linear momentum of a 2.00×10^7 kg ship moving west at 22.0 km/h?

35 Calculate the linear momentum of an electron, which has a mass of 9.1×10^{-31} kg, if it moves due east at 6.00×10^5 m/s.

36 A 145 kg pile-driver hammer is released from a height of 6.25 m above a pile. Determine the magnitude of the linear momentum of the hammer just before it strikes the pile. (*Hint:* Find the velocity just before the hammer strikes, using the principles of freely falling objects.)

37 A 475 lb pile-driver hammer is released from a height of 32.0 ft above a pile. Determine the magnitude of the linear momentum of the hammer just before it strikes the pile. (*Hint:* Find the velocity of the hammer just before it strikes the pile using the equations of motion for freely falling objects.)

38 A 4.75 kg rifle fires a 16.0 g bullet with a muzzle velocity of 775 m/s toward a target. If the rifle was initially at rest, what is its recoil velocity?

SUMMARY

Newton's three laws govern motion. The *first law* states that: An object at rest will remain at rest, and an object in motion will continue the motion in a straight line at a constant speed, unless it experiences an action called a *force*.

Mass is a measure of *inertia* (resistance to change in motion), and *weight* is due to the force of gravity.

The *second law* states: The net force is equal to the product of the mass and the acceleration: net $\mathbf{F} = m\mathbf{a}$. The *newton* (symbol N) is the SI unit for force, and the *slug* is the USCS unit for mass. The *weight w* of an object of mass *m* is found from $w = mg$, where *g* is the acceleration due to gravity.

The *third law* states: For every action there is an equal and opposite reaction. Friction tends to oppose any motion. The *friction force* is proportional to the normal force: $F \propto N$. The constant of proportionality μ is called the *coefficient of friction*.

$$F = \mu N \quad \text{and} \quad \mu = \frac{F}{N}$$

The coefficient of sliding or kinetic friction is less than that of static friction.

Linear momentum is the product of the mass and the velocity $\mathbf{p} = m\mathbf{v}$. *Impulse* is the product of the force and the time interval; it is also equal to the change in linear momentum:

$$\mathbf{F}\Delta t = \Delta \mathbf{p} = \Delta(m\mathbf{v}) = \frac{\Delta(w\mathbf{v})}{g}$$

Momentum is conserved.

QUESTIONS

1. When an aircraft performs aerobatics, the pilot is subjected to different forces. Describe some of the forces in terms of Newton's laws.
2. With practice, and many broken dishes, it is possible to jerk a tablecloth from a table without significantly disturbing the dishes. Explain how this trick works.
3. When a car accelerates, the passengers feel pushed backward into their seats, and when a car turns a corner, they tend to feel pushed to one side. Explain these effects.
4. Can the mass of a material object ever be zero? Can its weight ever be zero? Why?
5. Explain, in terms of Newton's laws of motion and friction, why it is difficult to walk on ice.
6. Why are lubricants and bearings used in machinery?
7. In space, astronauts are able to move with the aid of small rockets. Explain the principles involved.
8. Once a spaceship has cleared the gravitational pull of a celestial body, it will continue in a straight line through space without the use of additional rocket power. Explain why this is possible.
9. If you are on a frictionless surface, how could you move?
10. Seat belts normally reduce the possibility of injury during an automobile accident. Explain why.

REVIEW PROBLEMS

1. What is the acceleration when **(a)** a 15 kg mass is acted upon by a 12 N net force; **(b)** a 4.75 slug mass is acted on by a 25.0 lb net force?
2. If a certain net force accelerates an object at 96 ft/s^2, what acceleration would the same force produce if the object had four times the mass?
3. **(a)** Determine the mass of a person who weighs 803 N on the earth. **(b)** What is that person's weight on the moon, where the acceleration due to gravity is 1.65 m/s^2?
4. **(a)** What is the mass of a person who weighs 180 lb on the earth? **(b)** Determine that person's weight on the moon, where the acceleration due to gravity is 5.35 ft/s^2.
5. Determine the weight of a 75.0 kg person on the following planets, with the given accelerations due to gravity: **(a)** Earth; **(b)** Venus, $g = 8.63$ m/s^2; **(c)** Mars, $g = 3.83$ m/s^2; **(d)** Jupiter, $g = 26$ m/s^2; **(e)** Mercury, $g = 3.73$ m/s^2; **(f)** Saturn, $g = 11.5$ m/s^2.

6   If an electron is accelerated at 3.0×10^{10} m/s² by some force, what is its acceleration if the force is tripled?

7   If the rockets of a spaceship deliver a constant net thrust and it accelerates initially at 36.5 ft/s², what is its acceleration after it has reduced its mass to one-third of its initial value by burning fuel?

8   Determine the force in the supporting cable of an elevator if its total mass is 2700 kg and it accelerates upward at 1.20 m/s².

9   An 18.2 t aircraft accelerates uniformly from rest, at 2.50 m/s², to its takeoff speed of 192 km/h. Determine **(a)** the length of the runway used; **(b)** the time required; **(c)** the average net force on the aircraft.

10 A 1620 lb net force acts on a 10 500 lb truck for 15.0 s. If the truck was initially at rest, determine **(a)** its average acceleration; **(b)** its final speed; **(c)** the distance that it travels in the 15.0 s?

11 **(a)** What minimum force is required to start a 135 lb steel block sliding over a horizontal concrete surface? **(b)** What minimum force would keep the block sliding at a constant velocity?

12 **(a)** What minimum force is required to start a 34.0 kg wooden crate sliding over a horizontal concrete floor? **(b)** What minimum force would keep the crate sliding at a constant velocity?

13 What mass of steel could be kept sliding at a constant velocity over a horizontal concrete floor by an applied force of **(a)** 575 N; **(b)** 795 lb?

14 Determine the linear momentum of a 3500 lb car that travels due south at 55.0 mi/h.

15 Determine the magnitude of the linear momentum of an 225 g hockey puck traveling at 162 km/h.

16 Calculate the impulse given a 68.0 lb projectile if it is accelerated uniformly from 60.0 ft/s to 90.0 ft/s.

17 A 36.0 g golf ball is hit off a tee with an initial speed of 75.0 m/s. What magnitude impulse was it given?

18 A 6.00 oz ball travels at 155 ft/s after it has been struck with a bat. If the ball was initially at rest and the duration of the impact was 9.80 ms, **(a)** what was the impulse given to the ball; **(b)** what was the average force on the ball during the impact?

19 A 3.75 kg rifle fires a 16.0 g bullet with a muzzle velocity of 554 m/s. If the rifle is initially at rest, what is its recoil velocity?

20 An 8.00 lb rifle fires a 0.560 oz bullet with a muzzle velocity of 1800 ft/s. If the rifle is initially at rest, what is its recoil velocity?

6
ADDITION OF VECTORS

We must frequently "add" two or more vector quantities in order to find a single **resultant.** This resultant can then replace all of the original vectors because it represents their combined effect. The rules for the addition of vectors can be obtained by applying geometry and trigonometry to our knowledge of displacements and forces.

6-1 VECTOR ADDITION BY SCALE DIAGRAMS

If two or more similar vectors are not collinear, their vector sum is not equal to the algebraic sum of their magnitudes because the directions must be considered.

For example, if a man walks 4 km east and then 3 km north, his final position is only 5 km in a straight line from his starting position. Even though he walked a total distance of 7 km (Fig. 6-1), he could have accomplished the same result by walking 5 km in the direction shown in the figure. Therefore, the resultant displacement has a magnitude of 5 km in a direction 37° north of east.

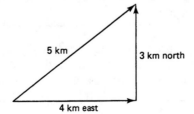

Figure 6-1

The scale diagram in Fig. 6-1 actually represents one method of adding two or more vectors to obtain a resultant. This method is called the **polygon method.** To add two or more vectors (*that have the same units*), proceed as follows:

1. Select a suitable scale and determine the lengths of all vectors in that scale.
2. Draw the first vector to scale.
3. Draw all other vectors in any order "tail to head" so that the origin of one vector is located at the terminal point of the preceding vector.
4. The resultant vector is drawn *from* the origin of the first vector *to* the terminal point of the last vector.
5. Measure the length of the resultant and use the scale to find the magnitude.
6. Measure the direction of the resultant with respect to some known or reference direction.

EXAMPLE 6-1

What is the total displacement of a car that travels 26 mi due north and then 32 mi due east?

Solution: Using a scale of 1.0 cm ≡ 10 mi, the magnitudes of the two displacement vectors represented by lengths of

$$s_1 = 26 \text{ mi} = 26 \text{ mi} \times \frac{1.0 \text{ cm}}{10 \text{ mi}} = 2.6 \text{ cm}$$

$$s_2 = 32 \text{ mi} = 32 \text{ mi} \times \frac{1.0 \text{ cm}}{10 \text{ mi}} = 3.2 \text{ cm}$$

Draw the vector (preferably on graph paper) to represent s_1, and from the terminal point of s_1, draw the vector s_2 (see Fig. 6-2). Note that the angle between the directions of the two vectors is 90° in this example.

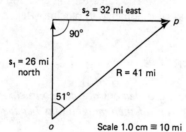

Figure 6-2 Scale diagram showing vector addition of displacements.

The resultant vector is drawn *from* the origin O of s_1 to the terminal point P of s_2. The magnitude of the resultant vector **R** (in the same scale) is the length of the line $OP = 4.1$ cm. Therefore, converting to the proper units, the magnitude of the resultant $R = 4.1 \text{ cm} \times \frac{10 \text{ mi}}{1.0 \text{ cm}} = 41$ mi. The direction of the resultant can be measured directly from the scale diagram as 51° east of north or 39° north of east.

Even though we used scale diagrams to add vectors, such as displacements, that occur in sequence (i.e., one after the other), the same process can also be used to add vectors, such as forces and velocities, that act simultaneously.

EXAMPLE 6-2

A boat traveling at a speed of 8.0 km/h heads directly across a river where the current speed is 4.0 km/h. Determine the speed and direction of the boat's path with respect to the bank of the river.

Sec. 6-1 / Vector Addition by Scale Diagrams

Solution: Let us choose a scale of 1.0 cm ≡ 2.0 km/h. Then the two lengths of the two velocity vectors are

$$v_b = 8.0 \text{ km/h} \times \frac{1.0 \text{ cm}}{2.0 \text{ km/h}} = 4.0 \text{ cm}$$

$$v_r = 4.0 \text{ km/h} \times \frac{1.0 \text{ cm}}{2.0 \text{ km/h}} = 2.0 \text{ cm}$$

Draw a line to represent the direction of the river bank and a line of length 2.0 cm parallel to the bank representing the river velocity (see Fig. 6-3). From the terminal point and at 90° to the river velocity vector draw the boat's velocity vector. The resultant velocity of the boat is drawn *from* the origin of the first vector (the river velocity) *to* the terminal point of the last vector (the boat's velocity). With a ruler we see that in this case the length of the vector in the diagram is 4.5 cm. This corresponds to a speed of

$$4.5 \text{ cm} \times \frac{2.0 \text{ km/h}}{1.0 \text{ cm}} = 9.0 \text{ km/h}$$

The direction can be measured with a protractor from the diagram as 63° from the bank.

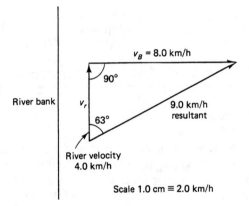

Figure 6-3 Scale diagram showing vector addition of velocities.

EXAMPLE 6-3

An aircraft flies 80 mi in a direction 45° south of east, then 110 mi 30° south of west, and finally 60 mi 10° east of north. What is the total displacement of the aircraft?

Solution: Let us choose a scale of 1 cm ≡ 20 mi; then the lengths of the vectors for the scale diagram are

$$A = (80 \text{ mi})\left(\frac{1 \text{ cm}}{20 \text{ mi}}\right) = 4.0 \text{ cm}$$

$$B = (110 \text{ mi})\left(\frac{1 \text{ cm}}{20 \text{ mi}}\right) = 5.5 \text{ cm}$$

$$C = (60 \text{ mi})\left(\frac{1 \text{ cm}}{20 \text{ mi}}\right) = 3.0 \text{ cm}$$

Draw a line to represent "north-south" and draw the first vector **A** on the graph paper (Fig. 6-4). From the terminal point of **A** draw the line to represent "east-west" and draw the second vector **B**. Finally, from the terminal point of **B** draw the "north-south" and then the vector **C**. The resultant **R** is drawn *from*

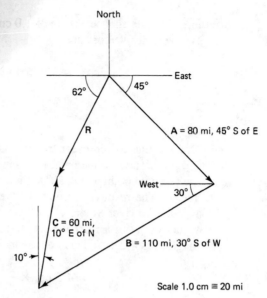

Figure 6-4 Vector addition of displacements.

the origin. The length of this vector on the diagram is 3.0 cm, which corresponds to

$$(3.0 \text{ cm})\left(\frac{20 \text{ mi}}{1 \text{ cm}}\right) = 60 \text{ mi}$$

Also from the diagram, the direction is 62° south of west.

Parallelogram Method

This method is often useful when we wish to add two vectors; it is actually quite similar to the polygon method. Suppose that we are given two vectors **A** and **B** with the same units. To find their vector sum, we draw **A** and **B** to scale from a common origin and complete a parallelogram using **A** and **B** as the sides (Fig. 6-5). The diagonal drawn from the common origin *O* represents the resultant **R** to scale.

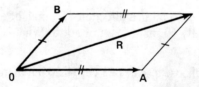

Figure 6-5 Parallelogram method of vector addition.

EXAMPLE 6-4

Two forces with magnitudes of 80 lb and 70 lb act at a common point. If the angle between the forces is 60°, what is the magnitude and direction of the resultant?

Solution: Choose a suitable scale (1.0 cm ≡ 20 lb), and draw arrows **OA** and **OB** from a common origin *O* to represent the 80 lb and 70 lb forces, respectively. From point *A* draw a line parallel to **OB**, and from point *B* draw a line parallel to **OA**. The intersection of these lines is the terminal point of the resultant drawn from *O* (Fig. 6-6). The magnitude of the resultant is the length of this diagonal to scale, 6.5 cm ≡ 130 lb, and the direction may be measured from either **OA** or **OB** (28° from **OA**).

Sec. 6-1 / Vector Addition by Scale Diagrams

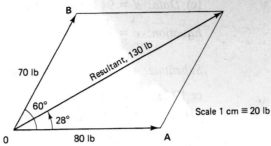

Figure 6-6 Vector addition of two forces by the parallelogram method.

EXAMPLE 6-5

A ship travels due east at 18 km/h for 3.0 h and then in a direction 40° south of east at 20 km/h for 2.0 h. Determine (a) the total distance traveled; (b) the total displacement; (c) the average speed; (d) the average velocity of the ship.

Solution: In this example we must first find distances and displacements for each leg of the journey, and then determine the average total speed and velocity. *We cannot add the velocities directly.*

(a) Data: $v_1 = 18$ km/h, $v_2 = 20$ km/h, $t_1 = 3.0$ h, $t_2 = 2.0$ h, $d = ?$

Equation: $s = vt$

Substitute: $s_1 = (18 \text{ km/h})(3.0 \text{ h}) = 54$ km
$s_2 = (20 \text{ km/h})(2.0 \text{ h}) = 40$ km

The total distance is therefore

$$d = s_1 + s_2 = 54 \text{ km} + 40 \text{ km} = 94 \text{ km}$$

(b) Choose a suitable scale (1.0 cm ≡ 20 km), then the magnitudes of the displacement vectors are represented by the lengths:

$$s_1 = (54 \text{ km})\left(\frac{1.0 \text{ cm}}{20 \text{ km}}\right) = 2.7 \text{ cm}$$

$$s_2 = (40 \text{ km})\left(\frac{1.0 \text{ cm}}{20 \text{ km}}\right) = 2.0 \text{ cm}$$

Draw the vector s_1 and from the terminal point draw s_2 (Fig. 6-7). The resultant displacement **s** is drawn from the origin O of s_1 to the terminal point P of s_2. Measuring the direction of the resultant, we obtain 17° south of east. The length of the resultant in the scale is 4.4 cm, which corresponds to

$$(4.4 \text{ cm})\left(\frac{20 \text{ km}}{1.0 \text{ cm}}\right) = 88 \text{ km}$$

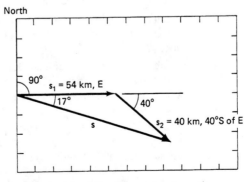

Figure 6-7 (To scale shown)

(c) *Data:* $d = 94$ km [from part (a)], $t = 3.0$ h $+ 2.0$ h $= 5.0$ h, $v = ?$

Equation: $v = \dfrac{d}{t}$

Substitute: $v = \dfrac{94 \text{ km}}{5.0 \text{ h}} = 18.8$ km/h

or 19 km/h to two significant figures.

(d) *Data:* $\mathbf{s} = 88$ km $17°$ south of east, $t = 5.0$ h, $\mathbf{v} = ?$

Equation: $\mathbf{v} = \dfrac{\mathbf{s}}{t}$

Substitute: $\mathbf{v} = \dfrac{88 \text{ km}}{5.0 \text{ h}}$, $17°$ S of E $= 17.6$ km/h, $17°$ S of E

or $\mathbf{v} = 18$ km/h in a direction $17°$ S of E (two significant figures).

PROBLEMS

1. Determine the total displacement of an aircraft that flies (a) 45 km due north and then 65 km due east; (b) 350 mi south and then 275 mi west.

2. What are the total displacements when a woman drives a car (a) 28 km due south and then 15 km due west; (b) 32 mi due west and then 45 mi north?

3. Two forces of magnitudes 36 N and 45 N are applied to a common point. If the angle between them is 48°, determine the magnitude and direction of their resultant.

4. Two forces are applied at a common point. If the forces have magnitudes of 60 lb and 110 lb and the angle between them is 36°, determine the magnitude and direction of the resultant.

5. Determine the speed and direction of the boat's path with respect to the bank of the river if it heads directly across the river at a speed of (a) 12 km/h, where the current speed is 4.0 km/h; (b) 18 mi/h, where the current speed is 8.0 mi/h.

6. What is the resultant of a 50.0 ft northward displacement and an 85.0 ft eastward displacement?

7. What is the resultant of a 70.0 N horizontal force and a 90.0 N vertical force?

8. An aircraft flies 330 km in a direction 45° south of east, and then 250 km in a direction 25° west of south. What is the total displacement?

9. Determine the resultant sum of the following vectors in standard position: (a) $\mathbf{A} = 50.0$ m at 30°, $\mathbf{B} = 35$ m at 120°; (b) $\mathbf{A} = 25$ km at 235°, $\mathbf{B} = 35$ km at 310°; (c) $\mathbf{A} = 250$ ft at 330°, $\mathbf{B} = 325$ ft at 25°; (d) $\mathbf{A} = 125$ N at 235°, $\mathbf{B} = 75.0$ N at 165°; (e) $\mathbf{A} = 5.00$ ft/s at 320°, $\mathbf{B} = 8.00$ ft/s at 45°; (f) $\mathbf{A} = 2.50$ m/s^2 at 190°, $\mathbf{B} = 3.25$ m/s^2 at 235°; (g) $\mathbf{A} = 75.0$ lb at 38°, $\mathbf{B} = 58$ lb at 315°.

10. Determine the resultant sums of velocities that are (a) 150 km/h in a direction 25° east of north and 250 km/h at 55° east of south; (b) 65.0 mi/h at 33° north of west and 55.0 mi/h at 52° west of south.

11. A man drives 110 km north, 60 km west, and then 85 km in a direction 45° north of east in a total time of 5.0 h. (a) Determine his displacement from the starting point. (b) What distance did he travel? (c) What was his average speed? (d) What was his average velocity?

12. An aircraft flies 160 mi due east, 120 mi due south, and then 145 mi in a direction 30° east of south in a total time of 3.0 h. Determine (a) the total distance traveled; (b) the total displacement; (c) the average speed; (d) the average velocity.

13. Determine (i) the total distance traveled, (ii) the total displacement, (iii) the average speed, and (iv) the average velocity of an aircraft that flies (a) due north at 250 km/h for 1.0 h and then due west at 325 km/h for 1.5 h; (b) due west at 180 mi/h for 2.50 h and then due south at 250 mi/h for 3.00 h.

Sec. 6-2 / Components of Vectors

14 Determine (i) the total distance traveled, (ii) the total displacement, (iii) the average speed, and (iv) the average velocity of a ship that sails (a) due south at 15 km/h for 5.0 h, then in a direction 35° west of south at a speed of 20 km/h for 3.0 h; (b) due west at 12 mi/h for 4.5 h and then in a direction 30° north of west at 10 mi/h for 3.0 h.

6-2 COMPONENTS OF VECTORS

We have seen that two or more vectors may be added to produce a single vector, the **resultant**. The reverse is also true. A single vector can be replaced by two or more vectors called **components** by a process called **vector resolution.** It is convenient to choose components that are perpendicular (at right angles to each other) because they are independent (i.e., a vector in one direction has no effect in a perpendicular direction). For example, a displacement toward the east does not produce any change in the perpendicular directions north or south, while a north-east displacement produces changes toward both north and east since it has components in those directions.

Consider any vector **OA** that makes an angle θ* with a specified direction (the x-axis, for example), and whose origin O is drawn at the center of a coordinate system (Fig. 6-8). From the terminal point A of the vector **OA**, construct a line AP that is perpendicular to the x-axis at the point P. The vector **OA** is the resultant of the sum of the vectors **OP** and **PA**. Thus

$$\mathbf{OA} = \mathbf{OP} + \mathbf{PA}$$

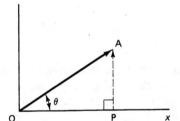

Figure 6-8 Vector resolution.

But **OP** and **PA** are in the directions of the axes of the coordinate system, and their magnitudes are the lengths of their arrows. Therefore, the magnitudes of the components can be determined by measuring the lengths of **OP** and **PA**, and converting these lengths back to the appropriate units using the scale. It should be noted that these components may be negative valued.

A component in the negative x or the negative y direction is given a negative value.

EXAMPLE 6-6

Determine the x and y components of the vectors illustrated in Fig. 6-9.

Solution

(a) $R = 4.0$ lb, $R_y = 2.0$ lb
(b) $R_x = 2.0$ m, $R_y = -3.0$ m (negative valued since it is downward)
(c) $R_x = -4.0$ N (negative since to the left), $R_y = -4.0$ N (negative valued since downward)

---

* θ is the lower case Greek letter theta.

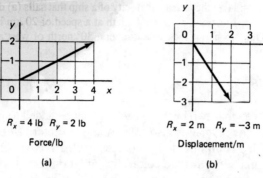

$R_x = 4$ lb $R_y = 2$ lb
Force/lb
(a)

$R_x = 2$ m $R_y = -3$ m
Displacement/m
(b)

$R_x = -4$ N $R_y = -4$ N
Force/N
(c)

Figure 6-9

We may also use trigonometry to find the magnitudes of components. Consider the vector in Fig. 6-10. Since the x and y axes are perpendicular, the triangle OAP has a right angle at P; thus

$$\cos \theta = \frac{OP}{OA} \quad \text{and} \quad OP = OA \cos \theta \tag{6-1}$$

Similarly,

$$\sin \theta = \frac{AP}{OA} \quad \text{and} \quad AP = OA \sin \theta \tag{6-2}$$

Thus the vector **OA** may be written as the sum of a component vector of magnitude $OA \cos \theta$ along the x-axis, and a component vector of magnitude $OA \sin \theta$ along the y-axis. Each component vector is the effective value of the original vector in the direction of the component.

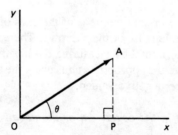

Figure 6-10

We can use the Pythagorean theorem to check answers since the perpendicular components (**OP** and **PA**) and the original vector **OA** form a right-angled triangle.

$$OP^2 + PA^2 = OA^2 \tag{6-3}$$

EXAMPLE 6-7

A boy pulls a wagon by exerting a 60.0 N force at an angle of 30° with the horizontal (Fig. 6-11). What are the **(a)** horizontal and **(b)** vertical components of this force?

Solution

Data: $F = 60.0$ N, $\theta = 30°$

(a) *Equation:* Horizontal component: $F_x = F \cos \theta$
Substitute: $F_x = (60.0 \text{ N}) \cos 30° = 52.0$ N

Sec. 6-2 / Components of Vectors

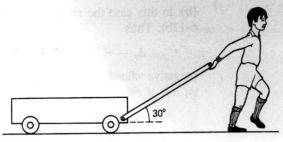

Figure 6-11

(b) *Equation:* Vertical component: $F_y = F \sin \theta$
Substitute: $F_y = (60.0 \text{ N}) \sin 30° = 30.0 \text{ N}$
Check: From the Pythagorean theorem,

$$(30.0 \text{ N})^2 + (52.0 \text{ N})^2 = (60.0 \text{ N})^2$$

EXAMPLE 6-8

Resolve the following vectors in standard position into their x and y components:
(a) 45.0 N at 66°; **(b)** 75.0 m at 190°; **(c)** 35.0 lb at 130°; **(d)** 80.0 ft at 325°.

Solution

Sketches: See Fig. 6-12.

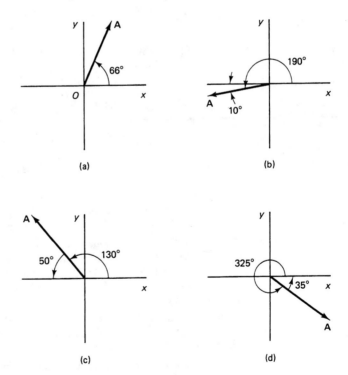

Figure 6-12

Equations: In each case $A_x = A \cos \theta$ and $A_y = A \sin \theta$, where θ is the angle between the x-axis and the vector.

(a) *Substitute:* $A_x = 45.0 \text{ N} \cos 66° = 18.3 \text{ N}$
$A_y = 45.0 \text{ N} \sin 66° = 41.1 \text{ N}$

Check: Using the Pythagorean theorem, we have

$$A^2 = A_x^2 + A_y^2 \quad \text{or} \quad (45.0 \text{ N})^2 = (18.3 \text{ N})^2 + (41.1 \text{ N})^2$$

(b) In this case the angle between the x-axis and the vector is $10°$ (see Fig. 6-12b). Thus

$$A_x = -75.0 \text{ m cos } 10° = -73.9 \text{ m} \quad (\text{or } 75.0 \text{ m cos } 190°)$$

(negative valued since the vector is directed toward the negative x-direction, that is, left of center).

$$A_y = -75.0 \text{ m sin } 10° = -13.0 \text{ m} \quad (\text{or } 75.0 \text{ m sin } 190°)$$

(negative valued since the vector is directed toward the negative y-axis, that is, downward).

Check: $(75.0 \text{ m})^2 = (-73.9 \text{ m})^2 + (-13.0 \text{ m})^2$

There is a slight discrepancy due to the rounding off.

(c) The angle between the x-axis and the vector is $50°$ (see Fig. 6-12c). Therefore,

$$A_x = -35.0 \text{ lb cos } 50° = -22.5 \text{ lb} \quad \text{(negative because toward the}$$
$$(\text{or } 35.0 \text{ lb cos } 130°) \quad \text{negative } x\text{-direction)}$$

$$A_y = 35.0 \text{ lb sin } 50° = 26.8 \text{ lb} \quad \text{(positive because toward the}$$
$$(\text{or } 35.0 \text{ lb sin } 130°) \quad \text{positive } y\text{-direction)}$$

Check: $(35.0 \text{ lb})^2 = (22.5 \text{ lb})^2 + (26.8 \text{ lb})^2$

(d) The angle between the x-axis and the vector is $35°$ (see Fig. 6-12d).

$$A_x = 80.0 \text{ ft cos } 35° = 65.5 \text{ ft} \quad \text{(positive valued)}$$
$$(\text{or } 80.0 \text{ ft cos } 325°)$$

$$A_y = -80.0 \text{ ft sin } 35° = -45.9 \text{ ft} \quad \text{(negative valued because the}$$
$$(\text{or } 80.0 \text{ ft sin } 325°) \quad \text{component is in the negative}$$
$$y\text{-direction)}$$

Check: $(80.0 \text{ ft})^2 = (65.5 \text{ ft})^2 + (-45.9 \text{ ft})^2$

PROBLEMS

15 A boy pulls a wagon by exerting a 35.0 N force at an angle of 30° with the horizontal. What are the horizontal and vertical components of the force?

16 Determine the x and y components of the following vectors in standard position: **(a)** 30.0 N at 75°; **(b)** 24 km at 38°; **(c)** 350 lb at 72°; **(d)** 45 m/s at 15°; **(e)** 16 km/h at 135°; **(f)** 750 N at 240°; **(g)** 66 lb at 345°; **(h)** 8.00 mi at 229°; **(i)** 34.0 m at 162°; **(j)** 778 lb at 148°; **(k)** 90.0 ft/s at 34°; **(l)** 234 ft at 180°.

17 Resolve the following into their east and north components: **(a)** 30.0 km at 45° east of north; **(b)** 20 km/h at 22° east of north; **(c)** 15.0 m/s² at 20° west of north; **(d)** 25.0 ft/s at 25° west of south; **(e)** 18.0 mi/h at 20° east of south; **(f)** 18 m at 15° south of east; **(g)** 350 ft at 29° south of west.

18 Resolve the following into horizontal and vertical components: **(a)** 72.0 N at 20° above the horizontal; **(b)** 28.0 lb at 45° above the horizontal; **(c)** 35.0 N at 60° *below* the horizontal; **(d)** 85 lb at 62° *below* the horizontal.

19 Resolve the following forces into their perpendicular x and y components: **(a)** 46.0 N at 35° with respect to the x-axis; **(b)** 38.0 lb at 84° with the x-axis; **(c)** 55.0 N at 145° with the x-axis; **(d)** 63.0 lb at 290° with the x-axis.

20 A weight at 50.0 N rests on an inclined plane that makes an angle of 30° with the horizontal. Find the components of the weight parallel and perpendicular to the plane.

6-3 VECTOR ADDITION BY COMPONENTS

Once vectors have been resolved into their perpendicular components, they are easy to add because the components are independent. In addition, the x-components are collinear and their algebraic sum is the x-component of the resultant. Similarly, the algebraic sum of the y-components is the y-component of the resultant.

EXAMPLE 6-9

Three displacements have the following components: **A** = (25.0 mi east, 20.0 mi north), **B** = (15.0 mi west, 12.0 mi north), and **C** = (30.0 mi east, 18.0 mi south). What are the components of their vector sum?

Solution

$$\mathbf{A} + \mathbf{B} + \mathbf{C} = (25.0 \text{ mi E}, 20.0 \text{ mi N}) + (15.0 \text{ mi W}, 12.0 \text{ mi N})$$
$$+ (30.0 \text{ mi E}, 18.0 \text{ mi S})$$
$$= (25.0 \text{ mi E} - 15.0 \text{ mi E} + 30.0 \text{ mi E}),$$
$$(20.0 \text{ mi N} + 12.0 \text{ mi N} - 18.0 \text{ mi N})$$

since west is the negative of east and south is the negative of north.
Thus **A** + **B** + **C** = (40.0 mi E, 14.0 mi N).

Once we have found the resultant of a vector in component form we can convert it into a magnitude and direction in the standard position (**polar form**) using the following procedure:

1. Sketch the components R_x and R_y, and the resultant **R**. Be sure to draw them in the correct directions given by the signs.

2. The components and the resultant form a right-angled triangle; therefore, we can use the Pythagorean theorem to find the magnitude R of the resultant **R**:

$$R^2 = R_x^2 + R_y^2 \quad \text{or} \quad R = \sqrt{R_x^2 + R_y^2} \qquad (6\text{-}4)$$

3. The direction of the resultant can be expressed in terms of the angle that it makes with the x-axis. This is found by using the tangent in the right-angled triangle. Care must be taken to state or show the correct location of the resultant vector.

$$\tan \phi^* = \frac{R_y}{R_x} \qquad (6\text{-}5)$$

Note that it is not necessary to use the signs in Eq. 6-5 as long as a diagram clearly illustrates the final direction.

EXAMPLE 6-10

If a vector has an x-component of 75.0 N and a y-component of −68.0 N, determine the magnitude and direction in standard position.

Solution

Sketch: See Fig. 6-13.

---

*ϕ is the lowercase Greek letter phi.

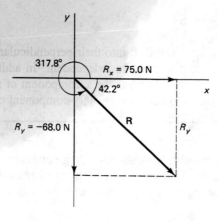

Figure 6-13

Data: $R_x = 75.0$ N, $R_y = 68.0$ N, $R = ?$, $\phi = ?$

Equations: $R = \sqrt{R_x^2 + R_y^2}$, $\tan \phi = \dfrac{R_y}{R_x}$

Substitute: $R = \sqrt{(75.0 \text{ N})^2 + (-68.0 \text{ N})^2} = 101$ N

$$\tan \phi = \frac{68.0 \text{ N}}{75.0 \text{ N}} = 0.907$$

Therefore, the angle $\phi = 42.2°$ below the x-axis. This corresponds to an angle of $360° - 42.2° = 317.8°$ in standard position.

Using *most* calculators, we would key this part of the solution to find the angle 42.2° as follows:

$$\boxed{68} \boxed{\div} \boxed{75} \boxed{=} \boxed{\tan^{-1}} \boxed{=}$$

We can use the component method to find the resultant of any number of vectors in the standard position by combining the techniques previously discussed. The procedure is as follows:

1. Sketch the vectors.
2. Resolve each vector into its components. It is often convenient to tabulate the results.
3. Algebraically add the components to find the components of the resultant.
4. Sketch the components and the resultant. Be sure to include the signs.
5. Use the Pythagorean theorem to find the magnitude of the resultant.

$$R = \sqrt{R_x^2 + R_y^2}$$

6. Find the angle that the resultant makes with the x-axis using the tangent function in the right-angled triangle formed by the vectors in step 4.

$$\tan \phi = \frac{R_y}{R_x}$$

EXAMPLE 6-11

Find the resultant sum of the following force vectors in standard position: $\mathbf{A} = 35.0$ N at 60°, $\mathbf{B} = 65.0$ N at 160°.

Solution

Sketch: See Fig. 6-14.

Sec. 6-3 / Vector Addition by Components

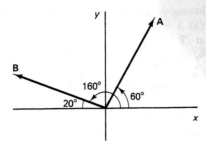

Figure 6-14

Resolve:

Vector	x-component	y-component
A	35.0 N cos 60° = 17.5 N	35.0 N sin 60° = 30.3 N
B	−65.0 N cos 20° = −61.1 N	65.0 N sin 20° = 22.2 N
	Sum: $R_x = -43.6$ N	$R_y = 52.5$ N

Sketch: See Fig. 6-15.

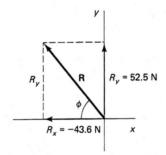

Figure 6-15

By the Pythagorean theorem,

$$R = \sqrt{R_x^2 + R_y^2}$$
$$= \sqrt{(-43.6 \text{ N})^2 + (52.5 \text{ N})^2} = 68.2 \text{ N}$$

Find the tangent:

$$\tan \phi = \frac{R_y}{R_x} = \frac{52.2 \text{ N}}{43.6 \text{ N}} = 1.20$$

so $\phi = 50.2°$. Note the direction in Fig. 6-15. This corresponds to an angle of $180° - 50.2° = 129.8°$ in standard position.

PROBLEMS

21 Determine the components of the vector sum of the following velocity vectors, which are written in terms of their components: **A** = (80.0 km/h E, 35.0 km/h S), **B** = (65.0 km/h W, 80.0 km/h S), **C** = (120.0 km/h E, 50.0 km/h N), and **D** = (90.0 km/h E, 60.0 km/h S).

22 Find the components of the resultant of the following force vectors, which are written in terms of their x and y components: **A** = (26.2 N, 19.4 N), **B** = (−15.3 N, −41.0 N), **C** = (−17.4 N, 25.5 N), and **D** = (22.1 N, −4.8 N).

23 Find the components of the vector sum of the following force vectors, which are given in terms of their components: **A** = (45.0 lb, −22.5 lb), **B** = (−35.0 lb, 15.0 lb), **C** = (−35.5 lb, −25.5 lb), and **D** = (−22.0 lb, 12.0 lb).

24. Given the following vectors in standard position, find in each case the components of their vector sum: (a) **A** = 25.0 m at 60°, **B** = 35.0 m at 135°; (b) **A** = 75.0 ft at 120°, **B** = 60.0 ft at 220°; (c) **A** = 45.0 N at 320°, **B** = 35.0 N at 200°; (d) **A** = 26.0 lb at 145°, **B** = 34.0 lb at 170°; (e) **A** = 15.0 m/s at 235°, **B** = 18.0 m/s at 160°.

25. In each case, find the sum of the vectors in standard position: (a) **A** = 8.00 N at 45°, **B** = 10.0 N at 80°; (b) **A** = 245 m at 235°; **B** = 175 m at 165°; (c) **A** = 24.0 m/s² at 315°, **B** = 18.0 m/s² at 36°; (d) **A** = 55.0 lb at 340°, **B** = 48.0 lb at 250°; (e) **A** = 35.0 ft at 155°, **B** = 48.0 ft at 215°; (f) **A** = 45.0 mi at 35°, **B** = 38.0 mi at 125°, **C** = 42.0 mi at 210°; (g) **A** = 26.0 N at 345°, **B** = 17.5 N at 165°, **C** = 22.5 N at 233°.

6-4 VECTOR SUBTRACTION

It is sometimes necessary to subtract one vector quantity from another. This process is equivalent to adding the negative value (a vector of the same magnitude but opposite direction), of the vector being subtracted. For example, to subtract a vector **B** from a vector **A**, form the negative of **B**, written as −**B**, and then add −**B** to **A** (Fig. 6-16). This is written as

$$\mathbf{A} - \mathbf{B} = \mathbf{A} + (-\mathbf{B})$$

To subtract a vector that is written in component form, we must change the sign of each of its components and then add them algebraically as before.

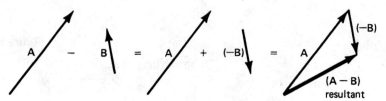

Figure 6-16 Vector subtraction.

EXAMPLE 6-12

A car traveling at an initial velocity with components $\mathbf{v}_o$ = (6.9 m/s E, 9.7 m/s N) accelerates to a final velocity with components $\mathbf{v}_1$ = (13.9 m/s E, 4.2 m/s S) in 50.0 s. Determine the components of (a) the change in velocity of the car; (b) the average acceleration.

Solution

Data: $\mathbf{v}_o$ = (6.9 m/s E, 9.7 m/s N), $\mathbf{v}_1$ = (13.9 m/s E, 4.2 m/s S), t = 50.0 s, $\Delta \mathbf{v}$ = ?, **a** = ?

(a) Equation: $\Delta \mathbf{v} = \mathbf{v}_1 - \mathbf{v}_o$

Substitute: $\Delta \mathbf{v}$ = (13.9 m/s E, 4.2 m/s S) − (6.9 m/s E, −9.7 m/s S)
  = (7.0 m/s E, 13.9 m/s S)

(b) Equation: $\mathbf{a} = \dfrac{\Delta \mathbf{v}}{t}$

Substitute: $\mathbf{a} = \dfrac{(6.9 \text{ m/s E, } 13.9 \text{ m/s S})}{50.0 \text{ s}}$ = (0.14 m/s² E, 0.278 m/s² S)

PROBLEMS

26. If **A** = 25.0 km N and **B** = 35 km W, find (a) **A** − **B**; (b) **B** − **A**.
27. Two forces combine to produce a resultant of 750 N north. If one of the forces is 450 N at 30° east of north, what is the other force?
28. If two forces combine to produce a resultant of 35.0 lb at 125° in the standard position, and one of the forces was 16.5 lb at 75°, what is the other force?

SUMMARY

We may add two or more vectors to find the resultant using scale diagrams. There are two common methods (Fig. 6-17):

Polygon method: The vectors are drawn to scale tail to head, and the resultant is drawn *from* the origin of the first vector *to* the terminal point of the last vector.

Parallelogram method: A parallelogram is constructed using two vectors drawn to scale from a common origin. The resultant is the diagonal drawn from the common origin.

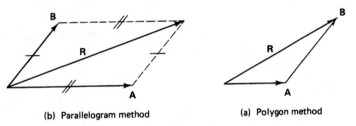

(b) Parallelogram method   (a) Polygon method

Figure 6-17

A vector **A** drawn in *standard position (polar form)* (Fig. 6-18) has perpendicular components:

$$A_x = A \cos \theta \qquad A_y = A \sin \theta$$

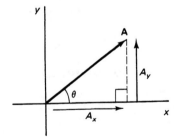

Figure 6-18

Also,

$$A = \sqrt{A_x^2 + A_y^2}$$

The algebraic sum of the x-components of a number of vectors is equal to the x-component of the resultant, and the algebraic sum of the y-components equals the y-component of the resultant:

$$R_x = A_x + B_x + C_x + \ldots, \quad R_y = A_y + B_y + C_y + \ldots$$

In standard position (or polar form) the magnitude of the resultant

$$R = \sqrt{R_x^2 + R_y^2}$$

and the direction of **R** is found from

$$\tan \phi = \frac{R_y}{R_x}$$

To *subtract a vector* we reverse its direction and then add.

QUESTIONS

1. Define the following terms: (a) resultant; (b) components; (c) collinear vectors; (d) negative vector.
2. Can the magnitude of a vector ever be less than the magnitude of its components? Explain your answer.
3. Is it possible to add two vectors and obtain a resultant with a magnitude less than the magnitudes of the original vectors? Explain the answer.
4. Give three reasons why you would want to subtract vectors.
5. Describe how you would subtract one vector from another using a scale diagram.

REVIEW PROBLEMS

1. What is the vector sum of a 15 lb horizontal force and a 12 lb vertical force?
2. The sum of two perpendicular forces has a magnitude of 380 N. If one of the forces has a magnitude of 225 N, what is the magnitude of the other?
3. Two forces are applied to a common point. If the forces have magnitudes of 80.0 lb and 50.0 lb, and the angle between them is 63°, what is the magnitude of the resultant?
4. An aircraft flies the following paths over the ground: **A** = 34° west of south for 315 km, **B** = 36° south of east for 450 km, and **C** = 125 km due south. If the total flying time is 3.50 h, determine (a) the total displacement of the aircraft; (b) the average velocity of the aircraft.
5. A ship sails the following paths over the earth: **A** = 15° west of south for 55 mi, **B** = 60° west of south for 25 mi, and **C** = 25° west of south for 35 mi. If the total sailing time is 7.50 h, determine (a) the total displacement of the ship; (b) the average velocity of the ship.
6. Resolve the following into their perpendicular x- and y-components: (a) **A** = 50.0 N at 36° with the x-axis, **B** = 35.0 N at 156° from the x-axis, **C** = 45.0 N at 280° from the x-axis; (b) **A** = 25.0 mi/h at 24° with the x-axis, **B** = 35.0 mi/h at 192° from the x-axis, **C** = 30.0 mi/h at 315° form the x-axis.
7. Resolve the following displacements into their perpendicular east and north components: (a) **A** = 75.0 km 35° west of south, **B** = 45.0 km 58° south of east, **C** = 60.0 km 10° east of north; (b) **A** = 120 mi 65° east of north, **B** = 150 mi at 32° south of east, **C** = 100 mi at 45° south of west.
8. Determine the components of the vector sum of the following velocity vectors: (a) **A** = (55 km/h E, 45 km/h N), **B** = (63 km/h W, 35 km/h N), **C** = (80 km/h W, 65 km/h S); (b) **A** = (35 mi/h E, 26 mi/h N), **B** = (80 mi/h W, 38 mi/h S), **C** = (50 mi/h E, 15 mi/h S).
9. An aircraft flies 325 km due south in 1.5 h and then 275 km in a direction 35° west of south in 1.0 h. Determine (a) the total distance traveled; (b) the total displacement; (c) the average speed; (d) the average velocity.

Review Problems

10. Find the vector sums of the following vectors in the standard position: (a) **A** = 450 N at 45°, **B** = 525 N at 110°. (b) **A** = 67.0 lb at 225°, **B** = 85.0 lb at 305°. (c) **A** = 22.0 m/s at 130°, **B** = 18.0 m/s at 250°. (d) **A** = 65.0 mi/h at 15°, **B** = 80.0 mi/h at 330°. (e) **A** = 14.0 km at 125°, **B** = 21.0 km at 335° and **C** = 18.0 km at 22°.

11. What is the velocity of A relative to B ($v_A - v_B$) if (a) boat A moves due north at 25 km/h and boat B moves due east at 32 km/h; (b) boat A moves due south at 15 mi/h and boat B moves due west at 20 mi/h?

12. Determine (i) the position of A relative to B (**A** − **B**), and (ii) the position of B relative to A (**B** − **A**), (a) if an aircraft A is 15 km north of a hill and aircraft B is 25 km due west; (b) if the aircraft A is 25 mi due east and aircraft B is 30 mi due north.

7
EQUILIBRIUM

We have seen that forces produce accelerations and therefore changes in the motion of objects, but in some cases there is no acceleration, even though forces are present. This state of no acceleration is called **equilibrium;** it arises because the forces balance each other. For example, structures such as buildings and bridges are subjected to very large forces, yet the structure must remain in equilibrium. Engineers design these structures so that they deform only slightly and they may often be treated as undeformed or rigid.

7-1 FIRST EQUILIBRIUM CONDITION

Newton's second law of motion tells us that any unbalanced (net) force produces an acceleration:

$$\text{net } \mathbf{F} = m\mathbf{a}$$

But acceleration is the time rate of change of velocity, therefore, a rigid object remains stationary or moves at a constant velocity (speed and direction) only when there is no net force acting on it. Two or more forces are in **equilibrium** only if their vector sum is zero since this corresponds to a state of no acceleration. This **first equilibrium condition** may be written as

$$\text{net } \mathbf{F} = 0 \qquad (7\text{-}1)$$

An equivalent statement of this condition is that the algebraic sum of the components of the forces in any direction must also equal zero:*

$$\text{net } F_x = 0 \qquad \text{and} \qquad \text{net } F_y = 0 \qquad (7\text{-}2)$$
(sum of x-components) \qquad (sum of y-components)

---

*In three dimensions a third component is also required: net $F_z = 0$.

Sec. 7-1 / First Equilibrium Condition

The first condition applies to any perpendicular set of axes that we choose, since the equilibrium conditions imply that the algebraic sum of the components of the forces in any direction equals zero.

In one dimension, this implies that the sum of the forces in the positive direction is equal in magnitude to the sum of the forces in the negative direction:

$$\text{net } F_+ = \text{net } F_- \tag{7-3}$$

EXAMPLE 7-1

A cable is used to support a load of machinery that hangs freely. Determine the tension in the cable if the load is (a) 75.0 kg; (b) 350 lb.

Solution

Sketch: See Fig. 7-1. The force of gravity pulls downward on the object with a force equal to its weight **w**, and the cable pulls upward on the load with a force equal to the tension **T**. If this were not true, the load would fall toward the earth.

Data: Remember, kilograms are mass units and pounds are force (weight) units.
(a) $m = 75.0$ kg, $g = 9.81$ m/s^2, (b) $w = mg = 350$ lb, $T = ?$
$T = ?$

Equations: weight $w = mg$, net $F_{up} = F_{down}$
Rearrange: $T = w = mg$
Substitute:
(a) $T = (75.0 \text{ kg})(9.81 \text{ m/s}^2) = 736$ N   (b) $T = 350$ lb

Figure 7-1 Free-body diagram.

EXAMPLE 7-2

A horizontal bridge weighing 980 000 N is supported at each end. If the upward force on the bridge from one of the supports is 350 000 N, what is the force at the other support?

Solution

Sketch: See Fig. 7-2.

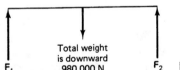

Figure 7-2

Data: Suppose that F_1 and F_2 are the magnitudes of the upward forces at the supports (the reaction forces). Then $F_1 = 350\,000$ N, $F_2 = ?$, $w = 980\,000$ N
Equation: Since the bridge is in equilibrium, the total upward force must balance the total downward force; therefore,

$$F_1 + F_2 = w$$

Rearrange: $F_2 = w - F_1$
Substitute: $F_2 = 980\,000$ N $- 350\,000$ N $= 630\,000$ N

PROBLEMS

1. A 275 lb sign is supported by two vertical wires. If the tension in one of the wires is 140 lb, what is the tension in the other?
2. A bridge is supported at each end. What is the weight of the loaded bridge if the reaction forces at the supports are (a) 180 000 N and 135 000 N; (b) 85 000 lb and 92 000 lb?

3   A 56.0 kg roof truss is supported at each end. If the load is distributed evenly, determine the reaction forces at the supports.

7-2 CONCURRENT FORCES IN EQUILIBRIUM

Forces whose lines of action pass through a common point are called **concurrent** (Fig. 7-3). These forces only tend to produce accelerations in straight lines in the direction of the net force. If the forces are concurrent, the first equilibrium condition is the only condition that must be satisfied for equilibrium to exist.

Concurrent forces are said to be in a state of **static equilibrium** if their resultant is zero and the system is at rest with respect to the observer. This type of equilibrium is present in most structures, such as buildings and bridges.

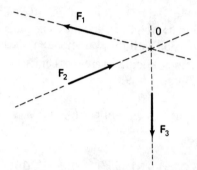

Figure 7-3   Three forces concurrent at 0.

Even moving objects can be in equilibrium as long as they do not accelerate. They must move in a straight line at a constant speed. For example, an aircraft in straight and level flight at a constant velocity is in equilibrium. In this case, the upward force on the wings (called **lift**) balances the weight, and the forward force (called **thrust**), due to the engines, balances the retarding forces due to the surrounding air (called **drag**) (Fig. 7-4). This is called **dynamic equilibrium.**

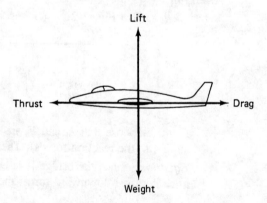

Figure 7-4   Aircraft in straight and level flight at a constant velocity.

Scale Diagrams

When two or more concurrent forces are in equilibrium, their net (**resultant**) force is zero; therefore, the scale diagram representing their vector sum must be **closed.** That is, since the resultant is zero, the terminal point of the last vector drawn must coincide with the origin of the first.

We may use this fact to solve equilibrium problems for concurrent forces. The procedure is as follows:

Sec. 7-2 / Concurrent Forces in Equilibrium

1. Draw the **free-body diagram** of the system, indicating all forces that act at the concurrence point (where they meet). Be sure to include all reaction forces in terms of their effect at that point, and any force of gravity (represented by the weight).
2. Construct the *closed* **scale diagram** representing the addition of the forces.
3. **Measure** the unknown values directly from the scale diagram and convert back to the appropriate units.

EXAMPLE 7-3

A crane supports a 7500 N load from a vertical cable attached to the top of a boom (see Fig. 7-5). Determine the horizontal tension in the cable and the compressive force in the boom.

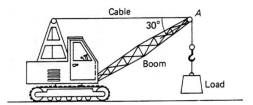

Figure 7-5

Solution: Draw the free-body diagram of the top of the boom where the forces are acting (point A, Fig. 7-6). The boom must be pushing as shown; otherwise, the system would fall downward and to the left. These reaction forces can usually be determined by asking the question: What would happen if they were not there?

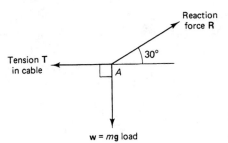

Figure 7-6 Free-body diagram of forces concurrent at point A.

Using a scale of 2000 N ≡ 1.00 cm, the weight of the load

$$7500 \text{ N} = (7500 \text{ N})\left(\frac{1.00 \text{ cm}}{2000 \text{ N}}\right) = 3.75 \text{ cm}$$

First, construct all vectors that we know completely in both magnitude and direction. In this case, the vertical weight vector (3.75 cm long). Although we do not know the magnitudes of the other vectors, we do know their directions. Since there are three forces in this case, we must construct a *closed* triangle (so that the resultant force is zero) by drawing all vectors tail to head. Therefore, construct a line *from* the terminal point of the weight vector in the direction of the reaction force **R**, and a line *to* the origin of the weight vector in the direction of the tension **T**. At the intersection of these lines we can now draw the terminal point of the reaction force **R** (see Fig. 7-7). This completes the closed triangle of vectors. The magnitudes of the vectors can now be measured directly from the diagram. Thus, on the diagram, the reaction force vector is 7.5 cm long and the tension vector is 6.5 cm long. These lengths correspond to magnitudes of

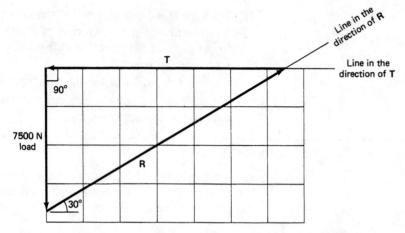

Figure 7-7 Scale diagram. Each division corresponds to 2000 N.

$$R = \left(\frac{7.5 \text{ cm}}{1.00 \text{ cm}}\right)(2000 \text{ N}) = 15\,000 \text{ N} \quad \text{and}$$

$$T = \left(\frac{6.5 \text{ cm}}{1.00 \text{ cm}}\right)(2000 \text{ N}) = 13\,000 \text{ N}$$

EXAMPLE 7-4

An 80.0 lb traffic light is suspended in equilibrium from two wires that make angles of 20° and 30° with the horizontal (see Fig. 7-8). Determine the tensions T_1 and T_2 in the wires.

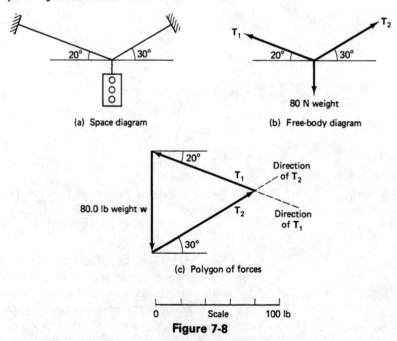

Figure 7-8

Solution: Draw the free-body diagram (Fig. 7-8b). Choose an appropriate scale, 1 cm ≡ 20 lb for example. In this scale,

$$80.0 \text{ lb} = (80.0 \text{ lb})\left(\frac{1.0 \text{ cm}}{20 \text{ lb}}\right) = 4.00 \text{ cm}$$

Therefore, construct the vertical vector representing the 80 lb weight (4.00 cm long in the chosen scale). From the terminal point, draw a line parallel to the

Sec. 7-2 / Concurrent Forces in Equilibrium

tension T_2. Since the triangle of forces must be closed, the terminal point of the remaining tension vector T_1 must coincide with the origin of the weight vector.

Construct a line parallel to the tension vector T_1 from the origin of the weight vector to intersect the line drawn parallel to T_2 (Fig. 7-8c). The intersection of these lines completes the triangle of forces. The magnitudes of T_1 and T_2 may be measured directly from the diagram (to the same scale as the weight vector). In this case the lengths of the vectors in the scale diagram are $T_1 \equiv 4.5$ cm and $T_2 \equiv 4.9$ cm; therefore, the magnitudes are

$$T_1 = (4.5 \text{ cm})\left(\frac{20 \text{ lb}}{1.0 \text{ cm}}\right) = 90 \text{ lb} \quad \text{and} \quad T_2 = (4.9 \text{ cm})\left(\frac{20 \text{ lb}}{1.0 \text{ cm}}\right) = 98 \text{ lb}$$

Trigonometric Solution

We may also use trigonometry to solve problems involving the equilibrium of concurrent forces. The procedure is as follows:

1. Draw a **free-body diagram** of the system for the point(s) where the forces are concurrent.
2. **Resolve** all forces into a *convenient* set of perpendicular components. Since the system is in equilibrium, we may choose any set of perpendicular axes. Horizontal and vertical axes are not always the most convenient, because a different choice will sometimes enable us to solve the problem with less work.
3. Apply the **equilibrium conditions** to obtain equations involving the unknown items.
4. **Solve** the equations obtained in step 3.

EXAMPLE 7-5

A sign is suspended in equilibrium from two wires, one horizontal and the other inclined at an angle of 60° with the horizontal. Determine the tensions in the wires if the sign has **(a)** a mass of 25.0 kg; **(b)** a weight of 50.0 lb.

Solution

Sketch: See Fig. 7-9. Since the weight of the sign acts vertically downward and one of the wires is horizontal, we shall use the horizontal and vertical directions as the x and y axes.

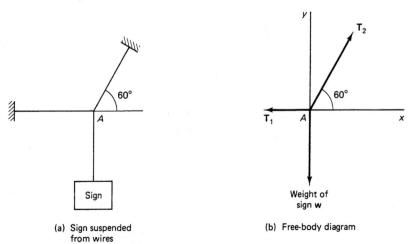

(a) Sign suspended from wires

(b) Free-body diagram

Figure 7-9

Data:
(a) $m = 25.0$ kg, $g = 9.81$ m/s², $T_1 = ?$, $T_2 = ?$
(b) $w = mg = 50.0$ lb, $T_1 = ?$, $T_2 = ?$

Equations: To resolve the forces into their components:

$$F_x = F \cos \theta \qquad F_y = F \sin \theta$$

Therefore, since down is considered as the negative of up:

Force	x-component	y-component
$w = mg$	0	$-w = -mg$
T_1	$-T_1$	0
T_2	$T_2 \cos 60°$	$T_2 \sin 60°$

Applying the equilibrium conditions: net $F = 0$ in any direction, therefore, adding the y-components,

$$\text{net } F_y = -mg + 0 + T_2 \sin 60° = 0$$

Rearrange: $T_2 \sin 60° = mg$ and $T_2 = \dfrac{mg}{\sin 60°}$

Substitute:

(a) $T_2 = \dfrac{(25.0 \text{ kg})(9.81 \text{ m/s}^2)}{\sin 60°}$
$= 283$ N

(b) $T_2 = \dfrac{50.0 \text{ lb}}{\sin 60°}$
$= 57.7$ lb

Also, adding the x-components yields

$$\text{net } F_x = 0 - T_1 + T_2 \cos 60° = 0$$

Rearrange: $T_1 = T_2 \cos 60°$

Substitute:

(a) Using $T_2 = 283$ N gives us
$T_1 = (283 \text{ N}) \cos 60° = 141$ N

(b) Using $T_2 = 57.7$ lb gives us
$T_1 = (57.7 \text{ lb}) \cos 60°$
$= 28.9$ lb

EXAMPLE 7-6

The boom illustrated in Fig. 7-10 supports a load. If the tie is horizontal and the boom makes an angle of 35° with the horizontal, find the tension in the tie and the compression in the boom when the load is (a) 3500 N; (b) 1250 lb.

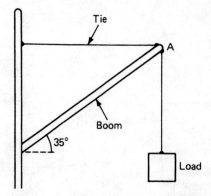

Figure 7-10

Solution

Sketch: See Fig. 7-11. Three forces are concurrent at point A. The boom must be pushing upward as shown (otherwise, the boom would fall to the left), the tie

Sec. 7-2 / Concurrent Forces in Equilibrium

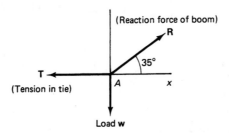

Figure 7-11 Free-body diagram.

pulls to the left with a force equal to the tension **T**, and the weight **w** acts vertically downward. Since the tie is horizontal and the weight acts downward, let us choose the horizontal and vertical as x and y axes.

Data:
(a) $w = 3500$ N, $\theta = 35°$, $T = ?$ (b) $w = 1250$ lb, $\theta = 35°$, $T = ?$
Equations: $F_x = F \cos\theta$, $F_y = F \sin\theta$
Resolve the forces into their components:

Force	x-component	y-component
w	0	$-w$
T	$-T$	0
R	$R \cos 35°$	$R \sin 35°$

net $F = 0$ in any direction

For the vertical axis,

$$\text{net } F_y = -w + 0 + R \sin 35° = 0$$

Rearrange: $R = \dfrac{w}{\sin 35°}$

Substitute:

(a) $R = \dfrac{3500 \text{ N}}{\sin 35°} = 6100$ N (b) $R = \dfrac{1250 \text{ lb}}{\sin 35°} = 2180$ lb

Similarly, for the x-axis,

$$\text{net } F_x = 0 - T + R \cos 35° = 0$$

Rearrange: $T = R \cos 35°$
Substitute:
(a) Since $R = 6100$ N (b) Since $R = 2180$ lb,
   $T = 6100$ N $\cos 35°$ $T = 2180$ lb $\cos 35° = 1790$ lb
   $= 5.0 \times 10^3$ N

EXAMPLE 7-7

Determine the friction force and the normal force on a block that is at rest on a 30° incline **(a)** if the block has a mass of 5.00 kg; **(b)** if the block weighs 12.0 lb.

Solution

Sketch: See Fig. 7-12. The friction force $\mathbf{F}_f$ must act up the incline (otherwise, the block would slide), and the normal force **N** acts perpendicularly to the incline.

Data: $N = ?$, $F_f = ?$
(a) $m = 5.00$ kg, $g = 9.81$ m/s² (b) $w = mg = 12.0$ lb
Equations: $w = mg$, $F_x = F \cos\theta$, $F_y = F \sin\theta$

In this case, since the friction force and the normal force are perpendicular to

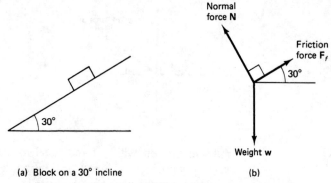

(a) Block on a 30° incline (b)

Figure 7-12

each other, it is convenient to choose the axes of the coordinate system along the directions of these forces (see the free-body diagram).

Resolve the forces into their components in these directions:

Force	x-component	y-component
F_f	F_f	0
N	0	N
$w = mg$	$-mg \cos 60°$	$-mg \sin 60°$

Equation: Using the equilibrium condition, net $F = 0$ in any direction, for the x-components:

$$\text{net } F = F_f - mg \cos 60° = 0 \quad \text{and therefore} \quad F_f = mg \cos 60°$$

For the y-components:

$$\text{net } F = N - mg \sin 60° = 0 \quad \text{and} \quad N = mg \sin 60°$$

Substitute: For the x-components:
(a) $F_f = (5.00 \text{ kg})(9.81 \text{ m/s}^2) \cos 60°$
    $= 24.5 \text{ N}$

(b) $F_f = 12.0 \text{ lb} \cos 60°$
    $= 6.00 \text{ lb}$

For the y-components:
(a) $N = (5.00 \text{ kg})(9.81 \text{ m/s}^2) \sin 60°$
    $= 42.5 \text{ N}$

(b) $N = 12.0 \text{ lb} \sin 60°$
    $= 10.4 \text{ lb}$

Equilibrant

If a system of concurrent forces is not in equilibrium, there is always some other force, called the **equilibrant,** that would give the system equilibrium if it were applied. To attain equilibrium (zero resultant force), the equilibrant must have equal magnitude but opposite direction to the resultant of the original forces.

EXAMPLE 7-8

A 30.0 N vertical force upward and a 40.0 N horizontal force are applied at a common point (see Fig. 7-13). Determine the resultant and the equilibrant.

Solution

(a) *Data:* $F_1 = 30.0$ N upward, $F_2 = 40.0$ N horizontal, resultant $R = ?$

Equation: Since the vectors are perpendicular, the magnitude of the resultant is

Sec. 7-2 / Concurrent Forces in Equilibrium

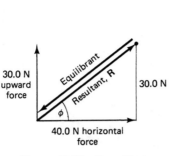

Figure 7-13 Resultant and equilibrant.

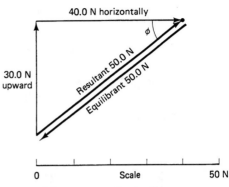

Figure 7-14 Scale diagram.

found using the Pythagorean theorem:

$$R = \sqrt{F_1^2 + F_2^2}$$

and the direction is found from $\tan \phi = \dfrac{F_y}{F_x}$.

Substitute: $R = \sqrt{(30.0 \text{ N})^2 + (40.0 \text{ N})^2} = 50.0 \text{ N}$

$$\tan \phi = \frac{30.0 \text{ N}}{40.0 \text{ N}} = 0.750 \quad \text{and} \quad \phi = 37°$$

This answer can also be obtained from the scale diagram (Fig. 7-14).

(b) The equilibrant is equal in magnitude to the resultant (i.e., is 50.0 N), but it is in the opposite direction. Therefore, the equilibrant is 50.0 N in a direction 37° below the horizontal, as shown (see Fig. 7-14).

PROBLEMS

4 A 20.0 lb horizontal force is applied to an object that weighs 50.0 lb. What is the additional force required to keep the object is equilibrium?

5 A sign is suspended in equilibrium from two wires. One of the wires is horizontal and the other is inclined at 45° with the horizontal. What are the tensions in the wires if the sign has **(a)** a mass of 32.0 kg; **(b)** a weight of 34.0 lb?

6 A lamp is suspended from two wires that make angles of 30° and 45° with the horizontal. Determine the tensions in the wires if the lamp has **(a)** a mass of 3.20 kg; **(b)** a weight of 6.50 lb.

7 Find the tension in the horizontal cable and the reaction force on the boom of a crane (Fig. 7-15) if the crane is supporting loads of **(a)** 2500 N; **(b)** 1750 lb; **(c)** 450 slug; **(d)** 1750 kg.

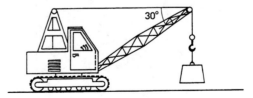

Figure 7-15

8 If the cable of a crane is inclined at 30° with the horizontal and 15° with the boom (Fig. 7-16), find the tension in that cable and the compressive force in the boom when the crane supports loads of **(a)** 22 500 N; **(b)** 1650 lb; **(c)** 345 slug; **(d)** 1250 kg; **(e)** 3750 kg.

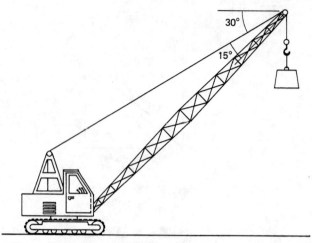

Figure 7-16

9. A block is held at rest on a 30° incline by a force (the friction force **F**) that is directed up the incline (Fig. 7-17). Determine the magnitudes of the normal force **N** (perpendicular to the incline) and the friction force if the block weighs (a) 35.0 N; (b) 8.25 lb. (*Hint:* Choose the axes of the coordinate system parallel and perpendicular to the incline.)

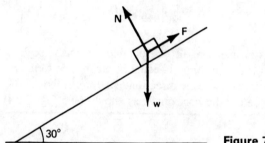

Figure 7-17

10. A load is suspended in equilibrium from two ropes, one horizontal and the other inclined at 56° with the horizontal. Determine the tensions in the ropes if the load is (a) 35.0 kg; (b) 125 lb.

11. Determine the tension in the horizontal tie wire and the compression in the boom illustrated in Fig. 7-18 (a) if the load is 22 500 N and the boom is inclined at 32° with the horizontal; (b) if the load is 3200 lb and the boom is inclined at 35° with the horizontal; (c) if the load is 2750 kg and the boom is inclined at 30° with the horizontal.

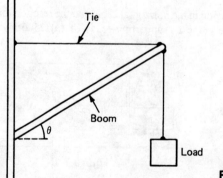

Figure 7-18

12. In the boom illustrated in Fig. 7-18, what is the load and the compression in the boom (a) if the tension in the horizontal tie wire is 12 000 N and the boom is inclined at 28°

Sec. 7-2 / Concurrent Forces in Equilibrium

with the horizontal; **(b)** if the tension in the horizontal tie wire is 7500 lb and the boom is inclined at 32° with the horizontal?

13 To pull a car from a ditch, the driver ties one end of a rope to the car and the other to a tree 18.0 m away. He then pulls sideways at the midpoint of the rope with a 525 N force (Fig. 7-19). How much force is exerted on the car when the man has pulled the rope 1.50 m to one side? (*Hint:* Find the angles from the diagram.)

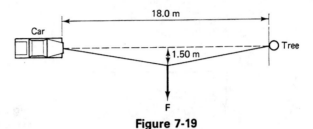

Figure 7-19

14 A load is applied at the junction of two members of a rigid structure. If the members are perpendicular to each other and inclined at 30° and 60° from the vertical (Fig. 7-20), determine the forces exerted in the members if the load is **(a)** 560 N; **(b)** 250 kg; **(c)** 425 lb; **(d)** 72.0 slug.

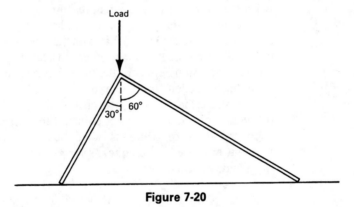

Figure 7-20

15 Determine the tension in the string and the magnitude of the horizontal force in the system illustrated in Fig. 7-21.

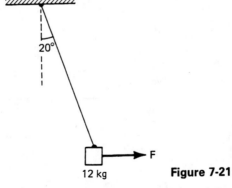

Figure 7-21

16 Determine the tension in the tie and the compressive force in the horizontal support of the boom in Fig. 7-22, when **(a)** the load is 725 N and $\theta = 30°$; **(b)** the load is 475 lb and $\theta = 28°$; **(c)** the load is 875 kg and $\theta = 32°$; **(d)** the load is 26.0 slug and $\theta = 35°$.

17 Determine the resultant and the equilibrant of **(a)** a 16.0 N upward force and a horizontal 12.0 N force to the right; **(b)** a 25.0 lb downward force and a horizontal 32.0 lb force to the left.

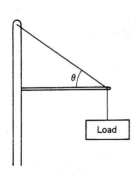

Figure 7-22

7-3 MOMENT OF FORCE AND TORQUE

Forces whose lines of action do not meet at a common point are called **nonconcurrent.** Two nonconcurrent forces have either parallel or antiparallel lines of action, and they tend to cause rotations, such as those in wheels, drive shafts, and gears (Fig. 7-23).

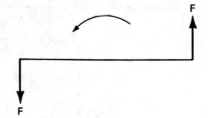

Figure 7-23 Nonconcurrent forces tend to cause rotations.

To describe the "turning effect" of a force, we define two terms, called **moment of force** and **torque**. Even though their definitions are identical, we normally use the term *torque* when there is an actual rotation. Moments of force are used in calculations when the system is in static equilibrium.

From experience, we know that less effort is required to open a door when we apply the force to the handle that is on the side away from the **axis of rotation,** or **pivot point** (the hinges). It is much more difficult to turn the door by pushing it near the hinges. The turning effect of the force depends on both its magnitude and the distance of its line of action from the axis of rotation.

Suppose that a force **F** acts on a rigid object that is free to rotate about some point O (the axis of rotation or pivot point). The perpendicular distance s from the line of action to the axis of rotation is called the **moment arm** (see Fig. 7-24). We define the **moment of force** (and **torque**) M (or τ) as the product of the moment arm s and the magnitude F of the force.

$$M_o = sF \qquad (\text{or } \tau = sF) \tag{7-4}$$

The subscript o indicates the **axis of rotation.** The units of torque and moment of force are the newton meter (N · m) in SI and pound foot (lb · ft) in USCS.

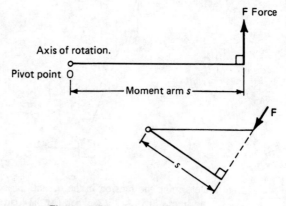

Figure 7-24 $M_O = sF$ and $\tau = sF$.

Torques tend to cause a system to rotate in either a clockwise or a counterclockwise direction. Clockwise torques oppose counterclockwise torques.

EXAMPLE 7-9

Determine the torques on the systems in Fig. 7-25 about the axes or rotation O.

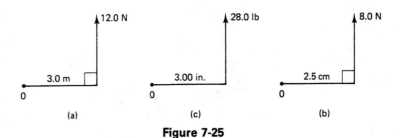

Figure 7-25

Solution

Data:
(a) $s = 3.0$ m, $F = 12$ N, $\tau = ?$
(b) $s = 2.5$ cm $= 2.5 \times 10^{-2}$ m, $F = 8.0$ N, $\tau = ?$
(c) $s = 3.00$ in. $= \dfrac{3.00 \text{ ft}}{12} = 0.250$ ft, $F = 28.0$ lb, $\tau = ?$

Equation: $\tau = sF$

Substitute:
(a) $\tau = (3.0 \text{ m})(12 \text{ N}) = 36$ N·m
(b) $\tau = (2.5 \times 10^{-2} \text{ m})(8.0 \text{ N}) = 0.20$ N·m
(c) $\tau = (0.250 \text{ ft})(28.0 \text{ lb}) = 7.00$ lb·ft

EXAMPLE 7-10

Determine the magnitude of the tangential force on the rim of a wheel **(a)** of radius 75.0 cm that would produce a torque of 225 N·m; **(b)** of radius 24.0 in. that would produce a torque of 166 lb·ft.

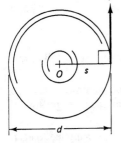

Figure 7-26
Tangential force on the rim of a wheel.

Solution

Sketch: See Fig. 7-26.
Data:
(a) $s = 0.750$ m, $\tau = 225$ N·m, $F = ?$
(b) $s = 24.0$ in. $= 2.00$ ft, $\tau = 166$ lb·ft, $F = ?$

Equation: $\tau = sF$

Rearrange: $F = \dfrac{\tau}{s}$

Substitute:
(a) $F = \dfrac{225 \text{ N·m}}{0.750 \text{ m}} = 300$ N
(b) $F = \dfrac{166 \text{ lb·ft}}{2.00 \text{ ft}} = 83.0$ lb

EXAMPLE 7-11

If there is no slipping, find **(i)** the average force, and **(ii)** the average torque supplied by the wheels when **(a)** a 1750 kg automobile with 40.0 cm diameter wheels accelerates from rest to 14.0 m/s in 12.0 s; **(b)** a 3650 lb automobile with 28.0 in. diameter wheels accelerates from rest to 60.0 mi/h in 10.0 s.

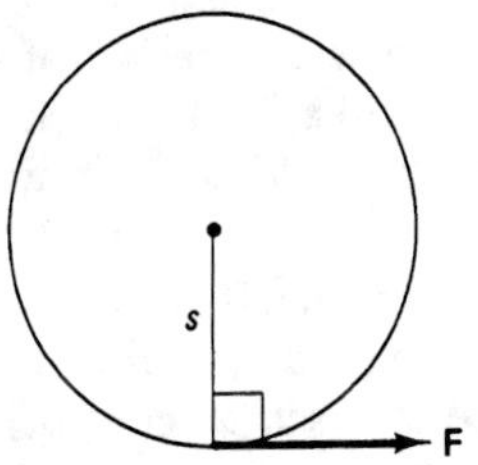

Figure 7-27

Solution

Sketch: The average force **F** is supplied at the rims of the wheel (Fig. 7-27).

Data:

(a) $t = 12.0$ s, $m = 1750$ kg,
$v_0 = 0$ m/s, $v_1 = 14.0$ m/s,
$s = \dfrac{d}{2} = 0.200$ m, $F = ?$,
$\tau = ?$

(b) $t = 10.0$ s, $w = 3650$ lb,
$s = \dfrac{14.0}{12}$ ft, $v_0 = 0$ ft/s,
$v_1 = 60.0$ mi/h $= 88.0$ ft/s,
$g = 32.2$ ft/s², $F = ?$, $\tau = ?$

Equations:

(i) $a = \dfrac{v_1 - v_0}{t}$, $F = ma$, $w = mg$

Rearrange:

(a) $F = \dfrac{m(v_1 - v_0)}{t}$

(b) $F = \dfrac{w}{g} \dfrac{(v_1 - v_0)}{t}$

Substitute:

(a) $F = (1750 \text{ kg})\left(\dfrac{14.0 \text{ m/s} - 0 \text{ m/s}}{12.0 \text{ s}}\right) = 2040$ N

(b) $F = \left(\dfrac{3650 \text{ lb}}{32.2 \text{ ft/s}^2}\right)\left(\dfrac{88.0 \text{ ft/s} - 0 \text{ ft/s}}{10.0 \text{ s}}\right) = 998$ lb

Equation:

(ii) $\tau = sF$

Substitute:

(a) $\tau = (0.200 \text{ m})(2040 \text{ N}) = 408$ N·m

(b) $\tau = \left(\dfrac{14.0}{12} \text{ ft}\right)(998 \text{ lb}) = 1160$ lb·ft

PROBLEMS

18 A string is wrapped around a cylinder with a radius of 6.00 in. If a force of 125 lb is applied to the string, what is the magnitude of the torque that is developed about the axis of the cylinder?

19 Determine magnitude of the torque that is exerted on a spark plug when (a) a 250 N force is applied perpendicular to the end of a 30.0 cm torque wrench; (b) a 35.0 lb force is applied to the end of an 18.0 in. torque wrench.

20 What tangential force on the rim would produce a 495 lb·ft torque on a solid cylindrical drive shaft that has a radius of 4.5 in.?

21 A mechanic can exert a maximum force of 1250 N. (a) What maximum torque can she exert at the end of a 52.0 cm long torque wrench? (b) What minimum length torque wrench can she use to produce a 575 N·m torque?

22 If there is no slipping, find (i) the net force, and (ii) the average torque supplied to each of the two rear wheels when (a) an 1800 kg car with 48.0 cm diameter wheels is accelerated from rest to 20.0 m/s in 16.0 s; (b) a 4250 lb car with 24.0 in. diameter wheels accelerates from 25.0 mi/h to 75.0 mi/h in 15.0 s.

23 If there is no slipping, determine (i) the net force, and (ii) the average torque supplied to the driving wheels of (a) a 2250 kg automobile with 38.0 cm diameter wheels that accelerates uniformly from rest to 15.0 m/s in a distance of 95.0 m; (b) a 5700 lb truck with 42.0 in. diameter wheels that accelerates uniformly from rest to 65.0 ft/s in a distance of 425 ft.

24 A 525 N·m average torque is applied to the wheels of a 1500 kg automobile. If the wheel diameters are 35.0 cm and the automobile starts from rest, determine (a) the net driving force on the wheels; (b) the acceleration; (c) the time required to reach a speed of 75.0 km/h; (d) the distance traveled in the first 8.00 s. Assume no slipping.

7-4 NONCONCURRENT FORCES IN EQUILIBRIUM

When the sum of the clockwise moments is not balanced by the sum of the counterclockwise moments, the system will start to rotate or change its rotation rate. The resultant torque is equal to the difference between the two torques.

Therefore, the first equilibrium condition,

$$\text{net } \mathbf{F} = 0 \quad \text{or} \quad \text{net } F = 0 \text{ in any direction}$$

is not sufficient for nonconcurrent forces since a rotational effect is also present. In Fig. 7-28, for example, the two forces are equal and opposite, but they are not in equilibrium because they cause a rotation or a change in rotation. A **second equilibrium condition** is required for nonconcurrent forces.

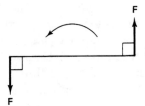

Figure 7-28 Nonconcurrent forces tend to produce rotations.

A system of two or more forces is in equilibrium if:

1. The vector sum of all the forces is zero.

$$\text{net } \mathbf{F} = 0$$

or

$$\text{net } F_x = 0 \quad \text{and} \quad \text{net } F_y = 0$$

2. The sum of the clockwise moments (torques) is equal to the sum of the counterclockwise moments *about any "pivot" point.*

$$\text{net } M_o \text{ clockwise} = \text{net } M_o \text{ counterclockwise} \qquad (7\text{-}5)$$

Using these conditions, we may solve problems involving systems in equilibrium. *Both* conditions must be satisfied for equilibrium to exist.

Careful selection of the pivot point frequently reduces the mathematical work required. Look for a point through which one of the unknown forces passes, and take moments about that point. This reduces the mathematical chore because that unknown force has no moment of force about the point through which it passes, and therefore it will not appear in the equation.

EXAMPLE 7-12

The system in Fig. 7-29 is in equilibrium. Determine **(a)** the magnitude of the force **F**; **(b)** the distance d.

Solution

Data: $\mathbf{F}_1 = 200$ N downward, $\mathbf{F}_2 = 50$ N downward, $s = 5.0$ m, $\mathbf{F} = ?$, $d = ?$

(a) *Equation:* net $F = 0$, upward taken as positive

Substitute: Net $F_{up} = F - 200 \text{ N} - 50 \text{ N} = 0$; therefore, $F = 250$ N.

(b) We may take moments about *any point*, but if we choose point O, the moment of the unknown force **F** about O is zero, because it passes through O and

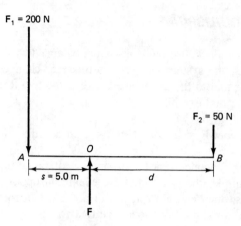

Figure 7-29

therefore the moment arm is zero. The point O becomes the pivot point, and we measure all distances to the lines of action of the forces from O. In other words, even though the system is in equilibrium, we consider the tendency of the forces to cause or change rotations. The 200 N force would tend to cause a counterclockwise rotation, and the 50 N force a clockwise rotation *about O*.

Equations: net M_o clockwise = net M_o counterclockwise, $M_o = sF$, M_o clockwise = dF_2, M_o counterclockwise = sF_1

Rearrange: $sF_1 = dF_2$, $d = \dfrac{sF_1}{F_2}$

Substitute: $d = \dfrac{(5.0 \text{ m})(200 \text{ N})}{50 \text{ N}} = 20 \text{ m}$

Check: The answer can be checked by taking moments about any other point (A, for example) with all distances measured from that point.

$$(5 \text{ m} + 20 \text{ m})(50 \text{ N}) = (5 \text{ m})(250 \text{ N})$$
$$\text{(clockwise)} \qquad \text{(counterclockwise)}$$

Center of Gravity

Every particle of matter of the surface of the earth is attracted toward the center of the earth by the force of gravity. The **weight** of the object is due primarily to the force of gravity on its particles, and it is usually represented by a single force vector **w**. This weight vector always passes through a center point (not necessarily in the object), called the **center of gravity** C. In fact, the object can be supported by a single upward force, which is equal in magnitude to the weight and acts at the center of gravity (see Fig. 7-30). The location of the center of gravity is extremely important in structures. When solving problems it is usually convenient to assume that:

The weight vector for any object always passes through the center of gravity, regardless of where it is supported or the location of the point about which we take moments.

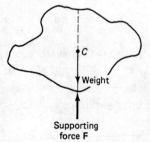

Figure 7-30 An object can be supported by a single force, **F**. The weight **w** and the supporting force **F** have lines of action that pass through the center of gravity C.

Sec. 7-4 / Nonconcurrent Forces in Equilibrium

If an object is uniform, we assume that the total weight acts at its center. This fact allows us to simplify the solution of many types of problems.

EXAMPLE 7-13

An 85.0 kg uniform beam AB is 7.50 m long and rests on two supports, one at A the other 1.50 m from B (see Fig. 7-31). Determine the reaction forces $\mathbf{R}_1$ and $\mathbf{R}_2$ of the supports.

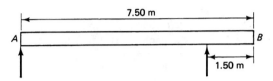

Figure 7-31

Solution

Sketch: See Fig. 7-32. The total weight $w = mg$ of the beam acts at its center, 3.75 m from the ends, because it is uniform.

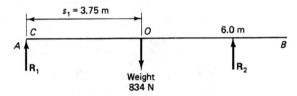

Figure 7-32 Free-body diagram.

Data: $m = 85.0$ kg, $g = 9.81$ m/s², and the weight of the beam $w = mg = (85.0 \text{ kg})(9.81 \text{ m/s}^2) = 834$ N. If we take moments about A, the moment arm for $\mathbf{R}_1$ is zero because it passes through A. This simplifies the solution. $s_1 = 3.75$ m, $s_2 = 6.0$ m, $w = 834$ N, $R_1 = ?$, $R_2 = ?$

Equations: net M_A (clockwise) = net M_A (counterclockwise), $M_o = sF$

Rearrange: $s_1 w = s_2 R_2$, $R_2 = \dfrac{s_1 w}{s_2}$

Substitute: $R_2 = \dfrac{(3.75 \text{ m})(834 \text{ N})}{6.0 \text{ m}} = 521$ N

Summing the forces in the vertical direction: net $F = 0$; thus

$$R_1 + R_2 - 834 \text{ N} = 0$$

Rearrange: $R_1 = 834 \text{ N} - R_2$

Substitute: $R_1 = 834 \text{ N} - 521 \text{ N} = 313$ N, since $R_2 = 521$ N

This answer can be checked by taking moments about any other point: C, for example.

Check: (6.0 m)(313 N) = (2.25 m)(834 N)
      (clockwise)   (counterclockwise)

Note that all distances must now be measured from the new pivot point C. The slight discrepancy is due to rounding off.

EXAMPLE 7-14

A 20.0 ft long uniform beam AB is supported at each end (Fig. 7-33). If the beam weighs 225 lb and a 325 lb load is located 6.0 ft from one end, find the reaction forces $\mathbf{R}_A$ and $\mathbf{R}_B$ at the supports.

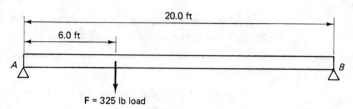

Figure 7-33

Solution

Sketch: See Fig. 7-34. The total weight of the beam (225 lb) is assumed to act at the center (10.0 ft from the ends) because the beam is uniform.

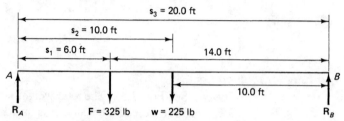

Figure 7-34 Free-body diagram.

Data: If we take moments about A, the moment arm for $\mathbf{R}_1$ is zero because it passes through A. This simplifies the problem. $s_1 = 6.0$ ft, $s_2 = 10.0$ ft, $s_3 = 20.0$ ft, $F = 325$ lb, $w = 225$ lb, $R_A = ?$, $R_B = ?$

Equations: net M_A (clockwise) = net M_A (counterclockwise), $M_o = sF$

Rearrange: $s_1 F + s_2 w = s_3 R_B$

Therefore,

$$R_B = \frac{s_1 F + s_2 w}{s_3}$$

Substitute: $R_B = \dfrac{(6.0 \text{ ft})(325 \text{ lb}) + (10.0 \text{ ft})(225 \text{ lb})}{20.0 \text{ ft}} = 210 \text{ lb}$

Equation: Since the beam is in equilibrium, summing the forces in the vertical direction, with net $F = 0$,

$$R_A + R_B = F + w$$

Rearrange: $R_A = F + w - R_B$

Substitute: $R_A = 325 \text{ lb} + 225 \text{ lb} - 210 \text{ lb} = 340 \text{ lb}$

Check: This answer can be checked by taking moments about any other point (B, for example). Using the values found, with all distances measured from the new pivot point B, we have

$$(20.0 \text{ ft})(340 \text{ lb}) = (14.0 \text{ ft})(325 \text{ lb}) + (10.0 \text{ ft})(225 \text{ lb})$$

EXAMPLE 7-15

Two men are carrying an 8.00 m long uniform beam that weighs 375 N. If one man A is 1.00 m from one end, and the other man B is 2.00 m from the other end, find the load carried by each man.

Solution

Sketch: See Fig. 7-35.

Data: Since the beam is uniform, the weight acts at its center, 4.00 m from the ends. If we take moments about point A, the upward force exerted by man A has no moment arm. Therefore, all distances are measured from A.

Sec. 7-4 / Nonconcurrent Forces in Equilibrium

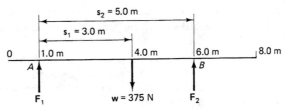

Figure 7-35 Free-body diagram.

$s_1 = 3.00$ m, $s_2 = 5.00$ m, $w = 375$ N, $F_1 = ?$, $F_2 = ?$

Equations: net M_A (clockwise) = net M_A (counterclockwise), $M_A = sF$

Rearrange: $s_1 w = s_2 F_2$, $F_2 = \dfrac{s_1 w}{s_2}$

Substitute: $F_2 = \dfrac{(3.00 \text{ m})(375 \text{ N})}{5.00 \text{ m}} = 225$ N

Equation: net $F = 0$ upward

Substitute: $F_1 + F_2 - 375 \text{ N} = 0$, $F_2 = 225$ N

Rearrange: $F_1 = 375 \text{ N} - 225 \text{ N} = 150$ N

Check: Taking moments about B, *with all distances measured from B*,

$s_1 = 5.00$ m, $s_2 = 2.00$ m, $w = 375$ N, $F_1 = 150$ N

$(5.00 \text{ m})(150 \text{ N}) = (2.00 \text{ m})(375 \text{ N})$

PROBLEMS

25 A 12.0 ft long uniform beam rests horizontally on two supports, 1.00 ft from each end. If the beam weighs 650 lb, determine the reaction forces at the supports.

26 A 375 kg horizontal uniform beam 8.0 m long is supported 50.0 cm from each end. Find the reaction forces at the supports.

27 A horizontal uniform beam 15.0 ft long weighing 250 lb is supported at each end, and a 150 lb load is hung 5.00 ft from one end. Determine the reaction forces at the supports.

28 Two men carry a 12.0 m long uniform beam on their shoulders. Man A is 1.0 m from one end, and man B is 2.0 m from the other end. If the beam has a mass of 82.0 kg, determine the load supported by each man.

29 If a 30.0 kg load is hung 3.0 m from man A in Problem 28, find the load supported by each man.

30 Each front wheel of a car with a 4.2 m wheelbase (the distance between the front and rear axles) presses on the ground with a 3500 N force, and each rear wheel with a 4000 N force. Determine the location of the center of gravity.

31 A horizontal 25.0 m long bridge weighing 225 000 N is supported at each end. If the center of gravity of the bridge is at its center, and a 2750 kg truck is on the bridge 8.00 m from one end, calculate the forces on the supports.

32 A horizontal 80.0 ft long bridge weighing 120 000 lb is supported at each end. If the center of gravity of the bridge is at its center and a 45 000 lb truck is on the bridge 30.0 ft from one end, determine the reaction forces at the supports.

33 A packing case containing machinery is 1.50 m high, 75.0 cm wide, 75.0 cm deep, and has a total mass of 175 kg. If the center of gravity is at the center of the case, what horizontal force applied to the top of the case will start it tipping? (*Hint:* It tips about the bottom front edge.)

34 When the front wheels of a truck are run onto a scale, the scale reads 1650 kg, and when the rear wheels are run onto the scale it reads 3450 kg. **(a)** If the truck has a wheelbase of 5.00 m, where is the center of gravity? **(b)** Where is the center of gravity if an additional 1850 kg load is located 1.50 m from the rear axle?

35 Two mechanics carry a 325 kg engine which is placed at the center of a 1.20 m long 15.0 kg uniform plank. If the mechanics support each end of the plank, how much weight does each carry?

36 A 185 lb construction worker stands 1.00 ft from the end of an 85.0 lb uniform scaffold, which is 12.5 ft long and supported at each end. If a 135 lb container of cement is located with its center 4.50 ft from the other end of the scaffold, determine the reaction forces at the supports.

SUMMARY

Equilibrium is a state of zero acceleration. Concurrent forces have lines of action that pass through a single point. *Moment of force (torque)* is the product of the moment arm s and the force F:

$$M_o = sF \quad \text{and} \quad \tau = sF$$

The units are newton meters (N·m) in SI and pound feet (lb·ft) in USCS.

The *two equilibrium conditions* are:

$$\text{net } F = 0 \text{ in any direction}$$

$$\text{net } M_o \text{ clockwise} = \text{net } M_0 \text{ counterclockwise, about any point } O$$

Center of gravity is the point through which all the weight is assumed to act, regardless of the locations of the supports.

QUESTIONS

1 Can moving objects ever be in equilibrium? Explain.
2 What is the relationship between a resultant force and the equilibrant?
3 Discuss the forces that are present when a block is at rest on an incline.
4 What is the purpose of a balance pole used by a tightrope walker?
5 Discuss the position of your body when you ride a bicycle in a straight line, and when you turn corners.
6 If a rider exerts a constant vertical force on the front pedals of a bicycle, discuss the variations in the torque produced (see Fig. 7-36).

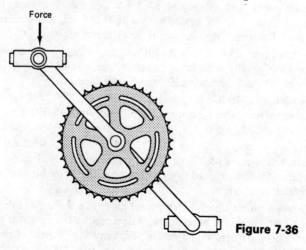

Figure 7-36

REVIEW PROBLEMS

1. A 750 000 lb horizontal bridge is supported at each end. If the upward force on the bridge at one support is 240 500 lb, what is the force at the other?
2. A 1250 kg uniform rectangular platform is supported by posts at each corner. What is the force at each support?
3. A 35.0 lb horizontal force is applied to an object that weighs 42.0 lb. What additional force will keep the object in equilibrium?
4. A 15.0 kg traffic light is suspended from two wires that are inclined at 30° and 45° with the horizontal. What are the tensions in the wires?
5. Determine the tension in the horizontal tie wire and the compression in the boom in Fig. 7-18, if (a) the load is 7800 N and the boom is inclined at an angle $\theta = 35°$ with the horizontal; (b) the load is 2850 lb and $\theta = 32°$; (c) the load is 875 kg and $\theta = 30°$.
6. A rope is tied to two fixed points that are 12.0 m apart and a 4.5 kg load is suspended at the midpoint, depressing the rope 2.5 m from the horizontal (see Fig. 7-37). Find the tension in the rope.

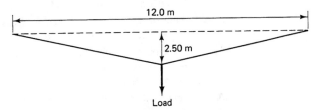

Figure 7-37

7. A load is applied at the rigid junction of a truss. Determine the forces in the truss members if they are inclined at 40° and 50° with the vertical (see Fig. 7-38) and the load is (a) 350 kg; (b) 175 lb; (c) 5750 N; (d) 85.0 slug.

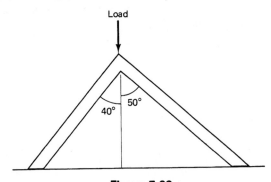

Figure 7-38

8. A mechanic can exert a maximum force of 680 N. (a) What maximum torque can she exert at the end of a 32.0 cm long torque wrench? (b) What minimum length torque wrench can she use to produce a 295 N·m torque?
9. A mechanic can exert a maximum force of 184 lb. (a) What maximum torque can he exert at the end of a 12.0 in. long wrench? (b) What minimum length wrench must he use to produce a 135 lb·ft torque?
10. A 1750 kg automobile with 35.0 cm diameter wheels accelerates from rest to 60.0 km/h in a distance of 125 m. Find (a) the net force; (b) the average torque supplied to the driving wheels if there is no slipping.
11. A 175 lb painter stands 5.25 ft from one end A of a 12.0 ft long, 65.0 lb uniform scaffold, which is supported at each end. If the painter has a 25.0 lb supply of paint 4.00 ft from end A, find the reaction forces at the supports.

12. A 35.0 kg uniform beam is 3.50 m long and is supported at one end A and 1.00 m from the other end B (see Fig. 7-39). How much can be added at a point 0.50 m from B before the beam tips?

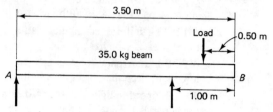

Figure 7-39

13. What tangential force would produce (a) 75.0 N·m torque on a drive shaft that has a 5.20 cm diameter; (b) a 175 lb·ft torque on a drive shaft that has a 2.50 in. diameter?

14. If there is no slipping, find (i) the net force, and (ii) the average torque supplied to the two rear wheels of (a) a 1500 kg car with 45.0 cm diameter wheels that accelerates from rest to 21.0 m/s in 15.0 s; (b) a 4450 lb car with 28.0 in. diameter wheels that accelerates from 20.0 mi/h to 55.0 mi/h in 12.5 s.

15. A 120 lb uniform beam that is 9.00 ft long rests horizontally on two supports 1.00 ft from each end. Determine the reaction forces at the supports.

16. The distance between the front and rear wheels of a truck is 6.50 m. If the front wheels each press on the ground with a 6250 N force, and each rear wheel with a 9750 N force, determine the location of the center of gravity.

17. A horizontal 25.0 m long bridge weighing 950 000 N is supported at each end. Calculate the forces at the supports if the center of gravity of the bridge is at its center and a 52 000 N truck is 5.0 m from one end.

18. One man carries a 2.75 m long uniform pole balanced horizontally with a 23.0 kg load hung 75.0 cm from one end. If the pole has a mass of 3.50 kg, at what point is the man carrying the pole?

19. One man carries an 8.0 ft long uniform pole balanced horizontally with a 12.0 lb load hung 2.0 ft from one end. If the pole weighs 8.0 lb, at what point is the man carrying the pole?

8
WORK, ENERGY, AND POWER

The words *work*, *power*, and *energy* are used in a variety of ways in everyday conversation. Normally, we associate the term *work* with mental or physical exertion that makes us tired. However, this is not precise enough for a physical definition.

Machinery is used frequently in many aspects of our lives, and the rate at which machines perform work is an important consideration. Different machines may be used to accomplish the same job, but they may do that job at different rates. The rate at which machines do work is described by the term *power*.

8-1 WORK

The application of a force may produce completely different effects. For example, if we push a heavy crate, we may not be able to move it even though we exert ourselves and become physically tired, or the crate may slide across the floor. Clearly, these results are quite different.

In physics, we define **mechanical work done** (or **work**) by a force as the product of the displacement of the object and the component of the force in the direction of the motion. Consider the simplest case, where the force **F** is in the same direction as the displacement, **s**. The work done by the force on the object is a scalar quantity that is equal to the product of the magnitudes of the force and the distance moved by the object.

$$W = Fs \quad (\textbf{F} \text{ and } \textbf{s} \text{ in the same direction}) \tag{8-1}$$

If the force and the displacement are not in the same direction, (Fig. 8-1), the work done is the product of the component of the force parallel to the displacement ($F \cos \theta$) and the magnitude of the displacement s. We must include the cosine of the angle between the force and the displacement.

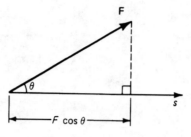

Figure 8-1 The work done by the force F in producing a displacement is $W = Fs \cos \theta$.

$$W = Fs \cos \theta \tag{8-2}$$

In SI, the unit of force is the newton (N) and displacements are in meters (m); therefore, the work unit is the newton meter (N · m), which is given the special name **joule** (J). One joule of work is done on an object when an applied force of 1 N displaces that object 1 m.

$$1 \text{ J} = (1 \text{ N})(1 \text{ m}) = 1 \text{ N} \cdot \text{m}$$

The USCS unit of work is the **foot pound** (ft · lb), which is the work done by a 1 lb force when it produces a displacement of 1 ft.

This definition of work is more restrictive than the common usage of the term. A person who spends all day trying, unsuccessfully, to move an object does no mechanical work *on the object*, but as far as the person is concerned, a great deal of work has been done because of the physical exertion.

EXAMPLE 8-1

A man applies a horizontal force of 180 N to a heavy crate on a horizontal floor. How much work is done by the man on the crate **(a)** if the crate does not move; **(b)** if the crate slides 4.0 m in the direction of the force?

Solution

Data:
(a) $F = 180$ N, $s = 0$ m, $W = ?$ **(b)** $F = 180$ N, $s = 4.0$ m, $W = ?$
Equation: $W = Fs$
Substitute:
(a) $W = (180 \text{ N})(0 \text{ m}) = 0$ J **(b)** $W = (180 \text{ N})(4.0 \text{ m}) = 720$ J

EXAMPLE 8-2

A crane lifts a girder from the ground to the third floor of a building at a constant velocity. How much work is done by the crane on the girder **(a)** if a 275 kg girder is raised a total vertical distance of 8.00 m; **(b)** if a 725 lb girder is raised 32.0 ft?

Solution

Sketch: See Fig. 8-2.
Data:
(a) $m = 275$ kg, $s = 8.00$ m **(b)** $w = mg = 725$ lb, $s = 32.0$ ft,
$g = 9.81$ m/s², $W = ?$ $W = ?$

Since there is no acceleration, the girder is in equilibrium even though the girder is in motion. thus $F_{up} = F_{down}$, or the magnitude of the applied force equals the force of gravity (weight), $F = w$. Also, $\theta = 0$ because **F** and **s** are both upward.
Equations: $w = mg$, $F = w$, $W = Fs$
Rearrange: $W = ws = mgs$

Sec. 8-1 / Work

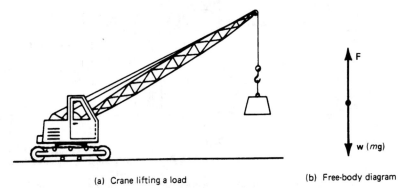

(a) Crane lifting a load   (b) Free-body diagram

Figure 8-2

Substitute:
(a) $W = (275 \text{ kg})(9.81 \text{ m/s}^2)(8.00 \text{ m})$
$= 2.16 \times 10^4 \text{ J}$

(b) $W = (725 \text{ lb})(32.0 \text{ ft})$
$= 2.32 \times 10^4 \text{ ft} \cdot \text{lb}$

EXAMPLE 8-3

A man moves a trolley of machinery over a horizontal surface at a constant velocity by pulling along the handle with a constant force. If the handle of the trolley is inclined at 20° with the horizontal, how much work does the man do on the loaded trolley (a) if he exerts a 151 N force and the trolley moves 12.0 m; (b) if he exerts a 45.0 lb force and moves the trolley 25.0 ft?

Solution
Sketch: See Fig. 8-3.

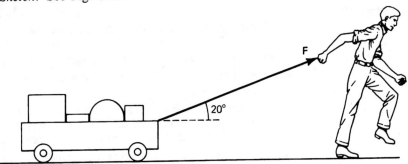

Figure 8-3 Man pulling a trolley.

Data: $\theta = 20°$
(a) $F = 151$ N, $s = 12.0$ m,
$W = ?$

(b) $F = 45.0$ lb, $s = 25.0$ ft,
$W = ?$

Equation: $W = Fs \cos \theta$
Substitute:
(a) $W = (151 \text{ N})(12.0 \text{ m}) \cos 20°$
$= 1700 \text{ J}$

(b) $W = (45.0 \text{ lb})(25.0 \text{ ft}) \cos 20°$
$= 1060 \text{ ft} \cdot \text{lb}$

It is important to appreciate that any force that is perpendicular to the motion does no work, since cos 90° = 0. In Fig. 8-4, a box is being pushed across a floor at a steady speed. The applied force and the friction force act horizontally in opposite directions, and the force of gravity acts vertically downward, while the reaction force of the floor (the normal force) balances the weight. The force of gravity does no work, because it has no component in the direction of the displacement.

When the smallest angle between the applied force and the displacement is greater than 90°, the cosine of that angle has a negative value, resulting in a negative

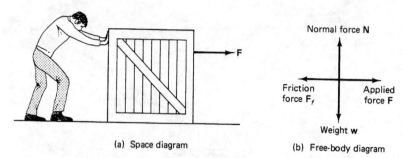

Figure 8-4 Man pushing a crate at a constant speed.

Figure 8-5 If an object is lowered at a constant speed, the applied force **F** is equal and opposite to the weight. Negative work is done since **F** and the displacement **s** are in opposite directions ($\theta = 180°$).

value for work. For example, the brakes of a car do negative work on the car because they slow the car down; the braking force is in the opposite direction to the motion. Similarly, the gradual lowering of an object to the floor represents negative work by the applied force, but positive work by the force of gravity (Fig. 8-5). Negative work actually results in a reduction of the energy of the object while positive work increases energy.

PROBLEMS

1 How much work is done when (a) a 45.0 N force acts through a distance of 7.50 m; (b) a 62.0 lb force acts through 9.60 ft?

2 Determine the work done by a force of (a) 75.0 N if it acts through 24.5 m; (b) 18.0 lb that acts through 5.00 ft.

3 How much work is done on an elevator (a) when a 1500 kg elevator is lifted 85.0 m at a constant speed; (b) When a 3800 lb elevator is lifted 125 ft at a constant speed?

4 A machine does 125 J of work to lift a 25.0 kg motor vertically at a constant velocity. What is the vertical displacement of the motor?

5 A machine does 2850 ft·lb of work and lifts a 75.0 lb load vertically from the ground at a constant velocity. How high above the ground is the final position of the load?

6 If a machine does 775 J of work, what load can it lift vertically through 5.25 m at a constant velocity?

7 A machine is used to lift a 57.5 kg engine vertically from the ground at a constant velocity. (a) If the machine does 1750 J of work, what is the final height of the engine above the ground? (b) How much work is required to lift the engine 4.50 m?

8 A crane lifts a 625 lb steel girder at constant speed from the ground floor to the top floor of a building, a total vertical distance of 132 ft. How much work is done on the girder?

9 How much work is done by a mechanic's hoist when (at constant speed) it raises (a) a 1250 kg car through a vertical distance of 2.75 m; (b) a 3800 lb car 6.00 ft?

10 A girl pulls a 7.50 kg sleigh by means of a rope over her shoulder. The ground is horizontal, and the angle between the rope and the horizontal displacement is 35°. If the tension in the rope is 65.0 N and the sleigh is pulled at a constant speed a total distance of 25.0 m, (a) how much work is done on the sleigh by the girl; (b) how much work is done against the force of friction; (c) how much work is done against the force of gravity?

11 A man pushes a 175 lb crate across a horizontal floor in a straight line at a constant speed for a distance of 22.0 ft by applying a horizontal 75.0 lb force. (a) How much work is done on the crate by the man? (b) What is the friction force? (c) How much work is done against the friction force? (d) What is the weight of the crate? (e) How much work is done against the force of gravity?

12 Determine the work done by a water pump if it raises 25.0 L of water vertically through a height of 25.0 m.

13 How much work is done by a conveyor belt on a 135 lb box if it pulls the box at a constant velocity 12.0 ft up a 25° incline?

14 A man pulls a 125 kg crate up a ramp which makes an angle of 30° with the horizontal

Sec. 8-2 / Energy

by applying a 725 N force parallel to the ramp. If the crate is pulled 4.75 m up the ramp at a constant speed, (a) how much work is done by the man on the crate; (b) how much work is done against the force of gravity; (c) how much work was done against the friction force; (d) what was the friction force?

15 Determine the applied horizontal force that would do 2500 J of work in order to displace a load of 25.0 kg a distance of 3.00 m up a 25° incline.

8-2 ENERGY

The concept of energy is fundamental to physics. Many aspects of science are actually concerned with the conversion of energy from one form into another. For example, the mechanical energy possessed by water may be converted into electrical energy in a hydroelectric power plant; the electrical energy is then transmitted to a consumer, where it may be converted into heat, light, and so on.

The concepts of work and energy are very closely related and they are even expressed in the same units: joules (J) in SI and foot pounds (ft · lb) in USCS. Energy may be used to do work, and also, work may be done to change the energy of a system. For example, we may do work to lift a hammer, but the hammer then has additional energy because it can do work (displacing a nail, for example) as it falls. Most types of energy may be classified as belonging to one of two main categories: kinetic energy or potential energy.

Kinetic Energy

Kinetic energy is the energy that an object possesses as a result of its motion. Moving cars, falling objects, and rotating wheels all have kinetic energy, because they may exert forces and do work on any object that tends to stop their motion. Also, if work is done on an object, its kinetic energy may increase, although part of the work done is frequently converted into other energy forms, such as thermal energy.

If an object of mass m is displaced a distance s in the direction of a force $\mathbf{F}$, the work done on the object by that force

$$W = Fs \tag{1}$$

According to Newton's second law, if there are no other forces,

$$\text{net } \mathbf{F} = \mathbf{F} = m\mathbf{a} \tag{2}$$

where $\mathbf{a}$ is the acceleration of the object. Combining Eqs. (1) and (2) yields

$$W = mas \tag{3}$$

If the force accelerates the object from an initial speed v_0 to a final speed v_1, then, according to the equations of uniformly accelerated motion,

$$v_1^2 - v_0^2 = 2as$$

or

$$as = \frac{v_1^2 - v_0^2}{2} \tag{4}$$

Therefore, substituting Eq. (4) into Eq. (3), we find that the work done

$$W = mas = \frac{mv_1^2 - mv_0^2}{2} \tag{8-3}$$

We define the **kinetic energy** E_k of a mass m moving at a speed v as the scalar quantity

$$E_k = \tfrac{1}{2} mv^2 = \tfrac{1}{2}\left(\frac{w}{g}\right) v^2 \tag{8-4}$$

where $w = mg$ is the weight of the object and g is the acceleration due to gravity.

Therefore, in Eq. 8-3 we see that the work done by the net force equals the change in the kinetic energy of the object:

$$W = E_{k1} - E_{k0} = \Delta E_k = \tfrac{1}{2} mv_1^2 - \tfrac{1}{2} mv_0^2 \tag{8-5}$$

where E_{k0} and E_{k1} are the initial and final kinetic energies, respectively, and ΔE_k is the change in kinetic energy. This is the energy possessed by a body because of its translational motion (it may, in addition, have some kinetic energy of rotation), and this energy may be used to do work. This is an important concept known as the **kinetic energy-work theorem.**

EXAMPLE 8-4

Calculate the kinetic energy of (a) a 275 kg iron wrecking ball if its speed is 1.20 m/s; (b) a 4200 lb car moving at 60.0 mi/h.

Solution

Data:
(a) $m = 275$ kg, $v = 1.20$ m/s, $E_k = ?$

(b) $w = 4200$ lb, $g = 32.2$ ft/s², $v = 60.0$ mi/h $= 88.0$ ft/s, $E_k = ?$

Equation:

(a) $E_k = \tfrac{1}{2} mv^2$

(b) $E_k = \dfrac{1}{2}\dfrac{wv^2}{g}$

Substitute:

(a) $E_k = \tfrac{1}{2}(275 \text{ kg})(1.20 \text{ m/s})^2$
$= 198$ J

(b) $E_k = \dfrac{1}{2}\dfrac{(4200 \text{ lb})(88.0 \text{ ft/s})^2}{(32.2 \text{ ft/s}^2)}$
$= 5.05 \times 10^5$ ft·lb

EXAMPLE 8-5

A 18.0 g bullet moving at 775 m/s strikes a target and penetrates horizontally a distance of 12.2 cm before stopping. Find the average retarding force exerted on the bullet by the target.

Solution

Data: $m = 0.0180$ kg, $v = 775$ m/s, $s = 0.122$ m, $F = ?$ In this case, the work done by the target in stopping the bullet is equal to the change in kinetic energy of the bullet. Also, since the final speed of the bullet is zero, the final kinetic energy is zero.

Equations: $W = \Delta E_k$, $W = Fs$, $\Delta E_k - E_{k0} = \tfrac{1}{2} mv^2 - 0$

Rearrange: $W = Fs = \tfrac{1}{2} mv^2$ or $F = \dfrac{mv^2}{2s}$

Substitute: $F = \dfrac{(0.0180 \text{ kg})(775 \text{ m/s})^2}{2(0.122 \text{ m})} = 44\,300$ N

The figures for this example have been chosen to represent a realistic situation. The force generated would be equivalent to the weight of more than three full-sized cars.

EXAMPLE 8-6

Determine **(i)** the initial kinetic energy, **(ii)** the final kinetic energy, **(iii)** the change in kinetic energy, and **(iv)** the average braking force **(a)** if a 1250 kg car brakes uniformly from 108 km/h to 36.0 km/h in a distance of 125 m; **(b)** if a 4250 lb car brakes uniformly from 88.0 ft/s to 44.0 ft/s in 325 ft.

Solution

Data:

(a) $v_0 = 108 \frac{km}{h} = \frac{108 \times 10^3 \text{ m}}{3600 \text{ s}}$
$= 30.0 \frac{m}{s}$, $m = 1250$ kg,
$v_1 = 36.0 \frac{km}{h} = \frac{36.0 \times 10^3 \text{ m}}{3600 \text{ s}}$
$= 10.0 \frac{m}{s}$, $s = 125$ m, $E_k = ?$,
$F = ?$

(b) $w = 4250$ lb, $g = 32.2$ ft/s^2,
$v_0 = 88.0$ ft/s, $v_1 = 44.0$ ft/s,
$s = 325$ ft, $E_k = ?$ $F = ?$

Equation:

(a) $E_k = \frac{1}{2} mv^2$ (b) $E_k = \frac{1}{2}\left(\frac{w}{g}\right)v^2$

(i) *Substitute:*
(a) $E_k = \frac{1}{2}(1250 \text{ kg})(30.0 \text{ m/s})^2 = 563\,000$ J
(b) $E_k = \frac{1}{2}\left(\frac{4250 \text{ lb}}{32.2 \text{ ft/s}^2}\right)(88.0 \text{ ft/s})^2 = 5.11 \times 10^5$ ft · lb

(ii) *Substitute:*
(a) $E_k = \frac{1}{2}(1250 \text{ kg})(10.0 \text{ m/s})^2 = 62\,500$ J
(b) $E_k = \frac{1}{2}\left(\frac{4250 \text{ lb}}{32.2 \text{ ft/s}^2}\right)(44.0 \text{ ft/s})^2 = 1.28 \times 10^5$ ft · lb

(iii) *Equation:* Change in kinetic energy $\Delta E_k = E_{k1} - E_{k0}$

Substitute:
(a) $\Delta E_k = 62\,500$ J $- 563\,000$ J $= -501\,000$ J
(b) $\Delta E_k = 1.28 \times 10^5$ ft · lb $- 5.11 \times 10^5$ ft · lb $= -3.83 \times 10^5$ ft · lb

(iv) *Equation:* $W = Fs = \Delta E_k$

Rearrange: $F = \frac{\Delta E_k}{s}$

Substitute:

(a) $F = \frac{-501\,000 \text{ J}}{124 \text{ m}}$
$= -4000$ N

(b) $F = \frac{-3.83 \times 10^5 \text{ ft} \cdot \text{lb}}{325 \text{ ft}}$
$= -1180$ lb

That is, 4000 N and 1180 lb in the **opposite directions** to the motions.

In many cases work is done on some object without increasing its energy. For example, we may do work by pushing a crate across a horizontal surface at a constant speed. Of course, we are essentially doing work to overcome the friction forces which tend to oppose the motion. This work is usually converted into other energy forms, usually thermal energy.

Potential Energy

Potential energy is the energy that an object possesses because of its position or state. A raised pile driver has the ability to do work because of its position; if it is allowed to fall, it can drive a pile into the ground. This is an example of **gravitational potential energy**. A compressed spring has **elastic potential energy** because of its state; if it is released, it can do work by pushing and displacing some other object. There are many other examples of potential energy. We will now examine the potential energy of an object in a uniform gravitational field. Again, the potential energy is an indication of the amount of work that could be done.

Gravitational Potential Energy. An object can do work if it is allowed to fall from some height. Therefore, when an object is raised, its potential energy increases.

Consider the work W that must be done in order to raise an object of mass m at a constant velocity through some vertical height Δh. We shall assume that there is no air resistance and that Δh is small enough that we can ignore the corresponding change in the force of gravity. Since the velocity is constant, the mass m is in equilibrium, and the applied force **F** must be equal and opposite to the force of gravity (weight) **w** = m**g** (see Fig 8-6). Therefore,

$$W = Fs \cos \theta = (mg) \Delta h \cos 0 = mg \, \Delta h$$

and this is known as the *change in the gravitational potential energy* ΔE_p of the object.

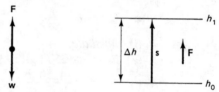

(a) Free-body diagram; no acceleration implies that the upward applied force F balances the force of gravity w

(b) The applied force **F** moves the mass through a vertical displacement $\Delta h = h_1 - h_0$

Figure 8-6

$$\Delta E_p = mg \, \Delta h = mg(h_1 - h_0) = w \, \Delta h \tag{8-6}$$

where h_0 and h_1 are the initial and final heights, respectively, and $w = mg$ is the weight of the mass m. Note that if the object is lowered, $(h_1 - h_0)$ is negative valued and there is a loss in potential energy.

EXAMPLE 8-7

Determine the change in potential energy of **(a)** a 255 kg steel beam **(i)** if it is raised from the ground to the top of a building 16.0 m high, and **(ii)** if it is lowered 8.00 m below the ground; **(b)** a 561 lb girder if it is **(i)** raised 52.2 ft, and **(ii)** lowered 26.2 ft.

Solution

Data:

(a) m = 255 kg, g = 9.81 m/s²,
(i) Δh = 16.0 m
(ii) Δh = −8.00 m (lowered)

(b) $w = mg$ = 561 lb ΔE_p = ?
(i) Δh = 52.5 ft
(ii) Δh = −26.2 ft (lowered)

Equation: $\Delta E_p = mg \, \Delta h$

Sec. 8-2 / Energy

(i) *Substitute:*
(a) $\Delta E_p = (255 \text{ kg})(9.81 \text{ m/s}^2)(16.0 \text{ m})$
$= 4.00 \times 10^4 \text{ J}$
(b) $\Delta E_p = (561 \text{ lb})(52.5 \text{ ft})$
$= 2.95 \times 10^4 \text{ ft} \cdot \text{lb}$

(ii) *Substitute:*
(a) $\Delta E_p = (255 \text{ kg})(9.81 \text{ m/s}^2)(-8.00 \text{ m}) = -2.00 \times 10^4 \text{ J}$
(b) $\Delta E_p = (561 \text{ lb})(-26.2 \text{ ft}) = -1.47 \times 10^4 \text{ ft} \cdot \text{lb}$

The minus sign means a loss in potential energy.

PROBLEMS

16. What is the kinetic energy of (a) a 1250 kg car that has a speed of 25.5 m/s; (b) a 5600 lb truck moving at 56.0 ft/s?

17. Determine the increase in kinetic energy when (a) a 1750 kg car increases its speed from 12.0 m/s to 24.5 m/s; (b) a 3400 lb car decreases its speed from 48.0 ft/s to 25.0 ft/s.

18. How much work must be done to throw (a) a 0.35 kg object at a speed of 25 m/s; (b) a 1.20 lb ball at 65.0 ft/s?

19. A net force of 250 N acts on a 45 kg mass which is initially at rest. (a) What is the kinetic energy after it has traveled 4.5 m? (b) What is its kinetic energy after 12 s?

20. A 132 lb beam is raised vertically through 85.0 ft at a constant velocity to the top of a structure. Determine (a) the work done on a beam; (b) the change in its potential energy.

21. What is the change in the height of a 225 kg girder if a crane does 7250 J of work on it to raise it vertically at a constant velocity?

22. A 625 lb pile-driver hammer is raised to a position 17.5 ft above the top of a pile. (a) What is the gravitational potential energy of the hammer with respect to the pile? (b) What is the kinetic energy of the hammer just before it strikes the pile?

23. A 32.0 g arrow traveling at 42.5 m/s strikes a stationary target and penetrates a distance of 7.80 cm. (a) What was the kinetic energy of the arrow before impact? (b) What was the average force exerted on the target?

24. A 2.50 lb rock is dropped from rest from a high bridge. What is its kinetic energy (a) after 4.50 s; (b) after it has fallen 161 ft?

25. A 75.0 kg crate is at rest on a horizontal floor. If a horizontal 1250 N force is applied, the opposing frictional force is 745 N, and the crate moves 4.00 m from rest, determine the work done by (a) the applied force; (b) the friction force; (c) the force of gravity; (d) the reaction force of the floor; (e) the net force when the crate moves 4.00 m from rest. (f) What is the final kinetic energy of the crate?

26. A 525 kg pile-driver hammer is raised to a position 12.0 m above the top of a pile. (a) What is the gravitational potential energy of the hammer with respect to the pile? (b) What is the kinetic energy of the hammer just before it strikes the pile?

27. What is the change in potential energy of 45.0 L of water when it falls 35.0 m over a waterfall?

28. Determine the change in the potential energy of a 35.0 kg load of bricks (a) when they are raised vertically through 25.0 m and (b) when they are lowered 12.5 m at constant velocities.

29. Determine the change in the potential energy of a 65.0 lb container of cement when it is (a) raised vertically through 73.5 ft and (b) lowered 41.0 ft at constant velocities.

30. How much work is done on a 1250 kg elevator when it is raised vertically 25.0 m at a constant velocity?

31. Determine the kinetic energy of an electron (mass 9.10×10^{-31} kg) moving at 6.50×10^6 m/s.

32. A 75.0 kg ski jumper starts from rest down a run, the end of which is 25.0 m below her starting point. Her touchdown point on the slope is 45.0 m below the end of the run. When she starts, (a) what is her potential energy relative to the end of the run; (b) what is her potential energy relative to the touchdown point; (c) how much potential energy does she lose during the flight?

8-3 CONSERVATION OF MECHANICAL ENERGY

There are many forms of energy, but after considerable observation, it has been discovered that these energy forms are related by a very useful physical law called the conservation of energy. This law may be stated in different ways, but they are all equivalent. Examples are:

1. *Energy cannot be created or destroyed.*
2. *The total energy in an isolated system is constant.*

It should be noted that work and energy are related. The energy may change if work is done either on or by a system. Energy may be transformed from one kind into another, but the total energy is constant in any closed system (i.e., where energy is not gained or lost from the outside).

There are many direct applications of the conservation of energy in mechanics, because systems often interchange only kinetic and potential energies. Therefore, if *mechanical energy is conserved* (i.e., is constant), this law may be expressed as

$$\Delta E_k + \Delta E_p + E_f = 0 \tag{8-7}$$

where ΔE_k and ΔE_p represent *changes* in kinetic and potential energies and E_f is the energy lost to friction (usually producing thermal energy). We may use Eqs. 8-5 and 8-6 in this expression. In general,

change in kinetic energy + change in potential energy
$$+ \text{ energy lost to friction} = zero$$

If there are no friction losses,

initial kinetic energy + initial potential energy
$$= \text{final kinetic energy} + \text{final potential energy} \tag{8-8}$$

Therefore, if a mass m moves with an initial speed v_0 at an initial height h_0 above some reference level, and its final speed is v_1 and its final height is h_1 above the same reference,

$$\tfrac{1}{2} m v_1^2 + m g h_1 = \tfrac{1}{2} m v_0^2 + m g h_0 \tag{8-9}$$

In many applications of this law an object moves from rest ($v_0 = 0$) and increases its speed to some value v_1 as it loses height by some amount Δh. If there are no frictional losses, the final kinetic energy equals the lost potential energy, that is,

$$\tfrac{1}{2} m v_1^2 = m g \, \Delta h \tag{8-10}$$

and

$$v_1 = \sqrt{2 g \, \Delta h} \tag{8-11}$$

EXAMPLE 8-8

What is the speed of the stone just before it hits the ground if it is dropped from the top of a building that is **(a)** 16.0 m high; **(b)** 78.0 ft high? Assume negligible air resistance.

Sec. 8-3 / Conservation of Mechanical Energy

Solution

Sketch: See Fig. 8-7.

Data: For convenience, take the ground as the reference level. Then $h_1 = 0$, and since the stone was dropped, the initial speed $v_o = 0$.

(a) $\Delta h = 16.0$ m, $g = 9.81$ m/s^2, $v_1 = ?$

(b) $\Delta h = 78.0$ ft, $g = 32.2$ ft/s^2, $v_1 = ?$

Equation: $v_1 = \sqrt{2g\,\Delta h}$

Substitute:

(a) $v_1 = \sqrt{2(9.81 \text{ m/s}^2)(16.0 \text{ m})}$
  $= 17.7$ m/s

(b) $v_1 = \sqrt{2(32.2 \text{ ft/s}^2)(78.0 \text{ ft})}$
  $= 70.9$ ft/s

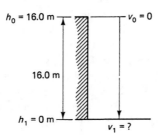

Figure 8-7 Stone dropped from a building.

EXAMPLE 8-9

A simple pendulum consists of a bob suspended from a long light string (Fig. 8-8). What is its speed at the lowest point of its swing if the bob is displaced through an arc and released from a point that is (a) 8.00 cm above its lowest position; (b) 3.00 in. above its lowest position? Assume negligible air resistance.

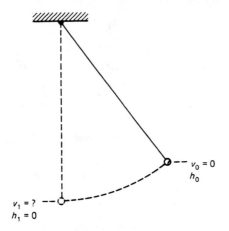

Figure 8-8 Simple pendulum.

Solution

Sketch: See Fig. 8-8.

Data: Again, potential energy is converted into kinetic energy. If the lowest point is taken as the reference level $h_1 = 0$, the initial conditions are $v_o = 0$,

(a) $\Delta h = 8.00$ cm $= 0.0800$ m, $g = 9.81$ m/s^2, $v_1 = ?$

(b) $\Delta h = 3.00$ in. $= 0.250$ ft, $g = 32.2$ ft/s^2, $v_1 = ?$

Equation: $v_1 = \sqrt{2g\,\Delta h}$

Substitute:

(a) $v_1 = \sqrt{2(9.81 \text{ m/s}^2)(0.0800 \text{ m})}$
  $= 1.25$ m/s

(b) $v_1 = \sqrt{2(32.2 \text{ ft/s}^2)(0.250 \text{ ft})}$
  $= 4.01$ ft/s

PROBLEMS

33. Determine the vertical speed at which water hits the bottom of a waterfall that is (a) 160 m high; (b) 315 m high; (c) 125 ft high; (d) 450 ft high.

34. What is the speed of the pile-driver hammer just before it strikes the pile if it falls (a) 12.5 m; (b) 35.0 ft?

35. A 20.0 kg rock is at rest at the top of a cliff 50.0 m high. (a) What is its potential energy with respect to the bottom of the cliff? (b) What would be its maximum kinetic energy if it fell freely to the bottom of the cliff?

36. A 38.0 lb rock is dropped from the top of a 135 ft high cliff. (a) What is its initial potential energy with respect to the bottom of the cliff? (b) At what speed does it strike the ground at the bottom of the cliff?

37. The bob of a simple pendulum is displaced through an arc so that it is 5.00 cm above its lowest position and then released. What is the speed of the bob as it passes through its lowest point? Assume no air resistance.

38. What is the maximum vertical height that the bob of a simple pendulum reaches above its lowest position if its speed at the lowest position is (a) 75.0 cm/s; (b) 2.50 m/s; (c) 30.0 in./s; (d) 8.20 ft/s?

39. How high is a waterfall if the water hits the bottom at a speed of (a) 35.0 m/s; (b) 62.5 m/s; (c) 95.0 ft/s; (d) 135 ft/s?

8-4 POWER

A particular amount of work may be performed by many different agents, but some will do that work faster than others. **Power** is defined as the time rate of doing work, or the time rate of expending energy. Therefore, if some agent does an amount of work W (or delivers an energy E) in some elapsed time t, the average power delivered

$$P = \frac{W}{t} \quad \text{or} \quad P = \frac{E}{t} \tag{8-12}$$

The SI unit for power is called a **watt** (symbol W). From the definition (Eq. 8-12), it can be seen that 1 W is the power of a source that delivers energy (or does work) at a rate of 1 J/s:

$$1 \text{ W} = 1 \frac{\text{J}}{\text{s}} = 1 \frac{\text{N} \cdot \text{m}}{\text{s}}$$

In USCS, power is expressed in units of either foot pounds per second (ft · lb/s) or horsepower (hp). The conversion is

$$1 \text{ hp} = 550 \frac{\text{ft} \cdot \text{lb}}{\text{s}}$$

The power of motors is frequently expressed in terms of **horsepower** (hp). another useful conversion factor is

$$1 \text{ hp} = 746 \text{ W}$$

EXAMPLE 8-10

What average power must a crane deliver to raise (a) a 175 kg container of concrete through a vertical height of 28.0 m in 12.0 s at constant speed; (b) a 325 lb girder through 26.5 ft in 8.00 s at a constant speed?

Sec. 8-4 / Power

Figure 8-9 Free-body diagram of the container, which is in equilibrium.

Solution

Sketch: See Fig. 8-9.

Data:
(a) $m = 175$ kg, $\Delta h = 28.0$ m, $t = 12.0$ s, $g = 9.81$ m/s^2, $P = ?$

(b) $w = mg = 325$ lb, $t = 8.00$ s, $\Delta h = 26.5$ ft, $P = ?$

and the magnitude of the applied force equals the force of gravity (weight).

$$F = w = mg$$

Equations: $P = \dfrac{W}{t}$, $W = Fs = mg\,\Delta h$

Rearrange: $P = \dfrac{mg\,\Delta h}{t}$

Substitute:

(a) $P = \dfrac{(175 \text{ kg})(9.81 \text{ m/s}^2)(28.0 \text{ m})}{12.0 \text{ s}}$

$= 4.01 \times 10^3$ W

(b) $P = \dfrac{(325 \text{ lb})(26.5 \text{ ft})}{8.00 \text{ s}}$

$= 1080 \dfrac{\text{ft} \cdot \text{lb}}{\text{s}} = \dfrac{1080}{550}$ hp

$= 1.96$ hp

EXAMPLE 8-11

If a pump delivers an average power of 0.250 hp; **(a)** how many liters of water does it raise from a 15.0 m deep well in 1.00 h; **(b)** how many U.S. gallons of water does it raise from a 32.0 ft deep well in 1.00 h? One U.S. gallon weighs 8.33 lb.

Solution

Data: The work done on the water is equal to the change in its potential energy $\Delta E_p = mg\,\Delta h$, $t = 1.00$ h $= 3600$ s.

(a) $\Delta h = 15.0$ m, $g = 9.81$ m/s^2,
$P = 0.250$ hp $= 0.250 \times 746$W,
$m = ?$

(b) $\Delta h = 32.0$ ft, $P = 0.250$ hp
$= 0.250 \times 550$ ft $\cdot$ lb/s,
$w = ?$

Equations: $P = \dfrac{\Delta E_p}{t}$, $W = mg\,\Delta h = \Delta E_p$

Rearrange: $P = \dfrac{mg\,\Delta h}{t}$; thus

(a) $m = \dfrac{Pt}{g\,\Delta h}$

(b) $w = mg = \dfrac{Pt}{\Delta h}$

Substitute:

(a) $m = \dfrac{(0.250 \times 746 \text{ W})(3600 \text{ s})}{(9.81 \text{ m/s}^2)(15.0 \text{ m})} = 4560$ kg

But a mass of 1.0 kg of water occupies a volume of 1.0 L; therefore, the pump can raise 4560 L of water in 1.00 h.

(b) $w = \dfrac{(0.250 \times 550 \text{ ft} \cdot \text{lb/s})(3600 \text{ s})}{32.0 \text{ ft}} = 15\,500$ lb

But 1 gal of water weighs 8.33 lb; therefore, 1 lb occupies $\dfrac{1}{8.33}$ gal and the pump raises $\dfrac{15\,500}{8.33}$ gal, or 1860 gal.

EXAMPLE 8-12

Determine the minimum output horsepower required by a motor of a crane that will raise (a) a 725 kg load vertically through 12.5 m in 52.0 s; (b) 425 lb load vertically through 26.0 ft in 34.0 s at a constant speed.

Solution

Data:
(a) $m = 725$ kg, $g = 9.81$ m/s^2,   (b) $w = mg = 415$ lb,
    $\Delta h = 12.5$ m, $t = 52.0$ s $\Delta h = 26.0$ ft, $t = 34.0$ s

Equations: $\Delta E = mg\,\Delta h$, $P = \dfrac{\Delta E}{t}$

Rearrange: $P = \dfrac{mg\,\Delta h}{t}$

Substitute:

(a) $P = \dfrac{(725 \text{ kg})(9.81 \text{ m/s}^2)(12.5 \text{ m})}{52.0 \text{ s}} = 1710$ W

But 1 hp = 746 W; therefore, 1 W = $\dfrac{1}{746}$ hp

Then 1710 W = $1710 \times \dfrac{1}{746}$ hp = 2.29 hp

(b) $P = \dfrac{(415 \text{ lb})(26.0 \text{ ft})}{34.0 \text{ s}} = 317$ ft·lb/s

But 1 hp = 550 ft·lb/s; therefore, 1 ft·lb/s = $\dfrac{1}{550}$ hp

Then 317 ft·lb/s = $317 \times \dfrac{1}{550}$ hp = 0.577 hp

Another convenient equation for finding power can be obtained from the definition. Since

$$\text{power} = \frac{\text{work done}}{\text{elapsed time}}$$

and if the applied force is in the same direction as the displacement,

$$\text{work} = \text{force} \times \text{distance}$$

Then

$$\text{power} = \frac{\text{force} \times \text{distance}}{\text{elapsed time}}$$

or

$$\text{power} = \text{force} \times \text{velocity} \qquad P = Fv \qquad (8\text{-}13)$$

provided, of course, that the force is constant.

EXAMPLE 8-13

What is the average power rating of the horse if it exerts a force of (a) 373 N to pull a loaded wagon at a steady speed of 2.00 m/s; (b) 83.8 lb to pull the wagon at 6.56 ft/s?

Sec. 8-4 / Power

Solution

Data:
(a) $F = 373$ N, $v = 2.00$ m/s,   (b) $F = 83.8$ lb, $v = 6.56$ ft/s,
    $P = ?$ $P = ?$

Equation: $P = Fv$

Substitute:
(a) $P = (373 \text{ N})(2.00 \text{ m/s})$ (b) $P = (83.8 \text{ lb})(6.56 \text{ ft/s})$
    $= 746$ W $= 550$ ft·lb/s $= 1.00$ hp

In fact, this value is commonly known as a horsepower (1 hp = 746 W). Horses have no trouble working at 20 or even 50 times this rate for short periods.

Efficiency

While the law of conservation of energy always applies to any machine, not all of the energy supplied is ever converted into useful work. Energy waste is frequently due to the friction between the moving parts of a machine, say a car engine, or as wasted heat in the exhaust gases, or even as light and sound. In many cases these losses may be quite large, and the useful work output is much less than the total energy or work input.

useful work output = energy input − energy losses

The ability of a machine to convert energy into useful work is described by a term called its **efficiency***, η, which is defined as the ratio of the useful work (or energy) output W_{out} to the total energy (or work) input W_{in}. It is usually expressed as a percentage.

$$\eta = \frac{\text{useful work output}}{\text{total energy input}} \times 100\%$$

$$\eta = \frac{W_{out}}{W_{in}} \times 100\% \qquad (8\text{-}14)$$

Also, if the elapsed time is the same for the input and output, the efficiency η of a device is the ratio of its useful output power P_{out} to the total power input P_{in}:

$$\eta = \frac{P_{out}}{P_{in}} \times 100\% \qquad (8\text{-}15)$$

EXAMPLE 8-14

Determine the efficiency of a laser if it draws 50 mW of electric power to produce a 2.0 mW output of laser light.

Solution

Data: $P_{in} = 50$ mW, $P_{out} = 2.0$ mW, $\eta = ?$

Equation: $\eta = \dfrac{P_{out}}{P_{in}} \times 100\%$

Substitute: $\eta = \dfrac{2.0 \text{ mW}}{50 \text{ mW}} \times 100\% = 4.0\%$

---

*Represented by the lower-case Greek letter eta

EXAMPLE 8-15

If the efficiency of a pump is 65%, what input power is required to lift (a) 60.0 L of water through 7.50 m in 1.00 min; (b) 15.0 U.S. gal of water through 23.2 ft in 1.00 min? One U.S. gallon of water weighs 8.33 lb.

Solution

Data: $\eta = 0.65$, $P_{in} = ?$

(a) $\Delta h = 7.50$ m, $t = 60.0$ s, and since 1.0 L of water has a mass of 1.0 kg, $m = 60.0$ kg

(b) $\Delta h = 23.2$ ft, $t = 60.0$ s, and 15.0 gal weighs
$w = 15.0 \times 8.33$ lb $= 125$ lb

Equations: $P_{out} = \dfrac{\Delta E}{t}$, $\Delta E = mg\,\Delta h$, $\eta = \dfrac{P_{out}}{P_{in}}$

Rearrange: $P_{in} = \dfrac{P_{out}}{\eta} = \dfrac{mg\,\Delta h}{\eta t}$

Substitute:

(a) $P_{in} = \dfrac{(60.0 \text{ kg})(9.81 \text{ m/s}^2)(7.50 \text{ m})}{(0.65)(60.0 \text{ s})} = 113$ W

(b) $P_{in} = \dfrac{(125 \text{ lb})(23.2 \text{ ft})}{(0.65)(60.0 \text{ s})} = 74.4$ ft·lb/s $= \dfrac{74.4}{550}$ hp $= 0.135$ hp

---

Perpetual motion is an idea that has intrigued people for centuries. A perpetual-motion machine would produce more energy than that supplied to it, that is, one whose efficiency would be greater than 100%. We now know that this is impossible, because energy can be neither created nor destroyed.

PROBLEMS

40. A 1500 kg elevator is raised 35.0 m in 6.50 s. (a) What useful power was required? (b) If the elevator draws 140 kW of input power, what is the efficiency?

41. What is the overall efficiency of the electric motor of an elevator that draws 80.0 hp of electric power when it raises (a) a 1750 kg elevator at a constant speed of 2.00 m/s; (b) a 3850 lb elevator at a constant 6.75 ft/s?

42. Determine the efficiency of an electric motor that draws 5.35 kW of power in order to raise a 250 kg object through a vertical height of 15.0 m in 12.0 s.

43. What is the average output power of a pump that raises (a) 45.0 L of water through a vertical height of 12.0 m in 3.00 min? (b) 12.0 U.S. gal of water through 35.0 ft in 3.00 min? One U.S. gallon weighs 8.33 lb.

44. What is the average power developed by a force of (a) 11 500 N if it pulls a loaded wagon at a steady speed of 1.75 m/s; (b) 2500 lb if it pulls the wagon at a steady 2.75 ft/s?

45. What input power is required to lift a loaded elevator, with a total mass of 1800 kg, at a steady speed of 2.50 m/s if the efficiency of the elevator system is 68%?

46. A 75.0 kg bricklayer carried 22.0 kg of bricks up a ladder through a vertical height of 5.80 m in 7.75 s. (a) What is the total average power required? (b) What is the average useful output power? (c) What is the efficiency of the process?

47. A 3750 lb elevator is raised at a constant speed through a vertical height of 65.0 ft in 8.00 s. What average power is required?

48. (a) How much work is done in 1.00 h by a source with an output power of 1.20 kW? (b) To what height would this lift a 1600 kg car?

49. Determine the minimum horsepower required by a motor that will raise an 850 kg elevator through 38.0 m in 8.00 s if the efficiency is (a) 100%; (b) 75%.

Summary

50 Determine the horsepower required by the motor of a crane if it is to lift a 1750 lb load vertically through 18.0 in. in 1 min 10 s and the efficiency is **(a)** 100%; **(b)** 85%.

SUMMARY

The *work done* W by a force $\mathbf{F}$ on an object is the product of the magnitudes of the force F, the displacement s, and the cosine of the angle θ between them:

$$W = Fs \cos \theta$$

Kinetic energy is the energy due to motion. For a mass m traveling at a speed v, the kinetic energy

$$E_k = \tfrac{1}{2}mv^2 = \tfrac{1}{2}\left(\frac{w}{g}\right)v^2$$

Potential energy is due to state or position. Near the earth's surface, the *change in the gravitational potential energy* of a mass m (or weight w) that is displaced vertically through Δh is

$$\Delta E_p = mg\,\Delta h = w\,\Delta h$$

The SI unit of work and energy is the *joule* (J), and the USCS unit is the *foot pound* (ft · lb).

Energy is conserved. That is, energy cannot be created or destroyed, but it can be converted from one type into another. If mechanical energy is conserved,

$$\Delta E_k + \Delta E_p + E_f = 0$$

where ΔE_k is the change in kinetic energy, ΔE_p the change in potential energy, and E_f the energy lost to friction.

$$\tfrac{1}{2}mv_1^2 + mgh_1 + E_f = \tfrac{1}{2}mv_o^2 + mgh_o$$

where v_o and v_1 are the initial and final speeds, and h_o and h_1 are the initial and final heights.

Power is the time rate of doing work or the time rate of using energy.

$$P = \frac{W}{t} = \frac{E}{t} \quad \text{and} \quad P = Fv$$

The units of power are watts or horsepower.

$$1 \text{ hp} = 746 \text{ W} = 550 \text{ ft} \cdot \text{lb/s}$$

Efficiency is

$$\eta = \frac{\text{work (or energy) output}}{\text{work (or energy) input}} \quad \text{or} \quad \eta = \frac{\text{output power}}{\text{input power}}$$

$$\eta = \frac{W_{\text{out}}}{W_{\text{in}}} \quad \text{and} \quad \eta = \frac{P_{\text{out}}}{P_{\text{in}}}$$

QUESTIONS

1. Define the following terms: **(a)** work; **(b)** kinetic energy; **(c)** power; **(d)** efficiency.
2. What is the difference between physical exertion and work?
3. Discuss the effects of friction forces on work.
4. When work is done on an object, that object does not necessarily accelerate. Explain why.
5. List five common examples of objects that have kinetic energy.
6. Describe, in terms of changes in energies, what happens during a roller coaster ride.
7. If an object was taken around the world and then returned to its original position, how much work was done on it?
8. How can two motors with different powers complete the same amount of work?
9. A kilowatt hour is the unit that is frequently used to measure electrical energy. What is the equivalent in joules? (*Hint:* Use the definition of power.)

REVIEW PROBLEMS

1. How much work is done when a 150 N force acts through a displacement of 32.0 m?
2. Determine the work done by a 350 lb force if it acts through 12.0 ft.
3. How much work is done by a crane on a 3350 lb beam if it lifts the beam through a vertical height of 32.0 ft at a constant speed?
4. A machine does 325 J of work on a 62.0 kg crate and raises it vertically at a constant velocity. What is the vertical displacement of the crate?
5. A machine does 2600 J of work on an engine and raises it vertically 3.50 m. What is the mass of the engine?
6. A man pulls a trolley of machinery at constant speed through a horizontal distance of 7.50 m by exerting a 145 N force at 42° with the horizontal. How much work does the man do on the trolley?
7. How much work is done by a construction worker if she raises 25.0 kg of bricks through a vertical distance of 16.0 m at a constant speed?
8. A girl pulls a sled along a horizontal surface by exerting a 35.0 lb force at an angle of 48°. How much work does she do on the sled in pulling it 8.50 ft horizontally **(a)** if there is no friction; **(b)** if there is a 15.0 lb friction force?
9. How much work is done by a pump in raising **(a)** 5.00 L of water vertically through a distance of 12.5 m; **(b)** 35.0 kg of water vertically through 15.0 m; **(c)** 495 lb of water vertically through 175 ft; **(d)** 24.0 U.S. gal of water vertically through 56.0 ft?
10. What is the kinetic energy of a 3450 lb car traveling at 55.0 mi/h?
11. Determine the kinetic energy of an electron traveling at 5.80×10^6 m/s if its mass is 9.1×10^{-31} kg.
12. How much work must be done to accelerate an 1800 kg car from rest to 80.0 km/h?
13. What is the kinetic energy of a 85.0 kg man if he is running at a speed of 11.5 km/h?
14. How much work must be done to throw a 1.20 lb ball with a speed of 82.0 ft/s?
15. Determine the change in the gravitational potential energy of a 180 kg crate if it is **(a)** raised vertically through 82.0 m; **(b)** lowered vertically through 22.5 m near the earth's surface.
16. Find the change in the gravitational potential energy of a 160 lb man if he descends a total distance of 42.0 ft vertically down a mine.
17. Determine the change in potential energy when a 2.50 kg object is raised from an altitude of 15.0 m to an altitude of 35.0 m above the earth's surface.
18. What is the change in the potential energy of 125 L of water when it falls down a 125 m waterfall?
19. Determine the vertical speed at which water hits the bottom of **(a)** a 180 m high waterfall; **(b)** a 285 ft high waterfall.

Review Problems

20. What is the speed of a pile-driver hammer just before it strikes if it falls (a) 18.0 m to a pile; (b) 26.0 ft to a pile?

21. If 15.0 kg of water is raised to a height of 35.0 m in 1.20 min by a pump, (a) how much work is done; (b) what is the average output power of the pump? (c) If the input power is 96.0 W, what is the efficiency?

22. A conveyor belt can pull a 122 lb box at a constant velocity 295 ft up a 30° incline in 5.25 s. (a) What is the work done? (b) What is the average output power developed? Ignore friction.

23. A train travels along a straight track at a constant speed of 88.0 km/h. If the total retarding force is 53 500 N, what is the average output power developed?

24. Determine the power required to lift a 285 lb load vertically through 28.0 ft in 22.0 s at a constant velocity.

25. What minimum horsepower must a pump deliver to raise 37.5 L of water per minute from the bottom of a 23.5 m well?

26. How many liters of water per minute can a 0.50 hp pump raise from the bottom of a 25.0 m well? Neglect power losses.

27. Determine the minimum output power in horsepower required by a motor that will raise an 850 kg elevator through 38.0 m in 8.00 s.

9 CIRCULAR AND ROTARY MOTION

Many objects, such as planets orbiting the sun, electrons orbiting an atomic nucleus, and aircraft, frequently accelerate and change the directions of their motion. If an object moves in a circle about a fixed point, it is said to be in **circular motion.**

Many rigid objects, such as drive shafts, wheels, cams, and gears, are designed to spin or rotate about some axis. A rigid object is said to move in pure rotary motion if all of its particles move in circles about an **axis of rotation.**

9-1 CIRCULAR MOTION

An object is said to perform **uniform circular motion** if it moves in a circle at a constant speed. However, it should be noted that even though the speed is a constant, an object in uniform circular motion continually changes its direction of motion, and therefore its velocity changes and it accelerates. This type of motion, which is repeated at regular intervals, is an example of a **periodic motion.** Other types of periodic motions are oscillations and vibrations. In each case the time required to complete one cycle of the motion (e.g., a circle, or a complete to-and-fro vibration or oscillation) is called the **period** T of the motion.

For an object in uniform circular motion of radius r at a constant speed v (Fig. 9-1), the period

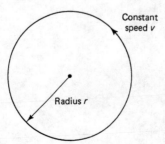

Figure 9-1 Uniform circular motion.

$$T = \frac{\text{distance traveled}}{\text{speed}} = \frac{\text{circumference}}{\text{speed}} = \frac{2\pi r}{v} \qquad (9\text{-}1)$$

The **frequency** f of the circular motion is the number of completed cycles per unit time. It is equal to the reciprocal of the period T.

$$f = \frac{1}{T} \qquad (9\text{-}2)$$

The unit for frequency for an object in circular motion is revolutions per second (rev/s) or revolutions per minute (rev/min or rpm).

EXAMPLE 9-1

Determine the speed of an aircraft if it completes a circular path of radius 775 m in 68.0 s.

Solution

Date: $r = 775$ m, $T = 68.0$ s, $v = ?$

Equation: $T = \dfrac{2\pi r}{v}$

Rearrange: $v = \dfrac{2\pi r}{T}$

Substitute: $v = \dfrac{2\pi (775 \text{ m})}{68.0 \text{ s}} = 71.6$ m/s

---

Objects in uniform circular motion are continually changing their direction of motion as they orbit a fixed point. Therefore, their velocity changes, and they must be continually accelerating. If we tie a ball to a string and whirl it in a circle, we must continually pull inward. This inward force, called a **centripetal force,** continually accelerates the ball, making it change its direction of motion. If we did not exert this centripetal force, the ball would fly off in one direction (Fig. 9-2). The faster the ball is whirled, the greater the inward (centripetal) force required to maintain the circular motion.

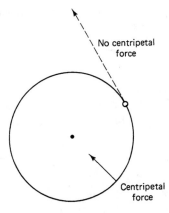

Figure 9-2 Without the centripetal force, the ball would fly off at a tangent to the circular path.

Both the centripetal force and the corresponding centripetal acceleration are directed toward the center of the circle. If an object of mass m and weight w moves at a constant speed v in a circle of radius r, its **centripetal acceleration** has a magnitude

$$a_c = \frac{v^2}{r} \tag{9-3}$$

and it is directed toward the center of the circle. From Newton's second law the corresponding centripetal force has a magnitude

$$F_c = \frac{mv^2}{r} = \left(\frac{w}{g}\right)\frac{v^2}{r} \tag{9-4}$$

and it is also directed toward the center of the circular path.

Note that when an object of mass m and weight $w = mg$ performs uniform circular motion, its kinetic energy

$$E_k = \frac{1}{2}mv^2 = \frac{1}{2}\left(\frac{w}{g}\right)v^2$$

is also constant, since its speed v is constant. The centripetal force always acts toward the center of the circle and perpendicular to the direction of motion; therefore, *centripetal forces do no work*.

Centripetal forces must be exerted on any object if it is to perform a circular motion. These centripetal forces may be provided by friction, gravitation, electric fields, magnetic fields, and many other sources.

Cars are able to turn corners on flat roads only because of the friction force between the wheels and the road surface (Fig. 9-3). If the magnitude of the friction force is not sufficient, the car will skid and not make the turn.

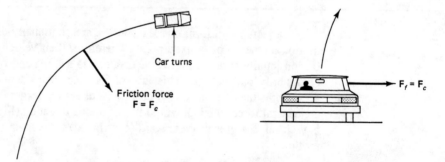

Figure 9-3 Friction provides the centripetal force that enables cars to turn corners on flat roads.

EXAMPLE 9-2

Determine the friction force between the tires and a flat horizontal road that enables **(a)** a 1650 kg car to make a turn of radius 135 m at a constant speed of 50.0 km/h; **(b)** a 3750 lb car to make a turn of radius 445 ft at a constant speed of 45.0 mi/h.

Solution

Data:

(a) $m = 1650$ kg, $r = 135$ m,
$v = 50.0 \dfrac{\text{km}}{\text{h}} = \dfrac{50.0 \times 10^3 \text{ m}}{3600 \text{ s}}$
$= 13.9$ m/s, $F = ?$

(b) $w = mg = 3750$ lb,
$r = 445$ ft, $g = 32.2$ ft/s^2,
$v = 45.0$ mi/h
$= 45.0 \times \dfrac{88}{60}$ ft/s $= 66.0$ ft/s,
$F = ?$

Sec. 9-1 / Circular Motion

Equation:

(a) $F_c = \dfrac{mv^2}{r}$ 	(b) $F_c = \left(\dfrac{w}{g}\right)\dfrac{v^2}{r}$

Substitute:

(a) $F_c = \dfrac{(1650 \text{ kg})(13.9 \text{ m/s})^2}{135 \text{ m}}$ 	(b) $F_c = \dfrac{(3750 \text{ lb})(66.0 \text{ ft/s})^2}{(32.2 \text{ ft/s}^2)(445 \text{ ft})}$

   $= 2360$ N 	 $= 1140$ lb

This must be provided by the friction force.

EXAMPLE 9-3

Determine the minimum coefficient of friction between the tires and a flat horizontal road that will enable cars to complete a turn (a) of radius 115 m at 80.0 km/h; (b) of radius 475 ft at 60.0 mi/h.

Solution

Sketch: See Fig. 9-4.

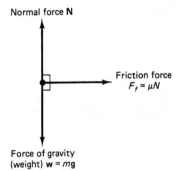

Figure 9-4 Free-body diagram.

Data:

(a) $r = 115$ m, $g = 9.81$ m/s², 	(b) $r = 475$ ft, $g = 32.2$ ft/s²,
    $v = 80.0$ km/h 	 $v = 60.0$ mi/h = 88.0 ft/s,
    $= \dfrac{80.0 \times 10^3 \text{ m}}{3600 \text{ s}} = 22.2$ m/s, 	 $\mu = ?$
    $\mu = ?$

Equations: If m is the mass of the car, N is the normal force of the road on the car and μ is the coefficient of friction, then since the car would be in equilibrium in the vertical direction (otherwise it would fall through or leap off the road), net $F = 0$ and $N - w = 0$. Therefore, $N - mg = 0$, and $N = mg$, since $w = mg$. In the horizontal direction, net $F = F_c = F_f$, the friction force. Also, $F_c = \dfrac{mv^2}{r}$ and $F_f = \mu N$.

Rearrange: $F_c = F_f = \mu N = \dfrac{mv^2}{r}$, but $N = mg$; therefore,

$$\mu mg = \dfrac{mv^2}{r} \quad \text{and} \quad \mu = \dfrac{mv^2}{mgr} = \dfrac{v^2}{rg} \tag{9-5}$$

Note that this expression for the coefficient of friction is independent of the mass or weight of the car!

Substitute:

(a) $\mu = \dfrac{(22.2 \text{ m/s})^2}{(115 \text{ m})(9.81 \text{ m/s}^2)} = 0.437$ 	(b) $\mu = \dfrac{(88.0 \text{ ft/s})^2}{(475 \text{ ft})(32.2 \text{ ft/s}^2)} = 0.506$

To reduce the possibility of cars skidding off a road because of a slippery surface or excessive speeds, roads are usually **banked** (Fig. 9-5). In this case the unbalanced horizontal component of the normal force **N** of the road on the car supplies at least a part of the centripetal force $\mathbf{F}_c$. The optimum angle of bank θ is given by

$$\tan \theta = \frac{v^2}{rg} \qquad (9\text{-}6)$$

where v is the speed of the car, r the radius of the curve, and g the acceleration due to gravity. At the optimum angle of bank, a car would be able to turn a curve without any assistance from friction. Note that this angle of bank is independent of the mass or weight of the car!

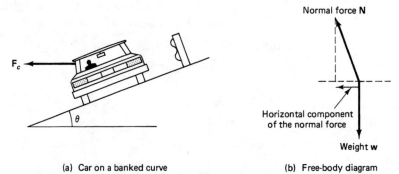

(a) Car on a banked curve   (b) Free-body diagram

Figure 9-5 Banking of a curve. The horizontal component of the normal force supplies the centripetal force.

EXAMPLE 9-4

At what optimum angle must a curve be banked so that cars can complete the turn without skidding **(a)** if the curve radius is 125 m and the car speed is 60.0 km/h; **(b)** if the curve radius is 485 ft and the car speed is 45.0 mi/h?

Solution

Data:

(a) $r = 125$ m, $g = 9.81$ m/s^2,
$v = 60.0$ km/h
$= \dfrac{60.0 \times 10^3 \text{ m}}{3600 \text{ s}} = 16.7$ m/s,
$\theta = ?$

(b) $r = 485$ ft, $g = 32.2$ ft/s^2,
$v = 45.0$ mi/h
$= 45.0 \times \dfrac{88}{60}$ ft/s $= 66.0$ ft/s,
$\theta = ?$

Equation: $\tan \theta = \dfrac{v^2}{rg}$

Substitute:

(a) $\tan \theta = \dfrac{(16.7 \text{ m/s})^2}{(125 \text{ m})(9.81 \text{ m/s}^2)}$
$= 0.227$
$\theta = 12.8°$

(b) $\tan \theta = \dfrac{(66.0 \text{ ft/s})^2}{(485 \text{ ft})(32.2 \text{ ft/s}^2)}$
$= 0.279$
$\theta = 15.6°$

A force called a **centrifugal force**, which acts away from the center of the circle, is often incorrectly introduced as a reaction force for objects in circular motion. In fact, the centrifugal force does *not* act on the object in circular motion, since these objects are not in equilibrium but continually accelerate due to the unbalanced centripetal force. The centrifugal force is fictitious force which is introduced to explain motion *in* an accelerating system. It is due to the tendency of the orbiting object to continue in a straight line.

PROBLEMS

1. A bicycle wheel of diameter 62.0 cm rotates 10 times in 6.00 s. Determine (a) the speed of a point on its rim; (b) the centripetal acceleration of its rim.

2. A 4.00 in. diameter drive shaft rotates 1500 times in 2.00 s. Determine (a) the speed of a point on its surface; (b) the centripetal acceleration of points on its surface.

3. Determine the time taken by an aircraft to complete a turn of radius 1250 m if its speed is 85.0 m/s.

4. A model of an atom pictures an electron in a circular orbit of radius 5.3×10^{-11} m about the atomic nucleus. If the speed of the electron is 2.2×10^6 m/s, determine the period of its orbit.

5. Determine the average speed of the earth about the sun if the average radius of its orbit is 1.50×10^{11} m and each orbit takes 365 days.

6. Find the speed of a train on a circular track of radius 1560 ft, if it takes 2.00 min to complete the circle.

7. Determine the centripetal force on a 5.80 kg object moving in a circle of radius 3.50 m at a constant speed of 2.75 m/s.

8. A 275 kg object moves in uniform circular motion. (a) If the centripetal force on the object is 345 N, what is its speed around a turn of radius 258 m? (b) If its speed is 24.0 m/s, what is the radius of its path?

9. An aircraft traveling at a constant speed of 328 mi/h turns a complete circle in 2.50 min. What is the radius of the turn?

10. Determine the friction force between the tires and a flat horizontal road that enables (a) a 1600 kg car to make a turn of radius 180 m at a constant speed of 55.0 km/h; (b) a 5800 lb car to make turn of radius 625 ft at 40.0 mi/h.

11. If the coefficient of friction is 0.52, determine the maximum speed that a car can travel around a curve of radius (a) 138 m and (b) 475 ft on a flat horizontal road.

12. What minimum coefficient of friction is required if there is a 65.0 km/h speed limit for a flat horizontal curve of radius 120 m?

13. Determine the minimum coefficient of friction that is required for cars to make turns of radius 482 ft at 60.0 mi/h on a flat horizontal road.

14. On a slippery day the coefficient of friction between the tires and the road is 0.28. On a flat horizontal road, at what maximum speed can a car make a turn of radius (a) 105 m; (b) 375 ft?

15. At what optimum angle must a curve of radius 142 m be banked so that cars can turn the curve at a constant speed of 85.0 km/h without skidding?

16. At what optimum angle must a curve be banked so that cars can turn the curve of radius 465 ft at a constant speed of 60.0 mi/h without skidding?

17. If a curve is banked at an angle of 11°, determine the optimum speed that cars can make a turn of radius (a) 160 m; (b) 525 ft.

9-2 MEASUREMENT OF ANGLES

There are three common ways that are used to indicate the amount through which an object rotates:

1. A **revolution** (rev) is a complete rotation of the object, so that it returns to its original position. This unit is frequently used in industry to measure the rotations of drive shafts, flywheels, and so on.

2. Degrees, minutes, and seconds of arc are frequently used to describe rotations. In this case the circle is divided into 360 equal parts called **degrees** (°); therefore, one revolution is equivalent to 360 degrees.

$$1 \text{ rev} = 360°$$

Each degree is also subdivided into 60 minutes (') of arc:

$$1° = 60'$$

and each minute is subdivided into 60 seconds ("):

$$1' = 60''$$

3. In SI, angles are measured in terms of a supplementary base unit called a radian (rad). A **radian** is defined as the angle subtended at the center of a circle by an arc length equal to the radius r (Fig. 9-6).

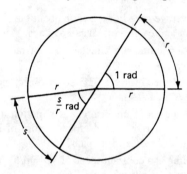

Figure 9-6 Definition of a radian. An arc length equal to the radius r subtends an angle of 1 rad at the center of the circle. An arc length s subtends s/r rad.

$$2\pi \text{ rad} = 360° \quad \text{or} \quad 1 \text{ rad} = \frac{360°}{2\pi} = \frac{180°}{\pi} \quad (9\text{-}7)$$

Note that the radian unit is really a ratio of two lengths, and therefore it has no physical dimensions. It is used to indicate that an angle is being considered.

Since an arc length equal to the radius r of a circle subtends an angle of 1 rad at the center of that circle, any other arc length s (s represents arc length and not displacement in this chapter) must subtend an angle:

$$\theta = \frac{s}{r} \text{ rad} \quad (9\text{-}8)$$

EXAMPLE 9-5

Express 3.29 rev in **(a)** degrees; **(b)** radians.

Solution: **(a)** By direct substitution, since 1 rev = 360°,

$$3.19 \text{ rev} = 3.19 \times 360° = 1150°$$

(b) 1 rev = 2π rad; therefore, 3.19 rev = 3.19 × 2π rad = 20.0 rad.

EXAMPLE 9-6

Convert 4.15 rad to **(a)** revolutions; **(b)** degrees.

Solution: **(a)** 1 rev = 2π rad; therefore, 1 rad = $\dfrac{1}{2\pi}$ rev. By direct substitution,

Sec. 9-2 / Measurement of Angles

$$4.15 \text{ rad} = 4.15 \times \frac{1}{2\pi} \text{ rev} = 0.660 \text{ rev}$$

(b) $360° = 2\pi$ rad; therefore, 1 rad = $360°/2\pi$. Hence

$$4.15 \text{ rad} = 4.15 \times \frac{360°}{2\pi} = 238°$$

EXAMPLE 9-7

If a 60.0 cm diameter car wheel rotates through 26.0 rev, how far does the car travel if the wheels do not slip?

Solution: The total distance traveled by a point on the rim (in a circular path) corresponds to the distance traveled.
Sketch: See Fig. 9-7.
Data: $r = 0.300$ m, $\theta = 26.0$ rev $= 26.0 \times 2\pi$ rad, $s = ?$
Equation: $\theta = \dfrac{s}{r}$
Rearrange: $s = r\theta$
Substitute: $s = (0.300 \text{ m})(26.0 \times 2\pi \text{ rad}) = 49.0$ m

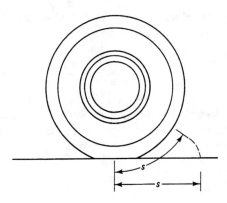

Figure 9-7 The distance traveled by the car is the distance traveled by a point on the rim.

PROBLEMS

18 Convert the following to radians and degrees: **(a)** 22.0 rev; **(b)** 12.0 rev; **(c)** 2300 rev; **(d)** 0.800 rev.

19 Convert the following to revolutions and degrees: **(a)** 78.0 rad; **(b)** 7530 rad; **(c)** 2.54 rad; **(d)** 3.50×10^4 rad.

20 Convert the following to revolutions and radians: **(a)** 18°; **(b)** 235°; **(c)** 775°; **(d)** 15°25'18".

21 How many radians are subtended by an arc length of 6.00 m at the center of a circle of radius 1.20 m?

22 How many radians are subtended at the center of a circle of radius 3.50 in. by an arc length of 5.60 in.?

23 If a 55.0 cm diameter bicycle wheel rotates through 56 revolutions without slipping, how far does the bicycle travel?

24 If a car has 28.0 in. diameter wheels, how far does it travel when its wheels rotate through 450 revolutions without slipping?

9-3 ANGULAR MOTION

Even though the motions are quite different, rotary motions and translational motions (in a straight line) are described by similar equations.

Angular Displacement

The angle through which an object rotates is known as the **angular displacement** θ; it is normally expressed in units of radians or revolutions. The time required for one revolution (one complete rotation) of a rigid object is called its **period T of rotation**. The number of revolutions per unit time is called the **frequency** f.

Angular Velocity

By analogy with linear velocity, the **angular velocity** ω* of a rotating object is defined as its time rate of change of angular displacement. Thus if an object rotates through an angular displacement θ in an elapsed time t, its average angular velocity is

$$\omega = \frac{\theta}{t} \tag{9-9}$$

Normally, the angular velocity is expressed in units of radians per second (rad/s) or revolutions per minute (rev/min or rpm).

$$\frac{1 \text{ rev}}{\text{min}} = \frac{2\pi \text{ rad}}{60 \text{ s}} \tag{9-10}$$

Suppose that a rigid object rotates through some angle θ in an elapsed time t. If some particle in that rigid object is located at a distance r from the axis of rotation and moves in an arc through a distance s, its average speed is

$$v = \frac{s}{t} = \frac{r\theta}{t} = r\omega \tag{9-11}$$

since $s = r\theta$ and the average angular velocity of the rigid object is

$$\omega = \frac{\theta}{t}$$

EXAMPLE 9-8

If the drive shaft of a motor rotates at a constant angular velocity of 1250 rpm, how many revolutions does it make in 25.0 s?

Solution

Data: $\omega = 1250$ rpm, $t = 25.0$ s $= \frac{25.0}{60}$ min, $\theta = ?$

Equation: $\omega = \frac{\theta}{t}$

Rearrange: $\theta = \omega t$

Substitute: $\theta = \left(1250 \frac{\text{rev}}{\text{min}}\right)\left(\frac{25.0}{60} \text{ min}\right) = 521$ rev

---

*Represented by the lower-case Greek letter omega

Sec. 9-3 / Angular Motion

EXAMPLE 9-9

A flywheel with a 1.50 ft diameter rotates through 125 rev in 10.0 s. Determine (a) the average angular velocity; (b) the average speed of a point on the rim.

Solution

Data: $\theta = 125$ rev $= 125 \times 2\pi$ rad $= 785$ rad, $t = 10.0$ s, $r = 0.750$ ft, $\omega = ?$, $v = ?$

(a) Equation: $\omega = \dfrac{\theta}{t}$

Substitute: $\omega = \dfrac{785 \text{ rad}}{10.0 \text{ s}} = 78.5$ rad/s

(b) Equation: $v = r\omega$

Substitute: $v = (0.750 \text{ ft})(78.5 \text{ rad/s}) = 58.9$ ft/s

---

Angular Acceleration

The angular acceleration α* of an object is defined as the time rate of change of its angular velocity. If an object changes from an angular velocity ω_0 to an angular velocity ω_1, in an elapsed time t, its average angular acceleration is

$$\alpha = \frac{\omega_1 - \omega_0}{t} = \frac{\Delta\omega}{t} \tag{9-12}$$

where $\Delta\omega$ is the change in angular velocity. Angular accelerations are usually expressed in units of rad/s².

If a particle in the rotating rigid object is at a distance r from the axis of rotation and it changes its velocity by an amount Δv in a very small time interval t, the magnitude of its tangential acceleration is given by

$$a_t = \frac{\Delta v}{t}$$

But $v = r\omega$, and r is a constant; therefore, $\Delta v = r \Delta\omega$, and

$$a_t = \frac{r \Delta\omega}{t}$$

or

$$a_t = r\alpha \tag{9-13}$$

All particles in a rotating rigid object experience a centripetal acceleration $\mathbf{a}_c$ directed toward the axis of rotation. At the same time, they may be subjected to an acceleration $\mathbf{a}_t$ tangential to their circular path (Fig. 9-8). Their total acceleration is the vector sum of $\mathbf{a}_c$ and $\mathbf{a}_t$. Since the centripetal and tangential accelerations are perpendicular to each other, they are independent. Tangential accelerations change the speed of the particles in the rotating object, and centripetal accelerations change the direction of motion.

---

*Represented by the lower-case Greek letter alpha

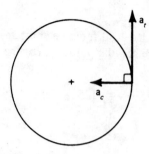

Figure 9-8 Centripetal and tangential accelerations.

EXAMPLE 9-10

A car starts from rest and accelerates uniformly to a final speed of 25.0 m/s in an elapsed time of 12.5 s. If the diameter of the wheels is 70.0 cm, determine (a) the acceleration of the car; (b) the final angular velocity of the wheels; (c) the angular acceleration of the wheels.

Solution

Data: $v_0 = 0$ m/s, $v_1 = 25.0$ m/s, $t = 12.5$ s, $r = \dfrac{d}{2} = 0.350$ m, $a_t = ?$, $\omega = ?$, $\alpha = ?$

(a) Equation: $a_t = \dfrac{v_1 - v_0}{t}$ since the acceleration of the car is equal to the acceleration of the rims of the wheels.

Substitute: $a_t = \dfrac{25.0 \text{ m/s} - 0 \text{ m/s}}{12.5 \text{ s}} = 2.00$ m/s²

(b) Equation: $\omega = \dfrac{v}{r}$

Substitute: $\omega = \dfrac{25.0 \text{ m/s}}{0.350 \text{ m}} = 71.4$ rad/s $= (71.4)\left(\dfrac{60}{2\pi} \text{ rpm}\right)$
$= 682$ rpm

(c) Equation: $\alpha = \dfrac{a_t}{r}$

Substitute: $\alpha = \dfrac{2.00 \text{ m/s}^2}{0.350 \text{ m}} = 5.71$ rad/s²

Four equations of rotational motion with a constant angular acceleration, which are quite similar to the equations of uniformly accelerated translational motion, may be obtained (Table 9-1). Note the similarity between the sets of equations. The rotational equations may be obtained by substituting angular displacement θ for linear displacement s, angular velocity ω for linear velocity v, and angular acceleration α for linear acceleration a in the translational equations.

TABLE 9-1

Equations of Motion with Constant Translational or Rotational Acceleration

Translational motion (straight line)	Rotational motion (fixed axis of rotation)
$s = \left(\dfrac{v_1 + v_0}{2}\right)t$	$\theta = \left(\dfrac{\omega_1 + \omega_0}{2}\right)t$
$v_1 = v_0 + at$	$\omega_1 = \omega_0 + \alpha t$
$s = v_0 t + \frac{1}{2}at^2$	$\theta = \omega_0 t + \frac{1}{2}\alpha t^2$
$v_1^2 = v_0^2 + 2as$	$\omega_1^2 = \omega_0^2 + 2\alpha\theta$

Sec. 9-3 / Angular Motion

EXAMPLE 9-11

A flywheel, which is initially rotating with an angular velocity of 50.0 rad/s, is subjected to an angular acceleration of 5.00 rad/s² for an elapsed time of 10.0 s. (a) What is its final angular velocity? (b) How many revolutions does it perform while it is being accelerated?

Solution

Data: $\omega_0 = 50.0$ rad/s, $\alpha = 5.00$ rad/s², $t = 10.0$ s, $\omega_1 = ?$, $\theta = ?$

(a) Equation: $\alpha = \dfrac{\omega_1 - \omega_0}{t}$

Rearrange: $\omega_1 = \omega_0 + \alpha t$
Substitute: $\omega_1 = 50.0$ rad/s $+ (5.00$ rad/s²$)(10.0$ s$)$
$= 100$ rad/s

(b) Equation: $\theta = \omega_0 t + \frac{1}{2}\alpha t^2$
Substitute: $\theta = (50.0$ rad/s$)(10.0$ s$) + \frac{1}{2}(5.00$ rad/s$)(10.0$ s$)^2$
$= 750$ rad $= \dfrac{750}{2\pi}$ rev $= 119$ rev

EXAMPLE 9-12

An aircraft engine idling at 1200 rpm accelerates uniformly at 60.0 rad/s² to a final angular velocity of 7800 rpm. Determine (a) the time taken; (b) the number of revolutions made while accelerating.

Solution

Data: $\omega_0 = 1200$ rpm $= 1200\left(\dfrac{2\pi}{60}\dfrac{\text{rad}}{\text{s}}\right) = 126$ rad/s

$\omega_1 = 7800$ rpm $= 7800\left(\dfrac{2\pi}{60}\dfrac{\text{rad}}{\text{s}}\right) = 817$ rad/s

$\alpha = 60.0$ rad/s²
$t = ?$, $\theta = ?$

(a) Equation: $\alpha = \dfrac{\omega_1 - \omega_0}{t}$

Rearrange: $t = \dfrac{\omega_1 - \omega_0}{\alpha}$

Substitute: $t = \dfrac{817 \text{ rad/s} - 126 \text{ rad/s}}{60.0 \text{ rad/s}^2} = 11.5$ s

(b) Equation: $\omega_1^2 = \omega_0^2 + 2\alpha\theta$

Rearrange: $\theta = \dfrac{\omega_1^2 - \omega_0^2}{2\alpha}$

Substitute: $\theta = \dfrac{(817 \text{ rad/s})^2 - (126 \text{ rad/s})^2}{2(60.0 \text{ rad/s}^2)}$

$= 5430$ rad $= \dfrac{5430}{2\pi}$ rev $= 864$ rev

PROBLEMS

25 Convert the following to revolutions per minute: (a) 4.50 rev/s; (b) 5.90 rad/s; (c) 25.0 rad/s; (d) 0.0520 rad/s.

26 Convert the following to radians per second: (a) 25.0 rpm; (b) 2300 rpm; (c) 2.30 rev/s; (d) 0.725 rev/s.

27. Determine the average angular velocity in radians per second and revolutions per minute of a wheel that makes (a) 35.0 rev in 2.50 min; (b) 1800 rev in 35.0 s; (c) 7800 rev in 58.0 s.

28. A wheel has an angular velocity of 258 rpm. Find the angular displacement in revolutions and radians after (a) 2.00 min; (b) 38.0 s; (c) 2.50 h.

29. A wheel has an angular velocity of 1250 rpm. (a) Determine the speed of a point on its rim if the wheel radius is 28.0 cm. (b) If the speed of a point on the rim is 25.0 m/s, what is its radius?

30. A drive shaft rotates at 1500 rpm. What is its angular displacement in radians after (a) 2.50 s; (b) 3.50 min?

31. A conveyor belt is wrapped around a 12.0 cm diameter pulley which drives it. If there is no slippage and the pulley rotates at 325 rpm, what is the linear speed of the belt?

32. The earth has a diameter of 1.276×10^7 m, and it completes one revolution about its axis in 24.0 h. (a) What is the average angular velocity of the earth? (b) What is the average speed of a point on the earth's equator?

33. A flywheel has a diameter of 32.0 in. and it makes 1250 rev in 1.50 min. Determine (a) the average angular velocity in rad/s; (b) the average speed of a point on the rim.

34. A car starts from rest and accelerates uniformly to a speed of 55.0 mi/h in 15.0 s. If the wheels of the car are 28.0 in. in diameter, determine (a) the acceleration of the car; (b) the final angular velocity of the wheels; (c) the angular acceleration of the wheels; (d) the number of revolutions of the wheels while the car accelerates.

35. The drive shaft of a motor starts from rest and accelerates uniformly to an angular velocity of 1500 rpm in 18 s. Determine (a) the angular acceleration; (b) the number of revolutions completed by the drive shaft while it accelerates.

36. A wheel is rotating with an angular velocity of 32.0 rad/s when it is subjected to an angular acceleration of 6.80 rad/s². (a) How long does it take the wheel to reach an angular velocity of 42.0 rad/s? (b) How many revolutions does it complete in the first 5.00 s that it accelerates?

37. Determine the speed of a point on the surface of a cylindrical piece of wood with a diameter of 2.00 in. if it rotates at 475 rpm on a lathe.

38. If the shaft of an electric motor slows from 1275 rpm to 350 rpm in 7.50 s, determine (a) the average angular acceleration; (b) the angular displacement in the 7.50 s.

39. At what angular velocity should a lathe turn to produce a tangential speed of 72.0 cm/s for a steel cylinder with a 6.50 cm diameter?

40. If a conveyor belt travels at a speed of 2.25 m/s, what is the angular velocity of its driving pulley, which has a diameter of 45.0 cm?

41. A wheel accelerates at 5.20 rad/s² to a final angular velocity of 1750 rpm in 5.00 s. Determine (a) the initial angular velocity; (b) the angular displacement of the wheel while accelerating.

42. An aircraft engine accelerates uniformly from 1500 rpm to 6000 rpm in 8.20 s. Find (a) the angular acceleration; (b) the number of revolutions completed while it accelerates.

43. The shaft of a generator is turning at 1500 rpm and decelerates at 5.00 rad/s² for 30.0 s. Determine (a) its final angular velocity; (b) the number of revolutions that it makes while decelerating.

9-4 ROTATIONAL ENERGY AND POWER

In addition to producing linear accelerations, forces may also cause objects to rotate or alter their rotation rate. For example, if we want to make a wheel rotate, we exert a tangential force along its rim (see Fig 9-9).

We describe the "turning effect" of a force by the term **torque** τ*, which is defined as the product of the applied force F and the perpendicular distance s from the

---

*Represented by the lower-case Greek letter tau

Sec. 9-4 / Rotational Energy and Power

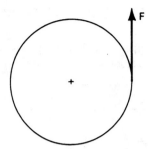

Figure 9-9 Tangential forces tend to cause rotations.

line of action to the point about which the object rotates (e.g., the hinge or an axle; Fig. 9-10):

$$\tau = Fs \qquad (9\text{-}14)$$

Remember, the SI unit for torque is the newton meter (N · m), and the USCS unit is the lb · ft.

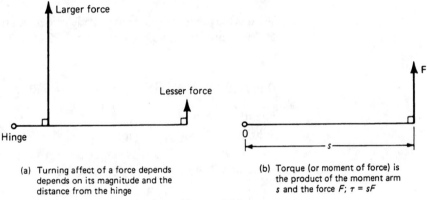

(a) Turning affect of a force depends depends on its magnitude and the distance from the hinge

(b) Torque (or moment of force) is the product of the moment arm s and the force F; $\tau = sF$

Figure 9-10

One of the most important applications of rotary motion is the development and transmission of power in rotary systems. In generating stations almost all power is generated as rotary power, and we use flywheels, drive shafts, drive trains, and so on, to transfer rotary power from one location to another.

If some torque τ produces an angular displacement of θ in a rotating object in some elapsed time t, the work done

$$W = \tau\theta \qquad (9\text{-}15)$$

The average power that is delivered to the rotating system is therefore

$$P = \frac{W}{t} = \frac{\tau\theta}{t} \qquad (9\text{-}16)$$

But the angular velocity of the rotating object $\omega = \theta/t$; therefore, if the torque τ is constant,

$$P = \tau\omega \qquad (9\text{-}17)$$

or

$$\text{power} = \text{torque} \times \text{angular velocity}$$

Note the similarity to $P = Fv$.

EXAMPLE 9-13

Determine the power delivered by a motor that supplies a torque of 62.5 lb·ft if its angular velocity is 15.5 rev/s.

Solution

Data: $\tau = 62.5$ lb·ft, $\omega = 15.5$ rev/s $= 15.5 \times 2\pi$ rad/s $= 97.4$ rad/s, $P = ?$

Equation: $P = \tau\omega$

Substitute: $P = (62.5 \text{ lb·ft})(97.4 \text{ rad/s}) = 6090$ ft·lb/s
But 1 hp = 550 ft·lb/s; therefore,

$$P = \frac{6090}{550} \text{ hp} = 11.1 \text{ hp}$$

EXAMPLE 9-14

What is the power developed by an electric motor that runs at an angular velocity of 480 rpm if it delivers a torque of 9.75 N·m?

Solution

Data: $\tau = 9.75$ N·m, $\omega = 480 \dfrac{\text{rev}}{\text{min}} = 480 \times \dfrac{2\pi \text{ rad}}{60 \text{ s}} = 50.3$ rad/s, $P = ?$

Equation: $P = \tau\omega$

Substitute: $P = (9.75 \text{ N·m})(50.3 \text{ rad/s}) = 490$ W

If we require the answer in horsepower, since 1 hp = 746 W, 1 W = $\frac{1}{746}$ hp and $P = 490 \times \frac{1}{746}$ hp = 0.657 hp.

EXAMPLE 9-15

Determine the torque delivered by a motor **(a)** in SI units and **(b)** in USCS units if it develops 35.0 hp at 4800 rpm.

Solution

Data:
(a) $P = 35.0$ hp
   $= 35.0 \times 746$ W
   $= 2.61 \times 10^4$ W, $\tau = ?$
(b) $P = 35.0$ hp $= 35.0 \times 550$ ft·lb/s
   $= 1.93 \times 10^4$ ft·lb/s, $\tau = ?$

$\omega = 4800 \dfrac{\text{rev}}{\text{min}} = 4800 \times \dfrac{2\pi \text{ rad}}{60 \text{ s}} = 503$ rad/s

Equation: $P = \tau\omega$

Rearrange: $\tau = \dfrac{P}{\omega}$

Substitute:

(a) $\tau = \dfrac{2.61 \times 10^4 \text{ W}}{503 \text{ rad/s}} = 51.9$ N·m   **(b)** $\tau = \dfrac{1.93 \times 10^4 \text{ ft·lb/s}}{503 \text{ rad/s}}$
$= 38.3$ lb·ft

PROBLEMS

44. Determine the magnitude of the torque that is exerted on a spark plug when a 280 N force is applied to the end of a 35.0 cm torque wrench.
45. Calculate the torque on a flywheel of diameter 32.0 in. by a tangential force of 85.0 lb along its rim.
46. What output power is developed by an electric motor that delivers a 6.50 N·m torque at 375 rpm?
47. What torque must be applied to produce 2.50 hp of power at 725 rpm?
48. Determine the power developed by an engine that delivers a torque of 35.0 lb·ft at 360 rpm.
49. What is the power in watts of an electric motor that delivers a torque of 6.20 N·m at 8.50 rad/s?
50. If a motor delivers 0.75 hp at 1500 rpm, what torque does it develop?
51. What is the angular velocity of a motor that develops 1.15 hp with a torque of 260 lb·ft?
52. Determine the power in horsepower that is delivered by a motor that develops a torque of 95.0 N·m at **(a)** 1200 rpm; **(b)** 1500 rpm; **(c)** 3200 rpm.
53. Determine the torque in newton meters developed by a drive shaft when it delivers 2.50 hp at 1200 rpm.
54. Determine the torque produced by a motor that develops an output power of 425 W at 4500 rpm.
55. What is the angular velocity of a motor that develops an output power of 0.750 hp and delivers a torque of 17.5 lb·ft?
56. Determine the torque produced at a drill bit by a drill that develops and output power of 375 W at 1250 rpm.

SUMMARY

For an object in uniform circular motion of radius r and speed v, the *orbital period*

$$T = \frac{2\pi r}{v}$$

The *centripetal acceleration*

$$a_c = \frac{v^2}{r}$$

and the *centripetal force*

$$F_c = \frac{mv^2}{r} \quad \text{and} \quad F_c = \frac{wv^2}{gr}$$

For flat horizontal roads, the coefficient of friction required for a car moving at a speed v to complete a turn of radius r without skidding:

$$\mu = \frac{v^2}{rg}$$

where g is the acceleration due to gravity. The *optimum angle of bank* θ is given by

$$\tan \theta = \frac{v^2}{rg}$$

$$2\pi \text{ rad} = 1 \text{ rev} = 360°$$

The *angular displacement* θ is the angle through which an object rotates:

$$\theta = \frac{\text{arc length}}{\text{radius}} = \frac{s}{r} \text{ rad}$$

Angular velocity ω is the time rate of change of angular displacement:

$$\omega = \frac{\theta}{t} \quad \text{and} \quad v = r\omega$$

Angular acceleration α is the time rate of change of angular velocity:

$$\alpha = \frac{\omega_1 - \omega_0}{t} = \frac{\Delta\omega}{t}$$

and the tangential acceleration $a_t = r\alpha$.

QUESTIONS

1. Define the following terms: **(a)** period of a circular motion; **(b)** frequency of the motion; **(c)** centripetal acceleration; **(d)** angular displacement; **(e)** angular velocity; **(f)** angular acceleration; **(g)** torque.
2. Describe the principles behind a chemical centrifuge.
3. If an object moves in a circle at a constant speed, does it accelerate? Explain your answer.
4. Show that 1 rpm = $\frac{2\pi}{60}$ rad/s.
5. Describe the forces that allow cars to turn curves.
6. Explain how aircraft turn. [*Hint:* Draw a free-body diagram and include a force (lift) perpendicular to the wings.]
7. Explain why centripetal forces do no work on objects in circular motion.

REVIEW PROBLEMS

1. A 1500 kg car moves at 22.0 m/s around a circular curve of radius 175 m. What is **(a)** the centripetal acceleration; **(b)** the centripetal force?
2. Determine the speed of a satellite in its circular orbit of radius 9.75×10^5 ft and period 1.38 h.
3. A satellite traveling at 27 800 km/h is in a circular orbit of radius 6680 km about the earth. **(a)** What is the period of its orbit? **(b)** Find the centripetal acceleration.
4. A curve of radius 125 m is banked at an angle of 11° with the horizontal. Find the optimum speed at which a car can make the turn without skidding.
5. A curve of radius 585 ft is banked at 12° with the horizontal. At what optimum speed can cars make the turn without skidding?.
6. At what optimum angle must a curve of radius 476 ft be banked so that cars can turn the curve at 56.0 mi/h without skidding?
7. At what optimum speed should a car complete a turn of radius 125 m on ice if the road is banked at an angle of 12°?

Chap. 9 / Review Problems

8. A 2750 kg aircraft completes a circular turn of radius 125 m at 190 km/h. Determine (a) the time required to complete a 360° turn; (b) the centripetal force on the aircraft.

9. An aircraft performs a loop at a constant acceleration of 4.5 g (i.e., 4.5 times the acceleration due to gravity). What is the radius of the loop if the aircraft has a speed of (a) 650 km/h; (b) 425 mi/h?

10. Express 2.18 rev in (a) degrees; (b) radians.

11. Convert 2.86 rad to (a) revolutions; (b) degrees.

12. A flywheel requires 2.80 s to rotate through 42.0 rev. If its angular velocity after the 2.80 s is 112 rad/s, find (a) the initial angular velocity; (b) the average angular acceleration.

13. A flywheel rotating at 1600 rpm decelerates uniformly to rest in 26.0 s. Calculate (a) the number of revolutions that it makes while decelerating to rest; (b) the angular acceleration.

14. The drive shaft of a car rotating at 475 rpm accelerates uniformly at 24.0 rad/s^2 for 3.20 s. What is the final angular velocity?

15. An engine idling at 1250 rpm accelerates uniformly to 6400 rpm in 8.00 s. Determine (a) the angular acceleration; (b) the number of revolutions made in the 8.00 s.

16. A flywheel is accelerated at 4.50 rad/s^2 for 12.0 s to a final angular velocity of 1950 rpm. Determine (a) the initial angular velocity; (b) the number of revolutions made in the 12.0 s.

17. A generator shaft decelerates at 2.00 rad/s^2 for 5.00 rev to a final angular velocity of 600 rpm. (a) What was the initial angular velocity? (b) How long did it decelerate?

18. What is the power generated by a drive shaft turning at 3500 rpm if it delivers a torque of (a) 2200 N · m; (b) 1480 lb · ft?

19. Determine the power delivered by an electric motor that exerts a torque of 1180 lb · ft on a drum, turning it at 14.5 rpm.

20. What is the tangential speed of the tip of a propeller that rotates at 4200 rpm in a circular arc with a diameter of (a) 3.20 m; (b) 7.50 ft?

21. Determine the power, in horsepower, that is delivered by a motor that develops a torque of 115 N · m at (a) 800 rpm; (b) 1600 rpm; (c) 2800 rpm.

22. What is the torque delivered by a motor that develops 620 W at 88.0 rpm?

23. Determine the torque developed by a drive shaft that delivers 1.80 hp at 1500 rpm.

10
SIMPLE MACHINES

Our ability to construct and use tools and machinery has enabled us to dominate many of the physically more powerful forces in our environment. The development of machinery has contributed considerably to the rapid advances of many technologies.

A **machine** is a device that may be used to transmit or modify applied forces and energies. With machinery, we can use our own power or the energy from some fuel to lift heavy objects, bend or break the strongest of materials, travel at great speeds, and perform many other tasks that we would otherwise find impossible.

Even though many machines in use today appear to be quite complex, they are all composed of one or more of three basic simple machines: the lever and the inclined plane (Fig. 10-1) and the hydraulic press (Chapter 11).

10-1 MECHANICAL ADVANTAGE AND EFFICIENCY

We can perform certain tasks more easily in some ways than in others. For example, we may find that it is easier to pull down than to pull up on a rope, and it may be easier to push rather than to pull objects. We also find it easier to do a certain amount of work by exerting small forces through large distances, rather than large forces through small distances. Machinery is used to change the magnitude, direction, and point of application of required forces in order to make tasks easier.

A measure of a machine's ability to assist the user is called the **mechanical advantage** of that machine. If a total **effort** (input force) F_i must be applied to the machine to overcome a particular **load** (output or resistance force) F_o on the machine, the **actual mechanical advantage** (AMA) of the machine is the ratio of the magnitudes of the output force to the input force.

$$\text{AMA} = \frac{\text{output force}}{\text{input force}} = \frac{\text{load}}{\text{effort}} = \frac{F_o}{F_i} \qquad (10\text{-}1)$$

The output of useful work (or energy) from any machine can never exceed the

Sec. 10-1 / Mechanical Advantage and Efficiency

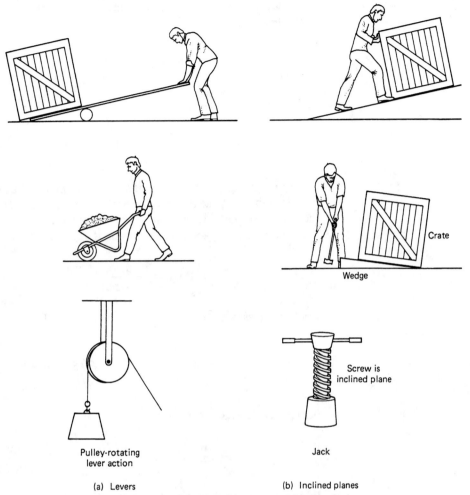

Figure 10-1 Examples of levers and inclined planes.

total input of work and energy. Even machines are not capable of perpetual motion! Thus the advantage gained by the difference between input and output forces is always lost elsewhere.

Friction between the moving parts of a machine always results in a loss of mechanical energy and power. In complex machines, which may have many moving parts, these losses may be very large. The effects of friction are accounted for in the **efficiency** (η) of the machine, which is defined as the ratio of the useful work or energy output W_o to the total work and energy supplied W_i.

$$\eta = \frac{\text{useful work output}}{\text{total work and energy input}} = \frac{W_o}{W_i} \qquad (10\text{-}2)$$

Thus if s is the distance moved by the effort (input force) F_i, h is the distance moved by the load (output force) F_o, and these forces act in the same directions as their displacements, then from the definition of work ($W = Fs \cos \theta$ with $\theta = 0$);

$$\eta = \frac{W_o}{W_i} = \frac{\text{load} \times \text{distance moved by load}}{\text{effort} \times \text{distance moved by effort}} = \frac{F_o h}{F_i s}$$

Rearranging this expression gives us

$$\eta = \frac{F_o/F_i}{s/h} \qquad (10\text{-}3)$$

For an ideal machine in which there are no frictional losses, the efficiency $\eta = 1$, and therefore in Eq. 10-3,

$$\frac{F_o/F_i}{s/h} = 1$$

and its mechanical advantage is

$$\text{MA} = \frac{F_o}{F_i} = \frac{s}{h}$$

Since the linkages and moving parts of a machine are usually rigid, the ratio of the distance moved by the effort to the useful distance moved by the load (s/h) represents the *maximum* possible mechanical advantage for that machine; it is called its **ideal mechanical advantage** (IMA) or **velocity ratio**.

$$\text{IMA} = \frac{\text{distance moved by effort}}{\text{distance moved by load}} = \frac{s}{h} \qquad (10\text{-}4)$$

Frictional losses are accounted for in the AMA. Thus the efficiency of any machine is

$$\eta = \frac{W_o}{W_i} = \frac{F_o/F_i}{s/h} = \frac{\text{AMA}}{\text{IMA}} \qquad (10\text{-}5)$$

or, expressed as a percentage,

$$\eta = \frac{W_o}{W_i} \times 100\% = \frac{F_o/F_i}{s/h} \times 100\% = \frac{\text{AMA}}{\text{IMA}} \times 100\% \qquad (10\text{-}6)$$

Remember, since the elapsed time is usually the same for input and output of work by the machine, the efficiency may also be expressed as the ratio of its output power P_o to the input power P_i:

$$\eta = \frac{P_o}{P_i} \quad \text{or} \quad \eta = \frac{P_o}{P_i} \times 100\% \qquad (10\text{-}7)$$

EXAMPLE 10-1

What is the efficiency of a machine that has an AMA of 6.45 and an IMA of 9.62?

Solution

Data: AMA = 6.45, IMA = 9.62, $\eta = ?$

Equation: $\eta = \dfrac{\text{AMA}}{\text{IMA}}$

Substitute: $\eta = \dfrac{6.45}{9.62} = 0.670$ or 67.0%

EXAMPLE 10-2

A simple machine is used to raise a 1710 lb load vertically through 12.0 ft at a constant velocity. If the input force (effort) of 281 lb is required to move through 92.0 ft, determine **(a)** the AMA; **(b)** the IMA; **(c)** the efficiency of the machine.

Sec. 10-1 / Mechanical Advantage and Efficiency

Figure 10-2
Free-body diagram of the load, which is in equilibrium since the velocity is constant.

Solution

Sketch: See Fig. 10-2.

Data: $F_o = w = 1710$ lb, $h = 12.0$ ft, $F_i = 281$ lb, $s = 92.0$ ft, AMA = ?, IMA = ?, $\eta = ?$

Equations: Since the load is in equilibrium (constant velocity), the output force balances the force of gravity (weight). Therefore, $F_o = w = mg$,

(a) AMA $= \dfrac{F_o}{F_i}$, (b) IMA $= \dfrac{s}{h}$, (c) $\eta = \dfrac{\text{AMA}}{\text{IMA}}$

(a) *Rearrange:* AMA $= \dfrac{F_o}{F_i} = \dfrac{w}{F_i} = \dfrac{mg}{F_i}$

Substitute: AMA $= \dfrac{1710 \text{ lb}}{281 \text{ lb}} = 6.09$

(b) *Substitute:* IMA $= \dfrac{92.0 \text{ ft}}{12.0 \text{ ft}} = 7.67$

(c) *Substitute:* $\eta = \dfrac{6.09}{7.67} = 0.794$ or 79.4%

EXAMPLE 10-3

A pump is used to raise water at a rate of 150 L/min from the bottom of an 80.0 m well. If the efficiency is 80%, what power is supplied to the pump?

Solution

Data: Since 1 L of water has a mass of 1 kg, the rate at which the water is raised

$$\frac{m}{t} = 150 \frac{\text{kg}}{\text{min}} = \frac{150 \text{ kg}}{60 \text{ s}} = 2.50 \text{ kg/s}$$

$g = 9.81$ m/s², $\Delta h = 80.0$ m, $\eta = 0.80$, $P_i = ?$

Equations: $P_o = \dfrac{W_o}{t}$, $\eta = \dfrac{P_o}{P_i}$

The work done equals the change in the potential energy of the water. Thus $W_o = \Delta E_P = mg\,\Delta h$.

Rearrange: $P_o = \dfrac{mg\,\Delta h}{t}$, $P_i = \dfrac{P_o}{\eta}$

Substitute: $P_o = (2.50 \text{ kg/s})(9.81 \text{ m/s}^2)(80.0 \text{ m}) = 1960$ W

Therefore,

$$P_i = \frac{P_o}{\eta} = \frac{1960 \text{ W}}{0.80} = 2450 \text{ W}$$

$$= \frac{2460 \text{ hp}}{746} = 3.29 \text{ hp}$$

since 746 W = 1 hp, and 1 W = $\dfrac{1 \text{ hp}}{746}$.

PROBLEMS

1. Determine the efficiency of a machine if its AMA is 3.5 and its IMA is 8.2.
2. If a machine has an AMA of 12.2, what load can it lift vertically with an input force of (a) 1650 N; (b) 375 lb?
3. A machine lifts a 225 kg load vertically through 85.0 cm. If the energy input to the machine is 2150 J, what is the efficiency?

4 How much input power is required to operate a laser with an output of 2.5 mW if its efficiency is 5.0%?

5 Find the missing quantities in Table 10-1.

TABLE 10-1

	Input force F_i	Load F_o	Distance s moved by effort	Distance h moved by load	Efficiency η	AMA	IMA
(a)	250 N	3500 N	35.0 m	1.50 m			
(b)	55.0 lb	775 lb	125 ft	4.50 ft			
(c)	75.0 N		26.0 m			4.50	7.20
(d)	15.0 lb		85.0 ft			4.75	8.10
(e)		8500 N		55.0 cm		3.60	5.45
(f)		1950 lb		11.5 ft		3.20	5.15
(g)	225 N			1.25 m	85%	7.50	
(h)	45.0 lb			4.00 ft	65%	6.10	
(i)		375 N	75.0 cm		65%		9.75
(j)		85.0 lb	16.0 ft		75%		8.70
(k)	450 N		8.75 m	55.0 cm	82%		
(l)	25 lb		285 ft	1.75 ft	83%		

6 A mechanical pump lifts 10.0 kg of water through a vertical height of 12.0 m in 18.0 s. If 110 W of input power is supplied to the pump, what is its efficiency?

7 How many U.S. gallons of water can a pump lift through 40.0 ft in 16.0 s if the pump efficiency is 68% and the input power to the pump is 0.250 hp?

8 A 1750 kg car climbs a 15° hill at a constant speed of 35.0 km/h. If only 55% of the total horsepower developed by the engine drives the wheels, determine the total horsepower developed by the engine.

9 A 5400 lb truck climbs a 12° hill at 30.0 mi/h. If only 65% of the total horsepower developed by the engine drives the wheels, determine the total horsepower developed by the engine.

10 If the efficiency is 65%, determine the minimum horsepower that must be supplied to a mechanical hoist to raise (a) a 1150 kg load through a vertical height of 1.20 m in 21.0 s; (b) a 3500 lb load through 5.00 ft in 25.0 s.

10-2 LEVERS

The **lever** is a very efficient, yet simple machine. It consists of a rigid bar that is free to rotate or pivot about a fixed point, called the **fulcrum**. Levers are classified according to the position P of the fulcrum with respect to the effort (input force) F_i and the load (output force) F_o (Fig. 10-3).

Friction losses are quite small in lever actions because of the small area of contact at the fulcrum. Therefore, when the lever is in equilibrium, the equilibrium conditions for nonconcurrent forces may be used to solve for unknown quantities.

If l_i and l_o are the lengths of the lever arms from the fulcrum to the effort and fulcrum to load, taking moments about the fulcrum for all three classes in Fig. 10-3, we obtain

$$\text{net } M_{\text{clockwise}} = \text{net } M_{\text{counterclockwise}}$$

or

$$l_o F_o = l_i F_i$$

Sec. 10-2 / Levers

Therefore,

$$\frac{F_o}{F_i} = \frac{l_i}{l_o} = \text{AMA} \tag{10-8}$$

Also, the IMA $= s/h$, and if friction is very small,

$$\text{IMA} = \frac{l_i}{l_o} = \text{AMA}$$

The weight of the lever itself is usually much less than the magnitudes of the other forces involved and it is usually ignored.

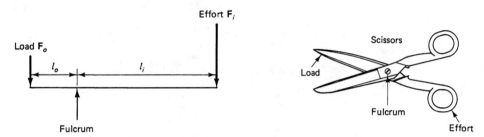

(a) First class: Fulcrum between the effort and the load; examples: scissors, pliers, valve rocker arms.

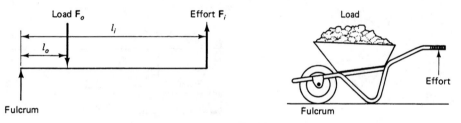

(b) Second class: load between fulcrum and effort; examples: wheelbarrow, nut crackers.

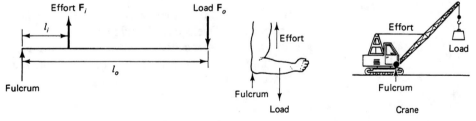

(c) Third class: effort between fulcrum and load; examples: human forearm, tweezers, boom of some cranes.

Figure 10-3 Classes of levers.

EXAMPLE 10-4

A man inserts one end of a 2.00 m long bar beneath a 325 kg rock and places a smaller rock, which acts as the fulcrum, 25.0 cm from the load (Fig. 10-4). **(a)** What is the minimum force that the man must apply to the bar to pry up the rock? **(b)** What is the IMA of this machine?

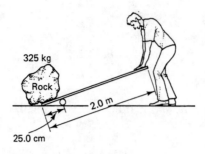

Figure 10-4

Solution

Sketch: See Fig. 10-5.

Data: $l_i = 1.75$ m, $l_o = 0.25$ m, $m = 325$ kg, $g = 9.81$ m/s², $F_i = ?$, IMA = ?

(a) Equations: net $M_{\text{clockwise}}$ = net $M_{\text{counterclockwise}}$, $M = sF$, $F_o = w = mg$

Rearrange: $l_i F_i = l_o F_o$, $F_i = \dfrac{l_o F_o}{l_i} = \dfrac{l_o mg}{l_i}$

Substitute: $F_i = \dfrac{(0.25 \text{ m})(325 \text{ kg})(9.81 \text{ m/s}^2)}{1.75 \text{ m}} = 455$ N

(b) Equations: IMA $= \dfrac{l_i}{l_o}$

Substitute: IMA $= \dfrac{1.75 \text{ m}}{0.25 \text{ m}} = 7.00$

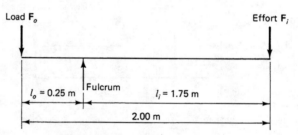

Figure 10-5 Free-body diagram.

EXAMPLE 10-5

A man uses an 8.00 ft long wooden stud as a second-class lever to pry a car free from the mud. One end A (the fulcrum) is placed under the car and the man then exerts a 168 lb perpendicular upward force on the other end (see Fig. 10-6). If the bumper of the car rests at a point 1.20 ft from the fulcrum, what is the force exerted on the car?

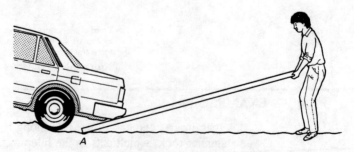

Figure 10-6

Sec. 10-2 / Levers

Solution

Sketch: See Fig. 10-7.
Data: $l_i = 8.00$ ft, $l_o = 1.20$ ft, $F_i = 168$ lb, $F_o = ?$
Equation: $l_i F_i = l_o F_o$
Rearrange: $F_o = \dfrac{l_i F_i}{l_o}$
Substitute: $F_o = \dfrac{(8.00 \text{ ft})(168 \text{ lb})}{1.20 \text{ ft}} = 1120$ lb

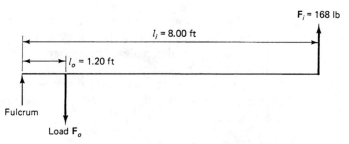

Figure 10-7 Free-body diagram.

PROBLEMS

11. A 2.00 m long crowbar of negligible weight is used as a first-class lever in order to raise a 125 kg load. If the fulcrum is located 32.0 cm from the load, **(a)** what minimum force must be applied to the other end of the crowbar; **(b)** what is the IMA?

12. A 7.50 ft long rod of negligible weight is used as a first-class lever in order to raise a 275 lb load. If the fulcrum is located 1.10 ft from the load, **(a)** what minimum force must be applied to the other end of the rod; **(b)** what is the IMA?

13. To lift the front of a tractor onto a block a farmer uses a 3.00 m long wooden stud as a first-class lever. He places one end A of the stud under the axle of the tractor and uses a rock as a fulcrum 42.0 m from A. If the farmer then exerts a force of 725 N on the other end of the stud, what is the upward force on the tractor?

14. An 82.5 kg man uses a piece of wood 2.65 m long, and of negligible weight, as a first-class lever to help raise a car that is stuck in the mud. One end of the wood is placed under the bumper of the car and a rock placed 35.0 cm from the bumper is used as the fulcrum. What upward force is exerted on the car when he stands on the other end of the wood?

15. A man uses an 8.00 ft long wooden stud of negligible weight as a first class lever to help raise a car that is stuck. One end of the wood is placed under the bumper of the car, and a rock, placed 1.00 ft from the bumper, acts as the fulcrum. If the man weighs 175 lb, what upward force is exerted on the car when he stands on the other end of the wood?

16. The center of gravity of a 125 lb load is located 1.35 ft from the axle (which acts as the fulcrum) of a wheelbarrow. **(a)** What minimum force must be exerted on the handles, 4.80 ft from the axle, in order to raise the load? **(b)** What is the IMA?

17. The center of gravity of a 65.0 kg load is located 42.0 cm from the axle (which acts as the fulcrum) of a wheelbarrow. **(a)** What minimum force must be exerted on the handles, 1.45 m from the axle, in order to raise the load? **(b)** What is the IMA?

18. A uniform steel rod with a mass of 30.0 kg and length 2.25 m is used as a second-class lever to lift a crate weighing 1250 N. What force must be exerted at the end opposite the fulcrum if the center of gravity of the crate is 32.0 cm from the fulcrum?

19. A 125 kg load is to be raised using a 2.50 m long rod of negligible weight. If the input force is to be 225 N, how far from the load should the fulcrum be located if the rod is

used as (a) a first-class lever; (b) a second-class lever? (c) Can a third-class lever be used in this case?

20 A 275 lb load is to be raised using an 8.00 ft long wooden stud of negligible weight. If the input force is 50.0 lb, how far from the load should the fulcrum be located if the stud is used as (a) a first-class lever; (b) a second-class lever? (c) Can a third-class lever be used in this case?

10-3 PULLEYS

A **pulley** is a wheel that is pivoted so that it is free to rotate about some axle at its center. The circumference of a pulley is often grooved or notched to give it greater hold on belts or other pulleys. If a pulley is fastened to the load, it is called **movable**; if it is fastened to a fixed point, it is called a **fixed pulley.** Systems of one or more pulleys can be used to change the direction and magnitude of required forces and also to transmit torques. Pulleys have many applications as hoists, belt drives, drive shafts, and gears.

Wheel and Axle

The **wheel and axle** consists of a large diameter pulley (the **wheel**) and a small-diameter pulley (the **axle**), which are joined and rotate together about the same axis. The effort F_i is supplied to a the wheel, and the load F_o is applied at the axle (Fig. 10-8). The ideal mechanical advantage is given by

$$\text{IMA} = \frac{R}{r} \tag{10-9}$$

where R and r are the radii of the wheel and axle, respectively.

Examples of devices employing the wheel and axle principle are winches, steering wheels, door knobs, and screwdrivers.

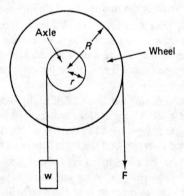

Figure 10-8 Wheel and axle.

EXAMPLE 10-6

A winch has an axle with a radius of 12.0 cm and a handle that turns in a radius of 32.0 cm. If the efficiency is 92%, determine (a) the IMA; (b) the AMA; (c) the force required to lift a 95.0 kg load.

Solution

Data: $r = 12.0$ cm, $R = 32.0$ cm, $\eta = 0.92$, $m = 95.0$ kg, $g = 9.81$ m/s^2, IMA = ?, AMA = ?, F = ?

(a) Equation: $\text{IMA} = \dfrac{R}{r}$

Substitute: $\text{IMA} = \dfrac{32.0 \text{ cm}}{12.0 \text{ cm}} = 2.67$

(b) Equation: $\eta = \dfrac{\text{AMA}}{\text{IMA}}$

Rearrange: $\text{AMA} = \eta(\text{IMA})$

Substitute: $\text{AMA} = (0.92)(2.67) = 2.45$

(c) Equations: $w = mg$, $\text{AMA} = \dfrac{F_o}{F_i} = \dfrac{w}{F_i}$

Rearrange: $\text{AMA} = \dfrac{mg}{F}$, $F = \dfrac{mg}{\text{AMA}}$

Substitute: $F_i = \dfrac{(95.0 \text{ kg})(9.81 \text{ m/s}^2)}{2.45} = 380$ N

A single fixed pulley is often used to change the direction of the input force so that the user can pull down rather than up on a belt (chain or rope). In this case, since equal distances are moved by the effort and the load, the IMA = 1. If friction effects and the mass of the pulley are negligible, the tension in the belt must be the same on each side of the pulley (Fig. 10-9). However, if a movable single pulley is attached to the load, two strings pull upward on the load. In this case, to raise the load through 1.0 m, the continuous string must be pulled through 2.0 m. Therefore, the input force must move twice as far as the load (see Fig. 10-10). Therefore, in this case the IMA = 2.

In general, this applies to all pulley systems:

IMA = number of strings pulling on the load

In the free-body diagram (Fig. 10-10), we see that two strings pull upward on the load. If the weight of the pulley is neglected and the system is in equilibrium, the tension in the belt is equal to half the weight of the load.

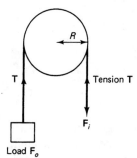

Figure 10-9 Single pulley.

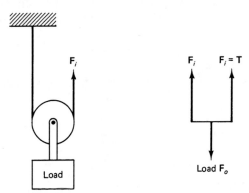

Figure 10-10 Two strings pull on the load if the load is attached to a single pulley.

Block and Tackle

The **block and tackle** is a more complex system of groups of pulleys, called **sheaves**, which are connected by the same belt or string. The load is suspended from the lower sheave (Fig. 10-11). Usually, it is desirable that the effort pull down. Therefore, if there are equal numbers of pulleys in each sheave, the belt is attached to the upper sheave. However, if the top sheave has more pulleys, the belt is attached to the lower sheave. It is also possible to have the effort pulling upward as in Fig. 10-12.

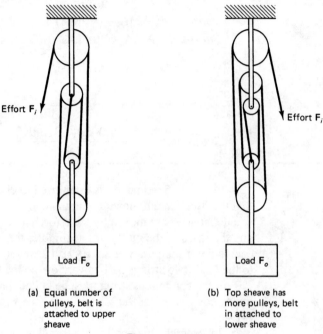

(a) Equal number of pulleys, belt is attached to upper sheave

(b) Top sheave has more pulleys, belt in attached to lower sheave

Figure 10-11

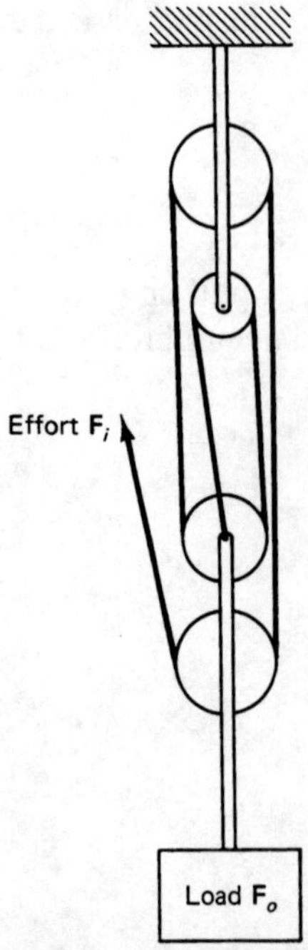

If friction and the weights of the strings and pulleys are neglected (since each pulley merely changes the direction of the force), the tension **T** is the same magnitude in each string and is numerically equal to the input force (effort) F_i. The IMA is determined by considering the tensions in the strings, which act on the lower sheave, or by the principle of work. For example, in the system in Fig. 10-11, each string that acts on the lower sheave pulls up on the load with a force equal in magnitude to the effort F_i. Therefore, in equilibrium, since four strings act on the lower sheave,

$$F_o = w = 4F_i$$

Since friction is ignored, the efficiency is 100% and

$$\text{AMA} = \frac{F_o}{F_i} = \text{IMA} = 4$$

Similarly, in Fig. 10-12, since five strings pull up on the load, the IMA = 5. In general,

$$\text{IMA} = \text{number of strings that pull on the lower sheave} \qquad (10\text{-}10)$$

Figure 10-12 Arrangement so that the effort pulls upward.

Sec. 10-3 / Pulleys

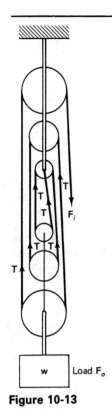

Figure 10-13

EXAMPLE 10-7

The block and tackle system illustrated in Fig. 10-13 is used to raise a 472 lb steel girder through 25.0 ft. If the efficiency is 94%, determine **(a)** the IMA; **(b)** the amount of belt that is pulled through the system; **(c)** the minimum effort required.

Solution

Data: $w = F_o = 472$ lb, $h = 25.0$ ft, $\eta = 94\%$, IMA = ?, s = ?, F_i = ?

(a) *Equation:* IMA = number of strings pulling on the load
Substitute: IMA = 6

(b) *Equation:* $\text{IMA} = \dfrac{s}{h}$

Rearrange: $s = h(\text{IMA})$
Substitute: $s = (25.0 \text{ ft})(6) = 150$ ft

(c) *Equations:* $\eta = \dfrac{\text{AMA}}{\text{IMA}}$, $\text{AMA} = \dfrac{F_o}{F_i} = \dfrac{w}{F_i}$

Rearrange: $\eta = \dfrac{w/F_i}{\text{IMA}}$; therefore, $F_i = \dfrac{w}{\eta(\text{IMA})} = \dfrac{mg}{\eta(\text{IMA})}$

Substitute: $F_i = \dfrac{472 \text{ lb}}{(0.94)(6)} = 83.7$ lb

Chain Hoist

A **chain hoist (Weston differential pulley)** consists of an upper block of two pulleys (one with a larger diameter than the other) and a single lower pulley to which the load F_o is attached (Fig. 10-14). An "endless belt" passes over the larger A of the upper

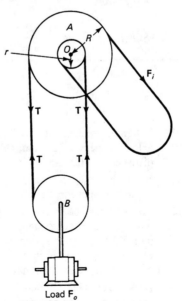

Figure 10-14 Chain hoist.

block pulleys, around the lower pulley B and then around the smaller upper block pulley C, leaving some slack. The effort F_i is applied as shown. If friction and the weight of the belt and pulleys are neglected and the system is in equilibrium, the tensions **T** in the belt support the load F_o. Thus

$$2T = F_o = w$$

If R and r are the radii of the larger and smaller pulleys of the upper block,

$$\text{IMA} = \frac{F_o}{F_i} = \frac{2R}{R - r} \qquad (10\text{-}11)$$

EXAMPLE 10-8

A chain hoist that has upper pulleys of diameters 22.0 cm and 18.0 cm and an efficiency of 95% is used to raise a 175 kg machine through 3.50 m. **(a)** How much of the belt must be pulled through the system by the applied force? **(b)** What minimum applied force is required?

Solution

Data: $D = 22.0$ cm or $R = 11.0$ cm, $d = 18.0$ cm or $r = 9.0$ cm, $\eta = 95\%$, $m = 175$ kg, $h = 3.50$ m, $s = ?$, $F_i = ?$

(a) *Equation:* $\text{IMA} = \dfrac{s}{h} = \dfrac{2R}{R-r}$

Rearrange: $s = h(\text{IMA})$

Substitute: $\text{IMA} = \dfrac{2(11.0 \text{ cm})}{11.0 \text{ cm} - 9.0 \text{ cm}} = 11.0$

Therefore, $s = (3.50 \text{ m})(11.0) = 38.5$ m.

(b) *Equations:* $\eta = \dfrac{\text{AMA}}{\text{IMA}} = \dfrac{F_o/F_i}{\text{IMA}}$, $F_o = w = mg$

Rearrange: $F_i = \dfrac{F_o}{\eta(\text{IMA})} = \dfrac{mg}{\eta(\text{IMA})}$

Substitute: $F_i = \dfrac{(175 \text{ kg})(9.81 \text{ m/s}^2)}{(0.95)(11.0)} = 164$ N

PROBLEMS

21 Determine the IMA of a wheel and axle **(a)** if the wheel has a diameter of 48 cm and the axle a diameter of 5.0 cm; **(b)** if the wheel has a diameter of 28.0 in. and the axle a diameter of 2.00 in.

22 A winch has a 15.0 cm long handle and a drum of radius 5.00 cm. Determine the minimum input force required to lift a 35.0 kg load if the efficiency is 100%.

23 **(a)** Determine the IMAs of the pulley systems in Fig. 10-15. **(b)** How much belt must be pulled through each system to raise a load through 8.00 m? **(c)** How much belt must be pulled through each system to raise a load through 12.5 ft? **(d)** What minimum effort must be applied in each case to lift a 125 kg load if the efficiency is 95%? **(e)** If the efficiency is 92%, what maximum loads can each system raise using a 135 lb effort?

24 Draw the possible combinations of the following pulley systems and find the IMA in each case: **(a)** two fixed and one movable; **(b)** two fixed and two movable; **(c)** two fixed and three movable; **(d)** three fixed and three movable; **(e)** four fixed and three movable. (*Hint:* There may be two answers for some depending on where the string is attached.)

Sec. 10-3 / Pulleys

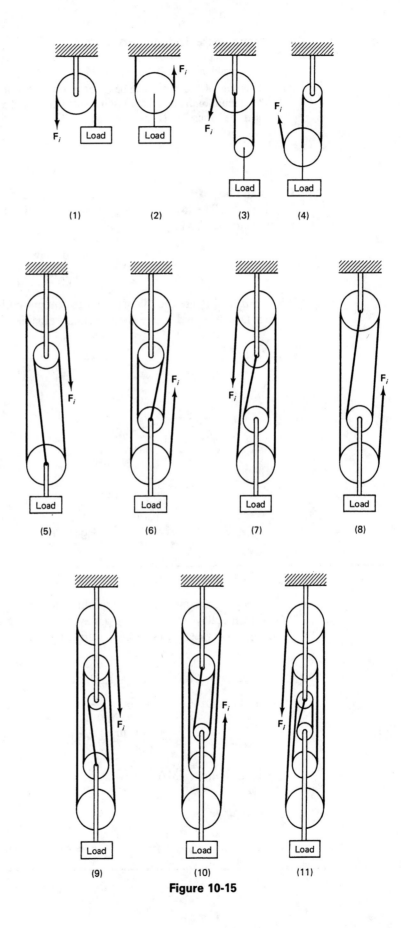

Figure 10-15

25. Determine the minimum force required to raise a 225 kg load through 35.0 cm and the amount of the belt that must be pulled through the system for each pulley in Problem 24 if the efficiency is (a) 100%; (b) 95%.

26. Determine the load that can be raised through 15.0 ft by an applied force of 75.0 lb and the amount of the belt that must be pulled through the system for each pulley in Problem 24 if the efficiency is (a) 100%; (b) 92%

10-4 GEARS

Gears and **belt-driven pulleys** (Fig. 10-16) are used extensively to transmit torques in machinery. The IMA in these systems is the ratio of the output torque τ_o to the input torque τ_i to the driving gear or pulley:

$$\text{IMA} = \frac{\text{output torque}}{\text{input torque}} = \frac{\tau_o}{\tau_i} \quad (10\text{-}12)$$

Also, if d_o and d_i are the diameters of the output pulley (or gear) and driving pulley (or gear), and ω_o and ω_i are their angular velocities,

$$\text{IMA} = \frac{d_o}{d_i} = \frac{\omega_i}{\omega_o} \quad (10\text{-}13)$$

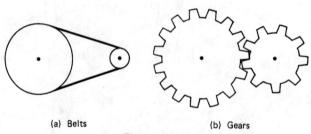

(a) Belts (b) Gears

Figure 10-16

EXAMPLE 10-9

A belt drive connects a 3.00 in. diameter driving pulley and a 15.0 in. diameter driven pulley. If the driving pulley rotates at 3200 rpm, determine (a) the IMA; (b) the angular velocity of the driven pulley.

Solution

Data: $d_i = 3.00$ in., $d_o = 15.0$ in., $\omega_i = 3200$ rpm, IMA = ?, $\omega_o = ?$

(a) Equation: $\text{IMA} = \dfrac{d_o}{d_i}$

Substitute: $\text{IMA} = \dfrac{15.0 \text{ in.}}{3.00 \text{ in.}} = 5.00$

(b) Equation: $\text{IMA} = \dfrac{\omega_i}{\omega_o}$

Rearrange: $\omega_o = \dfrac{\omega_i}{\text{IMA}}$

Substitute: $\omega_o = \dfrac{3200 \text{ rpm}}{5.00} = 640 \text{ rpm}$

Sec. 10-4 / Gears

Gears are essentially pulleys that have notches around their circumference to give them a greater hold on other gears (Fig. 10-16). They are capable of delivering large torques because the teeth of the gears intermesh, and there is no slippage of a belt or chain involved. The rate of rotation may be controlled by fixing the relative numbers of teeth in the interacting gears. The **gear ratio** is the ratio of the numbers of teeth in two interacting gears (Fig. 10-17):

$$\text{gear ratio} = \frac{\text{number of teeth in one gear}}{\text{number of teeth in the other gear}} \qquad (10\text{-}14)$$

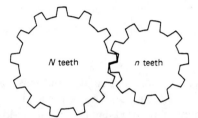

Figure 10-17 Gears. Gear ratio = N/n.

It should be noted that when two gears intermesh, they turn in opposite directions. If the output gear is required to rotate in the same direction as the driver gear, a third gear called the **idler** is inserted between them (see Fig. 10-18).

Figure 10-18 Idler.

If a gear with a large number of teeth is used to drive a gear with a smaller number of teeth, the rotation rate is increased by the gear ratio.

When gears, chains, and belt drives are used to transmit mechanical power from one rotating shaft to another, these shafts may be parallel or inclined at angles to each other. During the transmission of power, a gear may be used to increase or decrease the torque or angular velocity of the rotation. However, increases in torque result in decreases in the angular velocity, and vice versa.

If gear 1 has N_1 teeth, rotates at an angular velocity ω_1, and develops a torque τ_1, and gear 2 has N_2 teeth, rotates at ω_2, and develops a torque τ_2, the gear ratio between them

$$\text{gear ratio} = \frac{N_1}{N_2} = \frac{\omega_2}{\omega_1} = \frac{\tau_1}{\tau_2} \qquad (10\text{-}15)$$

EXAMPLE 10-10

A large gear with 85 teeth interacts with a smaller gear with 25 teeth. If the larger gear rotates at 125 rpm, determine **(a)** the gear ratio; **(b)** the angular velocity of the smaller gear.

Solution

Data: $N = 85$, $n = 25$, $\omega_{\text{larger}} = 125$ rpm, G.R. = ?, $\omega_{\text{smaller}} = ?$

(a) *Equation:* $\text{G.R.} = \dfrac{N}{n}$

Substitute: $\text{G.R.} = \dfrac{85}{25} = 3.4$

(b) *Equation:* $\text{G.R.} = \dfrac{\omega_{\text{smaller}}}{\omega_{\text{larger}}}$

Rearrange: $\omega_{\text{smaller}} = \text{G.R.}(\omega_{\text{larger}})$

Substitute: $\omega_{\text{smaller}} = (3.4)(125 \text{ rpm}) = 425$ rpm

---

Spur gears are cylindrically shaped and may have teeth on either the inner or outer surface (Fig. 10-19a and b). They are used to connect shafts that have parallel axes of rotation. Their main disadvantage is that the load is transferred suddenly from one tooth to another. This type of gear is used extensively in all kinds of machinery.

Helical gears have teeth that are not parallel to the shaft (Fig. 10-19c). As a result, there is no sudden transfer of the load. Consequently, these gears are quieter and stronger than spur gears, and they can operate at higher angular velocities.

Miter and **bevel gears** are used to connect shafts that are not parallel. They are basically cone shaped and have either straight or spiral teeth (Fig. 10-19d and e). The alignment of these cone-shaped gears specifies the angle of intersection of the shafts. The gear is called a **miter gear** when there is an equal number of teeth and the shafts are perpendicular.

Worm gears are mainly used to increase the torque (decrease the angular velocity) between perpendicular shafts (Fig. 10-19f and g). These gears may have a single or multithread worms. Their gear ratio is

$$\text{gear ratio} = \frac{\text{number of teeth on gear}}{\text{number of teeth on the worm}} \tag{10-16}$$

and may be very high. However, their efficiency is relatively low because sliding rather than rolling actions are involved. Worm gears are frequently used in hoists and winches.

Gear Trains

Gear trains (Fig. 10-20), which are composed of a number of interacting gears, are used to change rotation rates by even greater amounts. The total gear ratio of the gear train is the product of the individual gear ratios of the interacting pairs of gears in that train.

$$\text{G.R.} = \frac{N_A}{n_P} \times \frac{N_B}{n_Q} \times \frac{N_C}{n_R} \times \frac{N_D}{n_S} \times \cdots = \frac{\omega_o}{\omega_i} \tag{10-17}$$

where the N's are the numbers of teeth on the *driving* gears, the n's are the numbers of teeth on the *driven* gears, and ω_i and ω_o are the angular velocities of the first driver and last driven gears.

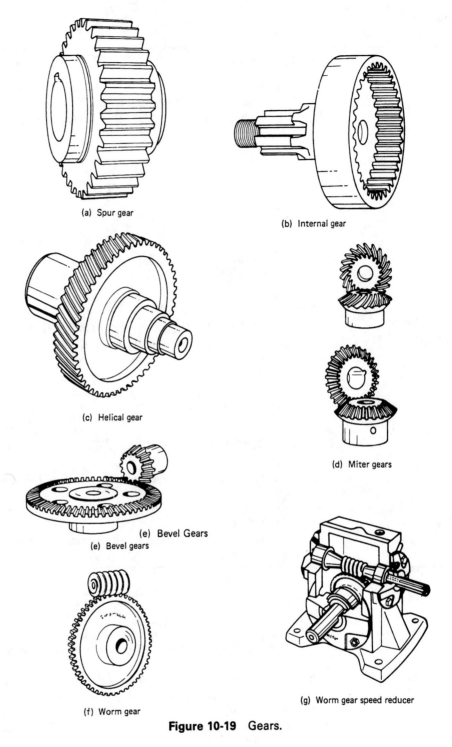

(a) Spur gear
(b) Internal gear
(c) Helical gear
(d) Miter gears
(e) Bevel Gears
(e) Bevel gears
(f) Worm gear
(g) Worm gear speed reducer

Figure 10-19 Gears.

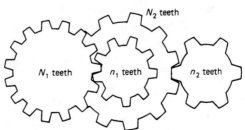

Figure 10-20 Gear train. Gear ratio of the gear train $= \dfrac{N_1}{n_1} \times \dfrac{N_2}{n_2}$.

EXAMPLE 10-11

In the gear train shown in Fig. 10-21, A has 15 teeth, B has 42 teeth, C has 12 teeth, D has 30 teeth, and E has 45 teeth. **(a)** If A rotates at 15.0 rpm, what is the angular velocity of E? **(b)** If A rotates clockwise, in which direction does E rotate?

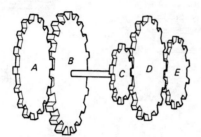

Figure 10-21

Solution

Data: Gears A, C, and D are drivers and B, D, and E are driven. Since D is both a driver and a driven gear, we shall see that it may be omitted from the calculation. $N_A = 15$ teeth, $n_B = 42$ teeth, $N_C = 12$ teeth, $n_D = N_D = 30$ teeth, $n_E = 45$ teeth, $\omega_A = 15.0$ rpm, $\omega_E = ?$

(a) Equation: $\dfrac{N_A \times N_C \times N_D}{n_B \times n_D \times n_E} = \dfrac{\omega_E}{\omega_A}$

Rearrange: $\omega_E = \omega_A \times \dfrac{N_A \times N_C \times N_D}{n_B \times n_D \times n_E}$

Substitute: $\omega_E = \dfrac{(15.0 \text{ rpm})(15)(12)(30)}{(42)(30)(45)} = 1.43 \text{ rpm}$

Note that D is both a driver and a driven gear, and its effect cancels in this expression.

(b) if A rotates clockwise, B with C rotate counterclockwise, D clockwise, and E counterclockwise.

PROBLEMS

27 A large gear with 65 teeth interacts with a smaller gear with 30 teeth. If the larger gear rotates at 150 rpm, determine **(a)** the gear ratio; **(b)** the angular velocity of the smaller gear.

28 A large gear with 75 teeth interacts with a smaller gear with 25 teeth. If the smaller gear rotates at 450 rpm, determine the gear ratio and the angular velocity of the larger gear.

29 A driver gear with 25 teeth rotates at 450 rpm and develops a torque of 1250 N·m. Determine the angular velocity and the torque developed by a driven gear if it has **(a)** 45 teeth; **(b)** 12 teeth; **(c)** 32 teeth.

30 In the gear train illustrated in Fig. 10-20, $N_1 = 65$, $n_1 = 12$, $N_2 = 45$, and $n_2 = 10$. If N_1 rotates with an angular velocity of 85.0 rpm, determine **(a)** the gear ratio; **(b)** the angular velocity of n_2.

31 In the gear train shown in Fig. 10-21, A has 20 teeth, B has 44 teeth, C has 10 teeth, D has 45 teeth, and E has 22 teeth. **(a)** If A rotates clockwise at 250 rpm, find the angular velocity and direction of C and D. **(b)** If C rotates clockwise at 875 rpm, find the angular velocitites and directions of rotation of A and E.

32 In the gear train shown in Fig. 10-22, A has 15 teeth, B has 48 teeth, C has 12 teeth, D has 32 teeth, E has 24 teeth, and F has 38 teeth. **(a)** If A rotates clockwise at 45.0 rpm, find the angular velocity and direction of rotation of F. **(b)** If C rotates clockwise at 875 rpm, find the angular velocities and directions of rotation of A and F.

Sec. 10-5 / Inclined Planes

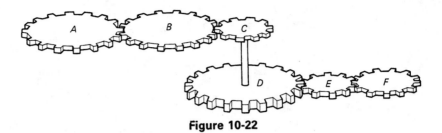

Figure 10-22

10-5 INCLINED PLANES

Normally, less force is usually required to push a load up a shallow incline than to lift it vertically; therefore, the incline itself is a form of a machine. Some common examples in which inclined planes are used as machines are ramps, staircases, wedges, screws, cams, and the hills of roads.

If a load F_o is pushed a distance s by a force F_i to a vertical height h (Fig. 10-23), then, from the definitions;

$$\text{AMA} = \frac{\text{load}}{\text{input force}} = \frac{F_o}{F_i}$$

and

$$\text{IMA} = \frac{\text{distance moved by the effort}}{\text{useful distance moved by the load}}$$

$$= \frac{s}{h} = \frac{1}{\sin \theta} \qquad (10\text{-}18)$$

where θ is the incline of the plane. Therefore, since the load is the weight of the object, the efficiency is

$$\eta = \frac{\text{AMA}}{\text{IMA}} = \frac{F_o/F_i}{s/h} = \frac{w/F_i}{s/h} = \frac{w \sin \theta}{F_i} \qquad (10\text{-}19)$$

Friction forces are always present to some degree between the incline and the load; they act parallel to the incline, resulting in efficiencies of less than 100%.

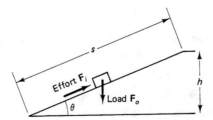

Figure 10-23

EXAMPLE 10-12

A person must exert a minimum force of 382 N in order to push a 145 kg load up a 12° incline. Determine **(a)** the AMA; **(b)** the IMA; **(c)** the efficiency of the inclined plane as a machine.

Solution

Data: $F_i = 382$ N, $m = 145$ kg, $\theta = 12°$, $g = 9.81$ m/s², AMA = ?, IMA = ?, $\eta = ?$

(a) *Equation:* $\text{AMA} = \dfrac{F_o}{F_i}$, $F_o = w = mg$

Rearrange: $\text{AMA} = \dfrac{mg}{F_i}$

Substitute: $\text{AMA} = \dfrac{(145 \text{ kg})(9.81 \text{ m/s}^2)}{382 \text{ N}} = 3.72$

(b) *Equation* $\text{IMA} = \dfrac{s}{h} = \dfrac{1}{\sin \theta}$

Substitute: $\text{IMA} = \dfrac{1}{\sin 12°} = 4.81$

(c) *Equation:* $\eta = \dfrac{\text{AMA}}{\text{IMA}}$

Substitute: $\eta = \dfrac{3.72}{4.81} = 0.774$ or 77.4%

Wedges, as machines, are commonly used as cutting instruments, such as knives and axes, or as inserts capable of raising large loads through small vertical displacements (Fig. 10-24). The IMA = s/h, but large friction losses between wedges and their loads usually give the wedge a relatively low efficiency as a machine; even so, it is still very useful. Rotary wedges find frequent uses (as **cams**) in complex machines. If r and R are the minimum and maximum distances, respectively, from the rim of the cam to the pivot, and the circumference of the cam is $2L$ (Fig. 10-25), then

$$\text{IMA} = \frac{\text{distance moved by effort}}{\text{useful distance moved by load}} = \frac{L}{R - r} \qquad (10\text{-}20)$$

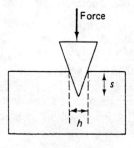

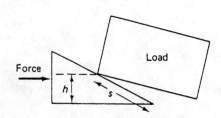

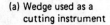

(a) Wedge used as a cutting instrument

(b) Wedge used as an insert to raise a large load

Figure 10-24 Wedges.

Screws and **screw jacks** consist of an inclined plane that spirals around a cylinder. The distance between successive crests of the thread is called the **pitch** p (Fig. 10-26). One complete rotation of the screw or screw jack moves the load through a distance equal to the pitch. Therefore, for a screw,

Figure 10-25 Cam.

$$\text{IMA} = \frac{\text{circumference of circle moved by screwdriver handle}}{\text{pitch of the thread}} \qquad (10\text{-}21)$$

Sec. 10-5 / Inclined Planes

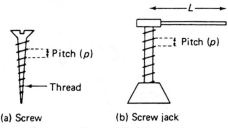

Figure 10-26

and for a jack,

$$\text{IMA} = \frac{2\pi L}{p} \tag{10-22}$$

where L is the length of the handle.

EXAMPLE 10-13

A screw jack with a thread pitch of 0.100 in. and a handle of length of 18.0 in. is used to lift a 3950 lb load. If the efficiency of the jack is 16%, what is the minimum force exerted at the end of the jack handle?

Solution

Data: $p = 0.100$ in., $L = 18.0$ in.; $F_o = w = mg = 3950$ lb, $\eta = 16\%$, $F_i = ?$

Equations: $\text{IMA} = \dfrac{2\pi L}{p}$, $\text{AMA} = \dfrac{F_o}{F_i} = \dfrac{w}{F_i}$, $\eta = \dfrac{\text{AMA}}{\text{IMA}}$

Rearrange: $\text{AMA} = \eta\,(\text{IMA})$, $\text{AMA} = \dfrac{w}{F_i}$

Therefore,

$$F_i = \frac{w}{\text{AMA}} = \frac{w}{\eta\,(\text{IMA})} = \frac{w}{\eta}\left(\frac{p}{2\pi L}\right)$$

Substitute: $F_i = \left(\dfrac{3950 \text{ lb}}{0.16}\right)\left(\dfrac{0.100 \text{ in.}}{2\pi \times 18.0 \text{ in.}}\right) = 21.8$ lb

---

To push a load up an inclined plane, the applied force must overcome not only the component of the load down the plane, but also the friction force between the load and that plane. Therefore, the efficiency of an inclined plane as a machine depends on the slope of the incline and the nature of the surfaces in contact.

PROBLEMS

33 A woman exerts a minimum force of 115 lb in order to push a 475 lb crate up a 10° incline. What is **(a)** the AMA; **(b)** the IMA; **(c)** the efficiency of the inclined plane as a machine?

34 A minimum force of 560 N is required to push a 115 kg load up a 20° inclined plane. Determine **(a)** the AMA; **(b)** the IMA; **(c)** the efficiency of the incline. If the load is raised vertically through 75.0 cm, determine **(d)** the work done on the load; **(e)** the energy lost to friction.

35 A screw jack with a thread pitch of 3.18 mm and a handle length of 55.8 cm is used to lift a 216 kg load. If the efficiency of the jack is 15%, determine the minimum force exerted at the end of the jack handle.

36 A screw jack with a thread pitch of 0.125 in. and a handle length of 22.0 in. is used to raise a 475 lb load. If the efficiency is 15%, determine the minimum force exerted at the end of the handle.

37 A screw with a pitch of 3.00 mm is driven into wood with a screwdriver that has a handle diameter of 2.20 cm. Determine the IMA.

10-6 COMPOUND MACHINES

Compound machines are constructed by combining two or more simple machines. The total efficiency of the compound machine is equal to the product of the efficiencies of its individual simple machine components.

EXAMPLE 10-14

A compound machine is constructed from a block and tackle which has an efficiency of 90%, and an inclined plane with an efficiency of 40%. What is the total efficiency of the combination?

Solution

Data: $\eta_1 = 90\%$, $\eta_2 = 40\%$, $\eta_{tot} = ?$
Equation: $\eta_{tot} = \eta_1 \times \eta_2$
Substitute: $\eta_{tot} = (0.90)(0.40) = 0.36$ or   36%

PROBLEM

38 A compound machine is constructed from a pulley system with an efficiency of 94% and an inclined plane with an efficiency of 35%. What is the total efficiency of the compound machine?

SUMMARY

A *machine* is a device used to transmit or modify forces and energies.

$$\text{AMA} = \frac{\text{output force}}{\text{input force}} = \frac{\text{load (or resistance)}}{\text{effort}} = \frac{F_o}{F_i}$$

Efficiency:

$$\eta = \frac{\text{work output}}{\text{work or energy input}} = \frac{W_o}{W_i}$$

Also,

$$\eta = \frac{\text{power output}}{\text{power input}} = \frac{P_o}{P_i}$$

$$\text{IMA} = \frac{\text{distance moved by effort}}{\text{distance moved by load}} = \frac{s}{h}$$

$$\text{Lever IMA} = \frac{l_o}{l_i} \qquad \text{Wheel and axle IMA} = \frac{\text{radius of wheel}}{\text{radius of axle}} = \frac{R}{r}$$

$$\text{Pulleys IMA} = \text{number of strings pulling on the load.}$$

$$\text{Chain hoist IMA} = \frac{2R}{R-r} \qquad \text{Belt drive IMA} = \frac{d_o}{d_i} = \frac{\omega_i}{\omega_o}$$

$$\text{Gears IMA} = \frac{d_o}{d_i} = \frac{\omega_i}{\omega_o} = \frac{N_o}{n_i}$$

$$\text{Gear trains: } \frac{\omega_o}{\omega_i} = \frac{N_A \times N_B \times N_C}{n_P \times n_Q \times n_R} \cdots$$

$$\text{Cam IMA} = \frac{L}{R-r} \qquad \text{Screw jack IMA} = \frac{2\pi L}{p}$$

QUESTIONS

1. Define the following terms: (a) machine; (b) actual mechanical advantage, (c) ideal mechanical advantage; (d) efficiency; (e) gear ratio; (f) gear train.
2. Discuss the advantages and disadvantages of attaching the belt to the upper or the lower sheave of a block and tackle pulley system.
3. What factors affect the efficiency of a machine? How are these factors reduced?
4. Discuss the advantages and disadvantages of large-diameter steering wheels in vehicles.
5. Why are gears and pulley drives used?
6. Explain what happens when you change gears in a car.
7. List the following simple machines in order of usual efficiency: inclined planes, levers, and pulleys.
8. Can a mechanical advantage ever be less than 1? Explain your answer.
9. If you had a pole to use as a lever to raise a load, describe the conditions for each class of lever that you could use. Which class of lever would give the greater IMA? Explain your answer.

REVIEW PROBLEMS

1. Determine the efficiency of a machine if its AMA is 4.8 and its IMA is 6.2.
2. What is the output power from an electric motor if its efficiency is 75% and the input power is (a) 1.80 kW; (b) 1.20 hp?
3. A pump raises 15.0 L of water through a vertical height of 4.65 m in 2.00 s. If the input power to the pump is 1.20 hp, what is its efficiency?
4. A pump raises 3.85 US gal of water through a vertical height of 15.0 ft in 2.00 s. If the input power is 1.20 hp, what is the efficiency of the pump?
5. A 3.75 m long steel rod is used as a first-class lever to raise a 175 kg load. If the load is located at one end of the rod and the fulcrum is 32.0 cm from the load, (a) what minimum force must be applied to the other end of the rod to lift the load; (b) what is the IMA? Ignore the weight of the rod.
6. A 12.0 ft long rod of negligible weight is used as a first-class lever to raise a 375 lb load. If the load is located at one end of the rod and the fulcrum is 1.00 ft from the load, (a)

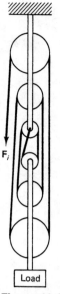

Figure 10-27

what minimum force must be applied to the other end of the rod to lift the load; **(b)** what is the IMA?

7 The center of gravity of an 85.0 kg load is located 45.0 cm from the axle of a wheelbarrow. **(a)** What minimum force must be exerted on the handles 1.50 m from the axle to raise the load? **(b)** What is the IMA?

8 A person uses a 3.20 m long pole as a second-class lever to pry a fallen tree from some location. One end A of the pole is placed under the tree which rests at a point 52.0 cm from A. If the person then exerts an upward 825 N force on the other end of the pole, what is the force exerted on the tree?

9 Determine the IMA of a wheel and axle if the wheel has a diameter of 78.0 cm and the axle diameter is 6.00 cm.

10 If the efficiency of the block and tackle in Fig. 10-27 is 95%, **(a)** what is the IMA; **(b)** what is the AMA; **(c)** how much work input is required to raise an 1800 kg car through a vertical height of 3.20 m at a constant speed; **(d)** how much work is required to raise a 5600 lb truck through a vertical height of 12.0 ft at a constant speed?

11 A large gear with 56 teeth interacts with a smaller gear with 12 teeth. If the smaller gear rotates at 3200 rpm, determine **(a)** the gear ratio; **(b)** the angular velocity of the larger gear.

12 Determine the gear ratio of a worm gear if the worm has 12 teeth and the gear has 35 teeth.

13 In the gear train shown in Fig. 10-28, A has 40 teeth, B has 12 teeth, C has 30 teeth, D has 40 teeth, E has 10 teeth, and F has 20 teeth. **(a)** What is the angular velocity and direction of rotation of F if A rotates clockwise at 120 rpm? **(b)** If E rotates clockwise at 1800 rpm, what are the directions of rotation and angular velocities of A and F?

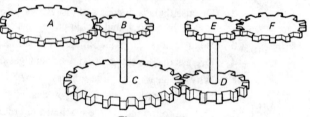

Figure 10-28

14 A compound machine is constructed from a lever system, with a total efficiency of 92%, and a series of inclined planes with a total efficiency of 28%. What is the efficiency of the compound machine?

15 Determine the IMA of a cam if its circumference is 16.0 cm, its larger radius is 3.20 cm, and its smaller radius is 1.40 cm.

16 Determine the IMA of a cam if its circumference is 7.20 in., its larger radius is 1.25 in. and its smaller radius is 0.620 in.

17 A screw jack has a thread pitch of 0.560 cm and a handle of length 75.0 cm. If the efficiency of the jack is 15%, what minimum force must be applied to the end of the handle to lift a 1500 kg car?

18 Determine the efficiency of a screw jack that has a pitch of 2.00 mm and a handle of length 72.0 cm if an applied force of 28.0 N is required to lift a 1450 kg car.

19 Determine the efficiency of a screw jack that has a pitch of 0.078 in. and a handle length of 28.0 in. if an applied force of 5.91 lb is required to lift a 3200 lb car.

20 A woman exerts a minimum force of 115 lb and pushes a 475 lb crate up a 12° incline. Determine **(a)** the AMA; **(b)** the IMA; **(c)** the efficiency of the inclined plane as a machine.

11 FLUIDS

The conditions of temperature and pressure determine whether the matter exists as a solid, liquid, or gas. All molecules of matter attract each other. The attraction between like molecules is called **cohesion;** the attraction between unlike molecules is called **adhesion.**

Solids tend to maintain a definite shape and volume because relatively strong cohesive forces fix the relative positions of the atoms. The cohesive forces between liquid molecules are not as strong and the molecules of a liquid are able to flow. A liquid will take the shape of any container to a height that depends on the volume. In general, a liquid will tend to have a level surface. It is not easily compressed because the molecules are quite close together. On the other hand, the molecules of a gas are relatively far apart and move freely between collisions. Consequently, a gas fully occupies any container and is easily compressed.

A **fluid** is defined as any matter that flows. Therefore, both liquids and gases are fluids. Many liquids and gases approximate perfect fluids because there are almost no forces retarding the flow of the molecules. Since we are surrounded by fluids, such as air and water, their uses and properties are of considerable importance. Aircraft, cars, and boats are streamlined so that they are able to move through fluids with less resistance. Pipelines and pumps are used to move fluids, such as gas, oil, and water, over large distances. In machinery, fluids, such as oil and air, reduce friction and wear between moving parts. Hovercraft even "float" on a cushion of air. Fluids are used as the agent to transfer heat in cooling and heating systems. Liquids are almost incompressible, and therefore they are often used in machinery, such as a hydraulic press, to transfer forces and gain a mechanical advantage.

11-1 PRESSURE

If a force is concentrated onto a small area, it usually has a greater effect than if it is spread out over a larger area. We describe the concentration of a force by a term called the **pressure** p, which is defined as the ratio of the magnitude F of the force to the perpendicular area A over which it acts.

$$p = \frac{F}{A} \tag{11-1}$$

Pressure is a scalar quantity that has units of newtons per square meter (N/m^2), which are given the special name **pascal** (symbol Pa) in SI ($1.0\ Pa = 1.0\ N/m^2$). The USCS units for pressure are pounds per square foot (lb/ft^2) and pounds per square inch (lb/in^2 or psi). Special units called **millibar** (mb) are also frequently used (1 mb = 100 Pa).

Pressures are caused by:

1. The application of a force to a particular area
2. The motion of the fluid molecules
3. The weight of a fluid above a surface area

EXAMPLE 11-1

A packing case weighing 1450 N is 1.75 m high and has a base with sides 1.00 m × 1.00 m. Determine the pressure that it exerts on the floor when it rests on **(a)** its base; **(b)** one of its sides.

Solution

Sketch: See Fig. 11-1.

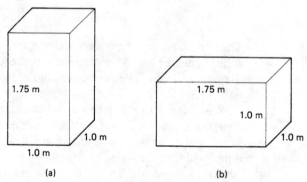

Figure 11-1

Data: $w = 1450$ N, $l = 1.00$ m, $b = 1.00$ m, $h = 1.75$ m, $p = ?$

Equations: $p = \dfrac{F}{A}$, **(a)** $A = lb$, **(b)** $A = bh$

Rearrange: **(a)** $p = \dfrac{F}{lb}$, **(b)** $p = \dfrac{F}{bh}$

Substitute: **(a)** $p = \dfrac{1450\ N}{(1.00\ m)(1.00\ m)} = 1450$ Pa

(b) $p = \dfrac{1450\ N}{(1.00\ m)(1.75\ m)} = 829$ Pa

Molecular Motion

The molecules of a gas are in continual and rapid motion, often colliding with each other and the walls of their container. It is the constant bombardment of the walls of the container that gives rise to pressure, because the molecules change their momentum as a result of these collisions. Thus there are reaction forces (according to Newton's third law) on the walls of the container.

Fluid Pressure

A fluid can exert significant forces which are distributed over the entire surfaces with which it is in contact. These forces also act perpendicular to the surfaces. A fluid cannot withstand any component of force parallel to its surface, because the fluid molecules would simply flow in response to such a force. For example, water exerts forces on the bottom and sides of any container, such as a bathtub, and these forces are perpendicular to the surfaces (Fig. 11-2); they are due to the weight of the water.

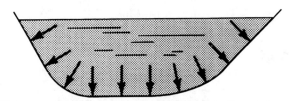

Figure 11-2 Forces act perpendicular to the surfaces.

EXAMPLE 11-2

Determine the total force exerted on an area of 8.65 ft² at a point in a liquid where the pressure is 13.8 lb/in².

Solution

Data: $p = 13.8 \dfrac{\text{lb}}{\text{in}^2} = \dfrac{13.8 \text{ lb}}{(\frac{1}{12} \text{ ft})^2} = 1990 \text{ lb/ft}^2$ since 1 ft = 12 in., $A = 8.65$ ft²,
$F = ?$

Equation: $p = \dfrac{F}{A}$

Rearrange: $F = pA$

Substitute: $F = (1990 \text{ lb/ft}^2)(8.65 \text{ ft}^2) = 17\,200$ lb

Atmospheric Pressure

Gases are easily compressed; therefore, their density varies with their depth, because the weight of the gas above compresses that below (Fig. 11-3). The weight of the earth's atmosphere produces a significant pressure over the entire surface of the earth. This atmosphere is "thinner" (less dense) at the higher altitudes than at the earth's surface. The air pressure at sea level on the earth's surface is approximately equal to 1 **standard atmosphere** (atm); that is, 101.325 kPa or 14.7 lb/in².

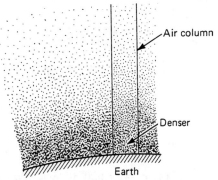

Figure 11-3 The weight of the atmosphere produces a pressure of about 101 kPa (14.7 lb/in²) at the earth's surface.

EXAMPLE 11-3

Determine the total force exerted by the earth's atmosphere on a perpendicular area of (a) 25.0 cm² and (b) 4.00 in², at a location where the pressure is 1.00 atm (101 kPa or 14.7 lb/in²).

Solution

Data:
(a) $p = 101 \times 10^3$ Pa, $A = 25$ cm² $= 25.0 \times 10^{-4}$ m², $F = ?$
[Since 1 cm $= 10^{-2}$ m, 1 cm² $= (10^{-2}$ m$)^2 = 10^{-4}$ m²]
(b) $p = 14.7$ lb/in², $A = 4.00$ in², $F = ?$

Equation: $p = \dfrac{F}{A}$

Rearrange: $F = pA$

Substitute:
(a) $F = (101 \times 10^3$ Pa$)(25.0 \times 10^{-4}$ m²$) = 253$ N
This is equivalent to the weight of a 25.8 kg mass at the earth's surface.
(b) $F = (14.7$ lb/in²$)(4.00$ in²$) = 58.8$ lb
Of course, we do not feel these forces because the pressures inside our bodies balance the atmospheric pressure.

Liquid Pressure

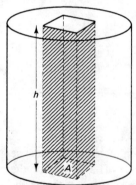

Figure 11-4 Pressure due to the liquid at a depth h. $p = \rho g h$; it is due to the weight of the liquid column.

Liquids are essentially incompressible, and their density remains approximately constant with depth. Consider a horizontal area A at a depth h in a liquid of density ρ (Fig. 11-4). The force F exerted on A by the liquid is equal to the weight w of the liquid directly above the area A. Since the volume of the liquid column $V = Ah$, and the density ρ is the ratio of mass m to the volume V, the mass of the liquid column:

$$m = \rho V = \rho A h$$

This column exerts a force of magnitude

$$F = w = mg = (\rho A h)g$$

on A; hence the pressure due to the liquid column is

$$p = \frac{F}{A} = \rho g h = Dh \qquad (11\text{-}2)$$

where D is the weight density and g is the acceleration due to gravity.

EXAMPLE 11-4

In 1960 the vessel *Trieste* carried a crew of two to the greatest known ocean depth of 11.0 km (6.83 mi). If it is assumed that the density of seawater was constant at 1040 kg/m³ (64.9 lb/ft³), what pressure due to the water (a) in SI units and (b) in USCS units did the *Trieste* withstand:

Solution

Data:
(a) $h = 11\,000$ m, $\rho = 1040$ kg/m³, $g = 9.81$ m/s², $p = ?$
(b) $h = 6.84$ mi $= 6.83 \times 5280$ ft, $D = 64.9$ lb/ft³, $p = ?$

Sec. 11-1 / Pressure

Equation (a) $p = \rho g h$ (b) $p = Dh$
Substitute:
(a) $p = (1040 \text{ kg/m}^3)(9.81 \text{ m/s}^2)(11\,000 \text{ m})$
    $= 1.12 \times 10^8$ Pa
(b) $p = (64.9 \text{ lb/ft}^3)(6.83 \times 5280 \text{ ft})$
    $= 2.34 \times 10^6 \text{ lb/ft}^2$ or $16\,300 \text{ lb/in}^2$

If there is also an external pressure p_o at the surface (Fig. 11-5), it must be added in order to obtain the total pressure p at any depth h in the liquid.

$$p = p_o + \rho g h = p_o + Dh \qquad (11\text{-}3)$$

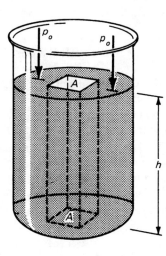

Figure 11-5 The total pressure at a depth h is the sum of the surface pressure p_o and the weight w per unit surface area A of the liquid column.

EXAMPLE 11-5

(a) If the surface of water in a reservoir is 85.0 m above a tap and the atmospheric pressure at the reservoir is 101 kPa, calculate the absolute water pressure at the tap. (b) If the tap nozzle has a diameter $d = 1.25$ cm, what is the force exerted on the tap?

Solution

Data: The density of water $\rho = 1.00 \times 10^3 \text{ kg/m}^3$, $d = 1.25$ cm, $h = 85.0$ m, $p = 101$ kPa $= 101 \times 10^3$ Pa, $g = 9.81 \text{ m/s}^2$, $p = ?$, $F = ?$
(a) Equation: $p = p_o + \rho g h$
Substitute: $p = 101 \times 10^3 \text{ Pa} + (1.00 \times 10^3 \text{ kg/m}^3)(9.81 \text{ m/s}^2)(85.0 \text{ m})$
$\qquad\qquad = 9.35 \times 10^5$ Pa
(b) Equations: $p = \dfrac{F}{A}$, $A = \dfrac{\pi d^2}{4}$
Rearrange: $F = pA = \dfrac{p \pi d^2}{4}$
Substitute: $F = (9.35 \times 10^5 \text{ N/m}^2) \dfrac{\pi (1.25 \times 10^{-2} \text{ m})^2}{4} = 115$ N

The height of the surface of water in the reservoir above the outlet is called the **pressure head**.

Pressure exists at all points in a liquid and has the same value at all points located at the same depth, regardless of the size of the container and the orientation of the surface. If this were not so, the liquid would simply flow until the pressure equalized. Stationary liquids tend to find a common surface level throughout any container (Fig. 11-6).

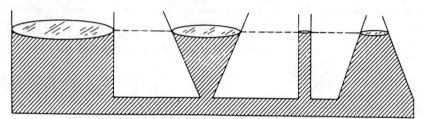

Figure 11-6 Stationary liquids tend to find a common surface level.

PROBLEMS

1. A 17 500 N packing crate has sides of lengths 1.50 m × 75.0 cm × 80.0 cm. Determine the pressure that it exerts on the floor when it rests on each of its three different sides.

2. Determine the pressure exerted on the floor by a 1950 lb packing crate if it rests on its base, which has dimensions of 3.95 ft × 2.80 ft.

3. Determine the total force exerted on a rectangular area of dimensions 75.0 cm × 50.0 cm, where the pressure is 135 kPa.

4. Determine the total force exerted on an area of 0.80 m^2 at a point in a liquid where the pressure is 95.0 kPa.

5. Determine the force exerted on an area of 5.40 in^2 at a point in a liquid where the pressure is 29.0 lb/in^2.

6. A nuclear submarine can operate at ocean depths of 700 m (2300 ft). If it is assumed that the density of seawater is constant at 1035 kg/m^3 (64.6 lb/ft^3), what is the total pressure on the hull (a) in SI units; (b) in USCS units?

7. What is the density of a liquid that exerts a pressure of 150 kPa at a depth of 23.0 m?

8. Determine the pressure at a point in a fluid where there is a total force of 151 lb on an area of 5.00 in^2.

9. Determine the total force on an area of 1.32 in^2 at a point in a fluid where the pressure is 45.9 lb/in^2.

10. A water container with vertical sides is 8.50 m high and has a bottom with an area of 1.20 m^2. Determine (a) the maximum mass of water that it could contain; (b) the maximum pressure that the bottom must withstand when it is filled with water.

11. Determine the pressure in SI and USCS units at the bottom of a 1.65 m (5.41 ft) deep tank due to the liquid if it is filled with (a) water; (b) mercury; (c) a liquid with a relative density (specific gravity) of 2.4.

12. Determine the pressure on the hull of a diving bell at a depth of 2500 m below the surface of seawater, which has an average relative density of 1.03. Assume that atmospheric pressure is 101 kPa.

13. The surface of water in a reservoir is 80.0 m above a tap, and the atmospheric pressure at the reservoir is 101 kPa. Calculate (a) the water pressure at the tap; (b) the force that the water exerts on the tap if the nozzle has a diameter of 1.20 cm.

14. The surface of water in a reservoir is 279 ft above a tap, and the atmospheric pressure at the reservoir is 14.7 lb/in^2. Determine (a) the water pressure at the tap; (b) the force that the water exerts on the tap if the nozzle has a diameter of 0.500 in.

15. Determine the density of a liquid that exerts a pressure of 725 kPa at a depth of 72.5 m.

16. What pressure would be required to raise a column of water to a height of (a) 1.25 m; (b) 3.00 ft?

17 Determine the height of a column of (a) water, (b) mercury, and (c) gasoline that can be supported by a pressure of 1.00 atm (101 kPa or 14.7 lb/in²).

18 A rectangular water tank with an open top has a depth of 2.00 m and a bottom with dimensions of 3.00 m and 2.50 m. If the atmospheric pressure is 101 kPa, determine (a) the pressure at the bottom of the tank; (b) the pressure on the sides of the tank 1.50 m below the water surface; (c) the total force on the bottom of the tank; (d) the total force on each vertical side of the tank.

19 Determine the water pressure at ground level required to supply water at a pressure of 240 kPa to the sixth floor of a building that is 23.5 m above the ground.

20 Determine the water pressure at ground level required to supply water at a pressure of 4.15 atm to the third floor of a building that is (a) 12.0 m above the ground; (b) 380 ft above the ground.

11-2 MEASUREMENT OF PRESSURES IN FLUIDS

Gas pressures may be measured with a variety of different devices. A **mercury barometer** measures pressures in terms of the height of a liquid column above an open surface (Fig. 11-7). The liquid used is usually mercury because of its high density and low vaporization at room temperatures. A simple mercury barometer may be constructed from a glass tube that is closed at one end. The tube is filled with mercury, the open end is sealed, the tube is inverted in an open mercury bath, and the seal is removed beneath the surface of the bath. The mercury column then drops until it exerts a pressure that is equal to the magnitude of the external pressure at the surface of the bath. Except for a small amount of mercury vapor, this leaves a vacuum between the closed end of the glass tube and the top of the mercury column. Thus the external pressure $p = \rho g h$, where ρ is the density of mercury, g the acceleration due to gravity, and h the height of the mercury column. The pressure may also be quoted in terms of the height h of the mercury column (1 atm or 101 kPa is equivalent to 760 mm of mercury).

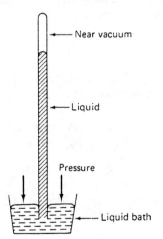

Figure 11-7 Barometer.

An **aneroid barometer** consists of one or more partially evacuated metallic capsules connected by levers to a cable that winds around a spindle to a spring (Fig. 11-8). A pointer is attached to the spindle and it moves over a calibrated scale. The capsules expand and contract as the air pressure varies. These changes are amplified by the system of levers so that there is a change in the tension of the cable, which produces a movement of the pointer over a scale.

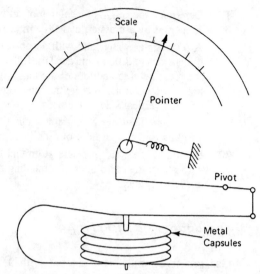

Figure 11-8 Aneroid barometer.

Since atmospheric pressure decreases with height above the earth's surface, a barometer may also be used to measure altitudes. A variation of the aneroid barometer, called a **pressure altimeter**, is calibrated in terms of altitudes rather than pressures, and it can easily be adjusted to the prevailing atmospheric pressure. These useful devices are found in all aircraft. However, they do not respond to rapid changes in altitude or to the terrain below the aircraft.

A **manometer** is a device that measures the difference between an unknown pressure and atmospheric pressure. This device consists of a U-shaped glass tube that is partially filled with mercury. One side of the tube is open to the unknown pressure p, and the other side is open to atmospheric pressure p_0. When the mercury is in equilibrium, the difference h in the height of mercury in both arms of the tube is related to the difference in pressure between p and p_0 (Fig. 11-9). If the mercury is higher on the side open to the atmospheric pressure,

$$p = p_0 + \rho g h = p_0 + Dh \qquad (11\text{-}4)$$

where ρ is the density, D the weight density, and g the acceleration due to gravity. Water is used instead of mercury in manometers that measure lower pressures, such as those in air ducts.

The value of atmospheric pressure p_0 may be determined with a barometer, and the unknown pressure may then be calculated.

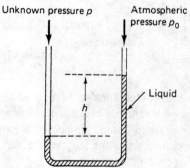

Figure 11-9 Manometer.

EXAMPLE 11-6

If the atmospheric pressure is 102 kPa, determine the pressure that would support a water column of 95.0 cm in a manometer.

Solution

Data: $p_0 = 102$ kPa, $\rho = 1000$ kg/m³, $g = 9.81$ m/s², $h = 0.950$ m, $p = ?$
Equation: $p = p_0 + \rho g h$
Substitute: $p = 102 \times 10^3$ Pa $+ (1000$ kg/m³$)(9.81$ m/s²$)(0.950$ m$)$
$= 111 \times 10^3$ Pa $= 111$ kPa

A **bourdon gauge** is frequently used to measure gas or steam pressures. It consists of a hollow tube that is bent into an arc; the end of the tube is connected via a linkage to a pointer (Fig. 11-10). When gas enters, the tube tends to straighten out under the pressure, and the motion of the tube is transmitted to the pointer.

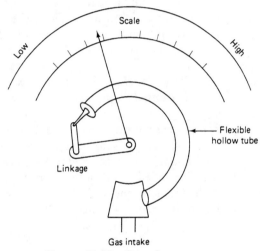

Figure 11-10 Bourdon gauge.

Gauge pressures indicate the magnitude of the pressure above atmospheric pressure. The total or **absolute pressure** is found by adding the atmospheric pressure to the gauge pressure:

$$\text{absolute pressure} = \text{gauge pressure} + \text{atmospheric pressure}$$
$$p = p_{\text{gauge}} + p_0 \tag{11-5}$$

EXAMPLE 11-7

What is the absolute pressure of a car tire if the gauge pressure is 245 kPa and atmospheric pressure is 103 kPa?

Solution

Data: $p_g = 245$ kPa, $p_0 = 103$ kPa, $p_{\text{abs}} = ?$
Equation: $p_{\text{abs}} = p_g + p_0$
Substitute: $p_{\text{abs}} = 245$ kPa $+ 103$ kPa $= 348$ kPa

PROBLEMS

21 What is the absolute pressure of a truck tire if the gauge pressure is 312 kPa and the atmospheric pressure is 105 kPa?

22 If the absolute pressure is 38.9 lb/in² and the atmspheric pressure is 14.8 lb/in², what is the gauge pressure?

23 A manometer is used to measure an unknown pressure. If the atmospheric pressure is 101 kPa, determine the unknown pressure if it supports a 120 cm column of water.

24 If the atmospheric pressure is 14.7 lb/in², determine the pressure that would support a water column of 37.4 in. in a manometer.

25 If the atmospheric pressure is 105 kPa, what is the magnitude of a pressure that supports a 65.0 cm column of water in a manometer?

26 Determine the unknown pressure in a manometer if the water level is 24.0 in. *below* the level on the side open to an atmospheric pressure of 15.0 lb/in².

11-3 PASCAL'S PRINCIPLE

Pressures in liquids may be produced by the weight of the liquid itself or by means of an external mechanical force. The basic principle regarding the transmission of pressures in liquids was discovered by Blaise Pascal (1623–1662). This principle is based on the fact that liquids are practically incompressible.

Pascal's principle: When a pressure is applied to a confined liquid, that pressure is transmitted throughout the extent of the liquid.

If this were not true, a pressure difference would exist, and the liquid would simply flow until the pressure equalized.

Liquids have the ability to transmit pressures, and they are used in a basic machine called a **hydraulic press** (Fig. 11-11). If a force F_1 is applied perpendicular to the area A_1 of piston 1, it produces a pressure p which is transmitted throughout the liquid and produces a force F_2 on the area A_2 of piston 2. Since the pressure is the same,

$$p = \frac{F_1}{A_1} = \frac{F_2}{A_2} \qquad (11\text{-}6)$$

Therefore,

$$\frac{F_2}{F_1} = \frac{A_2}{A_1} \qquad (11\text{-}7)$$

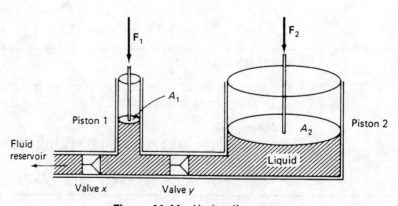

Figure 11-11 Hydraulic press.

Sec. 11-3 / Pascal's Principle

If a load F_2 is placed on the larger piston 2, it may be held in equilibrium by a smaller effort F_1 applied to the smaller piston 1, as long as the system can withstand the pressure. Therefore, the hydraulic press may be used to gain a mechanical advantage. The actual mechanical advantage is

$$\text{AMA} = \frac{\text{load}}{\text{effort}} = \frac{F_2}{F_1} = \frac{A_2}{A_1} \quad (11\text{-}8)$$

The efficiency of a hydraulic press is usually relatively high; it depends on the compressibility of the liquid and the losses in pressure due to leakage in the valves and along the sides of the pistons. If the effort F_1 moves through a distance s_1 and raises the load F_2 at the larger piston 2 a distance s_2, the ideal mechanical advantage is

$$\text{IMA} = \frac{\text{distance moved by effort}}{\text{distance moved by load}} = \frac{s_1}{s_2} \quad (11\text{-}9)$$

Therefore, for an **ideal machine** in which the efficiency is 100%,

$$F_2 s_2 = F_1 s_1$$

or

$$\frac{F_2}{F_1} = \frac{s_1}{s_2} = \frac{A_2}{A_1} \quad (11\text{-}10)$$

The load may be moved through greater distances by a series of "pumps" by the effort on piston 1 using valves to introduce more liquid to the confined area. To raise the load, valve x is closed and valve y is opened. The effort moves piston 1 down, forcing fluid through valve y as the load is raised by a relatively small amount. Valve y is then closed to prevent fluid from returning to the smaller cylinder, and valve x is opened. The smaller piston 1 a is then raised again, drawing fluid from the reservoir. Valve x is then closed, valve y is opened, and the process is repeated.

EXAMPLE 11-8

The smaller piston of a hydraulic press has a diameter $d_1 = 5.00$ cm, and the larger piston has a diameter $d_2 = 30.0$ cm. (a) If the hydraulic press has an efficiency of 100%, what minimum force is required to lift a 1600 kg car? (b) If the smaller piston can be pumped through 15.0 cm, how many times must it be pumped to raise the car 3.00 m?

Solution

Data: $d_1 = 5.00$ cm, $d_2 = 30.0$ cm, $s_2 = 3.00$ m. The mass of the car $m = 1600$ kg; therefore, the output force on the larger piston is the weight of the car, $F_2 = w = mg$, $F_1 = ?$

(a) Equations: $\dfrac{F_2}{F_1} = \dfrac{A_2}{A_1}$, $A = \dfrac{\pi d^2}{4}$

Rearrange: $F_1 = \dfrac{F_2 A_1}{A_2} = mg \left(\dfrac{\pi d_1^2 / 4}{\pi d_2^2 / 4} \right) = mg \left(\dfrac{d_1^2}{d_2^2} \right)$

Substitute: $F_1 = (1600 \text{ kg})(9.81 \text{ m/s}^2) \left[\dfrac{(5.00 \text{ cm})^2}{(30.0 \text{ cm})^2} \right] = 435$ N

(b) Equations: $\dfrac{s_1}{s_2} = \dfrac{A_2}{A_1}$, $A = \dfrac{\pi d^2}{4}$

Rearrange: $s_1 = s_2 \left(\dfrac{\pi d_2^2/4}{\pi d_1^2/4} \right) = s_2 \left(\dfrac{d_2^2}{d_1^2} \right)$

Substitute: $s_1 = (3.00 \text{ m}) \left[\dfrac{(30.0 \text{ cm})^2}{(5.00 \text{ cm})^2} \right] = 108$ m

But the length of each stroke of the smaller piston is 15.0 cm; therefore, it must be pumped

$$\dfrac{108 \text{ m}}{0.150 \text{ m}} = 720 \text{ times}$$

A **hydraulic jack** (Fig. 11-12) combines a hydraulic press with a second-class lever. For the lever,

$$\text{AMA} = \dfrac{F_1}{F} = \dfrac{l_i}{l_o}$$

Therefore, the total AMA is

$$\text{AMA}_{tot} = (\text{AMA})_{lever}(\text{AMA})_{press}$$

$$= \dfrac{F_1 F_2}{F F_1} = \dfrac{F_2}{F}$$

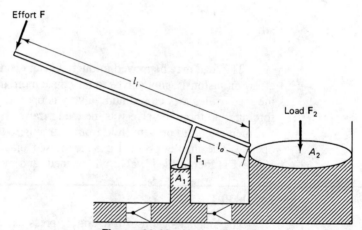

Figure 11-12 Hydraulic jack.

PROBLEMS

27. Determine the AMA of a hydraulic press if a 66.5 N force is required to lift a 205 kg load.
28. If the AMA of a hydraulic press is 28, what applied force is required to lift **(a)** a 225 kg load; **(b)** a 750 lb load?
29. If the AMA of a hydraulic jack is 280, what minimum applied force is required to lift **(a)** a 1500 kg car; **(b)** a 7500 lb truck?
30. What air pressure must be supplied to the intake pipe of a hydraulic lift to lift **(a)** a 2250 kg car if the diameter of the output piston is 25.0 cm; **(b)** a 5750 lb car if the diameter of the output piston is 12.0 in.?
31. Determine the AMA of a hydraulic press that raises a 1350 kg load when the input force is 525 N.
32. The AMA of a hydraulic jack is 225. **(a)** What is the load that can be raised by a 35.0 N input force? **(b)** What minimum input force is required to raise a 2500 kg load?
33. If the smaller piston of a hydraulic press has a diameter of 2.50 cm, what must the diameter of the larger piston be for a 75.0 N input force to raise a 1750 kg load?

Sec. 11-4 / Buoyancy and Archimedes' Principle

34 The smaller piston of a hydraulic press has a diameter of 3.00 in. and the larger piston diameter is 2.50 ft. If the efficiency of the press is 100%, determine the minimum force that will balance a 4000 lb car.

35 If the area of the small piston of a hydraulic press is 4.50 cm² and that of the larger piston is 35.0 cm², what minimum force must be applied to the small piston to lift a 1500 N load if the efficiency is 100%?

36 The smaller piston of a hydraulic press has a diameter of 2.25 in. and the larger a diameter of 1.80 ft. Determine the minimum force required to balance a 4000 lb car if the efficiency is 100%.

37 The smaller piston of a hydraulic press has a 2.00 cm diameter, and the larger piston has a 20.0 cm diameter. (a) If the press has a 100% efficiency, what minimum force is required to lift a 2100 kg load? (b) If the smaller piston can be pumped through 12.5 cm, how many times must it be pumped to lift the load 2.50 m? (c) How long does it take to raise the load if the pump delivers an average power 5.00 kW?

38 The smaller piston of a hydraulic press has a diameter of 4.00 cm, and the larger piston has a diameter of 80.0 cm. (a) If the hydraulic press has an efficiency of 100%, determine the minimum force required to lift a 1500 kg load. (b) If the smaller piston can be pumped through 20.0 cm, how many times must it be pumped to raise the load 2.50 m?

39 A hydraulic press has a smaller piston of diameter 2.50 in. and a larger piston with a diameter of 1.50 ft. (a) If the efficiency is 100%, determine the minimum force required to lift a 6000 lb load. (b) If the smaller piston can be pumped through 8.00 ft, how many times must it be pumped to raise the load 6.00 ft? (c) What minimum power should the pump deliver if the load must be raised 6.00 ft in 12.0 s?

11-4 BUOYANCY AND ARCHIMEDES' PRINCIPLE

Objects appear to weigh less in water than in air. Both air and water exert an upward **buoyant force** on the object, but the object has a greater buoyancy in the denser water.

The Greek mathematician Archimedes (287–212 B.C.) first discovered the law that governs buoyancy.

Archimedes' principle: When an object is in a fluid, it experiences a buoyant force (loss in apparent weight) equal to the weight of the fluid that is displaced. An object floats when it displaces its own weight in the fluid.

Consider a balloon filled with water and immersed in a pool of water. It will be in static equilibrium at any depth, and therefore its weight must be balanced by an upward (buoyant) force from the surrounding water. This buoyant force must still exist even if the contents of the balloon are replaced with some other substance. Similarly, a boat floats by displacing its own weight of water. If more loads are added to the boat, it sinks lower in order to displace more water to balance the extra load.

The relative density (specific gravity) of a material

$$\rho_r = \frac{\text{mass of the material}}{\text{mass of an equal volume of water}} \quad (11\text{-}11)$$

or

$$\rho_r = \frac{\text{weight of the material}}{\text{weight of an equal volume of water}} \quad (11\text{-}12)$$

If an object does not float in water,

$$\text{upward buoyant force } F = \text{weight of the water displaced}$$
$$= \text{loss of weight when immersed in water} \quad (11\text{-}13)$$

Therefore, if the object has a weight w in air, the relative density (specific gravity)

$$\rho_r = \frac{\text{weight of the substance}}{\text{loss of weight when immersed in water}} = \frac{w}{F}$$

EXAMPLE 11-9

A sample of brass weighs 10.800 lb in air and 9.552 lb when totally immersed in water. Determine (a) the buoyant force; (b) the relative density (specific gravity) of the metal; (c) the density of the metal.

Solution

Data: $w = 10.800$ lb, $w_w = 9.552$ lb, $D_w = 62.4$ lb/ft^3, $F = ?$, $\rho_r = ?$, $D = ?$

(a) Equation: $F = w - w_w$
Substitute: $F = 10.800$ lb $- 9.552$ lb $= 1.248$ lb

(b) Equation: $\rho_r = \dfrac{w}{F}$

Substitute: $\rho_r = \dfrac{10.800 \text{ lb}}{1.248 \text{ lb}} = 8.654$

(c) Equation: $D = D_{\text{water}} \rho_r$
Substitute: $D = (62.4 \text{ lb/ft}^3)(8.654) = 540$ lb/ft^3

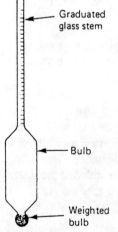

Figure 11-13
Hydrometer.

If the weight of an object is less than the buoyant force (the weight of fluid that its volume displaces), it rises. For example, a balloon filled with a low-density gas, such as helium, will rise in the air because its volume displaces a greater weight of air than its own weight. On the other hand, an object will sink when its weight is greater than the weight of the fluid that it displaces. A steel girder will not float on water.

Archimedes' principle is also used in submarines. In order to dive, a submarine allows water to enter special tanks, increasing its weight until the weight exceeds the buoyant force; the submarine then sinks. The submarine rises by forcing air into the same tanks, clearing the water and reducing the total weight until it is less than the buoyant force.

A **hydrometer** is a device that uses Archimedes' principle of flotation to measure relative densities of liquids. It consists of a graduated glass stem with two bulbs, the lower of which is weighted to ensure that the hydrometer floats vertically (Fig. 11-13). Since the weight of the hydrometer is fixed, the volume of liquid that it displaces depends on the relative density of that liquid. The hydrometer floats with its stem higher above the surface in denser liquids. These devices are commonly used to measure specific gravities in radiators to check the level of antifreeze. They are also used to test the charge on a storage battery; the specific gravity is higher (about 1.3) when charged and lower (about 1.2) when discharged. It should be noted however that the specific gravity depends on temperature.

PROBLEMS

40 A metal block weighs 80.0 lb in air and 67.0 lb when completely immersed in water. Determine (a) the buoyant force; (b) the relative density of the metal; (c) the metal density; (d) the volume of the metal; (e) the mass of the water that it displaces.

41 A 12.5 kg concrete block weighs 69.3 N when completely immersed in water. Find (a) the buoyant force; (b) the relative density of the concrete; (c) the volume of the block.

42 Determine the buoyant force on a boat that displaces (a) 725 m^3 of water; (b) 7.85 × 10^5 L of water; (c) 2500 ft^3 of water; (d) 1.95 × 10^5 U.S. gal of water.

43 Determine the buoyant force on a 525 cm³ metal cylinder when it is completely immersed in **(a)** water; **(b)** gasoline; **(c)** benzene.

44 Determine the volume of water displaced by **(a)** a floating 45.0 t ship; **(b)** a floating 2.50×10^6 lb ship.

45 A rectangular block of wood is 40.0 cm wide, 60.0 cm long, and 30.0 cm deep. If its relative density is 0.70, and it floats with its 30.0 cm side vertically, determine **(a)** the length of the block above the surface of the water; **(b)** the minimum force required to completely submerge the block.

11-5 FLUID FLOW

The flow of fluids may be controlled by pipes and channels, such as sewers, drains, household plumbing, and oil, gas, and water pipelines. Weather variations are caused by the flow of hot air masses relative to cold air masses. The stresses caused by the flow of fluids, such as the current of a river or the wind, must be considered in the design of structures, such as bridges, dams, and buildings.

Fluids may be set in motion by forces produced by hydrostatic pressure or by pumps. Once they are in motion, the fluid molecules possess momentum, and a force is required to stop the motion. Therefore, fluid molecules are able to exert **dynamic pressures** due to their motions. In addition, the energies of the individual molecules still produce **hydrostatic pressures**; this is an internal property of a fluid. For example, if we were to move with the fluid, we would still experience the hydrostatic pressure due to the molecular motion, but we would detect no dynamic pressure.

A fluid is in **steady flow** if its velocity past any fixed position is a constant in time. The velocity may vary at different positions in the fluid, but every molecule has the same velocity when it passes some particular position. The fluid molecules tend to follow each other in paths called **streamlines**. These streamlines do not cross and they are constant if the fluid is in a steady flow.

The **flow rate** R of a fluid is the rate of volume flow of the fluid. If the fluid has a constant density and is in steady flow at a velocity v through some area A, the flow rate (volume V per elapsed time t) is

$$R = \frac{V}{t} = Av \qquad (11\text{-}14)$$

This equation is used to size ducts and pipes through which fluids must flow. For air-conditioning systems, for example, the flow rate depends on the volume of the room and the number of occupants. Normally, the conditioned space should be completely replenished with air approximately every 10 minutes, but the air speed should be slow enough to avoid excessive noise and drafts.

EXAMPLE 11-10

What minimum duct size is required to deliver 810 ft³ of air per minute at 15.0 ft/s to an air-conditioned room? Assume that the air density is constant.

Solution

Data: $R = 810 \, \dfrac{\text{ft}^3}{\text{min}} = 810 \, \dfrac{\text{ft}^3}{60 \text{ s}} = 13.5 \text{ ft}^3/\text{s}, \, v = 15.0 \text{ ft/s}, \, A = ?$

Equation: $R = Av$

Rearrange: $A = \dfrac{R}{v}$

Substitute: $A = \dfrac{13.5 \text{ ft}^3/\text{s}}{15.0 \text{ ft/s}} = 0.900 \text{ ft}^2$

EXAMPLE 11-11

Fuel oil is pumped at 5.70 m/s through a pipe of diameter 8.00 cm. If it is delivered through a nozzle with a diameter of 3.00 cm, determine (a) the speed of the oil as it emerges; (b) the rate of flow in liters per minute.

Solution

Data: $v_1 = 5.70$ m/s, $d_1 = 8.00$ cm, $d_2 = 3.00$ cm, $v_2 = ?$, $R = ?$

Equations: $R = Av$, $A = \dfrac{\pi d^2}{4}$, $R = $ constant

(a) Rearrange: $A_1 v_1 = A_2 v_2$, $v_2 = v_1 \dfrac{A_1}{A_2}$

Therefore

$$v_2 = v_1 \frac{\pi d_1^2/4}{\pi d_2^2/4} = v_1 \left(\frac{d_1^2}{d_2^2}\right)$$

Substitute: $v_2 = (5.70 \text{ m/s}) \left[\dfrac{(8.00 \text{ cm})^2}{(3.00 \text{ cm})^2}\right] = 40.5$ m/s

(b) Rearrange: $R = \left(\dfrac{\pi d^2}{4}\right)v$

Substitute: $R = \dfrac{\pi(8.00 \times 10^{-2} \text{ m})^2 (5.70 \text{ m/s})}{4} = 2.87 \times 10^{-2}$ m³/s

But 1.0 m³ = 1000 L, and 1.0 s = $\tfrac{1}{60}$ min; therefore,

$$2.87 \times 10^{-2} \frac{\text{m}^3}{\text{s}} = 2.87 \times 10^{-2} (1000 \text{ L})\left(\frac{60}{\text{min}}\right) = 1.72 \times 10^3 \text{ L/min}$$

PROBLEMS

46 Determine the number of cubic meters of air that can pass through a 425 cm² duct at 5.75 m/s in 10.0 s. Assume that the air has a constant density.

47 Determine the minimum duct size required to change the air in a 15 000 ft³ room completely every 10.0 min if the air moves at 3.00 ft/s. Assume that the density of the air is constant.

48 Determine the minimum-size air duct that is required to change the air in a 435 m³ room completely every 10.0 min if the air moves at 92.5 cm/s and the air density is constant.

49 How long does it take to fill a 1.50 ft³ tank with water from a faucet that has a diameter of 0.500 in. if the water is emitted with a speed of 5.00 ft/s?

50 Oil flows with a speed of 3.00 m/s through a pipe with a diameter of 40.0 cm. Determine the speed of the oil through a constriction in the pipe where the diameter tapers to 10.0 cm.

51 Oil flows at 10.5 ft/s through a pipe with a diameter of 16.0 in. What is the speed of the oil through a constriction in the pipe where the diameter is 4.00 in.?

52 Water flows through a 5.00 cm diameter fire hose at 5.75 m/s. If the diameter of the nozzle is 1.50 cm, (a) what is the speed of the water as it emerges; (b) what is the flow rate in liters per minute?

53 Fuel oil exits through a hose at 25.0 m/s. If the delivery rate through a nozzle is 1.50 kL/min, what is the nozzle diameter?

SUMMARY

A *fluid* is matter that flows.

$$Pressure = \frac{\text{force}}{\text{perpendicular area}}, \quad p = \frac{F}{A}$$

The units are pascals (Pa) in SI and pounds per square foot (lb/ft^2) or pounds per square inch (lb/in^2 or psi) in USCS.

$$\text{Pressure due to a liquid column } p = \rho g h = Dh$$

$$\text{Absolute pressure } p = p_0 + p_{\text{gauge}}$$

Pascal's principle: When a pressure is applied to a liquid, that pressure is transmitted throughout the liquid.

For a hydraulic press,

$$\text{AMA} = \frac{F_2}{F_1} = \frac{A_2}{A_1} \qquad \text{IMA} = \frac{s_1}{s_2}$$

Archimedes' principle: When an object is completely or partially immersed in a fluid, it experiences an upward (buoyant) force F equal to the weight of fluid displaced.

If an object weighs w in air and the buoyant force is F in a liquid, the *relative density* (*specific gravity*) of the object

$$\rho_r = \frac{w}{F}$$

The *flow rate* (volume V per elapsed time t) of a fluid flowing at a speed v through a pipe where the cross-sectional area is A is

$$R = \frac{V}{t} = Av$$

QUESTIONS

1. Define the following terms: **(a)** pressure; **(b)** cohesive force; **(c)** adhesive force; **(d)** a pascal; **(e)** absolute pressure; **(f)** gauge pressure; **(g)** buoyant force; **(h)** pressure head.
2. Explain why a dam is constructed so that it is thicker at the bottom than at the top.
3. How does a gas exert pressure on the walls of its container?
4. Why must deep-sea divers wear weighted belts?
5. Describe why atmospheric pressure decreases with increases in altitude above the earth's surface.
6. Why does a balloon filled with helium rise in air, whereas a water-filled balloon falls?
7. How can a ship made of steel (or even concrete) float on water?
8. Describe what happens to the waterline of a cargo ship when it is loaded. Explain this in terms of the buoyant force.
9. Explain why steel floats on mercury.
10. A hydrometer can be used to test the antifreeze level in a radiator or the charge on a storage battery. Explain how.

REVIEW PROBLEMS

1. Determine the force on a 4.00 in^2 plate at a point in a liquid where the pressure is 175 lb/in^2.
2. Determine the total force on the shell of a spherical diving bell of radius 4.20 m at a depth where the pressure is 2.20 MPa.

3. Find the force on a 225 cm² surface at a point in a liquid where the pressure is 875 kPa.
4. What is the force on a 35.0 in² surface at a point in a liquid where the pressure is 128 lb/in²?
5. At a certain depth in a liquid, the force on an area of 77.5 cm² is 4390 N. What is the pressure at that depth?
6. At a certain depth in a liquid, the force on an area of 12.0 in² is 984 lb. What is the pressure at that depth?
7. At what depth in water is the pressure equal to 2.50 atm?
8. A 785 kg packing case is 1.45 m high, 75.0 cm wide, and 1.20 m long. Determine the pressure that it exerts on the floor when it rests on each of its three different sides.
9. A 365 lb packing case is 4.75 ft high, 2.50 ft wide, and 4.00 ft long. Determine the pressure that it exerts on the floor when it rests on each of its three different sides.
10. Determine the force exerted on an area of 75.0 cm² at the earth's surface due to atmospheric pressure of 105 kPa.
11. Determine the pressure that would be required to raise a column of water to a height of (a) 2.50 m; (b) 7.50 ft.
12. What is the absolute pressure of a tire if the gauge pressure is 44.1 lb/in² and atmospheric pressure is 14.7 lb/in²?
13. If the atmospheric pressure is 102 kPa, what pressure would support a 95.0 cm column of water in a manometer?
14. If the atmospheric pressure is 14.5 lb/in², what pressure would support a 30.0 in. column of water in a manometer?
15. The surface of water in a reservoir is 77.5 m above a tap, and the atmospheric pressure at the reservoir is 99.5 kPa. (a) Calculate the water pressure at the tap. (b) What force does the water exert on the tap nozzle if it has a diameter of 1.25 cm?
16. The surface of water in a reservoir is 250 ft above a tap, and the atmospheric pressure at the reservoir is 14.5 lb/in². (a) Determine the water pressure at the tap. (b) What force does the water exert on the tap nozzle if it has a diameter of 0.500 in.?
17. Water at a pressure of 565 kPa is used to operate a hydraulic lift. If the piston area is 17.5 cm and the efficiency is 100%, what maximum load can be raised?
18. Water at a pressure of 82.0 lb/in² is used to operate a hydraulic lift. If the piston diameter is 9.00 in. and the efficiency is 100%, what maximum load can be raised?
19. The smaller piston of a hydraulic press has a diameter of 5.0 cm, and the larger piston diameter is 75 cm. (a) If the efficiency is 100%, determine the minimum force required to lift a 2500 kg load. (b) If the smaller piston can be pumped through 15 cm, how many times must it be pumped to raise the load 3.2 m?
20. A sample of metal weighs 22.5 N in air and 19.6 N when immersed in water. Determine (a) the buoyant force; (b) the relative density of the metal; (c) the density of the metal; (d) the volume of the metal.
21. Determine the relative density of a block of material if it weighs 28.4 lb in air and 21.6 lb when it is totally immersed in water.
22. A barge has a bottom that is 6.00 m wide and 21.0 m long. How much deeper will it sink when its load is increased by 2.20 t?
23. Water flows at 6.15 m/s through a fire hose which has an inner diameter of 6.20 cm. Determine (a) the rate in liters per minute at which water flows through the hose; (b) the speed that water leaves the hose through a nozzle of diameter 2.00 cm.

12
HEAT

The concepts of heat, thermal energy, and temperature are very important in science and technology. Technicians are frequently concerned with the effects of heat on materials that they use.

12-1 TEMPERATURE AND HEAT

For now, we assume that temperature is an indication of how hot or cold an object is. To establish a temperature scale, a minimum of two reproducible standard points are required. A series of graduations, called **degrees,** are then made between them.

In two common temperature scales, the **Celsius** temperature scale and the **Fahrenheit** temperature scale (Fig. 12-1), the properties of pure water at a pressure of one standard atmosphere are used to define the two reproducible points. The

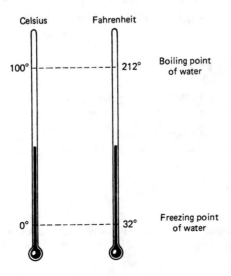

Figure 12-1

freezing point of water is taken as zero (0°C) on the Celsius scale and as thirty-two degrees (32°F) on the Fahrenheit scale. Similarly, the boiling point of water is taken as 100 degrees Celsius (100°C) and two hundred and twelve degrees Fahrenheit (212°F). The intervals between these points are then divided into even parts, which are extended both above the boiling point and below the freezing point. Even though degrees Celsius are not SI units, they are generally accepted for use with SI.

Since 0°C is equivalent to 32°F and 100°C is equivalent to 212°F, the difference between these points must be equal. Therefore, we have a relationship between **temperature changes**.

$$100°C - 0°C = 212°F - 32°F, \text{ or } 100°C = 180°F$$

Therefore,

$$5°C = 9°F \quad \text{(temperature changes)} \tag{12-1}$$

and

$$\Delta T_C = \frac{5}{9} \Delta T_F$$

where ΔT_C and ΔT_F are the changes in the Celsius T_C and Fahrenheit T_F temperatures, respectively. The actual conversion between the two temperature scales is given by

$$T_F = \tfrac{9}{5} T_C + 32° \tag{12-2}$$

and

$$T_C = \tfrac{5}{9}(T_F - 32°) \tag{12-3}$$

EXAMPLE 12-1

Convert 95°F to the equivalent Celsius temperature.

Solution

Data: $T_F = 95°F$
Equation: $T_C = \tfrac{5}{9}(T_F - 32)$
Substitute: $T_C = \tfrac{5}{9}(95 - 32)°C = 35°C$

EXAMPLE 12-2

Convert 75°C into the equivalent Fahrenheit temperature.

Solution

Data: $T_C = 75°C$
Equation: $T_F = \tfrac{9}{5} T_C + 32$
Substitute: $T_F = [\tfrac{9}{5}(75) + 32]°F = 167°F$

The SI unit for temperature is called the **kelvin** (symbol K);* it is a unit of **absolute temperature** (see Chapter 13). The relationship between kelvin and Celsius temperatures is approximately

$$T_K = T_C + 273 \tag{12-4}$$

---

*The kelvin scale is derived from thermodynamics. Note that the degree symbol is not used with kelvin temperatures; however, a single space should be left between the number and the unit.

Sec. 12-1 / Temperature and Heat

EXAMPLE 12-3

Convert **(a)** 20°C into a kelvin temperature; **(b)** 175 K into Celsius temperature.

Solution

Data:
(a) $T_C = 20°C$, $T_K = ?$ **(b)** $T_K = 175$ K, $T_C = ?$
Equation: $T_K = T_C + 273$
(a) *Substitute:* $T_K = (20 + 273)$ K $= 293$ K
(b) *Rearrange:* $T_C = T_K - 273$
Substitute: $T_C = (175 - 273)°C = -98°C$

Note that changes in temperature in kelvin are numerically equal to changes in temperature in Celsius.

$$\Delta T_K = \Delta T_C \tag{12-5}$$

For example, a temperature *change* from 300 K (27°C) to 312 K (39°C) is 12 K or 12°C.

The USCS unit for absolute temperature is called the **Rankine** temperature. It is related to the Fahrenheit temperature by the formula

$$T_R = T_F + 460 \tag{12-6}$$

EXAMPLE 12-4

Convert **(a)** 26°F into a Rankine temperature; **(b)** 385°R into a Fahrenheit temperature.

Solution

Data:
(a) $T_F = 26°F$, $T_R = ?$ **(b)** $T_R = 385°R$, $T_F = ?$
Equation: $T_R = T_F + 460$
(a) *Substitute:* $T_R = (26 + 460)°R = 486°R$
(b) *Rearrange:* $T_F = T_R - 460$
Substitute: $T_F = (385 - 460)°F = -75°F$

Changes in temperature in Rankine are numerically equal to changes in temperature in Fahrenheit.

$$\Delta T_R = \Delta T_F \tag{12-7}$$

For example, a temperature change from 520°R (60°F) to 600°R (140°F) is 80°R or 80°F.

Many physical quantities, such as volume, pressure, electrical resistance, and even color, vary with temperature. In fact, we often use these variations to measure the temperature. For example, the change in volume (due to thermal expansion) of mercury with temperature is used to measure temperature with the common mercury in glass thermometer.

All substances are composed of atoms. Even if an object is stationary, its atoms are in continual random motion. The atoms possess kinetic energy $E_k = \frac{1}{2}mv^2$ because of that motion, and potential energy due to their mutual interactions. The total energy of all the atoms in a mass of material is known as the **thermal energy** or **internal energy**.

Temperature is a measure of the average kinetic energy of the individual molecules in a substance. However, the thermal or internal energy of a substance is related to the total energy of all molecules. Therefore, the thermal energy depends on the mass and the nature of the material, as well as the temperature, since the energies of all the molecules must be included. For example, even if a cup of water and the water in a lake are at the same temperature, the water in the lake has the greater thermal energy because it contains more water molecules.

When two masses of materials with different temperatures are placed in contact, energy is transferred *from* the hotter (higher temperature) material *to* the cooler material. This energy that is transferred by means of a temperature difference is called **heat**.

Before it was known to be a transfer of energy, heat was measured in terms of its effect on pure water, and special units were defined to describe it. Some of these special units are still used.

A **calorie** (cal) is the amount of heat required to raise the temperature of 1 g of water by 1°C (from 14.5°C to 15.5°C).* The calorie used to describe food energy is really a kilocalorie (kcal).

The SI unit for heat and thermal energy is the same as that for all other energies [i.e., joule (J)]. The USCS unit for heat is the **British thermal unit** (Btu), which is the amount of heat required to raise the temperature of 1 lb of water by 1°F (from 58.5°F to 39.5°F).

Heat is an energy form, and therefore there is a direct conversion between heat units and the regular energy units.

$$1.0 \text{ cal} = 4.186 \text{ J} \quad \text{and} \quad 1.0 \text{ kcal} = 4186 \text{ J}$$

$$1.0 \text{ Btu} = 778 \text{ ft} \cdot \text{lb}$$

These relationships are known as **mechanical equivalents of heat**.

First Law of Thermodynamics

The **internal energy (thermal energy)** of a substance can be changed by some amount ΔE either by doing work W on it or by supplying an amount of heat Q to it. Therefore, since energy is conserved;

$$\Delta E = Q + W \tag{12-8}$$

This is known as the **first law of thermodynamics.** The following sign convention should be used: The work done W is positive valued if work is done on the substance, and negative valued when the substance does work (by expansion, for example). Similarly, the heat Q is positive valued if heat is supplied to the substance, but it is negative valued if the substance loses heat.

For example, when work is done by rubbing two objects together, the friction increases the internal energy and the temperature increases. We can also increase the internal energy and temperature of a pot of water by supplying it with some quantity of heat from a stove.

EXAMPLE 12-5

Determine the change in the internal energy of a substance if 3800 J of work is done by that substance when it receives 4500 J of heat.

---

*The temperature range is often ignored.

Solution

Data: $W = -3800$ J (negative because the substance does the work), $Q = 4500$ J, $\Delta E = ?$

Equation: $\Delta E = W + Q$

Substitute: $\Delta E = -3800$ J $+ 4500$ J $= 700$ J

PROBLEMS

1. Convert the following to Fahrenheit temperatures: (a) 28°C; (b) −38°C; (c) 158°C; (d) 1350°C; (e) −180°C; (f) 22°C.
2. Convert the following to Celsius temperatures: (a) 250°F; (b) −40°F; (c) 1780°F; (d) −135°F; (e) 73°F; (f) −5°F.
3. Convert the following to kelvin temperatures: (a) 35°C; (b) −48°C; (c) 380°C; (d) −125°C; (e) −210°C; (f) −273°C.
4. Find the equivalent Celsius temperatures of the following: (a) 345 K; (b) 1860 K; (c) 56 K; (d) 135 K; (e) 990 K; (f) 12 K.
5. Convert the following to Rankine temperatures: (a) 120°F; (b) 32°F; (c) −74°F; (d) −186°F; (e) 540°F; (f) −40°F.
6. Convert the following to Fahrenheit temperatures: (a) 186°R; (b) 602°R; (c) 445°R; (d) 1026°R; (e) 42°R; (f) 185°R.
7. Determine the temperature at which the Celsius and the Fahrenheit temperatures have the same magnitude.
8. How many calories are equivalent to (a) 6750 J; (b) 12 500 J; (c) 2.35×10^5 J; (d) 235 J?
9. How many joules are equivalent to (a) 455 cal; (b) 2.5 cal; (c) 3.55 kcal; (d) 0.225 kcal?
10. How many Btu are equivalent to (a) 45 000 ft·lb; (b) 7.50×10^6 ft·lb; (c) 860 000 ft·lb?
11. How many foot pounds are equivalent to (a) 5200 Btu; (b) 525 Btu; (c) 12 000 Btu; (d) 4.25×10^4 Btu?
12. Determine the change in the internal energy of a system if it does 3600 J of work and it loses 1800 J of heat.
13. Determine the work done *by* a system if it gains 2750 J of heat and its internal energy decreases by 3800 J.
14. How much work is done *on* a system if it loses 2750 Btu of heat and its internal energy increases by 1750 Btu?

12-2 HEAT TRANSFER

Heat flows from areas of higher temperature to areas of lower temperature at a rate that depends on the nature of the material (if any) through which it travels. Some materials, called **thermal insulators,** offer a large resistance to heat flow. They are used in clothing, refrigerators, buildings (as insulation to improve the air conditioning), and many other areas where hotter regions must be isolated from cooler regions. Other materials, called **thermal conductors,** readily allow heat to flow through them. These materials are used when heat must be transported from a hotter area to a cooler area. For example, in electronic circuits conductors called **heat sinks** are used to transfer heat away from devices to keep them cooler.

The control of heat flow has always been an important but often complex problem. In some cases, such as the air conditioning of buildings, large quantities of insulating materials are used to reduce the heat flow. In other cases, such as heating or cooling systems, heat flow is encouraged.

There are three basic mechanisms of heat transfer: **conducton, convection,** and **radiation.** Two or more of these mechanisms often occur simultaneously.

Conduction

The temperature of a material is related to the average kinetic energy possessed by its molecules. The higher the temperature, the greater their kinetic energy. This energy may be transferred from one molecule to its neighbors by a direct interaction process called **conduction.** The ability of a material to conduct heat depends largely on its structure. Since conduction requires the transfer of kinetic energy from one molecule to another, the distances between these molecules is important. Gases are good insulators because the average spacing between gas molecules is relatively large. Metals are the best conductors of heat since some of their electrons are free to move and can be used to transfer heat from molecule to molecule.

The rate at which heat is transferred by conduction through a material is given by

$$\frac{Q}{t} = \frac{KA\,\Delta T}{L} \tag{12-9}$$

where Q is the heat conducted through a cross-sectional area A in an elapsed time t, and L is the thickness (or length) between the points where the temperature difference is ΔT. K is called the **thermal conductivity** of the material; it is approximately constant for a particular material.

In SI, thermal conductivities have units of watts per meter kelvin [W/(m · K)] or watts per meter degree Celsius [W/(m · °C)]. However, other metric units, such as kilocalories per meter second degree Celsius [kcal/(m · s · °C)], are also frequently used.

In USCS, thermal conductivities have units of Btu per foot second degree Fahrenheit [Btu/(ft · s · °F)], but in many cases mixed units, such as Btu inches per foot squared hour degree Fahrenheit [Btu · in./(ft^2 · h · °F)], are used for convenience. Some typical values are listed in Table 12-1.

TABLE 12-1

Typical Thermal Conductivities

	Unit	
Substance	W/(m·K)	Btu·in./(h·ft^2·°F)
Aluminum	209	1460
Brass	108	750
Copper	385	2670
Iron	46	320
Asbestos	0.58	4.0
Brick	0.65	4.5
Concrete	1.08	7.5
Corkboard	0.043	0.30
Fiberglass	0.038	0.29
Glass	0.65	4.96
Gypsum board	0.17	1.17
Wood—pine (across grain)	0.113	0.78
Air	0.025	0.17
Hydrogen	0.18	1.3
Water	0.599	4.15

Sec.12-2 / Heat Transfer

EXAMPLE 12-6

A 0.250 in. thick glass window of a house is 6.00 ft wide and 4.50 ft high. If the outside temperature is −4°F and the inside temperature is 72°F, how much heat does it conduct in 12.0 h?

Solution

Data: $L = 0.250$ in., $A = 6.00$ ft × 4.50 ft, $\Delta T = 72°F - (-4°F) = 76°F$, $t = 12.0$ h, $K = 4.96$ Btu·in./(h·ft²·°F) (Table 12-1), $Q = ?$

Equation: $\dfrac{Q}{t} = \dfrac{KA\,\Delta T}{L}$

Rearrange: $Q = \dfrac{KA\,\Delta T t}{L}$

Substitute:

$$Q = \frac{[4.96 \text{ Btu·in./(h·ft}^2\text{·°F)}](6.00 \text{ ft} \times 4.50 \text{ ft})(76°F)(12.0 \text{ h})}{0.250 \text{ in.}}$$

$$= 4.89 \times 10^5 \text{ Btu}$$

EXAMPLE 12-7

An aluminum heat sink is 3.00 cm thick and has a cross-sectional area of 48.0 cm². How long does it take to conduct 1.00×10^6 J of heat away from an amplifier if one side is attached to the amplifier at 70°C and the other is maintained at 20°C?

Solution

Data: $L = 3.00$ cm $= 0.0300$ m, $A = 48.0$ cm² $= 48.0 \times 10^{-4}$ m², $\Delta T = 70°C - 20°C = 50°C$, $Q = 1.00 \times 10^6$ J, $K = 209$ W/(m·°C) from Table 12-1, $t = ?$

Equation: $\dfrac{Q}{t} = \dfrac{KA\,\Delta T}{L}$

Rearrange: $t = \dfrac{QL}{KA\,\Delta T}$

Substitute: $t = \dfrac{(1.00 \times 10^6 \text{ J})(0.0300 \text{ m})}{[209 \text{ W/(m·°C)}](48.0 \times 10^{-4} \text{ m}^2)(50°C)} = 598$ s

---

In heat load and air-conditioning calculations, manufacturers specify thermal properties of many standard insulating materials (which may or may not have uniform structures) in terms of a **thermal resistance** R or **R-value**, which is the ratio of the thickness L to the thermal conductivity K.

$$R = \frac{L}{K} \tag{12-10}$$

Therefore, Eq. 12-10 can now be rewritten as

$$\frac{Q}{t} = \frac{A\,\Delta T}{R} \tag{12-11}$$

The R-value has units of $m^2 \cdot °C/W$ in SI and $ft^2 \cdot h \cdot °F/Btu$ in USCS. The conversion between these units is

$$\text{USCS } R\text{-value} \times 0.176 = \text{SI } R\text{-value} \qquad (12\text{-}12)$$

For example, fiberglass batts rated at 20 $ft^2 \cdot h \cdot °F/Btu$ (which is usually written simply as R20) have an SI value of

$$20 \times 0.176 \text{ m}^2 \cdot °C/W = 3.52 \text{ m}^2 \cdot °C/W$$

This is also often written as: 3.52 (SI).

EXAMPLE 12-8

Determine the R-value of a 6.00 in. thick batt of fiberglass insulation.

Solution

Data: $L = 6.00$ in., $K = 0.29$ Btu $\cdot$ in./(h $\cdot$ ft² $\cdot$ °F) (Table 12-1), $R = ?$

Equation: $R = \dfrac{L}{K}$

Substitute: $R = \dfrac{6.00 \text{ in.}}{0.29 \text{ Btu} \cdot \text{in.}/(\text{h} \cdot \text{ft}^2 \cdot °F)} = 20.7 \text{ h} \cdot \text{ft}^2 \cdot °F/Btu$

EXAMPLE 12-9

Determine the heat lost in 3.00 h through a 7.75 m $\times$ 2.50 m wall that has a thermal resistance of 3.60 m² $\cdot$ °C/W if the inner and outer wall temperatures are 22°C and -30°C.

Solution

Data: $t = 3.00$ h $= 3.00 \times 3600$ s, $A = 7.75$ m $\times$ 2.50 m $= 19.4$ m²,
$\Delta T = 22°C - (-30°C) = 52°C$, $R = 3.60$ m² $\cdot$ °C/W, $Q = ?$

Equation: $\dfrac{Q}{t} = \dfrac{A \, \Delta T}{R}$

Rearrange: $Q = \dfrac{A \, \Delta T t}{R}$

Substitute: $Q = \dfrac{(19.4 \text{ m}^2)(52°C)(3.00 \times 3600 \text{ s})}{3.60 \text{ m}^2 \cdot °C/W} = 3.03 \times 10^6 \text{ J}$

Convection

Molecules of fluids are able to flow; therefore, heat may be transferred by the motion of the molecules. This heat flow is called **convection** (Fig. 12-2). Since most fluids expand when they are heated, the density of a warmer section of fluid is less than that

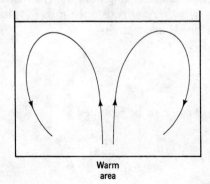

Figure 12-2 Convection currents: the warmer (higher energy) molecules rise while the cooler molecules sink.

Sec.12-2 / Heat Transfer

of the cooler areas, and the warmer fluid rises while the cooler fluid sinks. This circulation of fluid, which is due to density variations, is called **natural convection.** Many heating systems utilize natural convection in order to circulate hot air of water. Heat transfer inside most buildings is mainly the result of convection currents of air. Some heating systems use fans or pumps in order to force the higher-energy molecules to circulate. This is called **forced convection.**

Radiation

Heat transfer by conduction or convection requires a material medium, but heat reaches the earth directly from the sun through the near vacuum of space. In this heat flow process, called **radiation,** energy is transmitted at the speed of light as an electromagnetic wave. Although radiative heat transfer does not require a material medium, it can travel through materials. It passes through air and glass, for example, and it must be considered in air-conditioning load calculations.

Any object radiates heat at a rate that depends on its temperature and the nature of its surface. As it radiates heat, its molecules lose energy and the object cools. When an object absorbs radiation, the energy is transmitted to its molecules, increasing the internal energy and temperature.

PROBLEMS

15 How much heat is conducted in 8.00 h through a concrete wall that is 15.0 cm thick and has an area of 1.86 m² if the temperature difference is 28°C?

16 Determine the amount of heat that is conducted in 2.00 h through a concrete wall that is 6.00 in. thick and has an area of 160 ft² if the temperature difference is 50°F.

17 A glass window is 3.00 mm thick and has dimensions of 3.00 m × 2.00 m. If the outside temperature is −15°C and the inside temperature is 22°C, how much heat does it conduct in 6.00 h?

18 Determine the rate of heat flow in watts through a 1.00 cm thick corkboard that has an area of 35.0 cm² if the temperature difference between its surfaces is 30°C.

19 The inside dimensions of a refrigerator are 1.00 m × 90.0 cm × 2.30 m. If all the walls are constructed of a material 1.00 cm thick that has a thermal conductivity of 2.20×10^{-2} W/(m·°C) and the temperatures inside and outside are −5°C and 23°C, determine **(a)** the rate of the heat flow into the refrigerator; **(b)** the time required for 3600 J of heat to enter the refrigerator.

20 Determine the heat lost in 8.00 h through a 2.00 m by 1.20 m glass window pane that is 5.00 mm thick if the temperature difference is 56°C.

21 A wall is constructed from a layer of brick 4.00 in. thick and a layer of fiberglass 6.00 in. thick. If the total surface area of the wall is 100 ft² and the inside and outside temperatures are −10°F and 72°F, respectively, determine the rate of total heat flow through the wall.

22 Determine the R-value of a pane of glass that is **(a)** 4.00 mm thick; **(b)** 0.375 in. thick.

23 What is the R-value of a concrete wall that is **(a)** 20.0 cm thick; **(b)** 8.00 in. thick?

24 What is the R-value of a sheet of drywall (gypsum board) that is **(a)** 1.25 cm thick; **(b)** 0.500 in. thick?

25 What is the thickness of a glass plate that has an R-value of **(a)** 0.0750 m²·°C/W; **(b)** 0.450 ft²·h·°F/Btu?

26 An 8.00 m × 2.50 m wall has an R-value of 3.50 m²·°C/W. If the inside temperature is 23°C and the outside temperature is −38°C, how much heat escapes in 12.0 h through the wall?

27 Determine the minimum R-value required by a wall 6.00 m × 2.50 m if it is to limit the heat loss to a rate of 375 W when the temperature difference is 48°C.

28 A 120 ft² masonry wall of a house is 6.00 in. thick and has an R-value of 2.04 ft²·h·°F/Btu. Determine the rate of heat flow through the wall when the outside temperature is 20°F and the inside temperature is 70°F.

12-3 SPECIFIC HEAT CAPACITY

Even when different materials of the same mass and temperature are subjected to the same quantity of heat, their temperature changes may be quite different. The ability of a material to store or release heat is described by a term called the **specific heat capacity** c. It is defined by the relationship

$$c = \frac{Q}{m \, \Delta T} \quad \text{(in SI)} \tag{12-13}$$

and

$$c = \frac{Q}{w \, \Delta T} \quad \text{(in USCS)} \tag{12-14}$$

where Q is the quantity of heat that produces a temperature change ΔT in a mass m or weight w of the material. These equations may be rearranged to give

$$Q = mc \, \Delta T = mc(T_f - T_i) \quad \text{(in SI)} \tag{12-15}$$

and

$$Q = wc \, \Delta T = wc(T_f - T_i) \quad \text{(in USCS)} \tag{12-16}$$

where T_i is the intitial temperature and T_f is the final temperature.

The specific heat capacity is independent of the physical dimensions of the material.

In SI specific heat capacity has units of J/(kg · °C) or J/(kg · K). The USCS units are Btu/(lb · °F). Other units such as kcal/(kg · °C) are also still used. The specific heat capacities of some materials are listed in Table 12-2.

We may use these equations to determine the temperature changes produced by any quantity of heat Q, provided that the material does not change its phase between solid, liquid, and gas (or vapor) states.

EXAMPLE 12-10

How much heat is required to raise the temperature of 2.00 kg of water from 15°C to 58°C?

Solution

Data: $m = 2.00$ kg, $T_i = 15$°C, $T_f = 58$°C, $c = 4186$ J/(kg · °C) from Table 12-2, $Q = ?$
Equation: $Q = mc(T_f - T_i)$
Substitute: $Q = (2.00 \text{ kg})[4186 \text{ J/(kg} \cdot \text{°C)}](58\text{°C} - 15\text{°C})$
    $= 3.60 \times 10^5$ J

EXAMPLE 12-11

How much heat must be supplied to a 25.0 lb iron ingot to raise its temperature from 70°F to its melting point (2790°F)?

Solution

Data: $w = 25.0$ lb, $T_i = 70$°F, $T_f = 2790$°F, $c = 0.12$ Btu/(lb · °F) from Table 12-2, $Q = ?$
Equation: $Q = wc(T_f - T_i)$
Substitute: $Q = (25.0 \text{ lb})[0.12 \text{ Btu/(lb} \cdot \text{°F)}](2790\text{°F} - 70\text{°F})$
    $= 8160$ Btu

TABLE 12-2
Typical Heat Constants

Material	Melting point °C	Melting point °F	Boiling point °C	Boiling point °F	Specific heat capacity c ×10³ J/(kg·°C)	Specific heat capacity c kcal/(kg·°C) or Btu/(lb·°F)	Specific latent heat of fusion l_f kcal/kg	Specific latent heat of fusion l_f ×10⁵ J/kg	Specific latent heat of fusion l_f Btu/lb	Specific latent heat of vaporization l_v kcal/kg	Specific latent heat of vaporization l_v ×10⁶ J/kg	Specific latent heat of vaporization l_v Btu/lb
Air	—	—	—	—	1.0	0.24	—	—	—	51	0.21	92
Alcohol—ethyl	−130	−202	78	172	2.4	0.58	25	1.04	46	205	0.858	369
Aluminum	660	1220	1800	3270	0.921	0.22	77	3.2	139	1990	8.3	3591
Brass	900	1652	—	—	0.376	0.090	—	—	—	—	—	—
Copper	1083	1980	2296	4164	0.390	0.093	42	1.8	76	1750	7.3	3158
Glass	1100	2012	—	—	0.880	0.21	—	—	—	—	—	—
Hydrogen	−259	−434	−253	−432	14.3	3.4	—	—	—	108	0.45	195
Ice	0	32	—	—	2.13	0.51	—	—	—	—	—	—
Iron	1535	2790	3000	5430	0.50	0.12	5.5	0.23	9.9	—	—	—
Lead	328	622	1644	2990	0.135	0.032	5.5	0.23	9.9	—	—	—
Mercury	−39	−38	357	675	0.140	0.033	2.8	0.12	5.0	71	0.297	128
Oxygen	−218	−360	−183	−297	—	—	3.2	0.13	5.8	50	0.21	90
Steam	—	—	—	—	2.0	0.48	—	—	—	—	—	—
Water—liquid	0	32	100	212	4.186	1.00	80	3.35	144	540	2.26	970
Zinc	419	786	907	1664	0.377	0.090	28	1.2	51	—	—	—

EXAMPLE 12-12

An air-conditioning system supplies 3400 Btu to 280 lb of dry air at 22°F. What is the final air temperature?

Solution

Data: $Q = 3400$ Btu, $w = 280$ lb, $T_i = 22°F$, $c = 0.24$ Btu/(lb · °F) (Table 12-2), $T_f = ?$

Equation: $Q = wc(T_f - T_i)$

Rearrange: $T_f = \dfrac{Q}{wc} + T_i$

Substitute: $T_f = \dfrac{3400 \text{ Btu}}{(280 \text{ lb})[0.24 \text{ Btu/(lb} \cdot °F)]} + 22°F = 73°F$

Compared with most other substances, water has a relatively high specific heat capacity; it is also inexpensive and abundant. Therefore, many cooling and heating systems use water as a **heat transfer agent,** to "carry" heat from one place to another.

In general, a "hot" material loses heat to cooler materials and also to the surroundings. However, in a well-insulated container, called a **calorimeter,** heat losses to the surroundings are minimal. In an isolated system (such as a calorimeter) thermal energy may be conserved. Therefore, the heat lost by some components must equal the heat gained by the other components, as long as it is not converted into some other energy form.

total heat lost by some components = total heat gained by the other components

If a material loses heat, its final temperature T_f is less than its original temperature T_i. Thus in Eq. 12-15, the heat transfer:

$$Q = mc(T_f - T_i) \quad \text{or} \quad Q = wc(T_f - T_i)$$

is negative valued. On the other hand, if the final temperature T_f is greater than the original temperature T_i, the heat transfer is positive valued. We may therefore restate the **conservation of thermal energy** as follows:

If thermal energy is not transformed into some other kind of energy, the algebraic sum of all heat changes in an isolated system is equal to zero.

If the materials do not change phase between solid, liquid, and gas (or vapor), states, then

$$m_1 c_1 (T_f - T_{i1}) + m_2 c_2 (T_f - T_{i2}) + \cdots = 0 \qquad (12\text{-}17)$$

and

$$w_1 c_1 (T_f - T_{i1}) + w_2 c_2 (T_f - T_{i2}) + \cdots = 0 \qquad (12\text{-}18)$$

where the m's are the masses, w's are the weights, c's are the specific heat capacities, and T_i's are the initial temperatures of the components. Note that the final temperature T_f is the same for all components when thermal equilibrium is reached.

The specific heat capacity c of an unknown component or the final temperature T_f of a system may be determined from calorimetric experiments and the conservation of thermal energy equation.

Sec.12-3 / Specific Heat Capacity

EXAMPLE 12-13

Determine the final temperature when 2.80 lb of iron at 880°F is added to 18.0 lb of water at 68°F and there are no external heat influences.

Solution

Data: $w_i = 2.80$ lb, $w_w = 18.0$ lb, $T_i = 880°F$, $T_w = 68°F$, (Table 12-2), $c_i = 0.12$ Btu/(lb · °F) and $c_w = 1.00$ Btu/(lb · °F), $T_f = ?$
Equations: Total heat changes = 0, $Q = wc(T_f - T_i)$
Substitute: For the water,

$$Q = (18.00 \text{ lb})[1.00 \text{ Btu/(lb · °F)}](T_f - 68°F)$$
$$= 18.0 \text{ Btu/°F}(T_f - 68°F)$$

For the iron,

$$Q = (2.80 \text{ lb})[0.12 \text{ Btu/(lb · °F)}](T_f - 880°F)$$
$$= 0.34 \text{ Btu/°F}(T_f - 880°F)$$

Therefore,

$$18.0 \text{ Btu/°F}(T_f - 68°F) + 0.34 \text{ Btu/°F}(T_f - 880°F) = 0$$

Expanding the brackets and rearranging the units, we have

$$18.0 \, T_f + 0.34 \, T_f = (18.0)(68°F) + (0.34)(880°F) = 1520°F$$

Therefore,

$$T = \frac{1520°F}{18.34} = 83°F$$

EXAMPLE 12-14

A 350 g metal cylinder at 180°C is added to 750 g of water at 20°C. If the final temperature of the system is 27°C and there are no external heat influences, determine the specific heat capacity of the metal.

Solution

Data: $m_m = 0.35$ kg, $m_w = 0.75$ kg, $c_w = 4186$ J/(kg · °C) (Table 12-2), $T_{im} = 180°C$, $T_{iw} = 20°C$, $T_f = 27°C$, $c_m = ?$
Equations: Sum of the heat changes = 0, $Q = mc(T_f - T_i)$
In this case heat is lost by the metal and gained by the water.
Substitute: For the water,

$$Q = (0.75 \text{ kg})[4186 \text{ J/(kg · °C)}](27°C - 20°C) = 22\,000 \text{ J}$$

For the metal,

$$Q = (0.35 \text{ kg})c(27°C - 180°C) = -54c \text{ kg · °C}$$

Therefore, $-54c$ kg · °C + 22 000 J = 0, and

$$c = \frac{22\,000 \text{ J}}{54 \text{ kg · °C}} = 410 \text{ J/(kg · °C)}$$

PROBLEMS

Assume no heat influences from the surroundings.

29. Determine the quantity of heat required to raise the temperature of 15 kg of copper by 28°C from 20°C to 48°C.
30. How much heat is required to raise the temperature of 12.0 lb of air by 15°F?
31. Determine the heat required to raise the temperature of 6.25 kg of air by 8.0°C.
32. How much heat is emitted when 2.8 kg of water is cooled from 75°C to 40°C?
33. If 1600 Btu is added to a 1200 lb iron girder at 65°F, what is the final temperature?
34. When 15 g of a material is supplied with 68 cal of heat, its temperature rises from 38°C to 46°C. What is the specific heat capacity of the material?
35. How much heat is required to increase the temperature of a 480 kg iron rail from 20°C to its melting point (1535°C)?
36. Determine the amount of heat that is required to raise the temperature of 12.0 lb of copper from 68°F to its melting point.
37. A 100 g piece of metal at 150°C is added to a 75.0 g aluminum calorimeter that contains 200 g of water at 10°C. If the final temperature of the mixture is 15.5°C, determine the specific heat capacity of the metal.
38. How much heat must be added to a 380 kg copper boiler containing 360 kg of water at 22°C to raise the temperature of both to 100°C?
39. What is the final temperature when 3.00 lb of water at 38°F is mixed with 8.00 lb of water at 182°F?
40. When 280 g of a substance at 130°C is mixed with 420 g of water at 18°C, the final temperature is 28°C. What is the specific heat capacity of the substance?
41. How much water at 18°C is required to change the temperature of 2.5 kg of water at 82°C to a final temperature of 61°C?
42. What is the final temperature when 35.0 lb of zinc at 580°F is added to 120 lb of water at 58°F?
43. A 95 g block of aluminum at 260°C is added to a 210 g glass calorimeter that contains 400 g of water at 35°C. What is the final temperature of the system?

12-4 PHASE CHANGES

When a solid is heated, its internal energy and its temperature increase until the molecular vibrations in the solid become excessive. At this point, the application of more heat energy gradually breaks down the structure of the solid, and it "melts" at a constant temperature.

The heat input per unit mass (or weight) required to change the material from solid to liquid at a constant temperature and pressure is called the **specific latent heat of fusion** l_f of the material:

$$l_f = \frac{Q_f}{m} \quad \text{(SI)} \tag{12-19}$$

and

$$l_f = \frac{Q_f}{w} \quad \text{(USCS)} \tag{12-20}$$

where Q_f is the heat required to melt a mass m or weight w of the material at a constant temperature and pressure.

Sec.12-4 / Phase Changes

The specific latent heat of fusion l_f is usually expressed in units of J/kg in SI, but kcal/kg are also used. The USCS unit is Btu/lb.

If a material freezes from liquid to solid form, it reduces its energy by losing heat. This is indicated by giving the heat a negative value.

EXAMPLE 12-15

How much heat is liberated when 15.0 kg of liquid water solidifies into ice at 0°C at a constant pressure?

Solution

Data: $m = 15.0$ kg, $l_f = 3.35 \times 10^5$ J/kg (Table 12-2), $Q_f = ?$

Equation: $l_f = \dfrac{Q_f}{m}$

Rearrange: $Q_f = l_f m$

Substitute: $Q_f = -(3.35 \times 10^5 \text{ J/kg})(15.0 \text{ kg}) = -5.03 \times 10^6$ J

The water **loses** heat as it freezes; therefore, a minus sign will be required for the heat.

If, after all of a solid has been liquefied by the application of heat, the liquid is still heated, it gains more internal energy, and its temperature increases until the boiling or vaporization point is reached. At this point, some liquid molecules have sufficient kinetic energy to escape through the liquid surface, and continued application of heat energy causes the liquid to boil and vaporize at a constant temperature.

The amount of heat per unit mass (or weight) required to change a sample of material from liquid to vapor at a constant temperature and pressure is called the **specific latent heat of vaporization** l_v of that material.

$$l_v = \frac{Q_v}{m} \quad \text{(SI)} \tag{12-21}$$

and

$$l_v = \frac{Q_v}{w} \quad \text{(USCS)} \tag{12-22}$$

where Q_v is the heat required, m the mass, and w the weight of material vaporized. The specific latent heat of vaporization has units of J/kg in SI and Btu/lb in USCS. Other metric units such as kcal/kg are also used. If the material condenses from the vapor to the liquid state by absorbing heat, the heat "gained" is given a negative value. Far more heat is required to vaporize than to melt a fixed mass of the same material.

EXAMPLE 12-16

How much heat is required to vaporize 2.30 lb of water at 212°F?

Solution

Data: $w = 2.30$ lb, $l_v = 970$ Btu/lb (Table 12-2), $Q_v = ?$

Equation: $l_v = \dfrac{Q_v}{w}$

Rearrange: $Q_v = l_v w$

Substitute: $Q_v = (970 \text{ Btu/lb})(2.30 \text{ lb}) = 2230$ Btu

When a material undergoes both a change in temperature and a change in phase, we must consider both the specific heat and the latent heat involved. That is,

$$\text{total heat} = \text{specific heat} + \text{latent heat}$$

or

$$Q_{tot} = Q_{specific} + Q_{latent} \qquad (12\text{-}23)$$

EXAMPLE 12-17

How much heat is required to boil away 1.2 kg of water that is initially at 23°C?

Solution: We must first increase the temperature of the water from 23°C to the boiling point (100°C), and then the water boils at a constant temperature requiring latent heat.

Data: $m = 1.2$ kg, From Table 12-2, $c = 4186$ J/(kg·°C); $l_v = 2.26 \times 10^6$ J/kg, $T_i = 23$°C, $T_f = 100$°C, $Q_{tot} = ?$

Equation: $Q_{tot} = Q_{specific} + Q_{latent}$
$= mc(T_f - T_i) + l_v m$

Substitute:

$Q_{tot} = (1.2 \text{ kg})[4186 \text{ J/(kg·°C)}](100°C - 23°C) + (1.2 \text{ kg})(2.25 \times 10^6 \text{ J/kg})$

$= 3.1 \times 10^6$ J

In some cases, the phase change of one component may not be complete when the final equilibrium temperature is reached. The final temperature is then equal to the temperature at which that phase change occurs.

Under certain conditions, many substances, such as solid carbon dioxide (dry ice), change directly from solid to the vapor state. This is called **sublimation.** Also, liquids need not boil in order to vaporize; molecules at the liquid surface often receive extra energy from molecules below the surface. If their energy is sufficient, the surface molecules may escape. Since only the more energetic molecules escape, the remaining liquid has less thermal energy per molecule, and it is, therefore, cooler. This process is known as **evaporation.** Evaporation results in a loss of heat energy from the liquid, because the more energetic molecules escape. Therefore, evaporation cools a liquid. This effect is very noticeable when alcohol evaporates from skin, for example. The evaporation rate depends on the surface area, the temperature and pressure of the liquid.

Liquid molecules are continually evaporating from the surface, but some vapor molecules return to the liquid. When the rate of evaporation is equal to the rate of return of vapor molecules, the vapor is said to be **saturated.** If vapor molecules are continually removed from the region near the liquid surface (by ventilation), the net evaporation rate increases. Evaporation may occur even at relatively low temperatures, but the evaporation rate increases as the temperature of the liquid increases.

When a liquid boils, bubbles of saturated vapor form within the liquid body and rise to the surface. The internal pressure of these vapor bubbles must be equal to the total pressure at their depth in the liquid; otherwise, they would collapse. Near the surface of the boiling liquid, the internal vapor pressure of the bubbles is approximately equal to the surface pressure. Thus a liquid boils when its saturated vapor pressure is equal to the surface pressure. A reduction in the surface pressure causes the liquid to boil at a lower temperature. If the surface pressure is increased, the liquid will boil at a higher temperature, because more evaporation is required to increase the vapor pressure. Evaporation of water is responsible for many of our weather phenomena and is an important factor in air-conditioning systems.

Pressure also affects the freezing or fusion point of a material, although not to such a degree. When pressure is applied to a liquid, it tends to reduce the spacing between the molecules (Fig. 12-3). Therefore, most materials contract when they solidify, and an increase in pressure favors a change to the solid state.

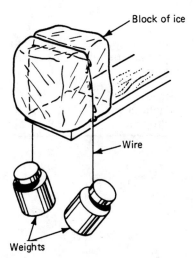

Figure 12-3 Pressure of the wire causes ice to melt.

Water expands when it freezes. Thus ice floats on water because its density is less than water. Therefore, an increase in pressure causes ice to melt. For example, an ice skater's skate blades exert pressure and melt the ice.

PROBLEMS

44 How much heat in joules is required to melt 85.0 kg of ice at 0°C?
45 How much heat in joules is evolved when 12.0 kg of steam is condensed at 100°C?
46 How much heat is required to melt 5.00 lb of ice at 32°F?
47 How much heat must be removed to freeze 12 kg of water at 0°C?
48 How much heat is required to melt 80.0 kg of iron at its melting point (1535°C)?
49 Determine the total heat required to boil away 3.5 kg of water initially at 20°C.
50 How much heat is required to change 2.00 lb of ice at 10°F to steam at 100°C?
51 If there are no external heat influences, determine the final temperature when 25 g of steam at 100°C is mixed with 6.0 kg of water at 10°C.
52 How much heat is required to melt 15.0 lb of aluminum if its initial temperature is 75°F?

12-5 HEAT OF COMBUSTION

One of the most common and useful sources of heat energy is the chemical energy that is released by the direct combustion of a fuel, such as coal, oil, or gasoline. The **heat of combustion** H of a fuel is the heat evolved per unit mass or volume of fuel when it is completely burned (Table 12-3). The heat evolved per unit mass is called the **specific heat of combustion**.

TABLE 12-3

Typical Heats of Combustion of Common Fuels

Material	Heat of Combustion	
	J/kg	Btu/lb
Coal	3.26×10^7	14 000
Wood	1.4×10^7	6 000
Gasoline	4.7×10^7	20 200
Diesel oil	4.5×10^7	19 400
Domestic fuel oil	4.6×10^7	19 800
Kerosene	4.6×10^7	19 800
Gases	J/m³	Btu/ft³
Natural gas	4.35×10^7	12 00
Propane	8.8×10^7	24 00

$$H = \frac{Q}{m} \quad \text{or} \quad H = \frac{Q}{V} \quad \text{(SI)} \tag{12-24}$$

and

$$H = \frac{Q}{w} \quad \text{(USCS)} \tag{12-25}$$

where Q is the heat liberated when a mass m, weight w, or volume V of fuel is burned. The heating capacity of a fuel is usually determined by its heating effects on water.

EXAMPLE 12-18

How much heat is liberated when 3.00 m³ of natural gas is completely burned?

Solution

Data: $V = 3.00$ m³, $H = 4.35 \times 10^7$ J/m³ (Table 12-3), $Q = ?$

Equation: $H = \dfrac{Q}{V}$

Rearrange: $Q = HV$

Substitute: $Q = (4.35 \times 10^7 \text{ J/m}^3)(3.00 \text{ m}^3) = 1.31 \times 10^8$ J

PROBLEMS

53. How much heat is evolved when (a) 12.0 kg of coal is completely burned; (b) 35.0 lb of wood is completely burned?
54. How much heat is emitted when (a) 0.14 m³ of natural gas is burned; (b) 2.75 ft³ of propane is burned?
55. (a) How much heat is required to heat 1.25 kg of water from 6°C to its boiling point? (b) If there are no heat losses to the surroundings, how much propane is required?
56. (a) How much heat is required to heat 5.00 gal of water from 55°F to its boiling point? (b) If there are no heat losses to the surroundings, how much natural gas is required? One U.S. gallon weighs 8.33 lb.
57. If 3.0 kg of wood is completely burned and 25% of the heat evolved is used to heat a 20 kg iron boiler contining 300 kg of water at 10°C, what is the final temperature of the water?

SUMMARY

$$T_F = \tfrac{9}{5}T_C + 32° \qquad T_C = \tfrac{5}{9}(T_F - 32)$$

$$T_K = T_C + 273 \qquad T_R = T_F + 460$$

Temperature is a measure of the average kinetic energy of the individual molecules. *Thermal (internal) energy* is related to the total energy of all molecules. *Heat* flows naturally from higher to lower temperatures.

A *calorie* (cal) is the quantity of heat required to raise the temperature of 1 g of water by 1°C (14.5°C to 15.5°C).

$$1 \text{ cal} = 4.186 \text{ J} \qquad 1 \text{ kcal} = 4186 \text{ J}$$

A *British thermal unit* (Btu) is the amount of heat required to raise the temperature of 1 lb of water by 1°F (58.5°F to 59.5°F).

$$1 \text{ Btu} = 778 \text{ ft} \cdot \text{lb}$$

The *first law of thermodynamics*,

$$\Delta E = W + Q$$

Heat is transferred by:

1. *Conduction:* from molecule to molecule. The rate of heat flow is

$$\frac{Q}{t} = \frac{KA\,\Delta T}{L}$$

and the *R-value* is

$$R = \frac{L}{K}$$

2. *Convection:* by the motion of the higher-energy molecules themselves.
3. *Radiation:* as an electromagnetic wave.

Specific heat capacity

$$c = \frac{Q}{m\,\Delta T} \text{ (SI)} \qquad \text{and} \qquad c = \frac{Q}{w\,\Delta T} \text{ (USCS)}$$

Specific latent heat of fusion

$$l_f = \frac{Q_f}{m} \text{ (SI)} \qquad \text{and} \qquad l_f = \frac{Q_f}{w} \text{ (USCS)}$$

Specific latent heat of vaporization

$$l_v = \frac{Q_v}{m} \text{ (SI)} \qquad \text{and} \qquad l_v = \frac{Q_v}{w} \text{ (USCS)}$$

In an isolated system, the total heat lost by some components is equal to total heat gained by the others.

Specific heat of combustion

$$H = \frac{Q}{m} \quad \text{or} \quad H = \frac{Q}{V} \quad \text{(SI)}$$

and

$$H = \frac{Q}{w} \quad \text{(USCS)}$$

QUESTIONS

1. Define the following terms: (a) calorie; (b) conduction; (c) convection; (d) radiation; (e) R-value; (f) specific heat capacity; (g) specific latent heat; (h) specific heat of combustion; (i) sublimation.
2. Explain why a drill bit becomes hotter when it is used to drill holes.
3. If a cup of water and the water in a bath tub are at the same temperature, what can you say about (a) the kinetic energies of the water molecules; (b) the thermal energies of the two bodies of water?
4. Describe the formation of clouds by the evaporation process.
5. What heat transfer processes are used to heat a house? Explain how each process is applied.
6. Give an example of a device that forces heat from a lower to a higher temperature area. Explain how this process works in terms of the first law of thermodynamics.
7. Why would you add ice rather than liquid water at 0°C to cool a drink?
8. Metal objects normally feel cooler to touch than wood. Explain why.

REVIEW PROBLEMS

Assume no external heat influences.

1. Convert the following to Celsius temperatures: (a) 63°F; (b) 196°F; (c) −48°F.
2. Convert the following to Fahrenheit temperatures: (a) 293°C; (b) 1200°C; (c) −86°C.
3. Convert the following to kelvin temperatures: (a) −28°C; (b) 89°C; (c) 1570°C.
4. Convert the following to Rankine temperatures: (a) 296°F; (b) −280°F; (c) 42°C; (d) −84°C.
5. Convert (a) 9600 kcal to joules; (b) 3.85 MJ to kilocalories; (c) 1580 Btu to foot pounds; (d) 9.75×10^8 ft · lb to Btu.
6. Determine the change in the internal energy of a system when (a) 3600 J of work is done on the system and it gains 1500 J of heat; (b) 1200 J of work is done by the system and it gains 1750 J of heat.
7. Determine the work done on a system when it gains 3400 Btu of heat and its internal energy increases by 1750 Btu.
8. Determine the heat lost in 8.00 h through a glass window pane 1.75 m × 2.50 m that is 4.50 mm thick if the temperature difference is 48°C.
9. How many Btu are lost in 10.0 min through a glass window 2.00 ft × 4.00 ft which is 0.250 in. thick if the inside and outside temperatures are 72°F and 30°F?
10. Determine the R-value of a batt of fiberglass that is (a) 15.0 cm thick; (b) 8.00 in. thick.
11. Determine the R-value of a sheet of drywall (gypsum board) that is (a) 2.00 cm thick; (b) 0.750 in. thick.

Review Problems

12. What thickness of concrete will give an R-value of (a) $2.50 \text{ m}^2 \cdot °\text{C/W}$; (b) $15 \text{ h} \cdot \text{ft}^2 \cdot °\text{F/Btu}$?

13. Determine the minimum R-value required by a wall $7.50 \text{ m} \times 2.50 \text{ m}$ if it is to limit the heat loss to a rate of 425 W when the temperature difference is 52°C.

14. How many Btu are conducted in 1.00 h through a 1.00 ft thick concrete wall of surface area 75.0 ft^2 if the outside temperature is $-28°\text{F}$ and the inside temperature is 72°F?

15. How much heat is required to raise the temperature of a 120 kg aluminum girder from 23°C to its melting point of 660°C?

16. How much heat in joules must be added to a 3.50 kg copper rod to raise its temperature from 20°C to 100°C?

17. How much water at 10°C must be mixed with 350 g of water at 80°C to bring the mixture to a final temperature of 48°C?

18. If 10.0 lb of steam at 212°F is passed into 500 lb of water at 40°F, what is the final temperature of the mixture?

19. If 455 g of ice at $-11°\text{C}$ is added to 1.60 kg of water at 5.6°C, how much ice melts?

20. How much ice melts if 1.00 lb of ice at 12°F is added to 3.50 lb of water at 42°F?

21. If 500 g of water and 100 g of ice are in equilibrium at 0°C, what is the final temperature if 200 g of steam at 100°C is added?

22. How much heat in joules is evolved when (a) 2.50 m^3 of propane is completely burned; (b) 25.0 ft^3 of natural gas is burned?

23. How much heat is given off when (a) 16.0 kg of coal is completely burned; (b) 75.0 lb of kerosene is burned?

13
THERMAL EXPANSION

Thermal expansion has many important consequences in structures, engines, electrical components, and many other areas. The properties of thermal expansion may be used to advantage in some devices, such as circuit breakers, thermometers, and thermal switches. It must also be allowed for in other areas, such as structures, engines, molds, and cables, which must operate within prescribed tolerances for all expected temperature ranges. While small variations in the temperature of liquids and solids cause only small changes in their dimensions, these expansions or contractions may result in enormous stresses. If these thermal stresses are not allowed for, they may cause mechanical failure.

13-1 LINEAR EXPANSION

It is a well-known fact that most substances expand when they are heated and that this expansion occurs in three dimensions. However, in solids, it is often necessary only to consider the consequences of the expansion in one direction. When a solid is heated, its temperature and the internal vibrational energy possessed by its atoms and molecules increase, and the material expands.

The amount that a solid expands depends on its initial length L_0, the change in temperature ΔT, and the nature of the material. Most solids expand relatively uniformly. Their change in length ΔL can be determined from

$$\Delta L = \alpha L_0 \Delta T \qquad (13\text{-}1)$$

The proportionality constant α is called the **coefficient of linear expansion;** its value depends on the material.

$$\alpha = \frac{\Delta L}{L_0 \Delta T} \qquad (13\text{-}2)$$

Sec. 13-1 / Linear Expansion

The coefficient of linear expansion is usually expressed in units of per kelvin ($1/K = K^{-1}$) or per degree Celsius ($1/°C$) in SI and per degree Fahrenheit ($1/°F$) in USCS. Some typical values are listed in Table 13-1. We may also rewrite Eq. 13-1 as

$$L = L_0[1 + \alpha(T - T_0)] \tag{13-3}$$

where L is the length at the temperature T and L_0 is the length at temperature T_0.

TABLE 13-1

Typical Coefficients of Linear Expansion of Solids at Room Temperature

	$\alpha/(10^{-5}/K)$ or $\alpha/(10^{-5}/°C)$	$\alpha/(10^{-5}/°F)$
Aluminum	2.3	1.3
Brass	1.9	1.1
Copper	1.7	1.1
Glass	0.91	0.51
Pyrex glass	0.32	0.18
Iron (soft)	1.2	0.67
Lead	2.9	1.6
Platinum	0.90	0.50
Quartz	0.055	0.031
Silver	1.9	1.1
Steel	1.2	0.65
Tungsten	0.45	0.25

EXAMPLE 13-1

If a steel I-beam in a bridge truss is 12.0000 m long at 5°C, what is its length when the temperature increases to 35°C?

Solution

Data: $L_0 = 12.0000$ m, $T = 35°C$, $T_0 = 5°C$, $\alpha = 1.2 \times 10^{-5}/°C$ (Table 13-1), $L = ?$
Equation: $L = L_0[1 + \alpha(T - T_0)]$
Substitute: $L = (12.0000 \text{ m})[1 + (1.2 \times 10^{-5}/°C)(35°C - 5°C)]$
$\qquad = 12.0043$ m

EXAMPLE 13-2

What minimum clearance should be allowed for if a steel pushrod is 10.0 in. long at 58°F and it is expected to operate at 220°F?

Solution

Data: $L_0 = 10.0$ in., $T_0 = 58°F$, $T = 220°F$, $\alpha = 6.5 \times 10^{-6}/°F$, (Table 13-1), $\Delta L = ?$
Equation: $\Delta L = \alpha L_0 \Delta T$
Substitute: $\Delta L = (6.5 \times 10^{-6}/°F)(10.0 \text{ in.})(220°F - 58°F)$
$\qquad = 0.0105$ in.

Even though thermal changes in length are usually relatively small compared with the original length, very large forces may arise because of them. These forces are quite capable of buckling even steel girders. Consequently, thermal elongation must be allowed for by special expansion joints in railway tracks and structures, such as bridges. In addition, U-shaped sections are included in pipes that experience large temperature changes (Fig. 13-1). Linear expansion also affects the fitting of engine parts, the length of a surveyor's measuring tape, the suspension of cables, the wire connections between electric circuit elements, and many other items.

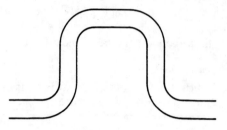

Figure 13-1 U-shaped pipe section to allow for expansion.

A **bimetallic strip** is composed of two different metal rods that are welded or riveted together lengthwise. Since each metal has a different coefficient of linear expansion, one metal rod expands more than the other when the strip is heated. If the composite rod is initially straight, it bends when it is heated. Thus a bimetallic strip may be used to measure temperature changes. For example, they may be used as the temperature-sensitive element of a thermostat (see Fig. 13-2).

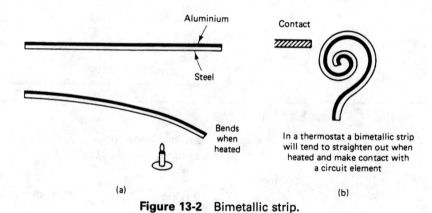

Figure 13-2 Bimetallic strip.

PROBLEMS

1. Determine the coefficient of linear expansion of a metal rod if it increases its length from 14.000 m to 14.014 m when the temperature increases from 0°C to 32°C.

2. Determine the coefficient of linear expansion of a metal rod if it increases its length from 58.000 ft to 58.025 ft when the temperature increases from 32°F to 118°F.

3. An aluminum girder is 6.750 00 m long at 18°C. What is its length when the temperature is **(a)** −32°C; **(b)** 34°C?

4. A steel girder is 15.000 00 ft long at 68°F. What is its length when the temperature is **(a)** 102°F; **(b)** 10°F?

5. What minimum gap should be left between steel girders that are 5.500 00 m long at 0°C if they will be subjected to a maximum temperature of 43°C?

6. A copper pipe 2.5000 m long at 20°C is used to carry hot water at 70°C. What is the length of the pipe at 70°C?

7. An aluminum tape is 25.000 00 ft long at 70°F. What is its length at **(a)** 105°F; **(b)** −15°F?

8. A steel I-beam is 7.2500 m long at 20°C. What is its change in length when the temperature changes to **(a)** 35°C; **(b)** −42°C?

9. A steel pushrod is 25.000 cm long at 23°C. What minimum clearance should be allowed for if it is expected to operate at 225°C?

10 A steel beam is 8.250 00 m long at 20°C. At what temperature is its length (a) 8.257 00 m; (b) 8.241 00 m?

11 If a glass rod is 3.5000 ft long at 72°F, at what temperature is its length (a) 3.5050 ft; (b) 3.4930 ft?

12 A surveyor has a steel tape that reads correctly at 20°C. If she measures a length of 38.000 m at −5°C, what is the actual length?

13 An engineer has a steel ruler that reads correctly at 65°F. If he measures a length of 12.0055 in. at 15°F, what is the actual length?

14 In a circuit, a copper wire is heated from 15°C to 45°C. Determine the percentage increase in its length.

15 An aluminum beam is 6.150 00 m long at 20°C. Determine the temperature at which its length is (a) 6.158 25 m; (b) 6.143 75 m.

16 A soft-iron plate has a circular hole with a diameter of 2.500 00 cm at 20°C. To what temperature must the plate be heated to fit a circular rod with a diameter of 2.505 85 cm into the hole?

17 The end of a 2.000 00 ft long steel rod is rigidly joined to the end of a 3.000 00 ft long aluminum rod with an equal cross-sectional area. Determine the total change in length when the temperature changes by (a) 50°F; (b) −60°F.

13-2 AREA EXPANSION

Solids expand in three dimensions when they are heated. Occasionally, we need to determine the area expansion (two dimensions) of a solid. In this case the change in area is given by

$$\Delta A = 2\alpha A_0 \Delta T = 2\alpha A_0 (T - T_0) \qquad (13\text{-}4)$$

or the area at the temperature T is

$$A = A_0[1 + 2\alpha(T - T_0)] \qquad (13\text{-}5)$$

where A_0 is the initial area at the temperature T_0, T the final temperature, and α the thermal coefficient of linear expansion.

EXAMPLE 13-3

A steel cylinder of an engine has a cross-sectional area of 82.000 00 cm² at 20°C. What is its cross-sectional area at 320°C?

Solution

Data: $A_0 = 82.000\,00$ cm², $T = 320°C$, $T_0 = 20°C$, $\alpha = 1.2 \times 10^{-5}/°C$ (Table 13-1), $A = ?$
Equation: $A = A_0[1 + 2\alpha(T - T_0)]$
Substitute: $A = (82.000\,00\text{ cm}^2)[1 + 2(1.2 \times 10^{-5}/°C)(320°C - 20°C)]$
 $= 82.590\,40$ cm²

PROBLEMS

18 A steel cylinder has a cross-sectional area of 78.539 82 cm² at 20°C. Determine the area at 185°C.

19 A steel rod has a cross-sectional area of 1.200 00 in² at 68°F. What is its cross-sectional area at (a) 100°F; (b) −20°F?

20 A copper pipe has a cross-sectional area of 1.227 18 cm² at 20°C. What is its cross-sectional area at 78°C?

21. An aluminum I-beam has a cross-sectional area of 12.000 00 in² at 70°F. At what temperature is its cross-sectional area (a) 12.002 50 in²; (b) 11.997 90 in²?

22. A steel beam has a cross-sectional area of 56.000 00 cm² at 20°C. At what temperature is its cross-sectional area (a) 56.005 25 cm²; (b) 55.998 75 cm²?

13-3 VOLUME EXPANSION

The maximum density of water occurs at about 4°C (39°F) under standard conditions. Water actually increases its volume, and therefore its density decreases, when it is cooled from 4°C (39°F) to 0°C (32°F). This is why ice can float on water. All other materials (and water in other temperature ranges) increase their volume when they are heated.

The **thermal coefficient of volume expansion** β is defined as the fractional change in volume per degree increase in temperature:

$$\beta = \frac{\Delta V}{V_0 \Delta T} = \frac{V - V_0}{V_0(T - T_0)} \tag{13-6a}$$

or

$$V = V_0[1 + \beta(T - T_0)] \tag{13-6b}$$

where V_0 is the initial volume at the temperature T_0, V is the volume at the temperature T, and ΔV is the change in the volume when the temperature increases by ΔT.

The coefficient of volume expansion β is also expressed in units of per kelvin (K^{-1}) or per degree Celsius (1/°C) in SI, and per degree Fahrenheit (1/°F) in USCS. Some typical values are listed in Table 13-2.

TABLE 13-2
Typical Volume Expansion Coefficients

	$\beta/(10^{-3} K^{-1})$ or $\beta/(10^{-3}/°C)$	$\beta/(10^{-4}/°F)$
Liquids		
Benzene	1.18	6.6
Ethyl alcohol	1.04	5.78
Gasoline	1.1	6.0
Mercury	0.182	1.01
Water (above 10°C)	0.37	2.0
Gases		
Air	3.67	20.4
Carbon dioxide	3.7	20.6
Hydrogen	3.66	20.3
Nitrogen	3.67	20.3
Oxygen	4.86	27
Solids		
Glass (Pyrex)	0.026	0.144

EXAMPLE 13-4

If a mass of gasoline occupies 70.0 L at 20°C, what is its volume at 35°C?

Solution

Data: $V_0 = 70.0$ L, $T_0 = 20°C$, $T = 35°C$, $\beta = 1.1 \times 10^{-3}/°C$ (Table 13-2), $V = ?$

Equation: $V = V_0[1 + \beta(T - T_0)]$

Sec. 13-3 / Volume Expansion

Substitute: $V = (70.0 \text{ L})[1 + (1.1 \times 10^{-3}/°C)(35°C - 20°C)]$
$= 71.2 \text{ L}$

EXAMPLE 13-5

If a mass of mercury occupies 250 in^3 at 95°F, what volume does it occupy at $-15°F$?

Solution

Data: $V_0 = 250$ in^3, $T_0 = 95°F$, $T = -15°F$, $\beta = 1.01 \times 10^{-4}/°F$ (Table 13-2), $V = ?$
Equation: $V = V_0[1 + \beta(T - T_0)]$
Substitute: $V = (250 \text{ in}^3)[1 + (1.01 \times 10^{-4}/°F)(-15°F - 95°F)]$
$= 247 \text{ in}^3$

Note that there is a drop in the temperature in this example; therefore, ΔT is negative valued and the final volume is less than the initial volume.

For a solid that has the same thermal coefficient of linear expansion α in all directions, the thermal coefficient of volume expansion $\beta \approx 3\alpha$. In this case the volume at some temperature T is given by

$$V = V_0[1 + 3\alpha(T - T_0)] \tag{13-7}$$

where V_0 is the volume at some temperature T_0.

EXAMPLE 13-6

A 475.00 g silver ingot has a volume of 45.2381 cm^3 and a density of 10 500 kg/m^3 at 20°C. Determine **(a)** the volume of the ingot; **(b)** the density at 35°C.

Solution

Data: $m = 475.00$ g, $V_0 = 45.2381$ cm^3, $\rho = 10\,500$ kg/m^3, $T_0 = 20°C$, $T = 35°C$, $\alpha = 1.9 \times 10^{-5}/°C$, $V = ?$, $\rho = ?$
(a) *Equation:* $V = V_0[1 + 3\alpha(T - T_0)]$
Substitute: $V = (45.2381 \text{ cm}^3)[1 + 3(1.9 \times 10^{-5}/°C)(35°C - 20°C)]$
$= 45.2768 \text{ cm}^3 = 45.2768 \times 10^{-6} \text{ m}^3$
(b) *Equation:* $\rho = \dfrac{m}{V}$
Substitute: $\rho = \dfrac{0.475\,00 \text{ kg}}{45.2768 \times 10^{-6} \text{ m}^3} = 10\,491$ kg/m^3

Note that the density decreases as the volume increases, the mass is constant.

Thermometers

There are many methods of measuring temperatures with some precision. Most of these methods depend on the variation of some other physical property as the temperature changes.

The most commom thermometer is the **liquid** (usually mercury) **in glass thermometer.** It consists of an evacuated glass capillary tube that is attached to a bulb full of the liquid. As the temperature rises, the volume of the liquid in the bulb expands, and the liquid moves up the capillary tube. Since the cross-sectional area of the capillary tube is approximately constant, the volume change of the liquid (and therefore the temperature change) is directly proportional to the distance that the liquid moves up or down the capillary.

PROBLEMS

23. If a mass of benzene occupies 86.0 L at 15°C, what is its volume at 38°C?
24. If a fixed mass of gasoline occupies 825 in³ at 68°F, what is its volume at (a) 103°F; (b) 32°F?
25. Determine the change in temperature of water if its volume changes from 350.68 mL to 355.27 mL.
26. Determine the change in volume of 525 in³ of benzene when its temperature increases by 45°F.
27. Determine the thermal coefficient of volume expansion of a liquid that occupies 86.000 L at 22°C and a volume of 87.860 L at 70°C.
28. What is the change in the temperature of water if its volume changes from 21.400 in³ to 21.680 in³?
29. If a mass of water occupies 136 mL at 93°C, what volume does it occupy at 27°C?
30. If a mass of water occupies 4.810 ft³ at 200°F, what volume does it occupy at 80°F?
31. If a mass of gasoline occupies 45.000 00 L at 20°C, what volume does it occupy at (a) 35°C; (b) −35°C?
32. If a U.S. gallon of water occupies 231 in³ at 68°F, at what temperature is its volume (a) 237 in³; (b) 230 in³?
33. If a mass of benzene occupies 12.000 L at 20°C, at what temperature is its volume (a) 12.288 L; (b) 11.483 L?
34. If the density of mercury at 0°C is 13 595.5 kg/m³, what is the density at 30°C? (*Hint:* The mass is constant. Consider the change in an initial volume of 1 m³.)
35. If the weight density of mercury at 32°F is 849.72 lb/ft³, what is its density at (a) 105°F; (b) −40°F?
36. A motorist buys 75.0 L of gasoline at 20°C. If gasoline costs 52.0 cents/L, how much would she pay for the same mass of gasoline at a temperature of (a) 35°C; (b) −42°C?
37. A steel ingot has a volume of 5250.00 cm³ at 20°C. Determine the volume of the ingot at (a) 42°C; (b) 275°C; (c) −38°C.
38. An aluminum ingot has a volume of 7520.000 cm³ at 20°C. At what temperature is the volume of the ingot (a) 7550.00 cm³; (b) 7475.000 cm³?
39. (a) Determine the volume of a 775 g bar of silver at 20°C if the density is 10 500 kg/m³. (b) What is the volume of the same mass of silver at 125°C? (c) What is the density at 125°C?

13-4 GAS LAWS

Gases are fundamentally different from liquids and solids. A fixed mass of gas has an indefinite volume since it occupies all of the available space. The spacing between gas molecules is often much greater than the spacing between molecules of liquids and solids. Therefore, gases are more easily compressed and are lighter (less dense).

Boyle's Law: Constant Temperature

Gas molecules are in continual rapid motion, colliding frequently with each other and the walls of any container. The pressure exerted by a confined gas is due mainly to this molecular bombardment of the container walls. If the distance between the walls of a container of gas is reduced, the volume of the gas is also reduced; however, if the temperature is constant, the molecules still have the same average speed, and they collide more frequently with the walls, producing an increased pressure.

The experimental investigation of the relationship between the pressure and volume of a confined gas was first performed by Robert Boyle (1627–1691) in 1662. His results, now called **Boyle's law**, are:

Sec.13-4 / Gas Laws

If a fixed mass of gas is maintained at a constant temperature, its absolute pressure p is inversely proportional to its volume V.

Thus

$$p \propto \frac{1}{V} \quad \text{or} \quad pV = \text{constant} \tag{13-8}$$

Therefore,

$$p_1 V_1 = p_2 V_2 \quad \text{(at constant temperature)} \tag{13-9}$$

where p_1 is the absolute pressure when the volume is V_1, and p_2 is the absolute pressure when the volume is V_2.

Remember: **Gauge pressures** must be converted into absolute pressures by the addition of atmospheric pressure before they can be used in Boyle's law (see Chapter 11).

EXAMPLE 13-7

What volume of air at atmospheric pressure 101 kPa is required to fill a 16.0 L cylinder to a gauge pressure of 606 kPa?

Solution

Data: $p_1 = 101$ kPa, $V_2 = 16.0$ L, $p_2 = 606$ kPa $+ 101$ kPa $= 707$ kPa absolute pressure, $V_1 = ?$

Equation: $p_1 V_1 = p_2 V_2$

Rearrange: $V_1 = \dfrac{p_2 V_2}{p_1}$

Substitute: $V_1 = \dfrac{(707 \text{ kPa})(16.0 \text{ L})}{101 \text{ kPa}} = 112$ L

EXAMPLE 13-8

Determine the final gauge pressure when 10.0 ft³ of oxygen at an atmospheric pressure of 15.0 lb/in² are pumped into a tank of volume 2.00 ft³ if the temperature remains constant.

Solution

Data: $V_1 = 10.0$ ft³, $p_1 = 15.0$ lb/in², $V_2 = 2.00$ ft³, $p_2 = ?$

Equation: $p_1 V_1 = p_2 V_2$

Rearrange: $p_2 = \dfrac{p_1 V_1}{V_2}$

Substitute: $p_2 = \dfrac{(15.0 \text{ lb/in}^2)(10.0 \text{ ft}^3)}{2.00 \text{ ft}^3} = 75.0$ lb/in²

But this is an absolute pressure; therefore, to find the gauge pressure, we must subtract the atmospheric pressure 15.0 lb/in². Thus

$$p_2 = 75.0 \text{ lb/in}^2 - 15.0 \text{ lb/in}^2 = 60.0 \text{ lb/in}^2 \text{ gauge.}$$

Like liquids and solids, gases expand when they are heated. However, gases are far more sensitive to pressure and volume changes, and their coefficients of thermal expansion are much larger.

Constant Pressure

If the pressure is constant, the change in the volume ΔV of a fixed mass of gas is directly proportional to its change in temperature ΔT:

$$\Delta V = \beta V_0 \, \Delta T$$

where β is the coefficient of volume expansion.

Since gas molecules are very small and relatively far apart, and molecules of different gases all behave in approximately the same way, the thermal coefficient of volume expansion at 0°C has the value of approximately 1/273°C for all low-pressure gases.

A graph of the volume V versus the temperature T of a fixed mass of gas under a constant low pressure yields a straight line (Fig. 13-3). The extension of the line through the temperature axis coincides with a temperature of -273.15°C or -460.27°F.

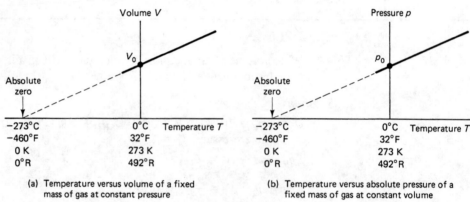

(a) Temperature versus volume of a fixed mass of gas at constant pressure

(b) Temperature versus absolute pressure of a fixed mass of gas at constant volume

Figure 13-3

If we take -273.15°C or -460.27°F as the origins of new temperature scales called **absolute temperatures,** the volume V is directly proportional to the absolute temperature T. Therefore

If a fixed mass of gas is maintained at a constant presssure, its volume V is directly proportional to its absolute temperature T.

$$V \propto T \quad \text{or} \quad \frac{V_1}{T_1} = \frac{V_2}{T_2} \quad \text{(constant pressure)} \tag{13-10}$$

The temperature -273.15°C (-460.27°F) is called **absolute zero.** Of course, in practice the gas would liquefy or even solidify before absolute zero is reached. The SI unit for absolute temperature is called a **kelvin** (symbol K); it is taken as a base unit. Absolute zero is taken as 0 K, and a temperature change of 1 K is equivalent to a temperature change of 1°C; therefore, the relationship between kelvin and Celsius temperatures is

$$T_K = T_C + 273 \tag{13-11}$$

The USCS unit for absolute temperature is the **Rankine** temperature. Absolute zero is taken as zero degrees Rankine, (0°R) and a temperature change of 1°R is approx-

Sec. 13-4 / Gas Laws

imately equal to a temperature change of 1°F; therefore, the relationship between Fahrenheit and Rankine temperatures is

$$T_R = T_F + 460 \tag{13-12}$$

EXAMPLE 13-9

If a mass of gas occupies 8.20 L at 23°C, what volume will it occupy at 80°C if the pressure is constant?

Solution

Data: $V_1 = 8.20$ L, $T_1 = (23 + 273)$ K $= 296$ K,
$T_2 = (80 + 273)$ K $= 353$ K, $V_2 = ?$

Equation: $\dfrac{V_2}{T_2} = \dfrac{V_1}{T_1}$

Rearrange: $V_2 = \dfrac{V_1 T_2}{T_1}$

Substitute: $V_2 = \dfrac{(8.20 \text{ L})(353 \text{ K})}{296 \text{ K}} = 9.78$ L

EXAMPLE 13-10

At 86°F a mass of gas occupies 28.0 ft³. At what temperature will it occupy 26.4 ft³ at the same pressure?

Solution

Data: $V_1 = 28.0$ ft³, $V_2 = 26.4$ ft³, $T_1 = (86 + 460)$°R $= 546$°R, $T_2 = ?$

Equation: $\dfrac{V_2}{T_2} = \dfrac{V_1}{T_1}$

Rearrange: $T_2 = \dfrac{T_1 V_2}{V_1}$

Substitute: $T_2 = \dfrac{(546 \text{ °R})(26.4 \text{ ft}^3)}{28.0 \text{ ft}^3} = 515$°R

Therefore, $T_2 = (515 - 460)$°F $= 55$°F

Constant Volume

If the volume of the fixed mass of gas is kept constant, as its temperature increases the kinetic energy and speeds of the molecules increase, and the gas molecules collide more frequently with any constraining walls.

If the volume of a fixed mass of gas is constant, its absolute pressure p is directly proportional to its absolute temperature T. Thus

$$p \propto T \quad \text{or} \quad \frac{p_1}{T_1} = \frac{p_2}{T_2} \quad \text{(constant volume)} \tag{13-13}$$

EXAMPLE 13-11

A car tire is inflated to a gauge pressure of 28.0 lb/in² when the temperature is 32°F. If the volume of the tire does not change and atmospheric pressure is 14.7 lb/in², what is the gauge pressure when the temperature is 99°F?

Solution

Data: The absolute conditions are $p_1 = 28.0$ lb/in² + 14.7 lb/in² = 42.7 lb/in², $T_1 = (32 + 460)°R = 492°R$, $T_2 = (99 + 460)°R = 559°R$; $p_2 = ?$

Equation: $\dfrac{p_1}{T_1} = \dfrac{p_2}{T_2}$

Rearrange: $p_2 = \dfrac{p_1 T_2}{T_1}$

Substitute: $p_2 = \dfrac{(42.7 \text{ lb/in}^2)(559°R)}{492°R}$

$= 48.5$ lb/in² (absolute)

Therefore, the final gauge pressure is

$$p = 48.5 \text{ lb/in}^2 - 14.7 \text{ lb/in}^2 = 33.8 \text{ lb/in}^2 \text{ (gauge)}$$

Ideal Gas Law

If we combine these relationships for temperature, pressure, and volume of gases, we obtain the more useful **ideal gas law:**

$$\frac{p_1 V_1}{T_1} = \frac{p_2 V_2}{T_2} \tag{13-14}$$

where p is the absolute pressure, V the volume, and T the absolute temperature (kelvin or Rankine).

EXAMPLE 13-12

If 4.20 L of gas at an absolute pressure of 120 kPa and a temperature of 0°C is heated until the temperature is 250°C and the absolute pressure is 360 kPa, what is the final volume of the gas?

Solution

Data: $V_1 = 4.20$ L, $p_1 = 120$ kPa, $p_2 = 360$ kPa, $T_1 = (0 + 273)$ K = 273 K, $T_2 = (250 + 273)$ K = 523 K, $V_2 = ?$

Equation: $\dfrac{p_1 V_1}{T_1} = \dfrac{p_2 V_2}{T_2}$

Rearrange: $V_2 = \dfrac{V_1 p_1 T_2}{p_2 T_1}$

Substitute: $V_2 = \dfrac{(4.20 \text{ L})(120 \text{ kPa})(523 \text{ K})}{(360 \text{ kPa})(273 \text{ K})} = 2.68$ L

PROBLEMS

40 A fixed mass of gas exerts a gauge pressure of 103 kPa when its volume is 339 L. If the temperature remains constant, what volume does it occupy when its gauge pressure is 261 kPa?

41 A fixed mass of gas exerts a gauge pressure of 15.0 lb/in² when its volume is 12.0 ft³. If the temperature remains constant, what volume does it occupy when its gauge pressure is 38.0 lb/in²?

42 Determine the volume of oxygen at an atmospheric pressure of 101 kPa that is required to fill a cylinder of volume 3.50 L to a gauge pressure of 800 kPa.

Summary

43 Determine the volume of nitrogen at 14.7 lb/in² that is required to fill a cylinder of volume 2.50 ft³ to a gauge pressure of 3.20 atm.

44 If 65.0 L of outside air at −15°C is drawn into a furnace and heated to 22°C at a constant pressure, what is the final volume of the air?

45 A car tire is inflated to a gauge pressure of 28.0 lb/in² when the temperature is 35°F. If the volume of the tire does not change, what is the gauge pressure when the temperature is **(a)** −35°F; **(b)** 82°F?

46 A car tire is inflated to a gauge pressure of 195 kPa when the temperature is 2.0°C. If the volume of the tire does not change, what is the gauge pressure when the temperature is **(a)** −37°C; **(b)** 28°C?

47 A gas occupies a volume of 25.0 L at 15°C. At what temperature will the gas occupy a volume of 72.0 L at the same pressure?

48 A tank contains 5.00 ft³ of oxygen at a temperature of 60°F and an absolute pressure of 35.0 lb/in². Determine the absolute pressure of the oxygen if its volume is increased to 25.0 ft³ and the temperature is raised to 450°F.

49 Helium has a density of 0.18 kg/m³ at 0°C and a pressure of 101 kPa. Determine the density at a pressure of 510 kPa and a temperature of 0°C. (*Hint:* The mass is constant; therefore, the volume varies inversely as the density.)

50 If 135 L of gas at atmospheric pressure 101 kPa and 0°C is used to fill a 15.0 L tank at 0°C, what is the gauge pressure of the gas in the tank?

51 A tank is filled with 35.0 ft³ of air at an atmospheric pressure of 14.7 lb/in² and a temperature of 65°F. Determine the final gauge pressure in the tank if the final volume is 2.50 ft³ and the temperature of the gas in the tank is 125°F.

52 If 8.50 L of gas at a gauge pressure of 720 kPa and a temperature of 0°C is heated until the gauge pressure is 1500 kPa and the temperature is 250°C, what is the final volume? Assume that atmospheric pressure is 101 kPa.

53 A gas at a gauge pressure of 525 kPa occupies 12.0 L at 0°C. What volume does it occupy if the gauge pressure is **(a)** 85 kPa; **(b)** 725 kPa; **(c)** tripled?

54 If 250 L of water is pumped at constant temperature into a pressure tank that contains 650 L of air at 101 kPa pressure, what is the final pressure in the tank?

55 A tank is filled with 325 ft³ of gas at an atmospheric pressure of 14.7 lb/in² and a temperature of 65°F. What is the final gauge pressure in the tank if the final volume is 25.0 ft³ and the temperature in the tank is 115°F?

56 A tank is filled with 975 L of air at an atmospheric pressure of 101 kPa and a temperature of 18°C. Determine the final gauge pressure in the tank if the final volume is 71 L and the temperature of the gas in the tank is 52°C.

57 A mass of gas at 60°F and an absolute pressure of 30.0 lb/in² occupies a volume of 6.00 ft³. Find **(a)** the volume of the gas at a temperature of 85°F and an absolute pressure of 50.0 lb/in²; **(b)** the pressure when the temperature is 32°F and the volume is 425 ft³; **(c)** the temperature of the gas if the absolute pressure is 32.0 lb/in² and the volume is 2.50 ft³.

58 A sample of gas at 15°C under an absolute pressure of 215 kPa occupies 27.5 L. Determine **(a)** the volume of the gas at a temperature of 35°C and an absolute pressure of 325 kPa; **(b)** the pressure when the temperature is 0°C and the volume is 35.0 L; **(c)** the temperature of the gas if the absolute pressure is 315 kPa and the volume is 22.0 L.

SUMMARY

Linear expansion:

$$\alpha = \frac{\Delta L}{L_0 \Delta T} \qquad L = L_0[1 + \alpha(T - T_0)]$$

Areas:

$$\Delta A = 2\alpha A_0 \Delta T \qquad A = A_0[1 + 2\alpha(T - T_0)]$$

Volumes:

$$\Delta V = 3\alpha V_0 \Delta T \qquad V = V_0[1 + 3\alpha(T - T_0)]$$

and

$$\Delta V = \beta V_0 \Delta T \qquad V = V_0[1 + \beta(T - T_0)]$$

where $\beta \approx 3\alpha$.

Gas laws:

$$T_K = T_C + 273 \qquad T_R = T_F + 460$$

$$p_{abs} = p_{gauge} + p_0$$

$$p_1 V_1 = p_2 V_2 = \text{constant} \qquad (T \text{ constant}) \; \textit{Boyle's law}$$

$$\frac{V_1}{T_1} = \frac{V_2}{T_2} = \text{constant} \qquad (p \text{ constant})$$

$$\frac{p_1}{T_1} = \frac{p_2}{T_2} = \text{constant} \qquad (V \text{ constant})$$

General gas equation:

$$\frac{p_1 V_1}{T_1} = \frac{p_2 V_2}{T_2}$$

Also, for a constant mass, the density $\rho \propto 1/V$.

QUESTIONS

1. Define the following terms: **(a)** coefficient of linear expansion; **(b)** coefficient of volume expansion; **(c)** gauge pressure.
2. Describe, in terms of thermal expansion, how you would fit a rim tightly over a wheel.
3. It is easier to remove the screw cap from a bottle after it has been warmed by placing it under hot water. Explain why.
4. It is possible to buckle a can of air by heating it and then sealing it when hot. As the can cools, it buckles. Explain why.
5. Describe the results of the gas laws in terms of the motions of the gas molecules.
6. Describe the operation of a thermostat.
7. Why must absolute temperatures and absolute pressures be used in the gas equations?
8. Explain the principles of operation of a mercury in glass thermometer.

REVIEW PROBLEMS

1. Determine the thermal coefficient of linear expansion of a metal rod if its length increases from 2.9841 m to 2.9881 m when the temperature increases from 4.5°C to 71.0°C.
2. What is the thermal coefficient of linear expansion of a metal rod if its length increases from 9.8000 ft to 9.8130 ft when the temperature increases from 40°F to 160°F?
3. A steel cable is 300.000 m long at 23°C. What is its length when the temperature drops to −40°C?

Review Problems

4. A steel wire is 2200 ft long at 40°F. What is the change in its length when the temperature changes to (a) 95°F; (b) −15°F?

5. A steel cable of a suspension bridge is 670.000 m long at 4.5°C. What is its increase in length if the temperature increases to 35.0°C?

6. An iron steam line is 125.000 m long at 32°C. What is its length when the temperature drops to −45°C?

7. If a mass of gasoline occupies 35.0 L at 5°C, what volume does it occupy at 30°C at the same pressure?

8. A steel tape measures the length of a copper rod as 1.000 000 ft at 68°F. What will it indicate the length of the copper rod to be at (a) 40°F; (b) 95°F? (*Hint:* The copper and the steel change lengths.)

9. Determine the change in the temperature of a container of mercury if its volume changes from 10.000 in^3 to 10.205 in^3.

10. If 50.0 m^3 of air is heated from 0°C to 75°C at constant pressure, what is its new volume?

11. What volume of air at absolute pressure of 15.0 lb/in^2 is required to fill a 1.50 ft^3 cylinder to a gauge pressure of 48.0 lb/in^2?

12. A cylinder of helium gas has a gauge pressure of 400 kPa at 15°C. What is its gauge pressure at 85°C?

13. The absolute pressure in a cylinder of a car engine rises from 401 kPa to 1.25×10^6 Pa during combustion of the gas and air mixture. If the initial temperature of the gas–air mixture was 200°C, what is the temperature after combustion?

14. If 4.00 L of helium is contained in a cylinder at 20°C, to what temperature must the helium be heated to triple the volume?

15. What is the final volume if 35.5 L of a gas at an absolute pressure of 101 kPa and a temperature of 5.6°C is heated to 31.0°C and an absolute pressure of 295 kPa?

16. A sample of gas occupies 15.0 L at 23°C and an absolute pressure of 125 kPa. Determine (a) the volume when the temperature is −38°C and the gauge pressure is 475 kPa; (b) the temperature when the gauge pressure is 535 kPa and the volume is 12.0 L; (c) the gauge pressure when the temperature is 42°C and the volume is 12.0 L.

17. A sample of gas occupies 8.00 L at 0°C under an absolute pressure of 115 kPa. Determine (a) its volume at 180°C under an absolute pressure of 212 kPa; (b) the temperature if the volume is tripled and the pressure is doubled.

18. What is the final volume if 1.25 ft^3 of a gas at an absolute pressure of 14.7 lb/in^2 and a temperature of 42°F is heated to 88°F and an absolute pressure of 43.0 lb/in^2?

14
VIBRATIONS AND WAVES

The transmission of mechanical energy from one place to another may be accomplished either by the direct transfer of matter, which has kinetic energy due to its motion, or by means of a vibratory disturbance called a **wave.**

The motion of a wave involves no transfer of matter, yet waves possess both kinetic energy and momentum. Sounds, light, and radio and television signals are all transmitted as waves. Some kinds of waves, such as sound waves, require a material medium (solid, liquid, or gas). Electromagnetic waves, such as light, radio, and television signals, do not require any material medium and are able to travel through a vacuum as well as through some materials. Even though many different kinds of waves exist, they behave alike in many respects.

14-1 VIBRATIONS

Objects will usually vibrate at some natural rate that may depend on the mass and the elastic properties. Some examples of mechanical vibrations are illustrated in Fig. 14-1. These are all examples of a **periodic motion,** in which the same motion is repeated at regular time intervals. One complete to-and-fro oscillation or vibration is called a **cycle.** The **period** T of the vibration or oscillation is the time taken to complete one cycle. Therefore, if the system completes N cycles in an elapsed time t, the period

$$T = \frac{\text{elapsed time}}{\text{number of cycles}} = \frac{t}{N} \qquad (14\text{-}1)$$

The frequency of the vibration or oscillation is the number of complete cycles per second.

$$f = \frac{\text{number of cycles}}{\text{elapsed time}} = \frac{N}{t} \qquad (14\text{-}2)$$

Sec. 14-1 / Vibrations

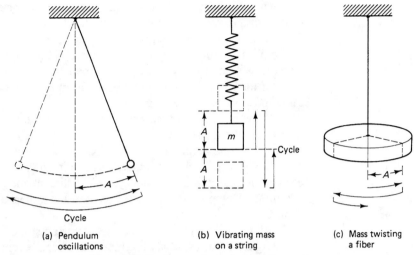

Figure 14-1 Examples of periodic motion.

Therefore,

$$f = \frac{1}{T} \tag{14-3}$$

The SI unit for frequency is called a **hertz** (symbol Hz); it is equal to the reciprocal of 1 second.

$$1 \text{ Hz} = \frac{1}{s}$$

The **amplitude** A of the vibration or oscillation is the maximum displacement from the rest position (see Fig. 14-1).

EXAMPLE 14-1

Determine (a) the period and (b) the frequency of a pendulum that completes 75.0 cycles in 30.0 s.

Solution

Data: $N = 75.0$ cycles (a cycle is not really a unit), $t = 30.0$ s

Equation:

(a) $T = \dfrac{t}{N}$ (b) $f = \dfrac{N}{t}$

Substitute:

(a) $T = \dfrac{30.0 \text{ s}}{75.0} = 0.400 \text{ s}$ (b) $f = \dfrac{75.0}{30.0 \text{ s}} = \dfrac{2.50}{\text{s}} = 2.50$ Hz

Check: $f = \dfrac{1}{T}$ or 2.50 Hz $= \dfrac{1}{0.400 \text{ s}}$

If energy is lost by the vibrating (or oscillating) system, it is said to be **damped**, and the amplitude decreases. However, we may supply additional energy by applying an oscillating (periodic) force. If the frequency of this force is not the same as the natural frequency of the vibration, the additional energy is usually dissipated and the resulting vibration is small.

On the other hand, when the frequency of the oscillating force is the same as the natural frequency, the energy and amplitude of the vibration may increase substantially. This is called **resonance.** For example, consider a child on a swing. To increase the amplitude of the swing and make the child go higher, the swing must be pushed with the same frequency as the natural frequency of the oscillation. We push and then wait for the child to complete a to-and-fro motion, and then we push the child at the same location as before (Fig. 14-2).

If the natural frequency is an integer (whole number) multiple of the applied frequency, the amplitude of the vibration may still increase. For example, we could still increase the amplitude by pushing the swing on every second complete cycle.

Resonance is a very important phenomenon that is used in electrical timing, oscillation and memory circuits, musical instruments, and a variety of mechanical vibrations. Damping is used to reduce unwanted mechanical vibrations and electrical "noise."

Vibrations may have disastrous effects on smooth-running machinery and relatively rigid structures. Many machines have been severely damaged or even destroyed because internal vibrations produced unwanted friction between moving parts and even caused some machines to break free of their mounts.

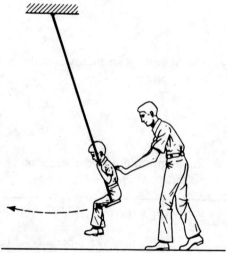

Figure 14-2 A swing has a natural frequency of oscillation. We push at that frequency to increase the amplitude.

PROBLEMS

1. What is the frequency of a sound if the period is 3.91 ms?
2. **(a)** What is the function of the pendulum in a grandfather clock? **(b)** If the pendulum completes 3.50 cycles in 7.00 s, determine its period and frequency.
3. If a stroboscope completes 630 flashes in 1.00 min, what is **(a)** the frequency; **(b)** the period?
4. How many times will a stroboscope flash in 28.0 s if its frequency is 14.0 Hz?
5. When tuning a car, a mechanic's timing light flashes 105 times in 60.0 s. Determine the frequency and period of the timing light.

14-2 WAVES

Waves are produced by vibrating or oscillating systems. A **wave** is a vibratory disturbance that transfers energy from one location to another without transfer of matter. A mechanical wave can exist only if the particles of the medium are close together so that the vibrational energy is transmitted from one particle to the next. Therefore, the medium must be relatively dense and elastic in order for the wave to travel very far.

The transfer of energy by a wave may be illustrated by creating a pulse in a spring (or rope) that is stretched between two points (Fig. 14-3). One end of the spring is quickly moved up and down, forming a pulse, which moves along the spring to the other end, where the pulse energy is evident to anyone holding the spring. Even though energy was transmitted along the spring, the coils of the spring were not transferred from one end to the other, but merely moved up and down as the pulse passed.

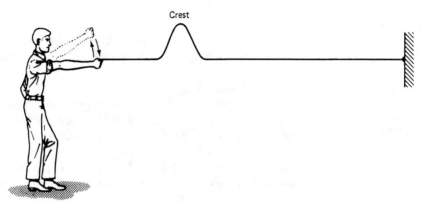

Figure 14-3 Creation of a pulse.

We can create a full cycle of a wave by quickly moving the spring in one direction, to create a **trough,** and then in the other direction to create a **crest** (Fig. 14-4). As a crest passes, the spring segments are displaced in one direction, and they are displaced in the opposite direction as the trough passes. This type of wave, in which the particles of the medium (spring coils in this case) vibrate in a direction perpendicular to the energy flow (the direction of the wave velocity), is called a **transverse wave** (see Fig. 14-5). Some examples of transverse waves are electromagnetic radiation, and the waves in vibrating strings.

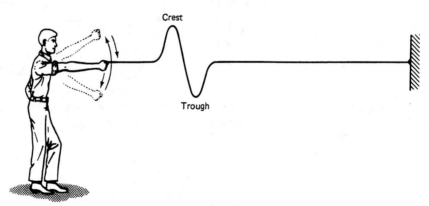

Figure 14-4 Creation of a wave.

Electromagnetic radiation consists of time-varying electric and magnetic fields. Even though these types of transverse waves do not require a material medium (they travel through a vacuum), they can also pass through some materials, such as air. Radio and television signals and visible light are all examples of electromagnetic radiation (see Chapter 16).

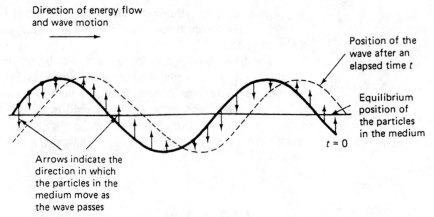

Figure 14-5 Transverse wave showing that the particles of the transmitting medium vibrate perpendicular to the wave velocity.

In a **longitudinal** or **compression wave** the particles of the transmitting medium vibrate along the same line as the energy flow (wave velocity; Fig. 14-6). An external compressive force causes the particles of the medium to "crowd together," forming a **condensation.** The crowded particles then repel each other as they attempt to regain their regular spacing, and since particles at *a* cannot move in the direction of the compressive force, particles at *b* are pushed closer to particles at *c*. The process is then repeated as region *c* is pushed away from *b*, and so on (Fig. 14-7).

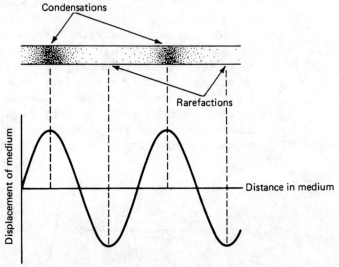

Figure 14-6 A compressive longitudinal wave causes condensations and rarifications in the medium.

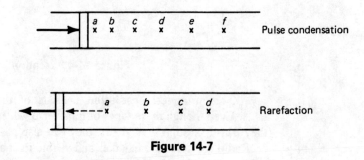

Figure 14-7

When the external force is moved in the opposite direction, it creates a region of reduced density as the particles move farther apart, producing a **rarefaction.** Particles at a move to fill the region of reduced density producing a rarefaction between a and b. This causes particles at c to move toward b, and so on (Fig. 14-7). Condensations correspond to higher-density regions, while rarefactions are the lower-density regions. By applying a series of fluctuating forces to an elastic medium, a series of condensations and rarefactions is transmitted.

Longitudinal waves may also be illustrated by the vibrations produced in a stretched spring when the coils are distorted longitudinally (compressed, for example) and then released. In this case the coils vibrate about their rest positions in a direction that is parallel to the energy flow or wave motion. Other important examples of longitudinal waves are sound waves, shock waves, and ultrasonic waves.

When a wave passes through a material medium, the particles of that medium vibrate about their rest positions. The **amplitude** A of the wave is the maximum displacement of these particles from their rest positions.

As a wave passes, the time interval between the arrivals of successive crests is the **period** T of the wave and the reciprocal of the period is the **frequency** f. These values are the same as the period and frequency of the vibrations of the particles in the medium; therefore,

$$f = \frac{1}{T}$$

Particles in the transmitting medium are said to have the same **phase of vibration** if they have the same displacement (magnitude and direction) and they are moving with the same velocity (speed and direction). Therefore, in Fig. 14-8, the particles at a and e are in phase, particles at b and f are in phase, and particles at c and g are in phase. Even though a and b have the same displacement, they are not in phase, because their velocities are different.

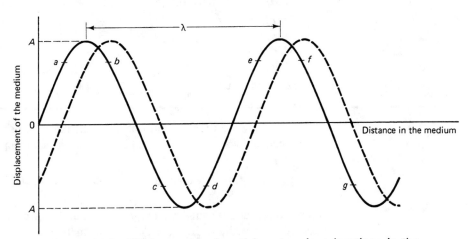

Figure 14-8 Displacement of particles at various locations in the medium.

The **wavespeed** v is the time rate at which a particular part of the wave (such as a crest or trough) travels through the medium.

The **wavelength** λ is the distance between the centers of two successive crests (or troughs) for transverse waves, or successive condensation (or rarefactions) for longitudinal waves.

Since a wave advances a distance equal to the wavelength in a time interval equal to the period, the wavespeed

$$v = \frac{\text{distance traveled}}{\text{time taken}} = \frac{\lambda}{T}$$

or since $T = 1/f$,

$$v = \frac{\lambda}{T} = f\lambda \qquad (14\text{-}4)$$

This is a very important relationship that is valid for all periodic waves that travel at a constant speed.

EXAMPLE 14-2
If the speed of sound in air is 1090 ft/s, what is the frequency of a sound that has a wavelength of 2.25 ft?

Solution

Data: $v = 1090$ ft/s, $\lambda = 2.25$ ft, $f = ?$

Equation: $v = f\lambda$

Rearrange: $f = \dfrac{v}{\lambda}$

Substitute: $f = \dfrac{1090 \text{ ft/s}}{2.25 \text{ ft}} = 484$ Hz

EXAMPLE 14-3
A source emits light with a frequency of 6.00×10^{14} Hz. If the speed of light is 3.00×10^8 m/s, what is the wavelength?

Solution

Data: $v = 3.00 \times 10^8$ m/s, $f = 6.00 \times 10^{14}$ Hz, $\lambda = ?$

Equation: $v = f\lambda$

Rearrange: $\lambda = \dfrac{v}{f}$

Substitute: $\lambda = \dfrac{3.00 \times 10^8 \text{ m/s}}{6.00 \times 10^{14} \text{ Hz}} = 5.00 \times 10^{-7}$ m

---

When a wave passes from one medium into another of different density, the wave speed and wavelength change, but their ratio, the frequency, remains the same.

$$f = \frac{v}{\lambda} = \text{constant} \qquad \text{(when waves change medium)}$$

Some waves, such as sound waves, speed up and their wavelength increases when they enter a denser medium. However, electromagnetic waves slow down, and their wavelength is reduced when they pass into a denser material.

PROBLEMS

6. Determine the frequency and period of a sound wave that has a wavelength of 21.9 cm and a speed of 331 m/s.
7. Determine the wavelength of a sound that has a frequency of 256 Hz if the speed of sound in air is **(a)** 330 m/s; **(b)** 1086 ft/s.

8. Light is an electromagnetic wave that has a speed of 3.0×10^8 m/s in a vacuum. (a) Determine the period of light that has a wavelength of 560 nm. (b) Calculate its frequency.

9. A source emits light at a frequency of 6.00×10^{14} Hz. If the speed of light is 9.82×10^8 ft/s, determine (a) its period; (b) its wavelength.

10. Microwaves, a form of electromagnetic radiation, travel at a speed of 3.00×10^8 m/s or 186 000 mi/s in a vacuum. What is the frequency of microwaves that have a wavelength of (a) 3.00 cm; (b) 3.45 cm; (c) 0.0750 ft; (d) 1.20 in.?

11. (a) Determine the frequency of a wavetrain that is moving at a speed of 26.0 m/s if the wavelength is 10.2 cm. (b) What is the period of the wave?

12. Determine the wavelength of a wave if its period is 15.0 ms and its speed is 750 m/s.

13. Determine the frequency and period of a sound wave that has a wavelength of 0.72 ft and a speed of 1100 ft/s.

14. Determine the frequency of a wavetrain that is moving at a speed of 85.0 ft/s if the wavelength is 4.00 in.

15. Radio waves travel at a speed of 3.00×10^8 m/s (9.82×10^8 ft/s) in air. What are the wavelengths of the radio waves transmitted by AM radio stations at frequencies of (a) 150 kHz; (b) 68.0 kHz?

16. Radio waves travel at 3.00×10^8 m/s or 9.82×10^8 ft/s (186 000 mi/s) in air and the frequency range of FM radio is 88.0 MHz to 108 MHz. Determine the range of wavelengths for FM radio waves.

17. Light travels at 3.00×10^8 m/s in air. Determine the frequency of the light emitted by a helium–neon laser if the wavelength is 633 nm.

14-3 WAVEFRONTS

In the previous discussion we considered waves and wave motion in one dimension when viewed from one side. However, to understand many other characteristics of waves, we must consider their motion in two dimensions by viewing the waves from the top. This is accomplished by means of a shallow tank with a glass bottom, which is known as a **ripple tank** (Fig. 14-9). The tank is filled with water and a light is

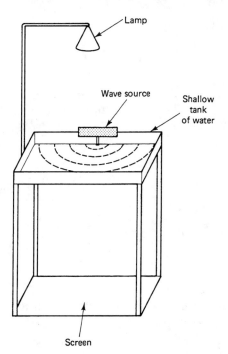

Figure 14-9 Ripple tank.

suspended above it. When waves are formed in the water, the crests and troughs act like lenses to concentrate and diverge the light, producing bright and dark images (Fig. 14-10). Bright areas correspond to crests, and dark areas to troughs. A wavefront is the continuous line representing a crest or a trough (or condensation or rarefaction in longitudinal waves).

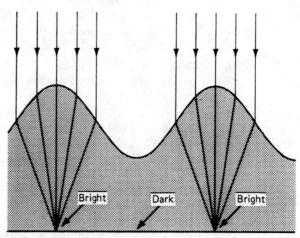

Figure 14-10 The crests and troughs of water in a ripple tank act like lenses to concentrate or diverge the light.

Wavefronts may have many different shapes. In a **plane wave,** the wavefronts are parallel (Fig. 14-11). They can be generated by moving a straight solid object, such as a length of wood, backward and forward in the water. **Circular waves** are generated by a small source, such as a vibrating ball (Fig. 14-12).

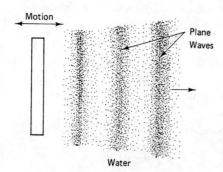

Figure 14-11 Plane waves are generated by moving a straight solid object from side to side in the water.

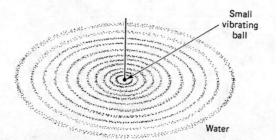

Figure 14-12 Circular waves are produced using a small vibrating source.

Sec. 14-3 / Wavefronts

Waves travel in directions that are perpendicular to their wavefronts. These "directions of energy flow" in a wave are represented by straight lines called **rays**. For convenience, in diagrams, we frequently draw rays instead of the corresponding wavefronts (Fig. 14-13). Waves may be completely reflected, or partially reflected and partially transmitted at a boundary between two different materials.

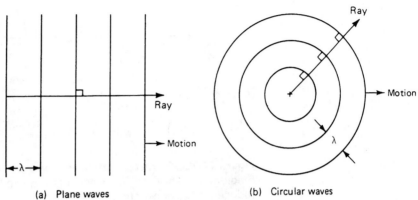

(a) Plane waves (b) Circular waves

Figure 14-13 Types of wavefronts.

Reflection

Suppose that a wave is incident at some angle on a smooth flat surface. The **angle of incidence** i is the angle between the normal **n** (perpendicular) to the reflecting surface and the incident ray. Similarly, the angle or reflection r is the angle between the normal **n** and the reflected ray (see Fig. 14-14). Then:

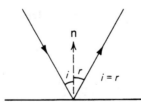

Figure 14-14 Laws of reflection.

1. The angle of incidence i is equal to the angle of reflection r.

$$i = r$$

2. The normal **n** to the surface and the incident and reflected rays all lie in the same geometric plane.

These statements are called the **laws of reflection**. They apply to both longitudinal and transverse waves.

Refraction

When a wave passes from one medium into another, its frequency remains the same, but its speed and wavelength change. As a result, the wavefront is bent at the boundary. This bending of the wavefront is called **refraction** (Fig. 14-15).

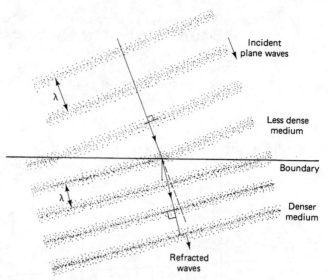

Figure 14-15 Refraction of plane waves when they pass from one medium into another. The wavelength, λ, and the wave speed, v, are reduced in the denser medium but the frequency f remains constant.

EXAMPLE 14-4

Light waves with a wavelength of 521 nm travel at 3.00×10^8 m/s in a vacuum and at 2.50×10^8 m/s in a transparent medium. Determine the wavelength in the medium.

Solution

Data: $v_1 = 3.00 \times 10^8$ m/s, $\lambda_1 = 521$ nm, $v_2 = 2.50 \times 10^8$ m/s, $\lambda_2 = ?$

Equations: $v = f\lambda$, $f =$ constant

Rearrange: $f = \dfrac{v_1}{\lambda_1} = \dfrac{v_2}{\lambda_2} =$ constant

Therefore, $\lambda_2 = \lambda_1 \dfrac{v_2}{v_1}$

Substitute: $\lambda_2 = \dfrac{(521 \text{ nm})(2.50 \times 10^8 \text{ m/s})}{3.00 \times 10^8 \text{ m/s}} = 434$ nm

Diffraction

Waves do not necessarily flow in straight lines from a source; they also have the ability to flow around barriers. This "bending" of the waves, delivering energy to regions behind barriers, is called **diffraction** (Fig. 14-16).

Even though waves can diffract around corners, the intensity of the wave is greatly reduced in regions directly behind obstacles. Therefore, while the center of the wave continues with relatively high intensity, the intensity of the curved portion behind the barrier is reduced.

PROBLEMS

18 Sound with a wavelength of 22.8 cm travels at 331 m/s in air and 5150 m/s in an iron railway track. What is the wavelength in the iron?

19 Sound with a wavelength of 9.00 in. travels at 1086 ft/s in air and at 16 900 ft/s in an iron railway track. What is the wavelength in iron?

Sec. 14-4 / Interference

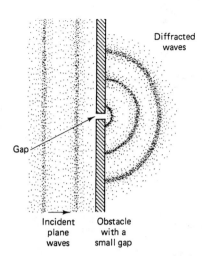

Figure 14-16 A small gap in an obstruction acts as a source of circular waves as the waves "bend" around the barrier.

20. Light at a frequency of 5.00×10^{14} Hz travels at 3.00×10^8 m/s in air. If the wavelength of the light in water is 462 nm, determine the speed of light in water.

21. In air a sound has a wavelength of 2.50 m and a speed of 330 m/s. What is the speed of that sound in helium if the wavelength is 7.31 m?

22. The speed of sound in air is 1086 ft/s and in helium it is 3170 ft/s. Determine the wavelength of sound in helium if its wavelength in air is 10.8 in.

23. Light with a frequency of 5.45×10^{14} Hz travels at 3.00×10^8 m/s (9.82×10^8 ft/s) in air and 1.94×10^8 m/s (6.35×10^8 ft/s) in a sample of glass. Determine (a) the wavelength in air; (b) the wavelength in the glass.

14-4 INTERFERENCE

Two or more waves may pass simultaneously through the same points in a medium, and each wave continues unaffected by the other waves that are present. However, each wave produces disturbances in the medium. This is called **interference**.

The total displacement of the particle in the medium is the vector sum of the displacements that the individual waves would produce alone at that instant. Therefore, if a wave A alone would produce a displacement s_A at some instant and a wave B alone would produce a displacement s_B at the same position and time, when they occur simultaneously at that position, their combined effect is to produce a displacement,

$$\mathbf{s} = \mathbf{s}_A + \mathbf{s}_B \tag{14-5}$$

This is called the **principle of superposition**.

If the waves arrive simultaneously in phase at some position (i.e., crest with crest and trough with trough), the waves are said to **interfere constructively** at that point. The resultant disturbance of the medium is increased (see Fig. 14-17a). In the ripple tank, constructive interference produces a bright region.

When two waves arrive simultaneously 180° out of phase (i.e., trough with crest) at some position, they are said to **interfere destructively** at that point. The resultant disturbance of the medium is reduced (see Fig. 14-17b). In the ripple tank this corresponds to darker areas.

The interference of waves has numerous applications. For example, precise measurements of lengths (with precisions approximately equal to half the wavelength of the light used) are possible using light sensors to count light fringes that are produced when two light waves interfere. Interference fringe patterns are also used to ensure that surfaces are flat (Fig. 14-18). Many other uses of interference will be discussed in later chapters.

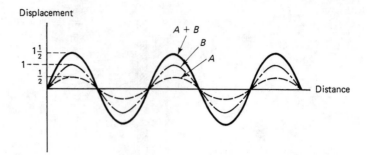

(a) Constructive interference between waves A and B which have the same speed, frequency, and phase, and emanate from the same point.

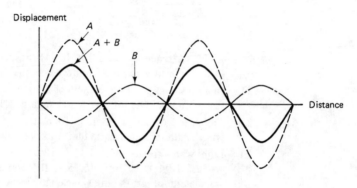

(b) Destructive interference between waves A and B which have the same frequency and speed, and emanate from the same point but differ in phase by 180°.

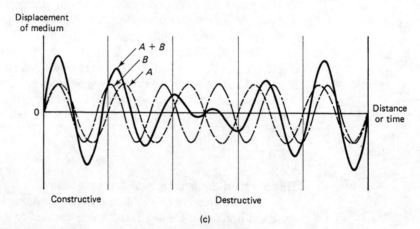

(c)

Figure 14-17

Figure 14-18 A fused quartz optical flat (which serves as a standard flat surface) is compared with a larger master flat under a light source. Interference between reflected light from the upper and lower surfaces and the narrow air wedge between them produces the pattern of interference fringes. Straight-line fringe patterns ensure that the smaller surface is flat. (Courtesy of the National Research Council of Canada.)

Sec. 14-4 / Interference

If the two waves have slightly different frequencies, a mixture of constructive and destructive interference patterns is produced (see Fig. 14-17c). When the two waves are in phase they support each other, and when they go out of phase they suppress each other, producing a phenomenon called **beats**. If sound waves of similar amplitudes are used, these beats are usually audible, since the intensity of the sound increases and a "beat" is heard when the waves interfere constructively. The intensity is reduced and the beat dies when destructive interference occurs. The frequency of the beats corresponds to the difference in the frequency between the two waves. For example, the "beat frequency" between a 505 Hz sound and a 500 Hz sound is the difference 5 Hz.

$$\text{beat frequency} = \text{difference in component frequencies} \quad (14\text{-}6)$$

In many electronic devices, such as radio receivers, incoming high-frequency signals are mixed with some other signal, producing a lower-frequency signal (the **beat frequency**) that is easier to amplify (make larger). This principle is used in heterodyne receivers.

Interference of waves constitutes the basis for radio and television communication systems. A constant high-frequency electromagnetic wave, which is called the **carrier wave** (radio, television, light, microwaves, etc.), is used to carry the signal from the source to the receiver. The sound or signal to be transmitted has a much lower variable frequency, which is used to modify the amplitude of the carrier wave (Fig. 14-19a). This process is called **amplitude modulation (AM)**. The modulated carrier signal is then transmitted to the receiver, where it is **demodulated** by a similar heterodyne process.

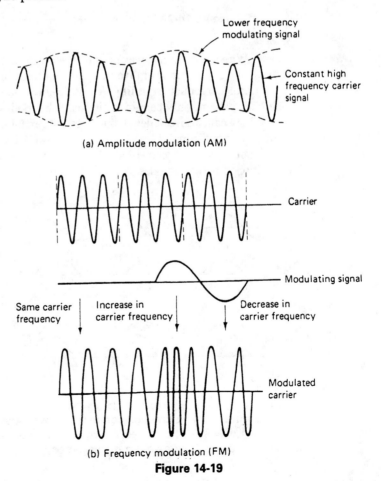

Figure 14-19

Frequency modulation (FM) is the variation of the carrier wave frequency while the amplitude of the carrier wave remains constant. In this case the magnitude of the frequency variation is related to the amplitude of the modulating signal, and the rate of carrier frequency variation corresponds to the frequency of the modulating signal (Fig. 14-19b). Frequency-modulated carrier waves are less susceptible to unwanted static, which produces fluctuations in the amplitude of the carrier.

PROBLEMS

24 If two similar waves have frequencies of 1500 Hz and 1505 Hz, what beat frequency will they produce?

25 If two sound sources produce a beat frequency of 35 Hz, and the frequency of one of the sources is 450 Hz, what can we say about the other frequency?

14-5 STANDING WAVES

If we stretch a spring between two points and generate a pulse in the spring, we see that when the pulse reaches the fixed end it is reflected, but the reflected pulse is displaced in the opposite direction (Fig. 14-20).

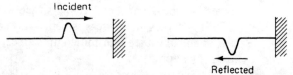

Figure 14-20 Reflection of a wave in a string from a fixed point.

When two similar waves with the same frequency and amplitude (or a single wave and its reflection) move in opposite directions through the same medium, they interfere to produce **standing** (or **stationary**) **waves.** In the standing-wave interference pattern, only certain positions called **nodes** N are never displaced, and other positions called **antinodes** A fluctuate with maximum amplitudes equal to twice the amplitude of the individual waves (Fig. 14-21). The distance between any two successive nodes is equal to one-half the wavelength of the standing wave. Antinodes are positioned between nodes.

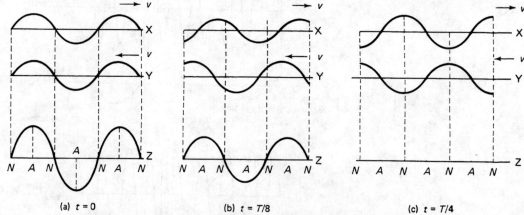

Figure 14-21 Superposition of two waves X and Y to form a standing wave Z. The nodes N are never displaced; the antinodes A fluctuate with maximum displacements.

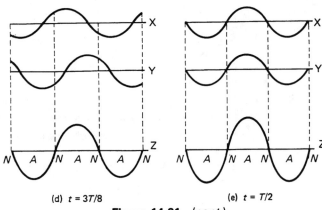

(d) $t = 3T/8$ (e) $t = T/2$

Figure 14-21 (cont.)

EXAMPLE 14-5

To determine the wavelength of microwaves, a reflector is placed 1.0 m in front of the source, and a diode is moved in a straight line between them. If the diode starts at a node and detects 28 additional nodes (null points) in a distance of 42 cm, what is the wavelength of the microwaves?

Solution

Data: 28 nodes in 42 cm, $\lambda = ?$

Equation: $\dfrac{\lambda}{2} = \dfrac{\text{distance}}{\text{number of nodes}}$

Rearrange: $\lambda = \dfrac{2(\text{distance})}{\text{number of nodes}}$

Substitute: $\lambda = \dfrac{2(42 \text{ cm})}{28} = 3.0 \text{ cm}$

In musical instruments, standing waves are produced in vibrating strings and air columns. Standing waves are also important in electronic devices, such as waveguides and coaxial cables.

PROBLEMS

26 Determine the distance between two successive nodes of an AM radio wave with a frequency of 1.05 kHz.

27 Determine the wavelength and frequency of an FM radio wave if a reflector produces a standing wave with nodes (a) 1.45 cm apart; (b) 0.380 in. apart.

SUMMARY

Periodic motion: $T = \dfrac{t}{N}$ $f = \dfrac{N}{t}$ $f = \dfrac{1}{T}$ $1 \text{ Hz} = \dfrac{1}{\text{s}}$

Waves: $v = f\lambda$ Changing medium $f = \dfrac{v}{\lambda} = \text{constant}$

Reflection: $i = r$; also, i, r, and the normal **n** are in the same plane.

Superposition: $\mathbf{s} = \mathbf{s}_A + \mathbf{s}_B$

Standing waves: $\dfrac{\lambda}{2} = \dfrac{\text{distance between nodes}}{\text{number of nodes}}$

QUESTIONS

1. Define the following terms: (a) periodic motion; (b) period; (c) frequency; (d) cycle; (e) wave; (f) wavespeed; (g) wavelength; (h) wave amplitude; (i) longitudinal waves; (j) transverse waves; (k) refraction; (l) diffraction; (m) interference of waves; (n) ray; (o) standing wave; (p) node; (q) antinode; (r) resonance.
2. Describe what is meant by (a) amplitude modulation; (b) frequency modulation.
3. Describe how you could measure the wavelength of a radio wave.
4. Describe what happens to a cork floating on the ocean as the waves pass.
5. (a) Describe what happens in terms of the disturbance of the medium when two waves interfere destructively at some point. (b) Is there any change in the energies of the two waves?

REVIEW PROBLEMS

1. Determine the period and wavelength of a light wave if its frequency is 5.74×10^{14} Hz and the wavespeed is 3.00×10^8 m/s.
2. Determine the period and the frequency of a light wave that has a speed of 9.82×10^8 ft/s and a wavelength of 2.08×10^{-6} ft.
3. Determine the number of flashes that a stroboscope completes in 1.00 min if its frequency is 186 Hz.
4. A spring vibrates with a period of 0.380 s. What is its frequency?
5. Find the wavelength and the period of a sound that has a frequency of 1200 Hz and a speed of (a) 332 m/s; (b) 1100 ft/s.
6. Light is a form of electromagnetic radiation that has a speed of 3.00×10^8 m/s in a vacuum. What is the frequency and the period of a light wave that has a wavelength of 550 nm?
7. If light travels as an electromagnetic wave at 3.00×10^8 m/s (9.82×10^8 ft/s) in a vacuum, determine (a) the period and (b) the wavelength of light that has a frequency of 7.50×10^{14} Hz.
8. Sound with a frequency of 850 Hz travels at 329 m/s in air and at 5050 m/s in steel. Find the wavelength in each medium.
9. Sound with a frequency of 1.20 kHz travels at 1086 ft/s in air and at 16 300 ft/s in aluminum. Find the wavelength in each medium.
10. Determine the beat frequency between two sound waves with frequencies of 480 Hz and 488 Hz.
11. Determine the wavelength and frequency of a radio wave if a reflector produces nodes (null points) that are (a) 82.0 m apart; (b) 245 ft apart.

15 SOUND

For a sound to be produced and heard, a source of vibratory energy (such as a speaker or vibrating string), a material medium (usually air), and a receiver (the ear) are required. Some of the energy from a vibrating source is transmitted as a longitudinal wave through the medium to the ear. Sound waves will not travel through a vacuum. The ear is sensitive to the small pressure changes caused by the sound wave and is able to convert these small pressure variations into electrical impulses. These are transmitted by auditory nerves to the brain, where they produce the sensation of sound.

Sound sources are present almost everywhere in our environment, and the control of sound is a major problem. Unwanted sounds are called **noise.** The control of noise is becoming an important consideration in our everyday lives. Excessive noise is known to produce drastic changes in our personalities, it is fatiguing, and it may cause deafness and even a reduction in our life span.

In recent years the production and recording of sound has developed into an important industry which utilizes vast quantities of electronic apparatus. **Acoustics** is the science of the production, transmission, reception, and effects of sound. Engineers must understand the acoustic properties of buildings and the building materials, which affect the reflection, transmission, and absorption of sound.

15-1 SOURCES OF SOUND

All sources of sound produce vibratory energy at frequencies within the audible range of the ear. Some of the most common sound sources are vibrating strings (stringed musical instruments), vibrating air columns (wind musical instruments), vibrating membranes (speakers and drums), and vibrating rods. This vibratory energy is transferred to the surrounding medium, normally air, where it is transmitted from particle to particle to the ear.

Many sound sources have several different **natural frequencies** or **resonant frequencies** of vibration. Each different vibratory state is called a **mode of vibration.** At these resonant frequencies, a small energy input results in a large amplitude.

Unlike springs and pendulums, which only have one resonant frequency, vibrating strings and air columns have an infinite number of different resonant frequencies.

Vibrating Strings

When a string is stretched under tension between two fixed points, we can make it vibrate by displacing it to one side and then releasing it. At most frequencies of vibration the energy is lost rapidly and the vibration dies out. However, when the string vibrates at certain **natural (resonant) frequencies**, **standing waves** are produced by the interference between the wave and its own reflection from the fixed ends (Fig. 15-1). These resonant frequencies depend on the length and mass of the string, and the tension in the string.

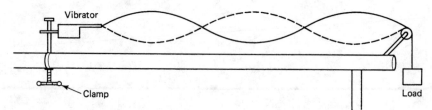

Figure 15-1 Standing waves in a vibrating string.

Nodes are points that are not displaced from their rest positions by the standing wave. Therefore, nodes must always occur at fixed points in vibrating strings, such as at the fixed ends.

Antinodes, where the displacements vary between extreme amplitudes, occur midway between the nodes. Therefore, since the distance between two successive nodes is equal to one-half the wavelength, the distance between a successive node and antinode is one-quarter of a wavelength.

A few of the possible resonant nodes of vibration of a string of fixed length L are illustrated in Fig. 15-2. In each case a node occurs at the fixed ends. The tension in the string is varied to produce the different **natural resonant frequencies.** Since the distance between node and nearest antinode is one-quarter of a wavelength $\lambda/4$ of the standing wave, we may easily determine the wavelengths of the natural modes of vibration.

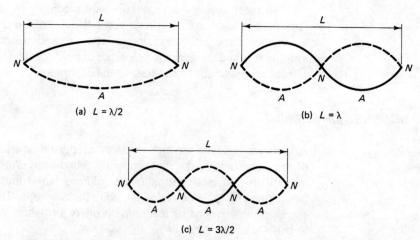

Figure 15-2 Three simplest natural modes of vibration of a string of length L which is fixed at each end, showing the positions of the nodes N and the antinodes A.

Sec.15-1 / Sources of Sound

The lowest possible natural (resonant) frequency of vibration of any source is called its **fundamental frequency.** For the string under tension in Fig. 15-2, this corresponds to the case where $L = \lambda/2$. Therefore, the fundamental frequency of the vibrating string:

$$f = \frac{v}{\lambda} = \frac{v}{2L} \quad \text{fundamental} \quad (15\text{-}1)$$

Harmonic frequencies are integer (whole-number) multiples of the fundamental frequency. For the vibrating string, these harmonic frequencies are given by

$$f_n = \frac{nv}{2L} \quad \text{harmonics} \quad (15\text{-}2)$$

where n is an integer (whole number). For example, the third harmonic is given by $n = 3$ in Eq. 15-2.

$$f_3 = \frac{3v}{2L} \quad \text{third harmonic}$$

EXAMPLE 15-1

A string is stretched under tension between two points that are 2.00 m apart. If the speed of a transverse wave in the string is 224 m/s, find **(a)** the fundamental frequency; **(b)** the frequency of the third harmonic; **(c)** the frequency of the sixth harmonic.

Solution

Data: $v = 224$ m/s, $L = 2.00$ m, $f_1 = ?, f_3 = ?, f_6 = ?$

Equation: $f = \dfrac{nv}{2L}$

Substitute:

(a) $f_1 = \dfrac{224 \text{ m/s}}{2(2.00 \text{ m})} = 56.0 \text{ Hz}$

(b) $f_3 = \dfrac{3(224 \text{ m/s})}{2(2.00 \text{ m})} = 168 \text{ Hz} = 3f_1$

(c) $f_6 = \dfrac{6(224 \text{ m/s})}{2(2.00 \text{ m})} = 336 \text{ Hz} = 6f_1$

Vibrating Air Columns in a Pipe That Is Closed at One End

Remember that in a longitudinal (compression) wave, the particles of the transmitting medium vibrate parallel to the direction of energy flow. Consider an air column in a tube of length L that is closed at one end only. The air in the tube can be made to vibrate with a natural frequency that depends on the length L and the speed of sound (which is related to the temperature). Longitudinal standing waves are formed at resonant frequencies as the wave reflected from the closed end of the pipe interferes with the original wave. A node exists at the closed end of the pipe because the air molecules there cannot vibrate longitudinally. Also, when the air column vibrates at a natural resonant frequency, an antinode exists approximately* at the open end of the pipe, where the air molecules receive a maximum variation in displacement. A few

---

*The antinode is slightly beyond the open end of the pipe by an amount that depends on the pipe diameter.

of the natural modes of vibration are illustrated in Fig. 15-3. Note that the displacement of the air is plotted vertically for convenience. Since longitudinal standing waves are produced, the air molecules are really displaced parallel to the length of the pipe. In this case, the fundamental frequency is given by

$$f = \frac{v}{\lambda} = \frac{v}{4L} \tag{15-3}$$

and the harmonics are

$$f_n = \frac{nv}{4L} \tag{15-4}$$

where n is an odd integer only. The even harmonics, corresponding to n being an even integer, do not exist.

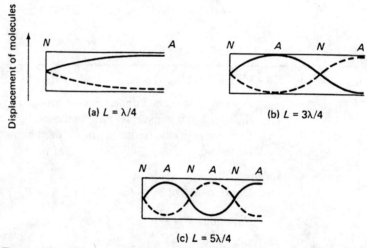

Figure 15-3 Three simplest natural modes of vibration of an air column in a pipe of length L that is closed at one end. (*Note:* Air molecules are really displaced parallel to the length of the pipe.)

EXAMPLE 15-2

If the speed of sound in air is 1100 ft/s and the fundamental frequency of a vibrating air column in a pipe that is open at one end only is 1024 Hz, find **(a)** the length of a vibrating air column; **(b)** the frequency of the seventh harmonic.

Solution

Data: $v = 1100$ ft/s, $f_1 = 1024$ Hz, $n = 1$ (fundamental), $L = ?$, $f_7 = ?$

(a) Equation $f_n = \dfrac{nv}{4L}$

Rearrange: $L = \dfrac{nv}{4f_n}$

Substitute: $L = \dfrac{(1)(1100 \text{ ft/s})}{4(1024 \text{ Hz})} = 0.27$ ft or 3.2 in.

(b) Equation $f_7 = 7f_1$

Substitute: $f_7 = 7(1024 \text{ Hz}) = 7168$ Hz

Even though we have discussed only sources of sound, the definitions of fundamental frequency and harmonics are of considerable importance in electronics. These terms are used extensively in wave shaping, frequency responses, and so on, and the operation of antenna systems are analogous to vibrating air columns.

There are many sources of sound, each of which is capable of transferring vibrational energy to a material medium, where it is transmitted as a longitudinal wave. In order to sustain the vibration, energy must be continually applied to the sound source.

In some cases, one vibrating object is able to transmit its vibrational energy to a second object, causing it to vibrate with the same frequency. This is called a **forced vibration.** For example, the air inside a speaker enclosure is forced to vibrate at the same frequency as the speaker. If the forced vibrations are at a resonant frequency of vibration, the amplitude of the vibration is greatly increased.

A simple experiment to show the effects of resonance is illustrated in Fig. 15-4. As the hollow pipe and vibrating tuning fork are moved vertically, the sound increases considerably when resonance occurs, that is, when the tuning fork delivers energy at one of the natural frequencies of vibration of the air column in the tube.

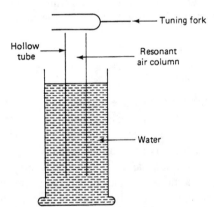

Figure 15-4 Simple experiment to show resonance in air columns.

While resonance effects are desirable in many musical instruments because of the larger amplitudes of vibration, in some cases they are quite irritating because they produce large distortions of normal sound. Resonance effects can also cause parts of a mechanical system to vibrate, and this can eventually lead to serious damage.

PROBLEMS

1. (a) Determine the wavelength of the fundamental frequency for a vibrating string that is 80.0 cm long. (b) What is the wavelength of the third harmonic?
2. A 3.00 ft long string is fixed at each end under a tension. If the speed of a transverse wave in the string is 35.0 ft/s, find (a) the fundamental frequency; (b) the wavelength and frequency of the third harmonic; (c) the wavelength and frequency of the fifth harmonic.
3. A string is 91.0 cm long and fixed at each end under a tension. If the speed of a transverse wave in the string is 10.7 m/s, find (a) the fundamental frequency; (b) the wavelength and the frequency of the third harmonic; (c) the wavelength and frequency of the fifth harmonic.
4. If the speed of a transverse wave in a string, which is attached at each end and under tension, is 15.0 m/s, (a) determine the length of the string that would vibrate with a fundamental frequency of 36.0 Hz. (b) What is the frequency of the fourth harmonic?
5. A string is attached at each end under tension. The speed of a transverse wave in the string is 52.0 ft/s. (a) Determine the length of the string if it vibrates in its fundamental frequency at a wavelength of 1.25 ft. (b) What is the frequency of the fifth harmonic?

6   If the speed of a transverse wave in a string, which is attached at each end and under tension, is 12.5 m/s and the frequency of the fifth harmonic is 212 Hz, what is the length of the string?

7   If the fundamental frequency of a vibrating string that is fixed at each end is 65.0 Hz, find the frequency of (a) the second harmonic; (b) the sixth harmonic.

8   If the speed of sound in air is 335 m/s, determine the length of a vibrating air column in a pipe that is open at one end if its fundamental frequency is 545 Hz.

9   If the speed of sound in an air column in a 2.50 ft long pipe that is closed at one end is 1100 ft/s, find the frequencies and wavelengths of (a) the fundamental; (b) the third harmonic; (c) the fifth harmonic.

10 If the speed of sound is 331 m/s in an air column in a 76.5 cm long pipe that is closed at one end, find the frequencies and wavelengths of (a) the fundamental; (b) the third harmonic; (c) the fifth harmonic.

11 If the speed of sound is 330 m/s and the frequency of the fifth harmonic of an air column in a pipe which is closed at one end only is 636 Hz, determine (a) the length of the pipe; (b) the fundamental frequency; (c) the frequency of the third harmonic.

12 A tuning fork produces resonance in a pipe closed at one end when the air column in the pipe is 15.0 cm long and then another resonance when the air column is 63.0 cm long. If the speed of sound in air is 336 m/s, determine (a) the frequency of the tuning fork; (b) the wavelength of the sound waves. (*Hint:* The two resonances occur half a wavelength apart. Why?)

15-2 MATERIAL MEDIUM

To pump

Figure 15-5 A material medium is required for the transmission of sound. If the bell jar is evacuated, we cannot hear the sound of the alarm.

Sound travels as a longitudinal wave and requires a material medium. Sound waves cannot travel through a vacuum. This can be demonstrated by placing a sound source, such as an electric bell or an alarm clock, in a bell jar (Fig. 15-5). When the air is pumped from the jar, there is no sound.

A vibrating object, such as a string or air column, makes the surrounding air molecules vibrate, creating the condensations and rarefactions (Fig. 15-6; see Chapter 14). Since the transmission of a longitudinal sound wave requires that the vibratory energy be transferred from molecule to molecule, the speed of sound depends on the density and the elastic properties of the material through which it travels. In liquids and solids the speed of sound is approximately independent of the temperature. However, in a gas such as air, the speed v of the sound varies as the square root of the **absolute** temperature T:

$$v \propto \sqrt{T}$$

or

$$\frac{v_2}{v_1} = \sqrt{\frac{T_2}{T_1}} \quad \text{and} \quad v_2 = v_1 \sqrt{\frac{T_2}{T_1}} \tag{15-5}$$

Some values for the speed of sound are given in Table 15-1.

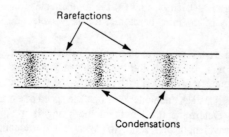

Figure 15-6 Condensations and rarefactions in the medium.

Sec. 15-2 / Material Medium

TABLE 15-1

Typical Speeds of Sound at 0°C(32°F)

Medium	Speed ft/s	Speed m/s
Fluids		
Air (dry)	1086	331
Carbon dioxide	853	260
Helium	3170	965
Hydrogen	4220	1285
Oxygen	1033	315
Solid rods		
Aluminum	16 400	5000
Iron	16 900	5150
Steel	16 600	5050

EXAMPLE 15-3

Determine the speed of sound in air at **(a)** 35°C; **(b)** 85°F.

Solution

Data: From Table 15-1 we have the speeds of sound in air.

(a) $v_1 = 331$ m/s, $T_1 = 0°C = (0 + 273)$ K $= 273$ K, $T_2 = 35°C = (35 + 273)$K $= 308$ K, $v_2 = ?$

(b) $v_1 = 1086$ ft/s, $T_1 = (32 + 460)°R = 492°R$, $T_2 = (85 + 460)°R = 545°R$, $v_2 = ?$

Equation: $v_2 = v_1 \sqrt{\dfrac{T_2}{T_1}}$

Substitute:

(a) $v_2 = (331 \text{ m/s}) \sqrt{\dfrac{308 \text{ K}}{273 \text{ K}}} = 352$ m/s

(b) $v_2 = (1086 \text{ ft/s}) \sqrt{\dfrac{545°R}{492°R}} = 1140$ ft/s

Objects that travel at speeds greater than the speed of sound are called **supersonic**. At subsonic speeds a longitudinal (compression) wave precedes the object. At supersonic speeds there is a sudden formation of a compression wave as the object passes. This compression wave is called a **shock wave**. When the shock wave from a supersonic aircraft reaches the ground, it is heard as a sonic boom; because of its high energy, it can cause damage by breaking windows, and so on.

Refraction

Sound waves are refracted as their speed changes when they pass into a different medium, or when they enter a region with different properties. Remember, even though the speed and wavelength of the sound change, the frequency remains constant.

The refraction of sound in air may be caused by a temperature variations above the earth's surface or by the effects of wind. Sounds usually carry farther over the earth's surface at night because the sound is refracted toward the cooler air near the earth. The reverse effect occurs in the daytime. At a fixed temperature the speed of sound is constant relative to the medium (air); therefore, wind velocities also change the speed of sound relative to the ground.

Interference

Two or more sound waves produce interference when they pass simultaneously through the same medium. The resultant sound is louder if the waves interfere constructively, and softer if there is destructive interference.

If the frequency of two sound waves is slightly different, the interference alternates between constructive and destructive producing amplitude variations called **beats.** The number of beats corresponds to the frequency difference between the sound waves. For example, if sounds with frequencies of 500 Hz and 504 Hz are produced simultaneously, there are four beats per second.

PROBLEMS

13 Determine the speed of sound in air at (a) −45°C; (b) −20°C; (c) 35°C; (d) −40°F; (e) 0°F; (f) 68°F.

14 Calculate the speed of sound in helium at (a) 5.0°C; (b) −25°C; (c) 70°F; (d) −12°F.

15 At what temperature is the speed of sound in air (a) 345 m/s; (b) 325 m/s; (c) 1060 ft/s; (d) 1100 ft/s?

16 If two sounds with frequencies of 1032 Hz and 1025 Hz are produced simultaneously, determine the frequency of the beats.

17 Two sound sources produce beats with a frequency of 4.0 Hz. If one of the sources has a frequency of 750 Hz, what are the possible frequencies of the other?

15-3 DOPPLER EFFECT

Sources and receivers of sound are not always stationary; as a result, the sound frequency detected by the receiver is not always the same as that transmitted by the source. The frequency difference depends on the relative motion between the source and the receiver. For example, there is a noticeable decrease in the pitch (frequency) of the sound we hear when a train or car moves first toward, then away, from us.

Frequency changes produced because of the relative motion of the source with respect to a receiver were predicted by Johann Doppler (1803–1853); the phenomenon is called the **Doppler effect.** It can be understood in terms of a swimmer in the ocean (Fig. 15-7). When the swimmer swims into the waves, the waves pass at a higher

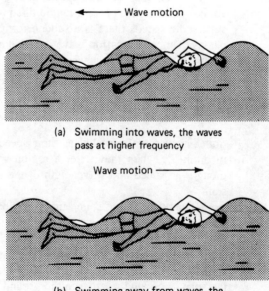

(a) Swimming into waves, the waves pass at higher frequency

(b) Swimming away from waves, the waves pass less frequently

Figure 15-7

frequency than when the swimmer is stationary. Similarly, the frequency at which the waves pass decreases when the swimmer tries to outrace them.

The Doppler effect has applications in detection and navigation devices, such as radar and sonar. The transmitter sends a wave to the target, where it is reflected. A part of the reflected wave then returns to the receiver, which is located at the transmitting station. Directional waves and the elapsed time between transmission and reception of the reflected wave give the location of the target. The change in frequency between the transmitted and the reflected wave is used to determine the velocity of the target.

15-4 HUMAN EAR

The time rate of emission of sound energy from a source is measured in watts. Different sources emit sound energy at different rates.

The **intensity** I of a sound at some point is the time rate of transmission of sound energy (or power) P per unit area A that is perpendicular to the direction of the energy flow.

$$I = \frac{\text{sound power}}{\text{perpendicular area}} = \frac{P}{A} \qquad (15\text{-}6)$$

As the wave spreads out from the source, its intensity decreases, and the wave is said to be **attenuated.**

The human ear is sensitive to variations in the intensity I and the pressure p of a sound, but the audio level or loudness of a sound to the human ear is not in direct proportion to either the intensity or the pressure. Our ears are approximately attuned to logarithmic intensity and pressure changes. The corresponding sound unit is called a **decibel.** Sound-level meters are usually calibrated to read sound pressure levels in decibels and effectively measure loudness (Fig. 15-8).

The response of a human ear to a pressure variation (caused by a particular sound) varies with the individual. In most cases, the human ear is sensitive only to a relatively small frequency range of approximately 20 Hz to 20 kHz. Outside this range a vibration is usually inaudible, regardless of its intensity (Fig. 15-9).

Figure 15-8 Sound meter.

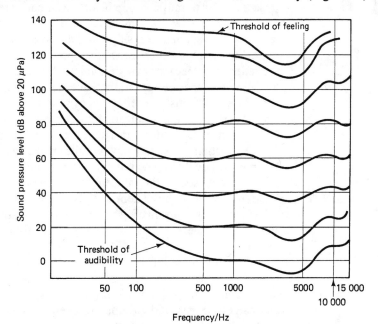

Figure 15-9 Frequency response of the human ear. The curves represent equal loudness of sound at different frequencies and pressure levels.

The sensation of loudness depends on the intensity and frequency of a sound. Human beings are usually most sensitive to sounds in the frequency range 2000 Hz to 4000 Hz. In this frequency range, the human ear can respond to pressure variations as small as 10 μPa. Low-intensity sounds in the most sensitive frequency range are judged to be as loud as sounds of higher intensity outside this range.

Within the audible range, the human ear is relatively sensitive to frequency variations. The term **pitch** is mainly associated with the frequency of a sound, but it also varies, to a lesser degree, with the sound pressure level. A sound with a high pitch has a greater frequency than a sound with a low pitch.

A single sound source may vibrate simultaneously in many different ways, producing a very complex waveform that contains the fundamental and many harmonics. The tone of a sound is related to the complexity of the total wave form. The human ear can distinguish between two sounds that produce the same basic note, because their tone is different. For example, a trumpet and a piano may both play the same note, but the two sounds are quite different because the waves produced have different complexities.

15-5 REVERBERATION

Rigid smooth surfaces are better reflectors of sound energy than soft irregular surfaces, but sound waves are reflected to some extent from most surfaces. As in the case of other waves, the maximum intensity of a reflected sound from a smooth surface occurs at an angle of reflection that is equal to the angle of incidence. The reflective properties of sound are used to measure distances and to focus and transmit sounds in a preferred direction.

Explorations for oil and minerals are aided by "soundings" of the earth. Explosions at the earth's surface produce pulses of sound, parts of which are reflected from the areas of different density below the surface of the earth, and the reflected sounds return to a series of receivers. By analysis of the elapsed times between the initial sounds and the reception of the reflections, scientists can estimate the earth's structure. The terrain and depths of ocean beds are measured by a similar process.

Sonar devices transmit a sound pulse through a body of water. This pulse is reflected from submarines, ships, and the ocean beds, and a part of the reflected sound returns to the transmitter. Since the speed of sound in water is known, the range of the reflecting object may be determined by measuring the elapsed time between the transmissions and the reception of the reflected signal.

In auditoriums, the reflection of sound may be used to advantage by locating a curved surface of good reflective properties behind the stage or behind the speaker. The curved surface directs the reflected sound to all parts of the audience.

Reflection of a sound may be quite critical in closed areas, such as in auditoriums and rooms. Usually, some reflection is required, but if the surfaces continually reflect a large percentage of the incident sound energy, the original sound will persist and interfere with any subsequent sounds. This continual reflection of audible sound is called **reverberation.** In effect, the reverberation time is the time required for a single sound to become inaudible. It usually varies from 0.5 s to 2.0 s in rooms and auditoriums.

The reverberation time of an enclosed space depends on the nature of the reflecting surfaces and the total volume. It may be reduced by covering the walls, floors, and ceiling with absorbing materials, such as draperies, cork, acoustic tiles, and carpets. Some auditoriums are even constructed with movable baffles, which control the volume of the space.

15-6 REPRODUCTION OF SOUND

Modern technology enables us to reproduce accurately all sounds within the audible range of the human ear. Stereo and multichannel systems even reproduce the relative locations of the individual sound sources.

A **transducer** is any device that is able to convert signals from one form of energy into another. In sound recording or reproduction processes, transducers, such as microphones and speakers, utilize crystals or electromagnets to convert mechanical vibrations into electrical impulses, and vice versa.

Records

In a recording process, a microphone converts sound vibrations into electrical impulses that are amplified (made larger) and then used to drive an electromagnetic cutting tool. The vibrating cutting tool makes a continuous spiral groove of variable dimensions in a rotating disk, and a master die is made from the finished disk. Plastic sheets are heated and then pressed to shape in the die to form records that are identical to the original disk.

Sound is reproduced from the record by the reverse process. The record is rotated on a turntable and a needle or stylus is located so that it vibrates due to the irregularities in the continuous groove. This vibration is transmitted to a special crystal or electromagnet in the pickup, where it is converted into small electric impulses. These impulses are amplified and used to drive an electromagnetic speaker cone that is attached to a diaphragm. The vibrating diaphragm produces longitudinal sound waves.

Tapes and Film Sound Tracks

Tapes are composed of a long plastic strip that contains small magnetic particles. During a recording process, the tape is passed through the recording head of the tape recorder (Fig. 15-10). Electric impulses from the microphone via an amplifier produce varying magnetic fields in the gap of the core of the electromagnetic recording head. These fields reorganize the magnetic particles into patterns.

Small electric impulses are reproduced when the magnetized tape is passed over the electromagnetic playback head. These impulses are amplified and used to drive a speaker.

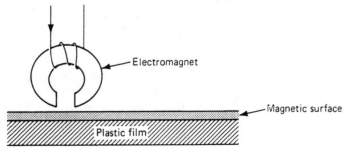

Figure 15-10 Head of a tape recorder.

SUMMARY

Sound and *ultrasonic waves* are longitudinal waves.

The *fundamental frequency* is the lowest natural resonant frequency of vibration. *Harmonic* frequencies are integer multiples of the fundamental frequency.

Vibrating strings fixed at each end: $f_n = \dfrac{nv}{2L}$

Vibrating air columns in a pipe closed at one end only: $f_n = \dfrac{nv}{4L}$ (n an odd integer)

Speed of sound: $v \propto \sqrt{T}$ and $v_2 = v_1\sqrt{\dfrac{T_2}{T_1}}$

The *Doppler effect* relates to the frequency changes that are due to relative motion between the source and the receiver.

Sound intensity: $I = \dfrac{P}{A}$

The *frequency range* of sounds audible to human beings is 20 Hz to 20 kHz.

QUESTIONS

1. Define the following terms: (a) noise; (b) fundamental frequency; (c) harmonic frequencies; (d) resonance; (e) intensity; (f) reverberation; (g) transducer.
2. Describe the operation of (a) records; (b) magnetic tape.
3. (a) If the reverberation time of sound in an auditorium is too long, how would you reduce it? (b) If it is too short, how would you increase it?
4. By measuring the shift in the light frequency from the stars when compared to stationary sources, we may determine the speed of the stars relative to the earth. Explain the theory involved.
5. The frequency shift of the Doppler effect depends on the relative velocity between the source and the receiver. What happens to the frequency shift if the relative speed between the source and the receiver (a) increases; (b) decreases?
6. Describe why different musical instruments have different sounds even if they play the same notes.

REVIEW PROBLEMS

1. A string is stretched between two fixed points 85.0 cm apart. If the speed of a transverse wave in the string is 125 m/s, find (a) the fundamental frequency; (b) the third harmonic frequency; (c) the wavelength of the sixth harmonic.
2. The speed of sound in an air column in a 12.5 cm long pipe that is closed at one end is 332 m/s. Find (a) the frequency of the third harmonic; (b) the frequency of the fifth harmonic.
3. A string is stretched between two fixed points. If the speed of a transverse wave in the string is 225 ft/s and the fundamental frequency is 35.0 Hz, find (a) the length of the string; (b) the wavelength of the third harmonic.
4. The speed of sound in an air column in a 97.5 cm long pipe that is closed at one end is 325 m/s. Determine (a) the wavelength of the fifth harmonic; (b) the funadmental frequency.
5. The speed of sound in an air column in a 3.20 ft long pipe that is closed at one end is 1080 ft/s? Determine (a) the wavelength of the fifth harmonic; (b) the fundamental frequency.
6. Determine the speed of sound in air at (a) 15°C; (b) −45°C; (c) 37°C; (d) −36°F; (e) 110°F.
7. At what temperature is the speed of sound in air (a) 342 m/s; (b) 320 m/s; (c) 1090 ft/s; (d) 1075 ft/s?
8. A 150 cm long string vibrates with five segments at a frequency of 120 Hz. (a) What is the wavelength? (b) Determine the wavespeed.

Review Problems

9 A 5.00 ft long string vibrates in three segments at a frequency of 165 Hz. Determine **(a)** the wavelength; **(b)** the wavespeed.

10 Two sound sources produce beats with a frequency of 45.0 Hz. If the frequency of one source is 825 Hz, what are the possible frequencies of the other source?

11 A 1240 Hz tuning fork produces resonance in a pipe closed at one end when the air column is 27.0 cm long and then another when the air column is 80.0 cm long. Determine **(a)** the wavelength; **(b)** the speed of the sound wave.

12 A 635 Hz vibration produces resonance in a pipe closed at one end when the air column is 10.7 in. long and then another when the air column is 31.5 in. long. Determine **(a)** the wavelength; **(b)** the speed of the sound wave.

16
LIGHT

The eye is sensitive only to a part of the total radiant energy that is emitted by matter. Light is the form of radiant energy that has the capacity to stimulate the sensation of vision. To be visible, light from an object must enter the eye. This light may be produced by the object, or merely reflected from it.

Bodies of matter that emit light as a part of their radiant energy are called **luminous bodies.** The sun and other stars are natural sources; tungsten lamps and fluorescent lamps are examples of artificial sources of light.

An object does not have to be a luminous body in order to be visible; most objects become visible because they reflect light. If light from a luminous body is incident on a surface, some of the light is reflected, and that surface is said to be **illuminated.**

Optics is the study of light and the phenomena and effects related to light. The science that is concerned with the measurement of light is called **photometry.**

TABLE 16-1
Typical Recommended Illumination Levels

Area or task	Illuminance	
	fc	lx
Chemical laboratory	50	540
Electrical testing	100	1100
Machine shops	100	1100
Medium bench and machine work	100	1100
Regular office work	90	1000
Store windows	180	2000
Outside building construction	10	107
Reading and writing	100	800

Sec. 16-1 / Nature of Light

Light is essential in many aspects of our lives because we rely on vision to detect and locate objects. Many measurement processes utilize light to transmit information about the relative size and shape of objects.

The improper illumination of a working area increases eye strain, brings on fatigue, and reduces the general efficiency of workers; therefore, the choice of a suitable lighting system is of great importance (Table 16-1). Adequate illumination of a working surface is essential, but excess illumination should also be avoided because it causes glare and shadows.

16-1 NATURE OF LIGHT

Light is a form of energy and can be transformed into other energy forms, such as thermal energy. We also know that energy may be transferred from one place to another by the motion of a particle (kinetic energy) or by means of waves. Therefore, for centuries, the actual nature of light was the subject of considerable controversy. Some people considered light to be a particle, and others thought that it was a wave. It was not until the twentieth century that we discovered that light possessed *both* particlelike and wavelike properties.

Wavelike Nature of Light

Light is a form of **electromagnetic radiation** that consists of time-varying electric and magnetic fields. These time-varying fields are perpendicular to each other and are also perpendicular to the velocity of the radiation forming a transverse wave (Fig. 16-1). Some other forms of electromagnetic radiation are radio waves, microwaves, television waves, gamma rays, X-rays, and cosmic rays. All forms of electromagnetic radiation exhibit the usual wavelike characteristics, such as diffraction and interference. They have a wavelength λ, a frequency f, and they all travel at the same enormous speed $c = 3.00 \times 10^8$ m/s or 9.82×10^8 ft/s (186 000 mi/s) in a vacuum. The **wave relation** is also valid for electromagnetic radiation.

$$c = f\lambda \tag{16-1}$$

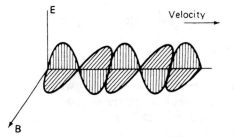

Figure 16-1 Electromagnetic wave. Time varying electric and magnetic fields and the velocity are mutually perpendicular.

EXAMPLE 16-1

Determine (a) the wavelength of microwaves that have a frequency of 15.0 GHz; (b) the frequency of yellow light that has a wavelength of 552 nm in a vacuum.

Solution

Data: $c = 3.00 \times 10^8$ m/s
(a) $f = 15.0 \times 10^9$ Hz, $\lambda = ?$ (b) $\lambda = 552 \times 10^{-9}$ m, $f = ?$
Equation: $c = f\lambda$

(a) *Rearrange:* $\lambda = \dfrac{c}{f}$

Substitute: $\lambda = \dfrac{3.00 \times 10^8 \text{ m/s}}{15.0 \times 10^9 \text{ Hz}} = 2.00 \times 10^{-2}$ m = 2.00 cm

(b) *Rearrange:* $f = \dfrac{c}{\lambda}$

Substitute: $f = \dfrac{3.00 \times 10^8 \text{ m/s}}{552 \times 10^{-9} \text{ m}} = 5.43 \times 10^{14}$ Hz

---

These electromagnetic waves may be organized into an ordered array of wavelengths called the **electromagnetic spectrum.** The different types of electromagnetic radiation are characterized by their wavelength or frequency. Light occupies only a very small region of the total electromagnetic spectrum, with wavelengths between about 380 nm and 760 nm. Within this range all the shades of color from violet to red are found. Each wavelength (or frequency) of light corresponds to a different color. For example, violet light has wavelengths around 400 nm, yellow about 550 nm, and red around 740 nm (see Fig. 16-2).

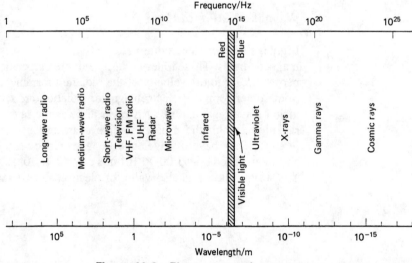

Figure 16-2 Electromagnetic spectrum.

Unlike other kinds of waves, electromagnetic waves do not require a material medium, because the time-varying electric and magnetic fields are self-sustaining. Therefore, light and all other electromagnetic waves can travel through a vacuum, although they can also travel through certain materials.

We use the wavelike characteristics such as reflection, refraction, diffraction, interference, and polarization to describe the way in which light travels from one place to another.

Particlelike Nature of Light

By the twentieth century most scientists believed the wave theory for light. However, a series of new experimental observations involving interaction between light and matter could not be explained in terms of wave theory. These observations led to the development of **quantum theory.**

Wave theory predicts a continuous range of energies for electromagnetic radiation. Quantum theory is based on the principle that the exchange of energy between electromagnetic radiation and matter is always in discrete amounts of energy called **quanta.** Each quantum of light has both particlelike and wavelike characteristics, and it is known as a **photon.** The existence of photons was proposed by Albert Einstein in 1905. He also suggested that each photon has an energy given by

$$E = hf \qquad (16\text{-}2)$$

where f is the frequency of the electromagnetic wave and $h = 6.63 \times 10^{-34}$ J·s is a constant called **Planck's constant.**

EXAMPLE 16-2

Determine the energy of a photon that has a wavelength of 540 nm.

Solution

Data: $\lambda = 540$ nm $= 5.4 \times 10^{-7}$ m, $h = 6.63 \times 10^{-34}$ J·s, $c = 3.0 \times 10^8$ m/s, $E = ?$

Equations: $E = hf$, $c = f\lambda$

Rearrange: $f = \dfrac{c}{\lambda}$; therefore, $E = hf = \dfrac{hc}{\lambda}$

Substitute: $E = \dfrac{(6.63 \times 10^{-34} \text{ J·s})(3.0 \times 10^8 \text{ m/s})}{5.4 \times 10^{-7} \text{ m}} = 3.7 \times 10^{-19}$ J

To study certain areas of science, we frequently develop a familiar model of the system to help us visualize our experimental observations. We then use the model to predict new results. For example, we could (crudely) picture an atom to be a miniature solar system with the electrons behaving (in some ways only) like the planets orbiting the nuclear "sun." The apparent contradiction concerning the wavelike and particlelike characteristics of electromagnetic radiation is merely due to our inability to compare it with a *single* model.

Electromagnetic radiation, such as light, possesses *both* particlelike and wavelike properties. Wave theory is used to describe how light travels, but particlelike characteristics explain the interactions between the radiation and matter. This is known as the **duality concept.**

Even though light exhibits wavelike properties, it is often convenient to consider its motion in terms of straight lines called **rays,** which represent the path of the light energy. The term **ray** is also used to describe the paths of other forms of electromagnetic energy, such as X-rays, gamma rays, and cosmic rays. **Geometrical optics** is founded on the assumption that light travels in straight lines represented by rays. Even though light really travels as a wave, the concepts of geometrical optics may be used to describe many optical phenomena, such as reflection, refraction, and shadows.

An object may transmit, reflect, and absorb some of the incident light energy. **Transparent** materials transmit light without significantly scattering the light energy. Objects can be seen with very little distortion through transparent materials. Materials that transmit and scatter light energy are called **translucent.** Objects cannot be clearly seen through translucent materials. **Opaque** materials do not transmit light but block the passage of light energy.

Shadows are produced when the path of light is blocked by an opaque object. If an opaque object blocks light from a small light source, it casts a sharp shadow on a screen (Fig. 16-3). The region on the screen where all the light rays from the source are excluded is called the **umbra.** When an extended light source, such as a fluorescent lamp, is used, an opaque object casts two types of shadow regions. All

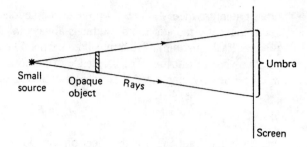

(a) The shadow cast by an opaque object when it blocks the light from a point source

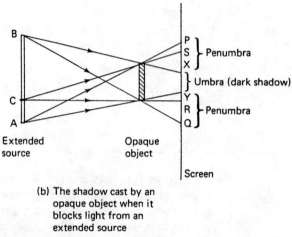

(b) The shadow cast by an opaque object when it blocks light from an extended source

Figure 16-3 Shadows that are cast when an opaque object is placed in the path of light from a source.

light rays are excluded from the umbra. In the **penumbra**, the light rays from *some* parts of the source are excluded. For example, light from point C of the source can illuminate all points of the screen except points between S and R. The illumination of the penumbra regions increases with the distance from the umbra region.

A **solar eclipse** is produced when the moon moves between the earth and the sun, and its shadow is cast on the earth's surface. The solar eclipse is said to be *total* in the umbra regions and *partial* in the penumbra regions of the moon's shadow. During a **lunar eclipse** the earth is located between the moon and the sun, and its shadow covers at least a part of the moon's surface.

PROBLEMS

1. Determine the wavelengths of light with the following frequencies: **(a)** 7.80×10^{14} Hz; **(b)** 5.20×10^{14} Hz; **(c)** 3.90×10^{14} Hz.

2. Determine the frequencies of light with the following wavelengths: **(a)** 425 nm; **(b)** 565 nm; **(c)** 685 nm; **(d)** 735 nm.

3. Determine the frequencies of electromagnetic waves with the following wavelengths, and determine the type of wave in each case from the electromagnetic spectrum: **(a)** 11.5 pm; **(b)** 3.00 cm; **(c)** 250 m; **(d)** 555 nm.

4. Determine the energies of the photons with the following frequencies: **(a)** 7.45×10^{14} Hz; **(b)** 5.35×10^{14} Hz; **(c)** 3.95×10^{14} Hz; **(d)** 630×10^{14} Hz.

5. Determine the energies of the photons with the following wavelengths: (a) 380 nm; (b) 425 nm; (c) 576 nm; (d) 752 nm.

6. Find the frequencies and wavelengths of the photons that have the following energies: (a) 3.20×10^{-19} J; (b) 1.64×10^{-19} J; (c) 5.25×10^{-18} J; (d) 1.70×10^{-18} J.

16-2 LIGHT SOURCES

Light energy constitutes only a fraction of the total electromagnetic radiation that is emitted from the sun. This electromagnetic radiation is the result of a continuous series of nuclear reactions, during which some of the sun's mass is converted directly into energy.

Many kinds of artificial light sources have also been produced. Most of these sources generate light by one of the following processes.

Incandescence

Incandescent lamps consist of a wire filament, usually made of tungsten, that is located in a glass bulb (Fig. 16-4). To reduce the evaporation of the filament at high temperatures, the bulb is either evacuated or contains a gas, such as argon or nitrogen, under low pressure. The mounting base also serves as the electrical connection between the bulb socket and the filament.

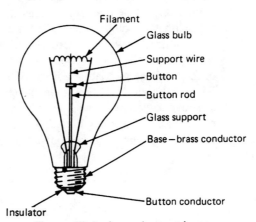

Figure 16-4 Incandescent lamp.

The filament has electric resistance, and when it carries an electric current, its temperature increases, and it glows, emitting thermal energy and light. In normal operation, at a temperature of approximately 2000°C, the filament emits light that is very similar to sunlight, but the efficiency of the incandescent lamp is quite low because most of the energy is emitted as heat.

Gas Discharge

A simple gas discharge tube consists of a glass tube containing a low-pressure gas and two electrodes. When an electric current is maintained through the low-pressure gas, the gas atoms become excited into high-energy states. The excited gas atoms return to their lower-energy states by emitting light with distinct wavelengths characteristic of the gas. Most discharge tubes emit light that is rather different from sunlight, so have only limited use in general lighting systems. However, gas discharge tubes are used for special types of lighting, such as neon signs and sunlamps.

Fluorescence

A **fluorescent lamp** consists of a long glass tube that contains mercury vapor and some argon gas under low pressure. The inner surface on the tube is coated with a special chemical phosphor. When there is an electric discharge through the mercury vapor, some of the mercury atoms are excited into a high-energy state, and they emit ultraviolet radiation as they return to their lower-energy states. The ultraviolet radiation then excites the phosphor atoms on the inner tube surface into higher-energy states, and they emit visible light when they return to the lower energy.

Fluorescent lamps have a relatively high efficiency and long lifetimes. The nature of the visible light that is emitted from a fluorescent lamp depends on the type of phosphor coating. Different types of fluorescent lamps are often used in factories, offices, schools, and houses.

Light Characteristics

All known sources emit significant amounts of other forms of electromagnetic radiation simultaneously with the light. The **radiant flux** (or **radiant power**) P of a source is defined as the time rate at which it radiates all forms of electromagnetic energy. The SI unit for radiant flux is the watt.

Initially, when a filament lamp is "turned on" (supplied with electrical power), its temperature increases rapidly, but it soon reaches an equilibrium state by dissipating energy at the same rate that it receives the electric energy. At this point the temperature of the filament remains almost constant, and the electrical power input is approximately equal to the radiant flux.

Normally, we are interested in the visible electromagnetic energy with wavelengths approximately between 380 nm and 760 nm, and this is usually only a fraction of the total energy that is radiated. The response of the human eye is not equal even for visible light of different wavelengths. Under normal conditions, the eye is most sensitive to yellow-green light with a wavelength of 555 nm. Relatively low power radiation at a wavelength of 555 nm produces a visual sensation that appears to be equally as strong as higher power radiations at other wavelengths (Fig. 16-5).

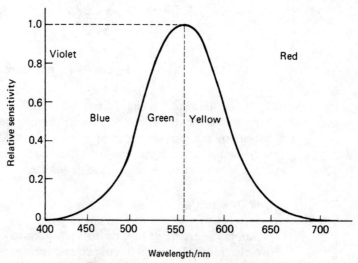

Figure 16-5 Eye sensitivity curve.

The **luminous flux** F is the time rate of flow of the electromagnetic energy that has the ability to stimulate the sensation of vision. It is normally expressed in units of **lumens** (symbol lm).

Sec. 16-2 / Light Sources

Light sources are rated in terms of their ability to stimulate vision. Since the sensitivity of the eye is involved, it is convenient to describe light in terms of special units.

If F is the luminous flux through a perpendicular area A, at a distance r from the source, we define the **luminous intensity** at that location as

$$I = \frac{Fr^2}{A} \tag{16-3}$$

The SI unit for luminous intensity is a base unit called a **candela** (symbol cd); it is defined in terms of the luminous intensity from a standard source (Fig. 16-6). The luminous intensity in a particular direction is called the **candle power** (cp) of the source.

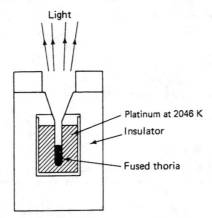

Figure 16-6 Standard light source used to define a candela.

EXAMPLE 16-3

Determine the luminous intensity of a source that produces a luminous flux of 24.5 lm on a surface area of 2.50 ft² at a distance of 3.50 ft.

Solution

Data: $F = 24.5$ lm, $A = 2.50$ ft², $r = 3.50$ ft, $I = ?$

Equation: $I = \dfrac{Fr^2}{A}$

Substitute: $I = \dfrac{(24.5 \text{ lm})(3.50 \text{ ft})^2}{2.50 \text{ ft}^2} = 120$ cd

If all directions in three dimensions are considered, the average luminous intensity of a source is called its **mean spherical luminous intensity** $\bar{I}$ in candela. The **total luminous flux** F_t of the source is given by

$$F_t = 4\pi\bar{I} \tag{16-4}$$

EXAMPLE 16-4

A small light source has a mean spherical luminous intensity of 31.0 cd, and a reflector concentrates all the light onto a perpendicular area of 3.75 m² at a distance 5.00 m from the source. Calculate (a) the total luminous flux from the source; (b) the luminous intensity of the light when viewed from the illuminated area.

Solution

Data: $\bar{I} = 31.0$ cd, $A = 3.75$ m^2, $r = 5.00$ m, $F_t = ?$, $I = ?$

(a) Equation: $F_t = 4\pi\bar{I}$

Substitute: $F_t = 4\pi(31.0 \text{ cd}) = 390$ lm

(b) Equation: $I = \dfrac{F_t r^2}{A}$

Substitute: $I = \dfrac{(390 \text{ lm})(5.00 \text{ m})^2}{3.75 \text{ m}^2} = 2600$ cd

---

Artificial light sources are often rated in terms of their ability to convert other energy forms into visible light energy (Table 16-2). The **luminous efficacy** (formerly luminous efficiency) K of a light source is defined as the ratio of its useful output power (i.e., the total luminous flux F_t), to the total input power P. In the case of electric lights, the input power is equal to the radiant flux P; therefore, the luminous efficacy

$$K = \frac{F_t}{P} \tag{16-5}$$

Luminous efficacies are not expressed as a percentage; they have units of lumens per watt.

TABLE 16.2
Typical Ratings of Some Electric Light Sources

Sources	Total luminous flux/(lm)	Luminous efficacy (lm/W)
40 W tungsten filament	470	11.8
60 W tungsten filament	830	13.8
100 W tungsten filament	1630	16.3
150 W tungsten filament	2700	18.0
100 W clear mercury lamp	3600	36
40 W fluorescent 48 in. long	2800	70
60 W fluorescent 48 in. long	4000	67
100 W fluorescent 48 in. long	5500	55

EXAMPLE 16-5

Determine the luminous efficacy of a small 200 W lamp if its mean spherical luminous intensity is 294 cd.

Solution

Data: $P = 200$ W, $\bar{I} = 294$ cd, $K = ?$

Equations: $F_t = 4\pi\bar{I}$, $K = \dfrac{F_t}{P}$

Sec.16-3 / Illumination of a Surface

$$\text{Rearrange: } K = \frac{4\pi \bar{I}}{P}$$

$$\text{Substitute: } K = \frac{4\pi(294 \text{ cd})}{200 \text{ W}} = 18.5 \text{ lm/W}$$

The luminous efficacies of most artificial light sources are quite small, because most of the input power is converted into thermal energy. Fluorescent lamps have much higher luminous efficacies than incandescent lamps.

PROBLEMS

7. Determine the luminous intensity of a source that produces a luminous flux of 250 lm on an area of 3.00 m^2 at a distance of 12.0 m.

8. If a source has a luminous intensity of 525 cd, what luminous flux does it produce on an area of 1.25 m^2 at a distance of 6.00 m?

9. Determine the luminous flux produced on an area of 3.25 ft^2 at a distance of 4.50 ft from a source that has a luminous intensity of 475 cd.

10. A small light source has a mean spherical luminous intensity of 85.0 cd, and a reflector concentrates all the light onto a perpendicular area of 14.0 cm^2 at a distance of 92.0 cm from the source. Determine (a) the total luminous flux; (b) the luminous intensity of the light when viewed from the illuminated area.

11. A small light source has a mean spherical luminous intensity of 92.0 cd, and a reflector concentrates all the light onto a perpendicular area of 1.50 ft^2 at a distance of 3.00 ft from the source. Determine (a) the total luminous flux; (b) the luminous intensity of the light when viewed from the illuminated area.

12. Determine the luminous intensity of a light beam if a reflector concentrates all the light from a small 45 cp source (a) onto the perpendicular area of 2.00 m^2 at a distance of 8.0 m from the source; (b) onto a perpendicular area of 25 ft^2 at a distance of 24.0 ft from the source.

13. A reflector concentrates all the light from a small source onto a perpendicular area of 2.32 m^2 at a distance of 5.50 m. If the luminous flux in the direction of the light is 1800 lm, determine the candlepower of the source.

14. Calculate the luminous efficacy of a 65.0 W lamp if its mean spherical luminous intensity is 350 cd.

15. A reflector concentrates all the light from a small 100 W source onto a perpendicular surface of area 4.50 ft^2 at a distance of 1.50 ft. If the luminous intensity in the direction of the light beam is 850 cd, calculate the luminous efficacy of the source.

16. A 40 W fluorescent lamp has a luminous efficacy of 70.0 lm/W. (a) Determine the total luminous flux. (b) What is its mean spherical luminous intensity?

17. A tungsten lamp emits a total luminous flux of 830 lm and has a luminous efficacy of 13.0 lm/W. Determine (a) the input power; (b) the mean spherical luminous intensity.

18. A 150 W tungsten lamp has a luminous efficacy of 18.0 lm/W. Find (a) the total luminous flux; (b) the mean spherical luminous intensity; (c) the luminous intensity of the light on a perpendicular area of 3250 cm^2 at a distance of 2.60 m; (d) the luminous intensity on a perpendicular area of 3.50 ft^2 at a distance of 8.50 ft.

16-3 ILLUMINATION OF A SURFACE

A surface is illuminated when it is irradiated with visible light from any number of sources. The **illuminance** E of the surface is defined as the total incident luminous flux F_t per unit surface area A.

$$E = \frac{F_t}{A} \qquad (16\text{-}6)$$

The SI unit for illuminance is called the **lux** (symbol lx); it is the illuminance when a surface receives one lumen of luminous flux per square meter.

$$1 \text{ lx} = 1 \frac{\text{lm}}{\text{m}^2}$$

In USCS, the unit of illuminance is the **footcandle** (fc); it is equal to one lumen of luminous flux per square foot.

$$1 \text{ fc} = 1 \frac{\text{lm}}{\text{ft}^2}$$

Many different light sources may contribute to the illumination of the same surface. The total illuminance of a surface A is equal to the *sum* of the illuminances due to the luminous flux that it receives from each light source:

$$E_{tot} = E_1 + E_2 + E_3 + \ldots \qquad (16\text{-}7)$$

and remember $E = \dfrac{F}{A}$.

EXAMPLE 16-6

Calculate the total illuminance of a 5.00 ft² surface if it receives 20.0 lm from one source and 80.0 lm from a second source.

Solution

Data: $A = 5.00$ ft² in each case, $F_1 = 20.0$ lm, $F_2 = 80.0$ lm, $E = ?$

Equations: $E = \dfrac{F}{A}$, $E_{tot} = E_1 + E_2$

Rearrange: $E = \dfrac{F_1}{A} + \dfrac{F_2}{A} = \dfrac{F_1 + F_2}{A}$

Substitute: $E = \dfrac{20.0 \text{ lm} + 80.0 \text{ lm}}{5.00 \text{ ft}^2} = 20.0$ fc

---

When an illuminated area is perpendicular to the direction from the source and there are no losses, the illuminance E varies inversely as the square of the distance r from the source.

$$E \propto \frac{1}{r^2}$$

and

$$\frac{E_1}{E_2} = \frac{r_2^2}{r_1^2} = \left(\frac{r_2}{r_1}\right)^2 \qquad (16\text{-}8)$$

where E_1 and E_2 are the illuminances at distances r_1 and r_2, respectively. This is another example of an inverse square law.

EXAMPLE 16-7

A point on a surface has an illuminance of 145 lx when irradiated with light from a small lamp at a distance of 1.25 m. How far from the same point should the lamp be located to reduce the illuminance to 115 lx?

Solution

Data: $E_1 = 145$ lx, $E_2 = 115$ lx, $r_1 = 1.25$ m, $r_2 = ?$

Equation: $\dfrac{E_1}{E_2} = \dfrac{r_2^2}{r_1^2}$

Rearrange: $r_2 = r_1 \sqrt{\dfrac{E_1}{E_2}}$

Substitute: $r_2 = (1.25 \text{ m})\sqrt{\dfrac{145 \text{ lx}}{115 \text{ lx}}} = 1.40$ m

---

Lighting systems are usually composed of a number of different light fixtures, and any point in a room may receive diffused and reflected light from its surroundings as well as direct light from each source. In practice the illuminance at some location in a room is often very difficult to calculate, but illuminances are easily measured with photoelectric devices.

PROBLEMS

19 Determine the illuminance of a 0.675 m² surface if it receives 42.0 lm.
20 Calculate the illuminance of a 5.00 ft² surface if it receives 30.0 lm from one source and 20.0 lm from a second source.
21 Determine the total illuminance of a 0.460 m² surface if it receives 30.0 lm from one light source and 42 lm from a second.
22 Calculate the total luminous flux that is incident on a surface with an area of 4.00 m² if it has a uniform illuminance of 375 lx.
23 Determine the total luminous flux that is incident on an 8.00 ft² surface that has a uniform illuminance of 62.0 fc.
24 A point on a surface has an illuminance of 180 lx when it is irradiated with light from a small lamp at a distance of 5.0 m. Calculate the illuminance at the same point if the lamp is moved **(a)** 4.0 m farther away; **(b)** 2.0 m closer.

SUMMARY

Emitters of light are called *luminous bodies*. Most objects are visible because they reflect light when they are illuminated. Light exhibits both particlelike and wavelike properties. Light is a form of electromagnetic radiation with wavelengths between approximately 380 nm and 760 nm. A *photon* is a *quantum* of light.

Wave relation $c = f\lambda$, and photon energy $E = hf = \dfrac{hc}{\lambda}$.

Radiant flux P is the time rate of emission of all radiant energy.

Luminous flux F is the time rate of emission of visible light energy. The unit of luminous flux is the *lumen* (lm).

Luminous intensity $I = \dfrac{Fr^2}{A}$. The SI base unit is the *candela* (cd).

The average luminous intensity of a source is called the *mean spherical luminous intensity* $\bar{I}$; then the total luminous flux

$$F_t = 4\pi\bar{I}$$

$$\text{efficacy } K = \dfrac{F_t}{P}$$

Efficacy has units of lumens per watt (lm/W).

Illuminance: $E = \dfrac{F}{A}$ $\quad E_{tot} = E_1 + E_2 + E_3 + \ldots$

The units are *lux* (lx) in SI and *footcandles* (fc) in USCS.

$$\dfrac{E_1}{E_2} = \dfrac{r_2^2}{r_1^2} = \left(\dfrac{r_2}{r_1}\right)^2$$

QUESTIONS

1. Define the following terms: (a) luminous body; (b) photometry; (c) photon; (d) spectrum; (e) ray; (f) translucent material; (g) opaque material; (h) umbra; (i) penumbra; (j) radiant flux; (k) luminous flux; (l) mean spherical luminous intensity; (m) luminous efficacy; (n) illuminance; (o) a lux.
2. List several phenomema that can be described in terms of (a) the wavelike and (b) the particlelike characteristics of light.
3. Are shadows always sharp? Explain.
4. List several undesirable effects of shadows in working areas.
5. Describe three kinds of artificial light sources.
6. What are the differences between light and other forms of electromagnetic radiation?
7. Describe the sensitivity of the human eye to different colors of light. To what color is the eye most sensitive?
8. Why do fluorescent lamps have greater luminous efficacies than incandescent lamps?
9. Why do objects appear to be different colors under sunlight and under fluorescent lamps?
10. What kind of artificial light source would you use (a) to resemble sunlight; (b) for economy?

REVIEW PROBLEMS

1. Determine the wavelengths of light with the following frequencies: (a) 4.56×10^{14} Hz; (b) 7.12×10^{14} Hz.
2. What are the frequencies of light with the following wavelengths: (a) 460 nm; (b) 575 nm; (c) 718 nm?
3. Determine the energies of the photons that have the following frequencies: (a) 4.68×10^{14} Hz; (b) 6.99×10^{14} Hz.
4. Determine the energies of the photons that have wavelengths of (a) 499 nm; (b) 632 nm; (c) 762 nm.
5. What are the frequencies and wavelengths of the photons that have energies of (a) 1.85×10^{-19} J; (b) 3.60×10^{-17} J; (c) 4.62×10^{-15} J?
6. Determine the luminous intensity of a light that radiates a luminous flux of 250 lm (a) on a 5620 cm^2 perpendicular surface that is 95.0 cm away; (b) on a 6.00 ft^2 perpendicular surface that is 3.00 ft away.
7. Determine the candlepower of a small light source in a direction where it emits a luminous flux of 350 lm (a) onto a perpendicular area of 10.0 m^2 that is 3.0 m away; (b) onto a perpendicular area of 25.0 ft^2 that is 8.25 ft away.
8. A small light source has a mean spherical luminous intensity of 24.0 cd, and a reflector concentrates all the light onto a perpendicular area of 1.12 m^2 at a distance of 67.0 cm. Calculate (a) the total luminous flux of the source; (b) the intensity of the light in the illuminated area.
9. A small light source has a mean spherical luminous intensity of 36.0 cd, and a reflector concentrates all the light onto a perpendicular area of 12.0 ft^2 at a distance of 2.20 ft. Determine (a) the total luminous flux of the source; (b) the intensity of the light in the illuminated area.

Review Problems

10. The bulb of a searchlight has a mean spherical luminous intensity of 75.0 cd, but a reflector and lens concentrate the light onto an area 25.0 ft^2, a distance of 900 ft from the source. Calculate the luminous intensity of the light in the direction of the light beam.

11. What is the luminous intensity of a spotlight with a 35.0 cd bulb if the beam is concentrated on a 125 m^2 area of a vertical wall at a distance of 35.0 m?

12. A 40.0 W tungsten lamp has a luminous efficacy of 11.8 lm/W. (a) What is its mean spherical luminous intensity? (b) At what distance from the lamp is the maximum illuminance 54.0 lx? (c) At what distance from the lamp is the maximum illuminance 65.0 fc?

13. Determine the total illuminance of a 3.00 m^2 surface if it receives 35 lm from one source and 25 lm from a second source.

14. Calculate the total luminous flux incident on a 5750 cm^2 area if it has a uniform illuminance of 456 lx.

15. Determine the total luminous flux incident on a 6.00 ft^2 area if it has a uniform illuminance of 25.0 fc.

16. A small lamp is 2.0 m above a table. (a) To what distance must it be lowered to increase the illuminance by a factor of 3? (b) To what distance must it be raised to reduce the illuminance to one-half the original value?

17 REFLECTION AND MIRRORS

In this chapter we consider the properties of light and optical instruments by assuming that light travels in straight lines represented by rays. Even though this assumption is not strictly correct, it enables us to describe many optical phenomena. We consider only light, but many of the principles of geometrical optics may also be applied to other forms of electromagnetic radiation.

Objects do not have to be luminous bodies in order to be visible; most objects are visible because they reflect light. Objects that have irregular surfaces reflect light in many directions and scatter the light; this type of reflection is called **diffuse** (Fig. 17-1). The reflection is perfectly diffuse if the reflecting surface has an equal luminance (or brightness) in all directions.

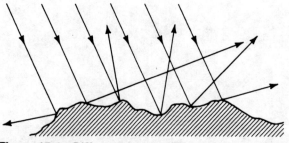

Figure 17-1 Diffuse reflection. The light is scattered.

A few objects, such as mirrors, have smooth surfaces, and there is a symmetrical relationship between the incident and the reflected light. This type of reflection is called **specular** or **regular** (Fig. 17-2). **Mirrors** and other **reflectors** are used extensively to control the direction and concentration of light and other forms of electromagnetic energy.

Sec.17-1 / Laws of Reflection

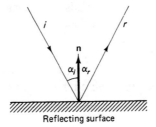

Figure 17-2 Specular reflection.

17-1 LAWS OF REFLECTION

The laws of reflection were discussed in Chapter 15. The angle of incidence α_i corresponds to the angle between the incident ray and the normal (perpendicular) to the surface. Similarly, the angle of reflection α_r is the angle between the reflected ray and the normal to the reflecting surface. Then the laws of reflection state:

1. The angle of incidence is equal to the angle of reflection. ($\alpha_i = \alpha_r$)
2. The incident ray i, the reflected ray r, and the normal **n** to the reflecting surface all lie in the same plane.

The Plane Mirror

We are all familiar with the images formed by a plane (flat) mirror. When we look at ourselves in the mirror, we see our likeness (**image**) behind the mirror surface.

Suppose that we locate some object in front of a plane mirror (Fig. 17-3). The light rays that are produced by (or reflected from) the object are incident on the mirror surface where they undergo specular reflection. To an observer, all the reflected light appears to come from the image, which is located behind the mirror surface.

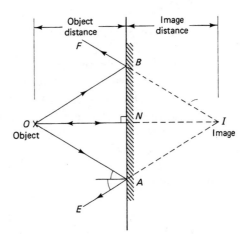

Figure 17-3 Image formation.

The location of the image I maybe determined from a **ray diagram,** which is a scale drawing (Fig. 17-3). Any number of rays may be drawn from the object to the surface of the mirror, and the laws of reflection give the directions of the reflected rays. For simplicity, we only need consider two light rays from each point (*OA* and *OB*, for example) and their reflections *AE* and *BF*. In each case the angle of incidence is equal to the angle of reflection. The intersection of the extensions of the reflected rays behind the mirror corresponds to the location of the image. If a ray ON is incident perpendicular to the mirror surface, it is reflected back along its own path. This type of ray is often convenient to use in ray diagrams.

Triangles ONA and INA are similar, and they have a common side NA; therefore, the distance ON from the object to the mirror (the **object distance**) is equal to the distance IN of the image I behind the mirror surface (the **image distance**). This result is always true for plane mirrors.

$$\text{object distance} = \text{image distance} \quad \text{(for plane mirrors)}$$

If we use an extended object, the images of the object points form an extended image behind the mirror surface (Fig. 17-4). From the ray diagram it can be seen that the image and the object are symmetrically located with respect to the mirror surface, and **the image is the same size as the object.**

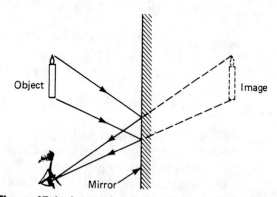

Figure 17-4 Image formed by an extended object.

Images

Two types of images can be formed:

1. If the light rays actually pass through the image, it is called a **real image**. This type of image can be projected onto a screen.
2. **Virtual images** cannot be projected onto a screen because light rays do not pass through them. Any image formed behind the surface of a mirror is a virtual image because the light rays do not penetrate the mirror.

For plane mirrors, the image is always virtual because it is always formed behind the mirror surface.

PROBLEM

1. A 1.75 m (5.74 ft) tall man stands in front of a vertical plane mirror. What is the minimum length of the mirror if the man sees his complete image?

17-2 SPHERICAL MIRRORS

In some optical instruments, curved mirrors are used to concentrate or disperse light. In many cases, the reflecting surfaces of these curved mirrors are shaped as if they were a part of a spherical surface, and they are called **spherical mirrors**.

Sec.17-2 / Spherical Mirrors

There are two types of spherical mirrors (see Fig. 17-5):

1. If the inner surface acts as the reflector, the mirror is called **concave.**
2. If the outer surface is the reflector, the mirror is called **convex.**

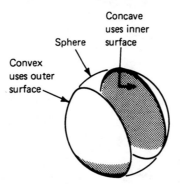

Figure 17-5 Concave and convex spherical mirrors.

The reflecting surface of a spherical mirror is a portion of an imaginary sphere. The center of the sphere is called the **center of curvature** C of the mirror, and the **radius of curvature** R of the mirror is the radius of sphere (Fig. 17-6).

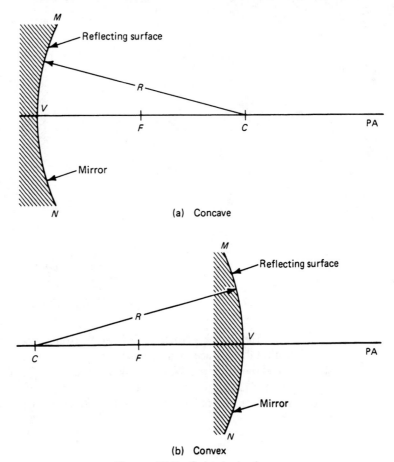

Figure 17-6 Spherical mirrors.

A spherical mirror normally has a circular boundary, the diameter MN of which is called the **aperture;** it is usually much smaller than the radius of curvature. The center of the mirror is known as the **vertex** V, and the straight line passing through the vertex V and the center of curvature C is called the **principal axis** PA.

All light rays that travel parallel to the principal axis of a concave spherical mirror are reflected (approximately) through a single point on the principal axis called the **principal focus** F. Similarly, for a convex mirror, light rays that travel parallel to the principal axis diverge after reflection as if they came from a single point, called the **principal focus** F (Fig. 17-7).

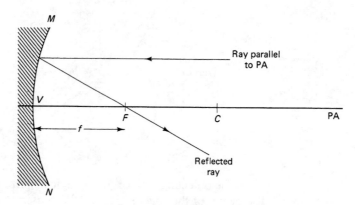

(a) Reflection from a concave spherical mirror. If the incident ray is parallel to the principal axis PA it is reflected through the principal focus F.

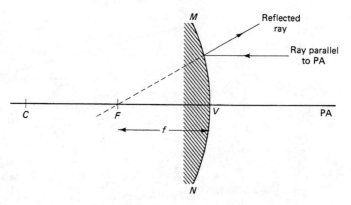

(b) Reflection from a convex spherical mirror. If the incident ray is parallel to the principal axis PA the reflected ray diverges as if it came from the principal focus F.

Figure 17-7

The principal focus is located in front of the surface of a concave mirror and behind the surface of a convex mirror. In each case the distance between the vertex V and the principal focus is called the **focal length** f for the mirror, and the principal focus F is approximately midway between the center of curvature C and the vertex. Therefore, the radius of curvature is approximately twice the focal length.

$$R \approx 2f \quad \text{(spherical mirrors)} \qquad (17\text{-}1)$$

Ray Diagrams

Ray diagrams may again be used to locate images formed by spherical mirrors; these images are usually different in size from the original objects. These diagrams must also be drawn to scale.

We generally draw two rays from each selected point on the object and use the laws of reflection to locate the images. The intersection of the reflected rays (or the extensions of the reflected rays) corresponds to the location of the image. In these diagrams we usually draw solid lines to represent the actual paths of the light rays, and dashed lines to indicate the directions from which the rays appear to originate.

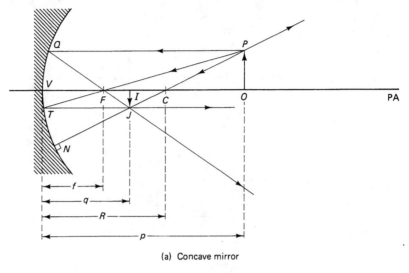

(a) Concave mirror

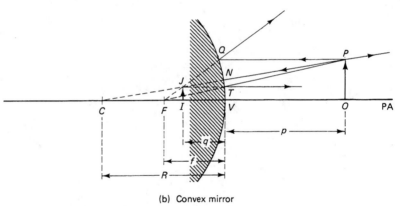

(b) Convex mirror

Figure 17-8 Ray diagrams for image location.

For convenience, any two of the following rays are normally used in ray diagrams (Fig. 17-8):

1. Any ray PQ parallel to the principal axis is reflected through the principal focus F of a concave mirror, or diverges as if it came from the principal focus of a convex mirror.
2. A ray PN that passes through the center of curvature C of a concave mirror (or is directed toward the center of curvature of a convex mirror) is reflected back along its original path.
3. Any ray PT that passes through the principal focus of a concave mirror (or is directed toward the principal focus of a convex mirror) is reflected parallel to the principal axis (along TS).

The location of the image J formed by the mirror is the intersection of the reflected rays (or the extensions of the reflected rays behind the mirror). The various types of ray diagrams are illustrated in Figs. 17-9 and 17-10.

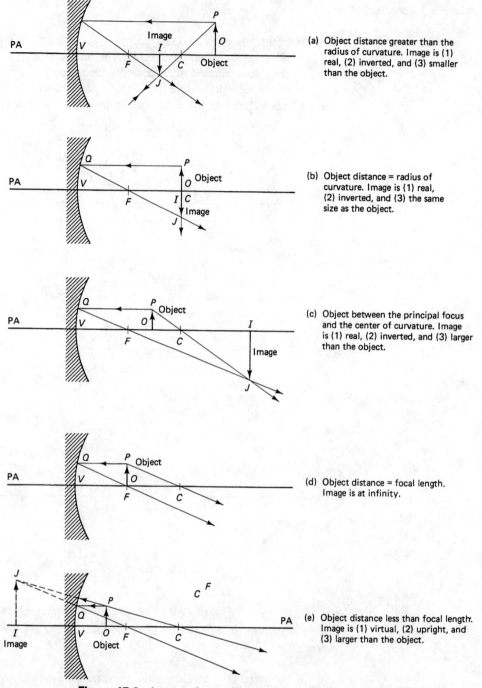

Figure 17-9 Images formed by concave spherical mirrors.

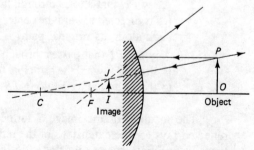

Figure 17-10 Images formed by convex spherical mirrors. The image is always (1) virtual, (2) upright, and (3) smaller than the object.

Sec.17-2 / Spherical Mirrors

The Mirror Equation

Suppose that an object is placed on the principal axis in front of a mirror of focal length f and radius of curvature R. If the object distance (measured from the vertex) is p and the mirror forms an image at a distance q from the vertex, then

$$\frac{1}{p} + \frac{1}{q} = \frac{2}{R} \approx \frac{1}{f} \qquad (17\text{-}2)$$

This expression is called the **mirror equation.** It is valid for both concave and convex spherical mirrors if the following **sign convention** is used.

All values for items that are located in front of the reflecting (shiny) side of the mirror are positive valued, and values for quantities behind the reflecting surface are negative valued.

For example, the center of curvature C and the principal focus F are located behind the reflecting surface of a convex mirror, and therefore the radius of curvature R and the focal length f are both negative valued. A concave mirror has its center of curvature and principal focus in front of the reflecting surface, and the radius of curvature R and focal length f are both positive valued. Similarly, real objects and images are located in front of the mirror where light rays can pass through them, and the corresponding object and image distances p and q are positive valued. Virtual images are located behind the mirror surface and their image distances are negative valued.

Linear Magnification

The ratio of the length (height or width) of the image H_i to the corresponding length H_o of the object is called the **linear magnification** M; by convention, it is positive when the image is erect and negative when the image is inverted.

It can be shown from the geometry:

$$M = \frac{\text{image length}}{\text{object length}} = \frac{H_i}{H_o} = \frac{-q}{p} \qquad (17\text{-}3)$$

where p is the object distance and q is the image distance.

EXAMPLE 17-1

An object 2.00 cm high is placed 20.0 cm in front of a convex spherical mirror that has a focal length of 30.0 cm. Find **(a)** the position; **(b)** the size of the image.

Solution: This problem may be solved by means of a ray diagram or the mirror equation.

Sketch: See Fig. 17-11.

Data: $H_o = 2.00$ cm, $p = 20.0$ cm, $f = -30.0$ cm (convex, and therefore the principal focus is behind the reflecting surface and negative valued), $q = ?$, $H_i = ?$

(a) *Equation:* $\dfrac{1}{p} + \dfrac{1}{q} = \dfrac{1}{f}$

Rearrange: $\dfrac{1}{q} = \dfrac{1}{f} - \dfrac{1}{p}$

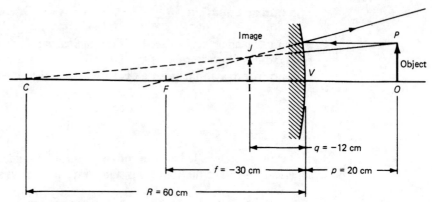

Figure 17-11 Ray diagram drawn to scale. (Image and object heights are in a different scale for convenience).

Substitute: $\dfrac{1}{q} = \dfrac{1}{-30.0 \text{ cm}} - \dfrac{1}{20.0 \text{ cm}} = \dfrac{-2-3}{60.0 \text{ cm}} = \dfrac{-5}{60.0 \text{ cm}}$

or $q = \dfrac{-60.0 \text{ cm}}{5} = -12.0$ cm. This may be found directly using a calculator.

The negative sign indicates that a virtual image is formed (i.e., 12.0 cm behind the reflecting surface).

(b) *Equation:* $M = \dfrac{H_i}{H_o} = \dfrac{-q}{p}$

Rearrange: $H_i = -H_o \dfrac{q}{p}$

Substitute: $H_i = \dfrac{-(2.00 \text{ cm})(-12.0 \text{ cm})}{20.0 \text{ cm}} = +1.20$ cm

The positive sign implies that the image is upright. The same answers were obtained with the ray diagram. *Even if a ray diagram is not used, it is always important to sketch the system.*

EXAMPLE 17-2

When an object is placed 10.0 in. in front of a spherical mirror, an erect virtual image is formed that is 1.50 times as large as the object. Find the focal length of the mirror.

Solution

Data: $M = 1.50$, $p = 10.0$ in., $f = ?$

Equations: $M = \dfrac{-q}{p}$, $\dfrac{1}{p} + \dfrac{1}{q} = \dfrac{1}{f}$

Rearrange: $q = -Mp$

Substitute: $q = -(1.50)(10.0 \text{ in.}) = -15.0$ in.

Then

$$\dfrac{1}{f} = \dfrac{1}{p} + \dfrac{1}{q} = \dfrac{1}{10.0 \text{ in.}} + \dfrac{1}{-15.0 \text{ in.}} = \dfrac{3-2}{30.0 \text{ in.}} = \dfrac{1}{30.0 \text{ in.}}$$

and $f = +30.0$ in. The mirror is concave because the focal length is positive valued (see Fig. 17-12).

Sec. 17-2 / Spherical Mirrors

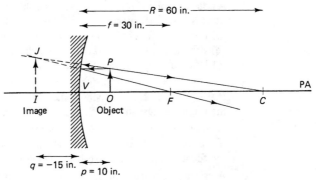

Figure 17-12 Ray diagram.

Applications of spherical mirrors include wide-angle driving mirrors (Fig. 17-13), which are used to increase the field of view, shaving mirrors, and reflectors for solar furnaces.

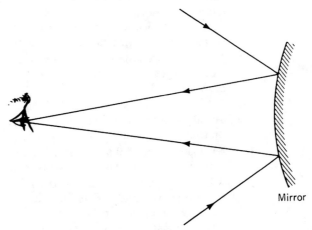

Figure 17-13 Wide-angle driving mirror.

PROBLEMS

2. A 3.2 cm high object is placed 8.00 cm in front of a concave spherical mirror that has a radius of curvature of 24.0 cm. (a) Determine the position and size of the image. (b) Describe the image. (c) Draw the ray diagram.

3. A 1.20 in. high object is located 4.50 in. in front of a concave spherical mirror that has a focal length of 5.00 in. (a) Determine the position and size of the image. (b) Describe the image. (c) Draw the ray diagram.

4. (a) Determine the position and size of an image that is formed when a 1.20 cm high object is located 8.00 cm in front of a convex spherical mirror that has a radius of curvature of 4.00 cm. (b) Draw the ray diagram.

5. A 2.50 in. high object is positioned 7.50 in. in front of a convex spherical mirror that has a focal length of 6.00 in. (a) Determine the position and size of the image. (b) Draw the ray diagram.

6. When an object is located 15.0 cm in front of a spherical mirror a virtual image is formed that is 0.600 times as large as the object. Determine the focal length of the mirror and draw the ray diagram.

7 A concave spherical mirror is used as a shaving mirror because it produces virtual upright images when the object distance is less than the focal length. Determine the magnification (a) if a man's face is 20.0 cm from a shaving mirror that has a focal length of 40.0 cm; (b) if a man's face is 12.0 in. from a shaving mirror that has a focal length of 36.0 in. (c) Draw the ray diagrams.

8 A convex spherical mirror forms a virtual image that is 0.800 times the size of the object. If the image is 22.0 cm behind the mirror, determine (a) the position of the object; (b) the radius of curvature of the mirror. (c) Draw the ray diagram.

9 A convex spherical mirror forms a virtual image that is 0.775 times the size of an object. If the image is 8.00 in. behind the mirror, determine (a) the position of the object; (b) the focal length of the mirror. (c) Draw the ray diagram.

10 A concave spherical mirror forms a real image that is 1.25 times the size of the object. If the focal length of the mirror is 50.0 cm, determine (a) the location of the object; (b) the location of the image. (c) Draw the ray diagram.

11 A concave spherical mirror has a radius of curvature of 88.0 cm. Determine the position and size of the image when a 1.85 cm object is located the following distances from the reflecting surface: (a) 125 cm; (b) 88.0 cm; (c) 60.0 cm; (d) 44.0 cm; (e) 18.5 cm.

12 A concave spherical mirror has a focal length of 17.5 in. Determine the position and size of the image that is formed when a 0.75 in. high object is located the following distances from the reflecting surface: (a) 50.0 in.; (b) 35.0 in.; (c) 25.0 in.; (d) 17.5 in; (e) 7.5 in.

13 A technician uses a concave mirror with a radius of curvature of 5.00 cm to check visually a circuit that is located in an awkward position. Determine the location of the image and the magnification when the circuit element is 1.50 cm from the mirror.

14 A concave spherical mirror with a radius of curvature of 5.00 ft is used to project an image onto a screen that is 15.0 ft from the mirror. Where must the object be located? Draw the ray diagram.

15 A 2.00 cm high object located 8.00 cm in front of a mirror forms an image 12.5 cm behind the mirror. Determine (a) the focal length of the mirror; (b) the image size. (c) Draw the ray diagram.

16 An object located 15.0 in. in front of a mirror forms an image 6.00 in. behind the mirror. Determine the focal length and radius of curvature of the mirror.

17 A 2.50 cm high object located 18.0 cm in front of a mirror forms a real image 24.0 cm from the mirror. Determine (a) the focal length of the mirror; (b) the size of the image. (c) Draw the ray diagram.

18 When an object is located in front of a concave spherical mirror with a focal length of 12.5 cm, a real image is formed that is 5.00 cm high and it is 30.0 cm from the mirror. Determine the position and size of the object. Draw the ray diagram.

19 Determine the position and size of the image that is formed when a 1.40 in. high object is placed 32.5 in. in front of a convex spherical mirror which has a radius of curvature of 14.0 in.

17-3 SPHERICAL ABERRATION AND PARABOLIC REFLECTORS

Rays that are parallel to the principal axis of a spherical mirror pass approximately through the same point (the principal focus) only if the aperture of the mirror is much smaller than its radius of curvature. If the aperture of the mirror is not relatively small, the parallel rays that are farthest from the principal axis are reflected through points on the principal axis that are closer to the vertex of the mirror (Fig. 17-14). This is called **spherical aberration;** it results in blurred images. The envelope of the reflected rays in the neighborhood of the principal focus is called the **caustic curve.**

Spherical aberration is eliminated if a **parabolic mirror** (Fig. 17-15) is used instead of a spherical mirror when a large aperture is required. Parallel rays always pass through the same point after they have been reflected from a parabolic mirror. Parabolic reflectors are used instead of spherical reflectors in some large telescopes (Fig. 17-16).

Sec. 17-3 / Spherical Aberration and Parabolic Reflectors   305

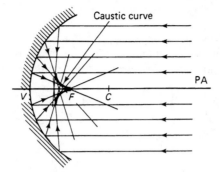

Figure 17-14 Spherical aberration.

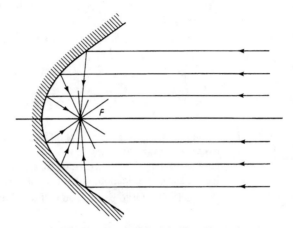

Figure 17-15 Parabolic mirror.

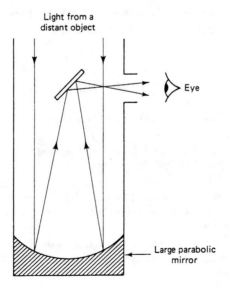

Figure 17-16 Reflecting telescope.

When a light source is placed at the principal focus of a parabolic mirror, parallel light rays emerge after reflection from the mirror's surface. This principle is used in automobile headlights and in searchlights.

SUMMARY

Plane mirror: Image distance = object distance

Light passes through *real images* and they can be projected onto a screen. Light does not pass through *virtual images* and they cannot be projected onto a screen.

Spherical mirrors: $R \approx 2f$, $\quad \dfrac{2}{R} \approx \dfrac{1}{f} = \dfrac{1}{p} + \dfrac{1}{q}$

Sign convention: Quantities on the shiny side are positive valued, and those behind the mirror surface are negative valued.

$$M = \frac{H_i}{H_o} = \frac{-q}{p}$$

Sign convention: Upright images have positive-valued linear magnifications. Inverted images have negative-valued linear magnifications.

Spherical aberration: Parallel rays of light are not reflected through a single point.

QUESTIONS

1. Define the following terms: **(a)** diffuse reflection; **(b)** specular reflection; **(c)** real image; **(d)** virtual image; **(e)** convex spherical mirror; **(f)** concave spherical mirror; **(g)** center of curvature; **(h)** principal focus; **(i)** focal length; **(j)** vertex; **(k)** aperture; **(l)** principal axis; **(m)** linear magnification; **(n)** spherical aberration.
2. Why is it difficult to perform some tasks, such as cutting your own hair, by looking into a mirror?
3. List a few common devices that use mirrors to form real enlarged images.
4. Describe what happens to the images formed by a concave spherical mirror when an object moves toward it from an initial distance greater than the radius of curvature.
5. Mirrors are used as reflectors for automobile headlights. Explain their function.
6. What kind of mirror would be used in a searchlight? Explain why.
7. Describe the advantages and disadvantages of spherical mirrors and parabolic mirrors.

REVIEW PROBLEMS

1. Sketch the image(s) that is (are) formed when an object is located in front of two mirrors that are perpendicular to each other.
2. An object that is 5.00 in. high is located 15.0 in. in front of a concave spherical mirror that has a focal length of 5.00 in. **(a)** Determine the position and size of the image. **(b)** Draw the ray diagram.
3. An object 1.50 cm high is placed 15.0 cm in front of a concave spherical mirror that has a radius of curvature of 10.0 cm. **(a)** Calculate the position and size of the image. **(b)** Draw the ray diagram.
4. An object 3.00 cm high is placed 8.00 cm in front of a concave spherical mirror of small aperture that has a focal length of 12.0 cm. **(a)** Calculate the position and size of the image. **(b)** Draw the ray diagram.
5. An object 1.20 cm high is placed 5.00 cm in front of a concave spherical mirror that has a radius of curvature of 28.0 cm. **(a)** Draw the ray diagram. **(b)** Determine the position and size of the image. **(c)** Describe the image.
6. An object 1.00 in. high is placed 12.0 in. in front of a concave spherical mirror that has a focal length of 6.00 in. **(a)** What is the position and size of the image? **(b)** Draw the ray diagram.

Review Problems

7 When an object is placed 18.0 in. in front of a spherical mirror, a real image 0.75 times as large as the object is formed. **(a)** Determine the focal length of the mirror. **(b)** Draw the ray diagram.

8 A 2.00 cm high object is located 12.0 cm in front of a convex spherical mirror that has a radius of curvature of 20.0 cm. Find the position and size of the image.

9 Determine the position and size of the image that is formed when a 18.0 mm high object is placed 6.80 cm in front of a convex spherical mirror that has a radius of curvature of 48.0 cm.

10 When a 2.50 in. high object is positioned 22.0 in. in front of a spherical mirror a virtual image that is 0.62 times as large as the object is formed. Find **(a)** the focal length of the mirror; **(b)** the position and size of the image. **(c)** Draw the ray diagram.

11 When an object is located 8.25 cm in front of a concave spherical mirror a virtual image 14.0 cm high is formed 17.5 cm behind the mirror. Determine **(a)** the focal length of the mirror; **(b)** the size of the object. **(c)** Draw the ray diagram.

18 REFRACTION AND LENSES

Transparent materials reflect some light from their surfaces, but they also transmit a fraction of any incident light energy. A **simple lens** is a piece of transparent material, such as glass, with one or two curved surfaces, that is used to change the directions of light rays.

18-1 REFRACTION

The speed of light depends on the nature of the medium through which it travels. Light travels at its greatest speed of 3.00×10^8 m/s or 9.82×10^8 ft/s (186 000 mi/s) in a vacuum. Its speed is reduced when it enters a material medium. The change in speed, when light passes from one material into another, may cause the paths of the light rays to bend, because one part of the wavefront travels faster than the other (Fig. 18-1). This bending process is called **refraction.**

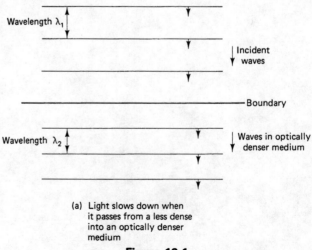

(a) Light slows down when it passes from a less dense into an optically denser medium

Figure 18-1

Sec. 18-1 / Refraction

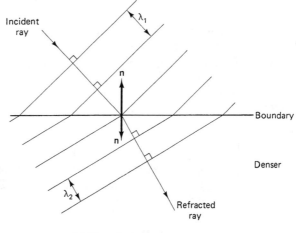

(b) The refracted rays are bent toward the normal to the boundary in the denser medium as the light slows down

Figure 18-1 (cont.)

Light rays are bent toward the normal (perpendicular) to the boundary as the waves slow down when they pass from an optically less dense material into an optically denser material. If light passes from an optically dense material into an optically less dense material, its speed increases, and the light rays are bent away from the normal to the boundary. For example, light slows down when it passes from air into optically denser water. This is why a straight stick appears to be bent when a part of it is placed in water (Fig. 18-2).

Figure 18-2 A straight stick appears to be bent in water because of refraction.

The **angle of incidence** α_i is the angle between the incident ray and the normal to the boundary, and the **angle of refraction** α_r is the angle between the normal and the refracted ray.

Note that the path of a refracted ray is reversible. That is, if the direction of the refracted light ray BC is reversed, it retraces its original path, and it is refracted (at the boundary) back along the path of the original ray, AB.

The ratio of the speed of light c in a vacuum to the speed of light v in a material medium is defined as the **index of refraction** n (or **optical density**) of the medium.

$$n = \frac{c}{v} \tag{18-1}$$

Since light travels fastest in a vacuum, the index of refraction of a material medium is always greater than 1. Some typical values for the indices of refraction of a few materials are listed in Table 18-1. These values depend on the wavelength of the incident light and the temperature of the system.

TABLE 18-1
Typical Indices of Refraction for Light with a Wavelength of 589.3 nm at 20°C

Substance	Index of refraction	Substance	Index of refraction
Vacuum	1.000 00	Solids	
Air	1.000 29	Crown glass	1.50
Liquids		Flint glass	1.58
Benzene	1.50	Dense crown glass	1.60
Ethyl alcohol	1.36	Dense flint glass	1.72
Water	1.33	Canada balsam	1.53
		Diamond	2.42
		Perspex	1.50
		Quartz (fused)	1.46

The **law of refraction** at a boundary between two media are as follows:

1. The ratio of the sine of the incident angle α_1 to the sine of the angle of refraction α_2 is a constant that is equal to the ratio of the speeds of light in the two media. If n_1 and n_2 are the indices of refraction of media 1 and 2, the corresponding speeds of light in these media are

$$v_1 = \frac{c}{n_1} \quad \text{and} \quad v_2 = \frac{c}{n_2}$$

Therefore, the **first law of refraction** states

$$\frac{\sin \alpha_1}{\sin \alpha_2} = \frac{v_1}{v_2} = \frac{n_2}{n_1} \tag{18-2}$$

or

$$n_1 \sin \alpha_1 = n_2 \sin \alpha_2$$

This relationship is called **Snell's law** (Fig. 18-3); it was discovered by W. Snell in the seventeenth century.

2. The incident ray, the refracted ray, and the normal to the boundary at the point of incidence lie in the same plane.

Sec. 18-1 / Refraction

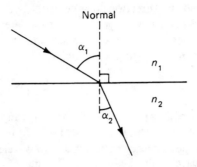

Figure 18-3 Snell's law: $n_1 \sin \alpha_1 = n_2 \sin \alpha_2$.

EXAMPLE 18-1

A narrow beam of light in air impinges on a flint glass plate at an angle of incidence of 52°. Determine (a) the angle of refraction; (b) the speed of light in flint glass.

Solution

Data: $\alpha_1 = 52°$, $n_1 = 1.000$, $n_2 = 1.58$ (Table 18-1), $c = 3.00 \times 10^8$ m/s or 186 000 mi/s, $\alpha_2 = ?$, $v_2 = ?$

(a) *Equation:* $\dfrac{\sin \alpha_1}{\sin \alpha_2} = \dfrac{n_2}{n_1}$

Rearrange: $\sin \alpha_2 = \dfrac{n_1 \sin \alpha_1}{n_2}$

Substitute: $\sin \alpha_2 = \dfrac{(1.00)(\sin 52°)}{1.58} = 0.50$

Therefore, $\alpha_2 = 30°$.

(b) *Equation:* $v_2 = \dfrac{c}{n_2}$

Substitute: $v_2 = \dfrac{3.00 \times 10^8 \text{ m/s}}{1.58} = 1.90 \times 10^8$ m/s

or

$$v_2 = \dfrac{186\,000 \text{ mi/s}}{1.58} = 118\,000 \text{ mi/s}$$

Light from distant stars and other celestial bodies reaches the earth after passing through the near vacuum of space. As this light penetrates the earth's atmosphere it is refracted downward (Fig. 18-4). For this reason the **angle of elevation** between the

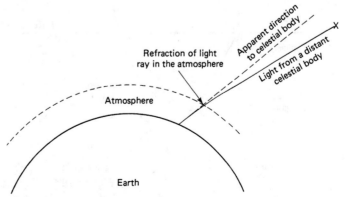

Figure 18-4 Refraction of light in the earth's atmosphere.

PROBLEMS

1. A narrow beam of light passes from air to water at an angle of incidence of 34°. Determine (a) the angle of refraction; (b) the speed of light in water.
2. (a) Determine the frequency of light that has a wavelength of 589.3 nm in a vacuum. (b) What is the wavelength of this light in flint glass?
3. A narrow beam of light is incident from air onto the surface of a transparent material. If the angle of incidence is 48° and the angle of refraction is 32°, calculate (a) the index of refraction of the material; (b) the speed of light in that material.
4. Determine the wavelength of light in a liquid that has an index of refraction of 1.42 if its wavelength in a vacuum is 550 nm.
5. A plate of crown glass is completely immersed in water parallel to the water surface, and a beam of light passes from air through the water to the glass plate. If the angle of incidence of the light at the boundary between the air and water is 38°, determine the angle of refraction in the glass plate.
6. A completely submerged diver shines a narrow beam of light toward the surface of the water. If the angle of incidence of the light beam is 28°, determine the angle of refraction in air.

18-2 TOTAL INTERNAL REFLECTION

When light passes from one transparent material into another, some of the light energy is reflected at the boundary between the two materials. The fraction of the total incident light energy that is reflected depends on the nature of the two transparent materials and the angle of incidence of the light at the boundary between them. As the angle of incidence is increased, a greater fraction of the incident light is reflected. For example, at large angles of incidence, a plate of glass behaves like a mirror.

Suppose that a beam of light passes from one transparent material, which has an index of refraction n_1, into another with a smaller index of refraction n_2. At small angles of incidence some of the light is reflected, and some is transmitted after it is refracted *away* from the normal at the boundary. As the angle of incidence α_i increases, a greater fraction of the incident light is reflected, and the angle of refraction α_r of the transmitted light approaches a maximum value of 90° (Fig. 18-5).

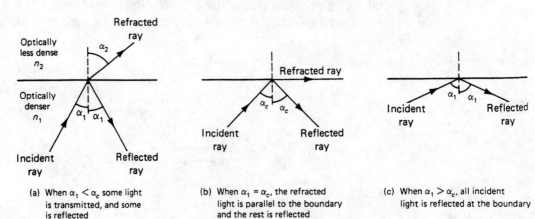

Figure 18-5 Internal reflection.

When the angle of refraction is 90°, the transmitted light emerges parallel to the boundary; the corresponding angle of incidence is called the **critical angle** α_c. Thus, according to Snell's law,

$$\frac{\sin \alpha_c}{\sin 90°} = \sin \alpha_c = \frac{n_2}{n_1} \tag{18-3}$$

since $\sin 90° = 1$.

If the angle of incidence is greater than the critical angle, all the incident light is reflected at the boundry between the two transparent materials. This **total internal reflection** occurs from a single surface, and it is a very efficient reflective process.

EXAMPLE 18-2

Determine the critical angle between crown glass and air.

Solution

Data: $n_2 = 1.00$, $n_1 = 1.50$ (Table 18-1), $\alpha_c = ?$

Equation: $\sin \alpha_c = \dfrac{n_2}{n_1}$

Substitute: $\sin \alpha_c = \dfrac{1.00}{1.50} = 0.667$ and $\alpha_c = 41.8°$.

The index of refraction of a transparent material depends on the amount of impurities that it contains; therefore, impurity concentrations may be determined by measuring the index of refraction. In liquids the index of refraction may be accurately and conveniently determined by measuring the critical angle.

Prisms

Total internal reflection from a polished glass surface with air is a very efficient process. There are only minute losses in light energy. The critical angle from glass to air is less than 45°, therefore, when a light ray in the glass is incident at an angle of 45° at the glass-air boundary, it undergoes total internal reflection. Many optical instruments contain glass **prisms** with polished sides and angles of 45°, 45°, and 90°. These prisms efficiently change the direction of the light or they invert images (Fig. 18-6).

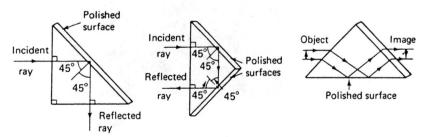

(a) Internal reflection produces a 90° change in the direction of a ray.

(b) A 180° change in the direction of a ray.

(c) Inversion of an image.

Figure 18-6 A 45°, 45°, 90° prism may be used to change the direction of light rays or to invert images by total internal reflection.

Optical Fibers

The recent development of special fibers has vastly improved our ability to control the path of light energy. These fibers are relatively flexible, and they can be bent into different shapes. Light is fed into one end of the fiber, and it undergoes multiple internal reflections as it is transmitted through the fiber (Fig. 18-7).

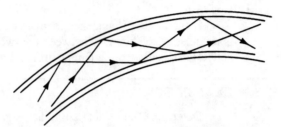

Figure 18-7 Total internal reflections in a fiber.

Special fibers of high-purity glass about as thick as a human hair are now widely used in telephone transmission (Fig. 18-8) and in light-wave communication systems as guides for light. These fibers are low cost and are able to carry high-frequency variations in the light with low losses. Optical fibers also have many other important applications. For example, in medicine, very fine fibers can be guided into the interior of a living organ, and the physician can then observe the internal operation of that organ. Scientists have already used fibers to observe the interior of a beating human heart.

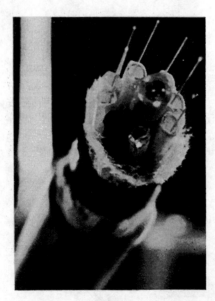

Figure 18-8 No thicker than a pencil, this cable can be used for simultaneous transmission of 300 000 telephone conversations. (Courtesy of Siemens.)

PROBLEMS

7. Determine the critical angle between (a) diamond and air; (b) crown glass and water.
8. (a) Determine the index of refraction of a material if the critical angle with air is 57°. (b) What is the speed of light in the material?

18-3 LENSES

When a ray of light passes from a medium through an optically denser prism, it is refracted towards the normal at the first surface, and then away from the normal at the second surface (Fig. 18-9). As a result, the ray is deflected from its original path.

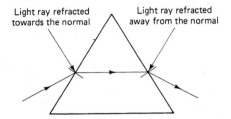

Figure 18-9 Double refraction.

Transparent objects that have two smooth surfaces that alter the shape of the wavefront are called **lenses**. Either one or both of the surfaces of a lens are curved, and the lens is named in terms of the shapes of these surfaces (Fig. 18-10).

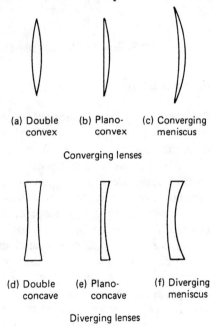

Figure 18-10 Lenses.

Most lenses have spherical surfaces, and they are made of materials, such as glass, that are optically denser than air. Light travels slower in glass than in air; therefore, when a light wave passes from air into a glass lens, the part of the wavefront that passes through the thicker portion of the lens spends more time in the lens. As a result, it lags behind other parts of the wavefront (Fig. 18-11). Therefore, the wavefront is changed by the lens. The amount of the change depends on the index of refraction and the shape of the lens.

A **converging lens** is usually thicker at its center;* its curved surfaces tend to refract incident parallel light rays through a common point. A **diverging lens** is usually thinner at its center; its curved surfaces tend to refract incident parallel light rays as if they originated at a common point.

---

*Unless the lens has a lower index of refraction than the surrounding medium, in which case the reverse is true.

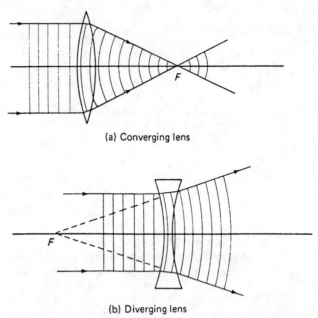

(a) Converging lens

(b) Diverging lens

Figure 18-11 Lenses change the shape of the wavefront.

Nomenclature

Refracting surfaces of **spherical lenses** are portions of spheres. The centers of these spheres are called the **centers of curvature** C_1 and C_2 of the lens, and the radii R_1 and R_2 of the spheres are called the **radii of curvature**.

The diameter of a lens is called the **aperture**; it is usually much smaller than the radii of curvature. The amount of light that passes through a lens depends on its aperture. Lenses with large apertures form bright images.

If a ray passes through a point called the **optical center** O of the lens, it always emerges from the lens in a direction that is parallel to the incident direction. In thin lenses, the optical center is located approximately at the center of the lens.

The **principal axis** PA of the lens is a straight line that passes through the centers of curvature and the optical center (Fig. 18-12). The distance between the object and the optical center is called the **object distance** p, and the distance between the image and the optical center is called the **image distance** q.

Rays that are initially parallel to the principal axis are refracted through (or diverge as if they came from) a common point called the **principal focus** F after they emerge from the lens. Note that there is a principal focus for each curved surface of the lens, and these foci are equidistant from the optical center. The focus that is located on the same side of the lens as the incident light is called the **first (near or virtual) principal focus** F_1. No light is refracted through the first principal focus. The other focus F_2 is called the **second (far or real) principal focus** and is located on the side of the lens where the refracted light emerges.

Ray Diagrams

The nature and location of an image that is formed by a thin lens may be determined with a ray diagram. Consider two or more rays that are drawn from the same point on an object. These rays are refracted as they pass through a lens, and the intersection of the refracted rays, or the extension of the refracted rays, corresponds to the image location. For convenience any two of the following rays are normally used (Fig. 18-13).

Sec. 18-3 / Lenses

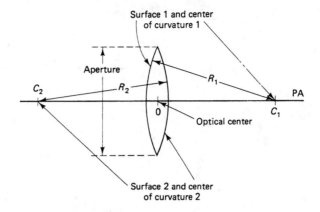

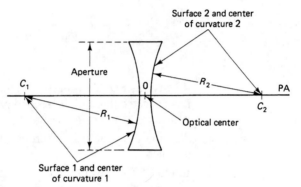

Figure 18-12

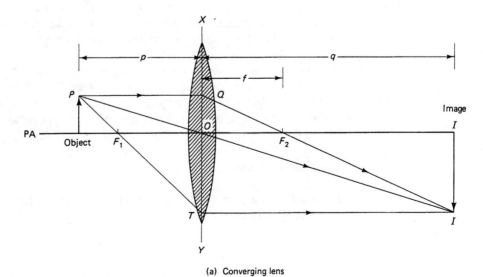

(a) Converging lens

Figure 18-13 Ray diagrams.

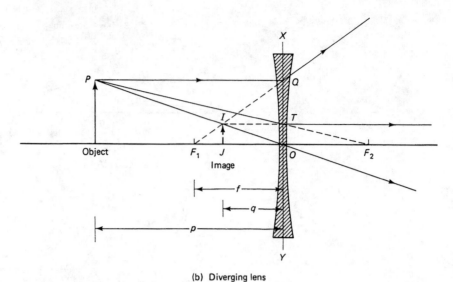

(b) Diverging lens

Figure 18-13 (cont.)

1. When a ray *PQ* that is parallel to the principal axis is incident on a thin lens, it is refracted through the second principal focus F_2 of a converging lens, or it diverges as if it originated at the first principal focus F_1 of a diverging lens.
2. Any ray *PO* that passes through the optical center *O* of any lens is not deviated.
3. If a ray *PT* passes through the first principal focus F_1 of a converging lens or if it is directed toward the second principal focus F_2 of a diverging lens, it is refracted by the lens, and it emerges parallel to the principal axis.

Refraction occurs at both surfaces of a lens; however, for convenience in ray diagrams, the direction changes are assumed to occur at the plane *XY* that is perpendicular to the principal axis and contains the optical center *O*.

Light rays actually pass through lenses; therefore, **real images** are formed on the side of the lens where the refracted light emerges, and they can be projected onto a screen. **Virtual images** are located on the same side of the lens as the incident light. These virtual images cannot be projected onto a screen, but the refracted light rays appear to originate there.

The images that are formed by diverging lenses are always virtual, upright, and diminished. Some of the types of images that are formed by converging lenses are illustrated in Fig. 18-14. Converging lenses may also be used to produce parallel rays of light by placing a light source at the first principal focus. The types of images formed by diverging lenses are illustrated in Fig. 18-15.

The Lens Equation

There is a simple equation that relates the object distance *p*, the image distance *q* and the focal length *f* of a lens:

$$\frac{1}{p} + \frac{1}{q} \approx \frac{1}{f} \qquad (18\text{-}4)$$

This is called the **lens equation.** It is valid for both converging and diverging lenses if the following **sign conventions** are used (Fig. 18-16):

Sec. 18-3 / Lenses

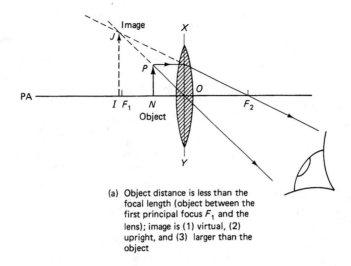

(a) Object distance is less than the focal length (object between the first principal focus F_1 and the lens); image is (1) virtual, (2) upright, and (3) larger than the object

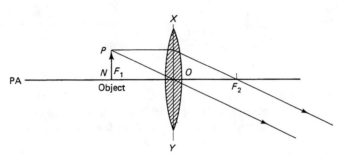

(b) Object at the first principal focus F_1; image is at infinity

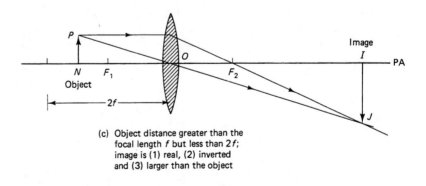

(c) Object distance greater than the focal length f but less than $2f$; image is (1) real, (2) inverted and (3) larger than the object

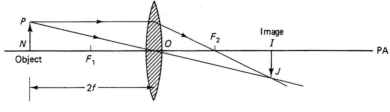

(d) Object distance equal to twice the focal length; image is (1) real, (2) inverted, and (3) same size as the object

Figure 18-14 Images formed by converging lenses.

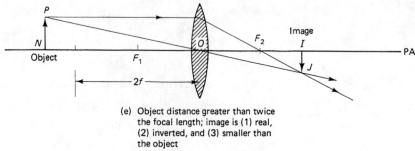

(e) Object distance greater than twice the focal length; image is (1) real, (2) inverted, and (3) smaller than the object

Figure 18-14 (cont.)

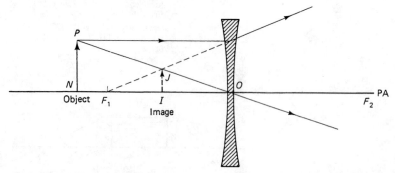

Figure 18-15 Image formed by a diverging lens is *always* (1) virtual, (2) upright, and (3) smaller than the object.

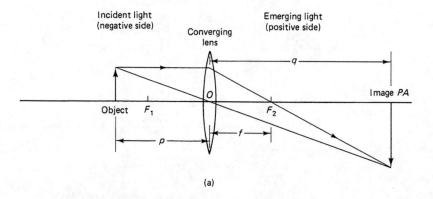

(a)

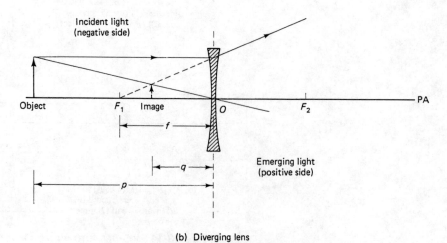

(b) Diverging lens

Figure 18-16 Sign convention for the lens equation.

Sec.18-3 / Lenses

1. The focal lengths f of converging lenses are positive valued, and the focal lengths of diverging lenses are negative valued.
2. Object distances p and image distances q of real objects and images are positive valued. They are negative valued for virtual objects and images.

Linear Magnification

As in the case of mirrors, the ratio of the length (height or width) of the image to the corresponding length (height or width) of the object is called the **linear magnification** M.

$$M = \frac{\text{image length}}{\text{object length}} = \frac{H_i}{H_o} = \frac{-q}{p} \qquad (18\text{-}5)$$

Power of a Lens

The **power** P of a lens in units of **diopters** is defined as the reciprocal of its focal length f in meters.

$$P \text{ (diopters)} = \frac{1}{f \text{ (meters)}} \qquad (18\text{-}6)$$

Lenses with short focal lengths have larger powers than those with long focal lengths, because they produce larger distortions of the wavefronts.

Before the thin-lens equation is used, it is advisable to sketch the ray diagram of the system where possible, and to apply the sign conventions to express the known quantities in terms of their symbols.

EXAMPLE 18-3

Determine **(a)** the position and **(b)** the height of the image that is formed when an object 3.00 cm high is placed 20.0 cm in front of a thin diverging lens that has a focal length of 5.00 cm.

Solution

Sketch: We may solve this problem with a ray diagram (Fig. 18-17).

Data: $p = 20.0$ cm, and, since the lens is a diverging lens, $f = -5.00$ cm; $H_o = 3.00$ cm, $q = ?$, $H_i = ?$

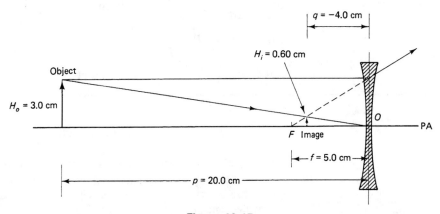

Figure 18-17

(a) Equation: $\dfrac{1}{p} + \dfrac{1}{q} = \dfrac{1}{f}$

Rearrange: $\dfrac{1}{q} = \dfrac{1}{f} - \dfrac{1}{p}$

Substitute: $\dfrac{1}{q} = \dfrac{1}{-5.00 \text{ cm}} - \dfrac{1}{20.0 \text{ cm}} = \dfrac{-4 - 1}{20.0 \text{ cm}} = \dfrac{-5}{20.0 \text{ cm}}$

or $q = \dfrac{-20.0 \text{ cm}}{5} = -4.00$ cm. The negative sign implies that the image is virtual and that it is located on the same side of the lens as the object. This type of calculation may be completed directly with a calculator.

(b) Equation: $\dfrac{H_i}{H_o} = \dfrac{-q}{p}$

Rearrange: $H_i = \dfrac{-H_o q}{p}$

Substitute: $H_i = \dfrac{-(3.00 \text{ cm})(-4.00 \text{ cm})}{20.0 \text{ cm}} = 0.600$ cm

These values can also be obtained from the ray diagrams.

EXAMPLE 18-4

The lens of a movie projector forms inverted images 7.00 m long on a screen that is 30.0 m away. If the film size is 35 mm, calculate (a) the focal length; (b) the power of the lens.

Solution

Sketch: This is difficult to draw to scale in this case because of the lengths involved.

Data: $H_o = 35$ mm, and since the images are real and inverted, $q = +30.0$ m, $H_i = -7.00$ m, $f = ?$, $P = ?$

(a) Equation $M = \dfrac{H_i}{H_o} = \dfrac{-q}{p}$

Substitute: $M = \dfrac{-7.00 \text{ m}}{0.035 \text{ m}} = -200 = \dfrac{-q}{p}$

Therefore, $p = \dfrac{q}{200}$ and

$$\dfrac{1}{f} = \dfrac{1}{p} + \dfrac{1}{q} = \dfrac{200}{q} + \dfrac{1}{q} = \dfrac{201}{q}$$

or $f = \dfrac{q}{201} = \dfrac{30.0 \text{ m}}{201} = 0.149$ m.

(b) Equation: $P = \dfrac{1}{f}$

Substitute: $P = \dfrac{201}{30.0 \text{ m}} = 6.70$ diopters

PROBLEMS

9 If a lens has a focal length of 15.0 cm, what is its power in diopters?

10 (a) Determine the position and height of the image that is formed when a 1.50 cm object is placed 15.0 cm in front of a diverging lens that has a focal length of 7.50 cm. (b) Draw the ray diagram.

Sec.18-4 / Simple Optical Instruments

11. **(a)** Calculate the position and size of the image that is formed when a 25.0 mm high object is located 125 mm in front of a thin diverging lens that has a focal length of 75.0 mm. **(b)** Draw the ray diagram.

12. When a 3.00 in. high object is placed 12.0 in. in front of a thin lens, an inverted real image 27.0 in. high is formed. Determine the focal length of the lens.

13. The lens of a movie projector has a focal length of 12.5 cm and is used to project an image from a 35 mm projector onto a screen 10.0 m away. Determine **(a)** the image size; **(b)** the linear magnification; **(c)** the power of the lens.

14. An upright virtual image 125 mm high is formed when an object 75.0 mm high is placed 65.0 mm from a lens. Determine **(a)** the location of the image; **(b)** the focal length of the lens.

15. A converging lens with a focal length of 22.5 in. is used to form an image that is three times as large as the object. Determine how far from the lens the object should be **(a)** if the image is real; **(b)** if the image is virtual. **(c)** Draw the ray diagrams.

16. A 5.00 in. high object is located 18.0 in. from a converging lens and a real image is formed 12.0 in. from the lens. Determine **(a)** the focal length of the lens; **(b)** the image height; **(c)** the power of the lens. **(d)** Draw the ray diagram.

17. If a 2.00 cm high object is placed 8.00 cm from a converging lens with a power of 5.80 diopters, determine **(a)** the position and size of the image; **(b)** the linear magnification. **(c)** Draw the ray diagram.

18. A camera that has a lens with a focal length of 75.0 cm is located in a satellite. If the camera takes photographs from an altitude of 150 km, how large is the image of a 125 m long object on the film?

18-4 SIMPLE OPTICAL INSTRUMENTS

The **human eye** is a very important optical instrument because it enables us to see, but it does have limitations. We frequently rely on other optical instruments to improve the visual abilities of the eye.

Cameras

In inexpensive cameras a single converging lens is used to produce small inverted real images of objects (Fig. 18-18). These images are recorded on a light-sensitive film. Light can only enter the camera through the lens. The duration of the exposure is

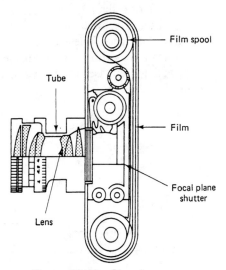

Figure 18-18 Simple camera.

controlled by a **shutter**, which opens and allows light to enter the camera for a predetermined **exposure time**. An adjustable **diaphragm** effectively changes the aperture of the lens, and it controls the amount of light that is allowed to enter the camera. The **aperture** or **f-number** of a camera lens is normally stated as a fraction of the focal length. For example, a setting of $f/16$ implies that the diameter of the aperture is one-sixteenth of the focal length of the camera lens.

Objects that are within a certain range, which is called the **depth of field**, are focused to relatively sharp images on the film; objects outside this range form blurred images. The depth of field may be increased by reducing the aperture of the lens, but this also reduces the illumination of the film, and the exposure time must then be increased. Expensive cameras usually contain a number of lenses to correct for spherical and chromatic aberrations (see Section 18-6).

Projectors

An **illuminating system** of a projector consists of an intense light source, a reflector, and a pair of **condensing lens** that direct the light onto a film. The film is located slightly farther than the focal length from a converging **projection lens**. This lens forms real enlarged and inverted images of the objects on the film (Fig. 18-19). Normally, the film is inverted so that the observer sees an upright image on the screen.

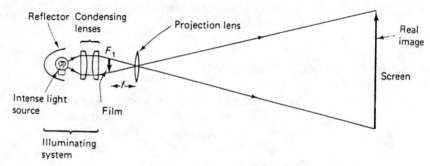

Figure 18-19 Projector.

Magnifiers

A normal human eye sees objects most distinctly when they are about 25 cm away. If an object is farther than 25 cm, the eye is unable to distinguish fine details; if the object is closer than 25 cm, blurred images are formed on the retina.

Virtual erect and enlarged images are formed by a single converging lens when the object distance is less than the focal length. Therefore, a single converging lens can act as a **magnifier** (or **magnifying glass;** Fig. 18-20). It enables us to see clearly objects that are closer to the eye than 25 cm by producing an image 25 cm or more from the eye.

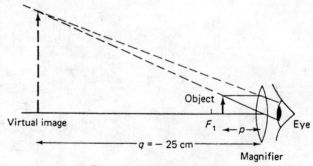

Figure 18-20 Magnifier.

Compound Microscopes

A **compound microscope** is a device that is used to produce very large magnifications. The basic instrument consists of two converging lenses (or corrected converging lens systems) that are arranged on a common principal axis (Fig. 18-21). The lens that is closest to the object is called the **objective**. It has a very short focal length, and it forms a real inverted and enlarged image I_1J_1 of any object PQ that is located slightly beyond the first principal focus F_1. A simple magnifier with a moderately short focal length acts as the **eyepiece** (or **ocular**) of the microscope. The eyepiece is arranged so that it forms a virtual enlarged image I_2J_2 of the real image I_1J_1; this final image is 25 cm (the distance of most distinct vision) from the eyepiece.

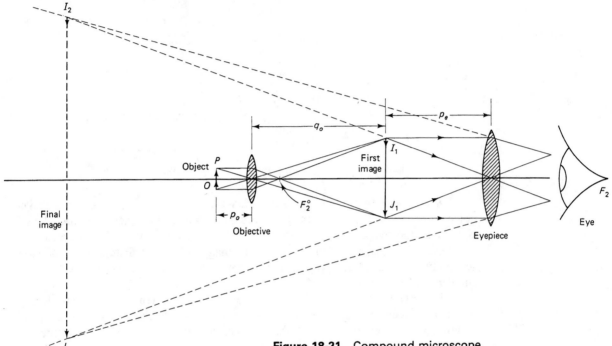

Figure 18-21 Compound microscope.

Microscopes normally have a **turret nose** with three compound objectives with different powers. These objectives are moved into alignment with the eyepiece by rotating the turret effectively changing the magnification.

Telescopes

A basic **refracting astronomical telescope** also consists of two converging lenses that are arranged on a common principal axis with the distance between the lenses approximately equal to the sum of their focal lengths (Fig. 18-22). The **objective** has a large

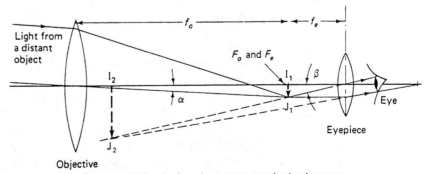

Figure 18-22 Refracting astronomical telescope.

aperture and a long focal length f_o; it produces a real inverted and diminished image I_1J_1 of a distant object. The eyepiece is a simple magnifier with a moderately short focal length f_e; it forms a virtual enlarged image I_2J_2 of the real image I_1J_1.

Telescopes are normally used to observe distant objects, and they form final virtual images that are inverted and smaller than the object.

Binoculars are also a form of astronomical telescope in which light is internally reflected in two 45° prisms. These prisms shorten the length of the telescope and rectify the give an upright final image (Fig. 18-23).

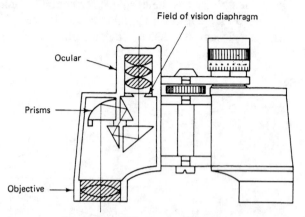

Figure 18-23 Prism binocular.

A **Galilean telescope** (or **opera glass**) also produces an erect image. The objective is a converging lens, but a diverging lens with a short (negative) focal length is used as the eyepiece. The distance between the objective and the eyepiece is approximately equal to the difference in the magnitudes of their focal lengths. Consequently, Galilean telescopes are shorter than astonomical telescopes, but they have smaller fields of view.

Terrestial telescopes consist of three converging lenses. The third lens is inserted between the objective and the eyepiece to rectify the image. However, the addition of the extra lens increases the length of the device, and it reduces the intensity of the final image.

A **surveyor's theodolite** essentially consists of a telescope mounted on graduated plates. These instruments may be used to measure accurately horizontal and vertical angles.

18-5 LENS DEFECTS

Spherical lenses are relatively easy to manufacture, but they are subject to a number of defects.

Spherical Aberration

Incident light rays that are parallel to the principal axis of a spherical lens pass through a common focus only if the aperture is small. If the aperture is not very small, the parallel rays that are incident nearer to the perimeter of the lens are refracted through larger angles, and they are brought to a focus at points that are closer to the lens than the principal focus (Fig. 18-24). This is called **spherical aberration**.

To minimize spherical aberrations, lenses are often designed so that the incident light is deviated by equal amounts at each refracting surface. A diaphragm may be used to limit the effective aperture of the lens, but it also reduces the amount of light that is transmitted.

Sec.18-5 / Lens Defects

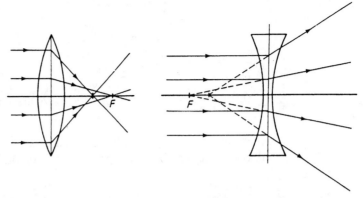

Figure 18-24 Spherical aberration. Incident rays near the perimeter of the lens are brought to focus at points closer to the lens than the principal focus.

Distortion

When a diaphragm is used to control spherical aberration by limiting the aperture, it also produces a variation in the linear magnification at different points on the image. This is called **distortion**; it results in distorted images.

Chromatic Aberration

The index of refraction of a material depends on the wavelength of the incident light. A material has a larger index of refraction for blue light than for red light. Therefore, a lens refracts different colors of incident light through different angles; this is called **chromatic aberration** (Fig. 18-25). The lens forms a separate image for each color of light, and the total image becomes blurred.

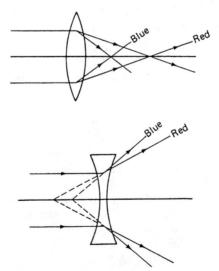

Figure 18-25 Chromatic aberration.

Special lenses that have been corrected for chromatic aberrations are called **achromatic lenses** (Fig. 18-26). These lenses are really two lenses in contact, a converging lens and a diverging lens, but the lenses are made from different materials, such as crown glass and flint glass. The achromatic lens is designed so that the color dispersion in the converging lens is equal and opposite to that in the diverging lens,

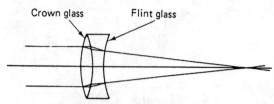

Figure 18-26 Achromatic lens.

but one of these component lenses has a greater power than the other. Consequently, an achromatic lens has a net power, but it produces no color dispersion.

SUMMARY

Refraction is the bending of light rays when they pass from one material into another. It is caused by a change in the speed of the light.

Index of refraction: $n = \dfrac{c}{v}$

Snell's law: $n_1 \sin \alpha_1 = n_2 \sin \alpha_2$

Critical angle for total internal reflection is given by

$$\sin \alpha_c = \frac{n_2}{n_1}$$

A *lens* is a device used to alter the shape of a wavefront. *Converging lenses* are normally thicker at the center. *Diverging lenses* are normally thinner at the center.

Lens equation: $\dfrac{1}{f} = \dfrac{1}{p} + \dfrac{1}{q}$ power $P = \dfrac{1}{f \text{ (in meters)}}$ diopters

Linear magnification: $M = \dfrac{H_i}{H_o} = \dfrac{-q}{p}$

Spherical aberration: light rays parallel to the principal axis do not pass through a common focus.

Distortion: a variation in the linear magnification distorts the image.

Chromatic aberration: the index of refraction depends on the wavelength of the light and the lens forms a separate image for each color.

QUESTIONS

1. Define the following terms: (a) refraction; (b) index of refraction; (c) critical angle; (d) converging lens; (e) diverging lens; (f) aperture; (g) principal focus; (h) optical center; (i) real image; (j) linear magnification; (k) power of a lens in diopters; (l) aperture (f-number) of a camera; (m) depth of field.
2. What are the differences among a refracting astronomical telescope, a Galilean telescope, and a terrestial telescope?
3. Describe the images formed by (a) a compound microscope; (b) prism binoculars; (c) a magnifier.
4. How would you increase the depth of field of a camera?
5. Describe the formation of an image on the retina of a human eye.
6. Describe what is meant by (a) spherical aberration; (b) distortion; (c) chromatic aberration.
7. How would you correct for (a) spherical aberration; (b) chromatic aberration?

REVIEW PROBLEMS

1. Determine the speed of light in water if the index of refraction is 1.33.
2. (a) Determine the frequency of laser light that has a wavelength of 633 nm in a vacuum. (b) What is the wavelength of this light in water?
3. A narrow beam of light is incident at an angle of incidence of 38° on a crown glass plate. Determine (a) the angle of refraction; (b) the speed of light in the glass.
4. Determine the critical angle between (a) air and dense crown glass; (b) benzene and water.
5. Determine the critical angle between water and air at 20°C.
6. A ray of light enters a sample of glass at an incident angle of 30°. If the index of refraction for the glass is 1.50, determine (a) the speed of light in the glass; (b) the angle of refraction.
7. An object 5.00 cm high is placed 30.0 cm in front of a converging lens of focal length 15.0 cm. (a) What is the position and size of the image? (b) Is it real or virtual?
8. A lens has a focal length of 20.0 cm. What is its power in diopters?
9. An object 2.00 cm high is placed 25.0 cm in front of a thin lens of focal length -10.0 cm. (a) What kind of lens is used? (b) Sketch the system. (c) Use the lens equation to determine the position and size of the image. (d) Is the image real or virtual?
10. At what distance from a converging lens of focal length 5.60 cm must an object be placed in order that a real image be formed twice the size of the object?
11. An object 5.00 in. high is placed 15.0 in. in front of a thin lens of focal length -10.0 in. (a) What kind of lens is used? (b) Draw the system. (c) Use the lens equation to determine the position and size of the image. (d) Is the image real or virtual?
12. A magnifier with a focal length of 1.80 cm forms a virtual image 25.0 cm from the eye. Determine (a) the object distance; (b) the linear magnification.
13. Determine the location of the image that is formed when an object is placed 9.25 cm from a lens that has a power of 6.20 diopters.
14. A thin diverging lens with a focal length of magnitude 12.0 cm forms a virtual image that is 28.0 mm high and 4.00 cm from the lens center. Determine the position and size of the object.

19 ELECTROSTATICS

All fundamental properties of electricity and magnetism can be traced to the state or motion of something called **electric charge.** Matter itself is composed of charged particles, and charges play an important role in holding matter together. Differences in the electrical structure of matter are responsible for many of the properties of matter.

Electrostatics is the study of these charges at rest. Stationary charges attract and repel each other because electric forces act over distances; the charges do not have to be in contact for these forces to act. In addition, electric forces are much stronger than gravitational forces.

19-1 ELECTRIC CHARGE

There are many common effects of static electricity. Our hair may be attracted to a brush or comb. On a dry day we may receive a mild shock when we turn on a lamp or the television after walking across a carpet. A balloon clings to a wall after it has been rubbed with a cloth. In each case the objects are said to be **electrified** or **charged**, and they possess a net electric charge.

Many fires and even explosions have been caused by static electricity, so precautions are often necessary to prevent the buildup of charge. Fuel trucks drag chains along the road, lightning rods are used to protect buildings from electrical storms, and so on. However, we shall also see that the motion of these charges and their accumulation are essential for the operation of electronic circuits and devices, which we now take for granted.

Even though the exact nature of **electric charge** is not known, we can describe many of its properties. A glass rod becomes electrified when it is rubbed with silk and a hard rubber (or ebonite) rod becomes electrified when it is rubbed with fur (Fig. 19-1). However, these rods have different kinds of charge. This can be seen by a simple experiment (Fig. 19-2). If an electrified glass rod is suspended from a light

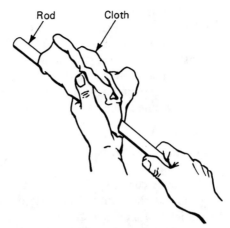

Figure 19-1 Electrification by rubbing suitable materials together. The rod and the cloth obtain opposite net charges.

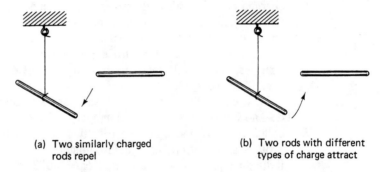

(a) Two similarly charged rods repel

(b) Two rods with different types of charge attract

Figure 19-2 Forces between electric charges.

thread, it is repelled by a second, similarly charged glass rod. But if the electrified rubber rod is brought close to the electrified glass rod, they attract each other. Finally, two of the electrified rubber rods repel each other. From this experiment we see one of the basic characteristics of charges:

Like charges repel each other and unlike charges attract each other.

In the eighteenth century Benjamin Franklin* introduced a sign convention that is still used to describe the two types of charge. He chose the term *positive* for the charge on a glass rod after it had been rubbed with silk, and the term *negative* for the charge on a rubber rod after it had been rubbed with fur. Today we use plastics instead of glass and rubber to demonstrate the two types of charges, because they can be given much larger charges.

Using this convention we determine the **net charge** on an object simply by finding the algebraic sum of the charges present. If the net charge is zero, the object is said to be **uncharged** or **electrically neutral**.

The SI unit for charge is a derived unit called a **coulomb** (symbol C). A charge of 1 C is actually a very large charge; objects charged by rubbing normally have only a very small fraction of this (i.e., only micro-, nano-, or picocoulombs).

---

*By flying a kite in a thunderstorm and collecting the charge, Franklin was also able to show that lightning was an electrical discharge. In fact, he was lucky that he was not electrocuted while performing this famous experiment.

Electric charges are generally associated with particles such as protons and electrons. We use charges to store or transfer energy. For example, work must be done to produce a net charge, such as on the plates of a capacitor, and energy can be transferred from one place to another by means of a flow of electric charge, which is called an **electric current**.

To explain many of the basic phenomena of electricity and magnetism, we must first investigate the structure of matter.

Atomic Structure

All matter consists of one or more fundamental substances called **elements** which cannot be chemically decomposed. An **atom** is the smallest complete subdivision (particle) of an element.

Atoms were originally thought to be indivisible, but in 1911, Ernest Rutherford discovered that each atom has a very small, dense core, called the **nucleus**, which also has a positive electric charge. Then in 1913, Niels Bohr proposed that an atom resembled a small-scale model of the solar system, in which even smaller, negatively charged particles called electrons orbited the nucleus at high speeds in the same way that the planets revolve around the sun. The modern model of an atom, developed about 1925, assumes that a small, very dense, positively charged nucleus is surrounded by a "cloud" of electrons in somewhat irregular orbits that are related to their energies. Individual **electrons** have a mass of 9.108×10^{-31} kg and a charge of -1.602×10^{-19} C.

The investigation of nuclear structure is an active area of research. However, we do know that the nucleus is composed of one or more positively charged particles called **protons** and other electrically neutral particles called **neutrons**. Individual protons and neutrons have approximately the same mass of 1.67×10^{-27} kg, and individual protons have a charge of $+1.602 \times 10^{-19}$ C. Therefore, electrons and protons have charges of equal magnitude but opposite signs, and in an electrically neutral atom the number of nuclear protons is equal to the number of orbiting electrons.

Electrons of the same atom may perform their orbits at different distances from the nucleus. Most of the atom is empty space (see Chapter 1). The electrons with the farthest orbits from the nucleus are not held as tightly to the atom as the electrons with the closer orbits.

When two materials, such as rubber and fur, are rubbed together, electrons in one material (the rubber) may be more strongly attracted to the atoms in the other material (the fur). As a result, some negatively charged electrons may move from the first material (rubber) leaving it with a deficiency of electrons and thus a positive charge. At the same time, the second material (fur) gains the electrons, giving it an excess of electrons and therefore a net negative charge.

Electric charges are not created or destroyed when we charge a material; instead, electrons are transferred from one material to the other. As the negative charge builds up on one material, a positive charge forms on the other, but the total net charge of the whole system remains the same. This fact is expressed in terms of the **law of conservation of charge**, which states:

The algebraic sum of all electric charges in any isolated system is a constant.

If an object has a net **negative charge**, it has an excess of electrons. It has a net **positive charge**, if it has a deficiency of electrons.

Sec.19-1 / Electric Charge

EXAMPLE 19-1

One object has a net charge of $+5.0$ C and a second has a net charge of -3.0 C. If the objects are isolated from external influences and the charge on one changes to $+8.0$ C, what is the charge on the other?

Solution

Data: Original charges: $Q_1 = 5.0$ C, $Q_2 = -3.0$ C; final charges: $Q_3 = +8.0$ C, $Q_4 = ?$

Equation: Total charge is conserved; therefore,

$$Q_1 + Q_2 = Q_3 + Q_4$$

Rearrange: $Q_4 = Q_1 + Q_2 - Q_3$

Substitute: $Q_4 = 5.0$ C $+ (-3.0$ C$) - (8.0$ C$) = -6.0$ C

Note that the net charge remains constant at 2.0 C.

Electrons and protons possess the smallest known quantity of charge, an *elementary charge* $e = 1.602 \times 10^{-19}$ C. Since an object becomes charged by either gaining or losing charged particles, normally electrons:

A net electric charge always exists in amounts that are integer multiples of an elementary charge e.

For example, net charges of $501e$ and $-201e$ might be carried by certain objects, but not net charges of $2.5e$ or $-1.9e$ because 2.5 and 1.9 are not integers.

EXAMPLE 19-2

Determine the number of elementary charges in a 320 μC net charge.

Solution

Data: Net charge $Q = 320\,\mu$C $= 3.2 \times 10^{-4}$ C, $e = 1.6 \times 10^{-19}$ C, $N = ?$

Equation: Number of elementary charges $= \dfrac{\text{total charges}}{\text{elementary charge}} = \dfrac{Q}{e}$

Substitute: $N = \dfrac{3.2 \times 10^{-4}\text{ C}}{1.6 \times 10^{-19}\text{ C}} = 2.0 \times 10^{15}$

The ease with which charges are able to move through a material depends on the structure of that material. Solids are composed of large numbers of atoms (approximately 10^{21} atoms/cm^3).

Some substances, called **conductors**, contain charges that are relatively free to move. These charges may be electrons which are not tightly bound to an individual atom, or charged atoms or charged groups of atoms called **ions** which are free to flow. In a copper wire, for example, the outermost electrons in the atoms are not tightly bound, and only a small amount of energy is required to move them from atom to atom through the wire. Ions are found in some solutions called **electrolytes**, which are important in devices such as batteries.

Other materials, such as rubber and wood, are poor conductors of electricity because they do not contain charges that are relatively free to move, even when relatively large amounts of energy are applied. These materials are called **insulators** or **dielectrics**.

Semiconductors have electrical properties between conductors and insulators. These materials normally have some "free" charges (that may be positive or negative) but not as many as in conductors.

PROBLEMS

1. Determine the net charge in a system of the following charges: 2.600 μC, -925 nC, and $-0.003\ 700$ mC.
2. How many electrons are required to produce a net charge of $-5.0\ \mu$C?
3. What is the net charge on the nucleus of a nitrogen atom if it contains seven protons?
4. One object has a net charge of 18.0 μC and a second has a net charge of $-13.0\ \mu$C. If the objects are isolated and the charge on one changes to $-6.0\ \mu$C, what is the charge on the other?
5. How many elementary charges are required to produce a net charge of 52.5 nC?
6. Determine the net charge transferred when 6.20×10^{16} electrons move from one material onto another.

19-2 CHARGING PROCESSES AND GROUND

A **leaf electroscope** is a simple device that is used to compare and study electric charges. It consists of a metal conducting rod with a plate or a knob at the top and a pair of metal foil leaves (such as aluminum or gold) that hang freely at the other end. The rod is suspended vertically through an insulating ring at the top of a container that has metal sides with a glass window (Fig. 19-3). If the electroscope is uncharged and there are no net charges nearby, the metal foil leaves hang limply. However, if a charged object is brought close to the plate, the leaves diverge to indicate the presence of that charge.

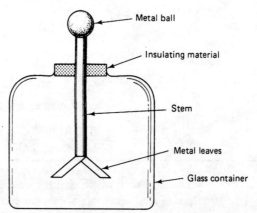

Figure 19-3 Leaf electroscope.

When a positively charged rod is brought close to an uncharged electroscope, electrons are attracted to the unlike positive charge. Therefore, electrons flow from the leaves and conducting rod onto the metal plate. A negative charge is then said to have been **induced** on the metal plate, even though there was no physical contact between the rod and the plate. However, as the plate becomes negatively charged, the metal leaves lose electrons and they are left with a net positive charge. Since like charges repel, the leaves diverge.

If the positively charged rod is now removed, the electrons on the metal plate repel each other. The "extra" electrons then redistribute themselves over the plate, conducting rod and metal foil leaves, neutralizing the net charge on the leaves, and the leaves collapse. During this whole process, the net charge on the electroscope remains constant at zero.

Similarly, when a negatively charged rod is brought close to an uncharged electroscope, electrons are repelled from the metal plate and they flow to the leaves, giving them a net negative charge. The leaves repel each other, and they diverge, but the metal plate now contains an induced net positive charge. When the negatively charged rod is removed, the electrons are redistributed and the leaves collapse.

An electroscope may be given a net charge by two different techniques:

1. Charge by conduction. Objects may be charged by touching them with some other charged object. This process, called charging by conduction, results in the same polarity (or sign) of net charge on each object. For example, to give a conductor a negative charge by a conduction process, we can simply touch it with a negatively charged object.

As the charged object approaches the conductor, it repels some electrons to the farthest surface, leaving the closer surface with a net positive charge, which actually attracts the opposite polarity charges on the rod. When the rod touches the conductor, some electrons flow from the rod, and they are distributed over the surface of the conductor. If the rod is now removed, the conductor is left with an excess of electrons and therefore a net negative charge. During the conduction process the net charge is shared between the rod and the conductor (not necessarily evenly) and the net charge on the rod decreases (Fig.19-4).

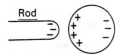
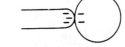
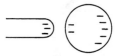

(a) As rod approaches, the electrons in the object are repelled away leaving the closer surface with a net positive charge

(b) When the rod touches, electrons are transferred to the object

(c) Object has a net negative charge when the rod is removed

Figure 19-4 Charging by conduction (contact).

The magnitude of the divergence of the metal foil leaves of an electroscope is related to the net charge on any object that approaches or makes contact with the metal plate. The larger the net charge, the greater the divergence of the leaves.

2. Charge by induction. Compared with objects on its surface, the earth is very large and it contains an extremely large number of atoms. When electrons flow between an object and the earth, the change in the charge of the object may be quite noticeable, but because of its size, the change in the earth's net charge is insignificant. Consequently, the earth may be considered as a source or drain of electrons.

An **electrical ground** or **earth** uses the earth as a reservoir of charge and it is represented schematically by the symbols

Earth ground   Chassis ground

We may also use our bodies to ground a charge (as long as it is not very large) merely by touching it.

When a charged object is connected to the earth by a conductor (**grounded**), electrons flow between the earth and that object in a direction which tends to neutralize the charge. If a negatively charged object is grounded, electrons flow from that object *to* the earth to neutralize the negative charge. Similarly, when a positively charged object is grounded, electrons flow *from* the earth to that object to neutralize the positive charge.

Using an electrical ground, we may also charge conductors by a process called **charging by induction**. If a negatively charged rod is brought close to an uncharged conductor, electrons are repelled to the farthest side of that conductor, leaving the front surface with an induced net positive charge (Fig. 19-5). When the rear surface of the conductor is grounded, some of the excess electrons flow from the conductor to the earth. If the ground is removed (while the rod is still nearby), the conductor is left with a net positive charge. The negatively charged rod can then be removed so that the remaining electrons can redistribute over the entire surface of the conductor. The conductor is still left with a net positive charge. An object may be given a net negative charge by a similar process.

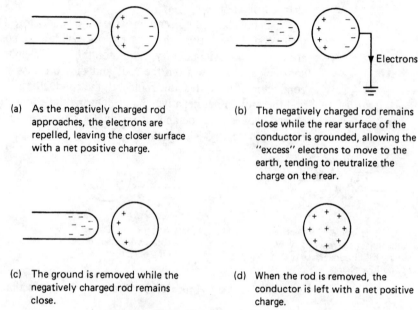

(a) As the negatively charged rod approaches, the electrons are repelled, leaving the closer surface with a net positive charge.

(b) The negatively charged rod remains close while the rear surface of the conductor is grounded, allowing the "excess" electrons to move to the earth, tending to neutralize the charge on the rear.

(c) The ground is removed while the negatively charged rod remains close.

(d) When the rod is removed, the conductor is left with a net positive charge.

Figure 19-5 Charging by induction.

Note that when a conductor is charged by induction, the net charge on the initial rod remains unchanged (except for leakage) during the charging process, because it is not shared with the conductor.

Determination of Charge Polarity

The **polarity** (sign) of the net charge on an object may be determined with an electroscope. For example, suppose that we give the electroscope a small negative charge and then bring a negatively charged rod close to the plate (Fig. 19-6). More electrons are repelled from the plate onto the leaves, increasing their net negative charge and causing then to diverge further. However, if a positively charged rod is brought close, electrons are attracted to the plate, reducing the charge on the leaves and allowing them to collapse.

In general, if the charge is the same on the rod and the electroscope, the divergence will increase. Care should be taken in using this technique, however, because a strongly charged object can overpower the relatively small charge on the electroscope. This will cause the leaves to collapse and then reseparate.

Charge Separation by Induction

Some devices use induction to generate relatively large net charges. The process can be illustrated by considering two conductors that are electrically neutral and connected by a conducting wire (Fig. 19-7).

Sec. 19-2 / Charging Processes and Ground

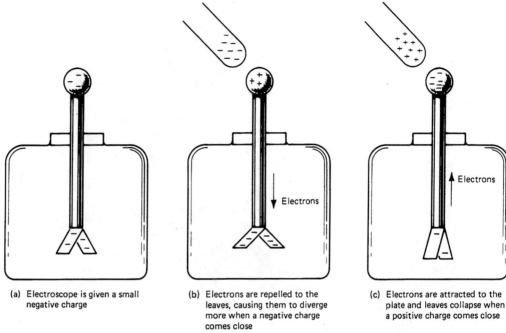

(a) Electroscope is given a small negative charge

(b) Electrons are repelled to the leaves, causing them to diverge more when a negative charge comes close

(c) Electrons are attracted to the plate and leaves collapse when a positive charge comes close

Figure 19-6 Determination of charge polarity.

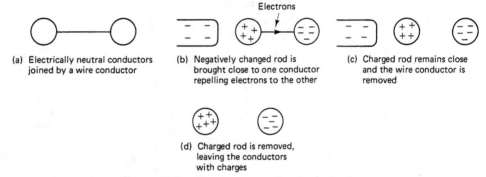

(a) Electrically neutral conductors joined by a wire conductor

(b) Negatively changed rod is brought close to one conductor repelling electrons to the other

(c) Charged rod remains close and the wire conductor is removed

(d) Charged rod is removed, leaving the conductors with charges

Figure 19-7 Charge separation by induction.

When a negatively charged rod is brought close to one of the conductors, it repels electrons to the other conductor. If first the conducting wire and then the negatively charged rod are now removed, the two conductors retain net charges, one positive and the other negative.

This principle is used in machines such as the **Wimshurst static machine** (Fig 19-8a). In operation, the two disks are rotated in opposite directions. If one foil obtains a net charge by friction, it induces opposite polarity charges on the foils in the other disk as they rotate past each other. These charges are removed with combs and stored in two capacitors.

Van de Graaf generators (Fig. 19-8b) are also used to produce very large net charges. In these devices a high-voltage source deposits electrons on a belt made of an insulating material. The belt carries the electrons to the top of a column, where they are removed and placed on a spherical conductor. Extremely high charges may be produced on the spherical conductor.

Many important and often spectacular phenomena are caused by static electricity and charge separation. For example, lightning is an electrical discharge between oppositely charged clouds, or between charged clouds and the earth. Static electricity can also be very hazardous, because it causes sparks that can ignite materials. We must be especially careful around flammable materials such as gasoline.

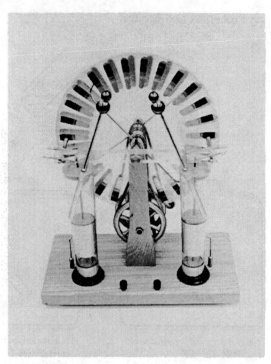

(a) Wimshurst static machine.

(b) Van de Graaf generator.
Figure 19-8

19-3 COULOMB'S LAW

Like charges repel and unlike charges attract each other. Also, electric forces always act along a line joining the centers of two charges. The magnitude of the electric forces depends on the nature of the medium, the magnitude of the charges, and the distance between their centers.

Sec. 19-3 / Coulomb's Law

Charles Augustin de Coulomb (1736–1806) determined that (Fig. 19-9):

The magnitude of the electric force F between two charges Q_1 and Q_2 is proportional to the product of the charges and inversely proportional to the square of the distance d between their centers.

$$F \propto \frac{Q_1 Q_2}{d^2} \quad \text{or} \quad F = \frac{kQ_1 Q_2}{d^2} \tag{19-1}$$

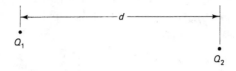

Figure 19-9 Coulomb's law: The force between the charges is $F = \frac{kQ_1 Q_2}{d^2}$.

This statement is known at **Coulomb's law**. The proportionality constant k is called the **Coulomb constant**; its value depends on the material medium around the charges. In a vacuum (and approximately in air), $k = 9.00 \times 10^9$ N·m²/C².

If two charges have opposite polarity, they attract each other, and in Coulomb's law, this gives rise to a negative value for the force. Similarly, like charges repel, and this corresponds to a positive value for the force in Coulomb's law. According to Newton's third law, in an isolated system the force experienced by one charge must be equal and opposite to the force experienced by the other.

EXAMPLE 19-3

What is the electric force between a $+3.80$ μC charge and a -2.40 μC charge if they are 4.60 cm apart in a vacuum?

Solution

Data: $Q_1 = 3.80 \times 10^{-6}$ C, $Q_2 = -2.40 \times 10^{-6}$ C, $d = 4.60 \times 10^{-2}$ m, $k = 9.00 \times 10^9$ N·m²/C², $F = ?$

Equation: $F = \dfrac{kQ_1 Q_2}{d^2}$

Substitute:

$$F = \frac{(9.00 \times 10^9 \text{ N·m}^2/\text{C}^2)(3.80 \times 10^{-6} \text{ C})(-2.40 \times 10^{-6} \text{ C})}{(4.60 \times 10^{-2} \text{ m})^2}$$

$$= -38.8 \text{ N} \quad \text{(the minus sign means attractive)}$$

EXAMPLE 19-4

If the force between an electron and a proton in a vacuum is -1.40×10^{-29} N, how far apart are they?

Solution

Data: $Q_1 = e = 1.60 \times 10^{-19}$ C and $Q_2 = -1.60 \times 10^{-19}$ C (Appendix E); $k = 9.00 \times 10^9$ N·m²/C², $F = -1.40 \times 10^{-29}$ N, $d = ?$

Equation: $F = \dfrac{kQ_1 Q_2}{d^2}$

Rearrange: $d^2 = \dfrac{kQ_1 Q_2}{F}, \quad d = \sqrt{\dfrac{kQ_1 Q_2}{F}}$

Substitute:

$$d = \sqrt{\frac{(9.00 \times 10^9 \text{ N·m}^2/\text{C}^2)(1.60 \times 10^{-19} \text{ C})(-1.60 \times 10^{-19} \text{ C})}{-1.40 \times 10^{-29} \text{ N}}}$$

$$= 4.06 \text{ m}$$

Coulomb's law is valid only for pairs of charges, but it can be used for systems containing more than two charges by determining the forces between the pairs and adding the force vectors.

Distribution of Charge

Charges cannot flow very easily in insulators; therefore, insulated areas of net charge tend to remain isolated. However, any net electric charge is always located on the outer surface of a conductor. This can be seen in terms of the electric forces between like charges (Fig. 19-10). The vector sum of the repulsive forces is always toward the outer surface of the conductor.

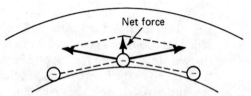

Figure 19-10 Like charges on an inner surface of a conductor repel each other producing net forces toward the outer surface.

By investigating conductors which have different shapes, we find that excess charges tend to concentrate in regions where the surface curvature is the greatest, especially at points (Fig. 19-11). Again, this can be seen in terms of the electric forces between like charges. The components of the repulsive forces tangent to the surface are smallest, and charges can move closest together, where the curvature is the sharpest. The components of the forces away from the surface are strongest where the curvature is the sharpest.

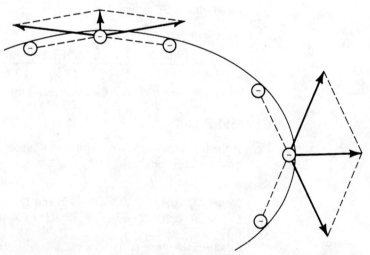

Figure 19-11 Repulsive forces between like charges. The components *along the surfaces* are largest where the curvature is least.

This net force between like charges may result in a leakage of charge from the sharply curved portion of the surface of a conductor to the surroundings. This phenomenon is used in some electrical devices to transfer a charge from one conductor to another.

Sec.19-4 / Potential Difference

PROBLEMS

7. Determine the electric force between two very small particles in a vacuum if they have charges of 8.00 C and −5.00 C and they are 2.50 m apart.

8. The force between two very small particles with equal charges is 3.60×10^{-18} N in a vacuum when they are 25.0 mm apart. What is the magnitude of the charge on each particle?

9. Determine the electric force between two very small particles in a vacuum if they have charges of 27.0 nC and 16.0 nC if they are 2.50 mm apart.

10. An electron and a proton in a hydrogen atom are 5.3×10^{-11} m apart. Determine the electric force between them.

11. If the force between two charges is 6.40×10^{-8} N, what is the force between them if the distance between their centers is halved?

12. Determine the electric force between two small particles with charges of 12.0 nC charge and 9.0 nC if they are 25.0 cm apart in a vacuum.

13. What magnitude charge exerts a 2.80×10^{-18} N force on an electron at a distance of 35.0 cm in a vacuum?

14. At what distance in a vacuum is the force between two electrons equal to 3.65×10^{-27} N?

15. Determine the distance between two electrons in a vacuum if there is a 1.15×10^{-25} N force between them.

16. If the force between two small charged particles is 9.60×10^{-12} N, what is the force between them if their charges are each tripled and the distance between their centers is reduced to half the original distance?

19-4 POTENTIAL DIFFERENCE

Two or more charges interact even if they do not make contact, because electric forces act over distances. Objects that possess net charges are affected by the presence of any other charge, and experience electrostatic (Coulomb) forces.

An **electric field** is said to exist in any region where a charge experiences an electric force. Since electric forces occur only between charges, electric fields are produced by charges.

Potential Difference

To accumulate a net charge, work must be done against electrostatic forces which tend to drive like charges apart. For example, work must be done to move a positive charge q toward some other positive charge Q. However, the charge q then has a greater potential energy because the electrostatic forces can do work on it as the charges repel each other. This is very similar to lifting an object against the force of gravity, thus increasing the potential energy of that object.

As in the case of gravitational potential energy, we normally consider changes in electric potential energy. We define the **electric potential difference** V between any two points A and B as the work W that must be done per unit positive charge Q when that charge is moved from A to B (Fig. 19-12):

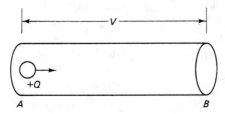

Figure 19-12 Work W must be done in order to move a charge Q between two points A and B. The electric potential difference between A and B is $V = \dfrac{W}{+Q}$.

$$\text{potential difference} = \frac{\text{work done}}{\text{charge moved}}$$

or

$$V = \frac{W}{Q} \tag{19-2}$$

Even though the Coulomb force between the charges and therefore the work done depends on the magnitude and sign of the charge Q, the potential difference V is independent of the charge Q.

The SI unit for potential difference is called the **volt** (abbreviated V). From Eq. 19-2 it can be seen that 1 V is 1 J/C.

$$1 \text{ V} = 1 \frac{\text{J}}{\text{C}}$$

One joule of work is required to move a charge of 1 C between two points that have a potential difference of 1 V.

Electric potential difference is a scalar quantity; it is also called the **voltage** or **potential drop**.

EXAMPLE 19-5

Determine **(a)** the work done by the electric field and **(b)** the final speed of an electron which is accelerated from rest through a potential difference of 250 V.

Solution

Data: $V = 250$ V; $m = 9.1 \times 10^{-31}$ kg and $Q = -1.6 \times 10^{-19}$ C (Appendix E); $W = ?$, $v = ?$

(a) *Equation:* $V = \dfrac{W}{Q}$

Rearrange: $W = QV$

Substitute: $W = (-1.6 \times 10^{-19} \text{ C})(250 \text{ V}) = -4.0 \times 10^{-17}$ J

We can usually ignore the sign; it only tells us that work is done by the electric field rather than against it.

(b) *Equation:* The work done must equal the gain in kinetic energy, and since the electron was initially at rest, its initial kinetic energy was zero, so

$$W = \tfrac{1}{2}mv^2$$

Rearrange: $v = \sqrt{\dfrac{2W}{m}}$

Substitute: $v = \sqrt{\dfrac{2(4.0 \times 10^{-17} \text{ C})}{9.1 \times 10^{-31} \text{ kg}}} = 9.4 \times 10^6$ m/s

---

Even though we have defined potential difference and voltage in terms of electrostatics, we shall see that it plays a very important role in the description and operation of electric circuits and devices. In practice, ground (the earth) is usually taken as the zero voltage reference level, and all other voltages are measured with respect to ground.

From the definition of potential difference we see that if the potential difference between two points is zero, no work must be done against the electric forces to move a charge between those points (although work must usually be done against other forces). A surface joining all points that are at the same electric potential (have no potential difference) is called an **equipotential surface.** The surface of any conductor is an equipotential surface, because if a potential difference existed, the charges would simply flow over the surface.

In the operation of many devices and circuits the electric field does work in order to displace charges from one point to another. If an electric field does work to move a charge Q between two equipotential surfaces a distance s apart that are at a potential difference V, the force exerted on the charge Q by the field

$$F = \frac{QV}{s} \tag{19-3}$$

The term V/s is known as the **potential gradient** of the field.

EXAMPLE 19-6

Determine the magnitude of the average force on an electron if it is accelerated from rest through a potential difference of 240 V between two electrodes that are 2.0 cm apart.

Solution

Data: $V = 240$ V, $s = 2.0 \times 10^{-2}$ m, $Q = 1.6 \times 10^{-19}$ C (we can ignore the sign in this case), $F = ?$

Equation: $F = \dfrac{QV}{s}$

Substitute: $F = \dfrac{(1.6 \times 10^{-19} \text{ C})(240 \text{ V})}{2.0 \times 10^{-2} \text{ m}} = 1.9 \times 10^{-15}$ N

When a large potential gradient exists in a region, it may cause the surrounding material to break down and a spark may jump through the material. For dry air this breakdown occurs when the potential gradient is about 30 kV/cm. However, some electrical discharge through air occurs at even lower potential gradients (about 10 kV/cm) and causes a zone of ionization of the air known as a **corona discharge.** This discharge is present around power lines and other high-voltage areas; it often appears as a greenish glow.

Corona discharges have many very important applications. In an **electrostatic precipitator** corona discharges are used to charge waste materials (in dirty exhaust gases), which can then be collected electrostatically so that only clean gases are discharged. Other important applications include paint sprayers and xerographic copying machines, using charges to attract a black dust which is used as the ink.

PROBLEMS

17 Calculate the work that must be done in order to move an electron between two points that differ in electric potential by 320 V.

18 An electron is accelerated through a potential difference of 45.0 kV in an X-ray tube. Determine **(a)** its gain in kinetic energy; **(b)** its final speed if it started from rest.

19 An electron in a cathode ray tube accelerates from rest to a final speed of 5.80×10^6 m/s in a distance of 2.50 cm. Find **(a)** the work done by the field; **(b)** the potential difference that was applied.

20 In what distance will an electron increase its kinetic energy by 4.80×10^{-17} J as it moves through a uniform potential gradient of 2500 V/m?

21 The anode and cathode of a vacuum tube are 2.80 cm apart and they have a potential difference of 180 V. Determine **(a)** the potential gradient; **(b)** the average electric force that would be exerted on an electron in the tube.

22 An electron is accelerated between two electrodes that are 1.50 cm apart and have a potential difference of 125 V. Determine (a) the average electric force on the electron; (b) the acceleration of the electron. (c) If the electron is initially at rest, how long would it take to move through 1.20 cm, and what maximum speed would it attain in the 1.20 cm?

SUMMARY

Electrostatics is the study of charges at rest. Like charges repel and unlike charges attract.

A *coulomb* (C) is the SI unit for charge.

An *atom* consists of a very dense core, called the *nucleus,* which contains positively charged particles called *protons* and electrically neutral particles called *neutrons*. The nucleus is surrounded by a cloud of negatively charged particles called *electrons*.

Electric charge is conserved. An object has a *negative* charge when it has an excess of electrons and a *positive* charge when it has a deficiency of electrons.

The *earth (ground)* is considered to be a reservoir of charge and it is represented by the symbol

Earth ground   Chassis ground

Coulomb's law: The force between two charges Q_1 and Q_2 varies directly as the product of their magnitudes and inversely as the square of the distance between their centers.

$$F = \frac{kQ_1Q_2}{d^2}$$

The *potential difference* between two points is the work W that must be done per unit positive charge Q in moving that charge between the two points.

$$V = \frac{W}{Q} \qquad \text{SI unit is the } volt \text{ (V)}$$

The force F on a charge Q in a *potential gradient* V/s is

$$F = \frac{QV}{s}$$

QUESTIONS

1 Define the following terms: (a) electrostatics; (b) conductor; (c) insulator; (d) semiconductor; (e) ion; (f) electric ground; (g) electric potential difference; (h) equipotential surface; (i) corona discharge; (j) volt; (k) potential gradient.

2 In a dry climate hair becomes unmanageable when it is combed or brushed with a nylon comb or brush. Explain why.

3 Mild electric shocks are common when a person walks across a nylon carpet and touches metal, such as a TV knob. Explain why.

4 Describe what happens when a positive charge is formed by conduction on a conductor.
5 Explain how you would give a conductor a positive charge by induction.
6 (a) How does a piece of plastic wrapping paper become charged when it is taken from a roll? (b) Why does it cling to glass but not to a metal pot?
7 Describe how you would determine the type of charge on a conductor.
8 Why must the nozzle of a hose be grounded if it is used to pump fuel into a tank?
9 Why are conductors with sharp points frequently used to transfer charges from one conductor to another?
10 Explain how Newton's third law (action and reaction) applies to the forces between charges.
11 If two objects attract each other can you conclude that they are both charged? Explain.

REVIEW PROBLEMS

1 What is the net charge on a conductor that has an excess of 3.8×10^{17} electrons?
2 Determine the net charge in a system of three charges that have magnitudes of 1.6 μC, -560 nC, and -0.045 mC.
3 How many excess electrons are required to produce a net negative charge of magnitude 2.38 mC?
4 If the electric force between two charges is 2.50×10^{-15} N when they are 32.0 cm apart, what is the force when the same charges are 75.0 cm apart?
5 Determine the electric force between a 650 nC charge and a 825 nC charge if they are 12.5 cm apart in a vacuum.
6 If the force between a 720 nC charge and a 460 nC charge in a vacuum is 4.80 mN, how far apart are they?
7 (a) Determine the average force on an electron if it is accelerated from rest through 5.20 cm, where the potential gradient is 3200 V/m. (b) What is the final speed of the electron? (c) How long does it take the electron to travel the 5.20 cm?
8 How much work is done by the electric field when a 2.50 nC charge is moved through a 38.0 V potential difference?
9 An electron initially at rest is accelerated through a potential difference of 12.0 V. (a) What is its final kinetic energy? (b) Determine its final speed.
10 An electron accelerates between two parallel plates 1.25 cm apart. If the potential difference between the plates is 850 V, determine (a) the potential gradient; (b) the average force on the electron.
11 The anode and cathode of a vacuum tube are 2.40 cm apart and they are maintained at a potential difference of 135 V. (a) What is the potential gradient? (b) What magnitude force would be exerted on an electron by the field between the electrodes? (c) How much work is done by the field in moving an electron from the cathode to the anode?

20
DIRECT-CURRENT CIRCUITS

Most applications of electricity involve **electric currents** which are flows of electric charges around closed paths called **circuits**. A circuit usually contains special devices (each having some basic function) which are interconnected by metallic wire conductors. The development of electrical and electronic devices has had a dramatic impact on our lives. We rely on electricity for many "necessities" and comforts such as heating, lighting, cooking, and even entertainment. Technicians and scientists frequently use electricity and electrical devices in their work.

Electronics is the study of the emission, properties, and effects of electrons in circuits, and in devices that operate directly or indirectly because of electron flow.

20-1 CHARGE AND ELECTRIC CURRENT

An **electric current** is a flow of electric charge. The magnitude of an electric current depends on the number of available charges and their ability to move through the material. Materials differ in their suitability as carriers of electric current.

Free positive charges move *from* the higher (more positive) electric potential *toward* the lower (more negative) potential. Negative charges (normally electrons) tend to move in the opposite direction. However, a negative charge moving in one direction is equivalent to the motion of a positive charge in the opposite direction. Both positive and negative charges can contribute to the total electric current.

A current may be a flow of positively charged particles, negatively charged particles, or both positive and negative charges simultaneously. By international agreement, the direction of **conventional electric current** is taken as the direction of positive charge flow. If electrons move one way in a material, the conventional current is in the opposite direction (Fig. 20-1).

The SI unit for electric current is the **ampere** (symbol A). It is defined in terms of the magnetic forces between two parallel electric currents.

Sec. 20-1 / Charge and Electric Current

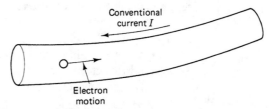

Figure 20-1 Negatively charged particles, such as electrons, move in the opposite direction to the conventional current I.

The unit of charge, the coulomb (C), is derived from the ampere. A **coulomb** is the net quantity of electric charge that flows past a fixed point in a conductor in each second if there is a steady (constant) electric current of 1 A. Therefore, an electric current is a time rate of flow of charge:

$$I = \frac{Q}{t} \tag{20-1}$$

where Q is the net charge in coulombs that flows past the fixed point in the elapsed time t in seconds and the current I is in amperes. Rearranging this expression, we find the net charge that flows:

$$Q = It \tag{20-2}$$

or $1 \text{ C} = (1 \text{ A})(1 \text{ s}) = 1 \text{ A} \cdot \text{s}$.

EXAMPLE 20-1

What is the average current if 18.0 C of charge flows past a fixed point in a copper wire in 12.0 s?

Solution

Data: $Q = 18.0$ C, $t = 12.0$ s, $I = ?$

Equation: $I = \dfrac{Q}{t}$

Substitute: $I = \dfrac{18.0 \text{ C}}{12.0 \text{ s}} = 1.50$ A

EXAMPLE 20-2

How many electrons flow past a given point in a copper wire in 1.00 min if the wire carries a steady current of 135 mA?

Solution

Data: $I = 135$ mA $= 0.135$ A, $t = 1.00$ min $= 60.0$ s, the charge on an electron has a magnitude $e = 1.60 \times 10^{-19}$ C, $N = ?$

Equations: Net charge that flows $Q = It$, $N = \dfrac{Q}{e}$

Substitute: $Q = (0.135 \text{ A})(60.0 \text{ s}) = 8.10$ C

Therefore,

$$N = \frac{Q}{e} = \frac{8.10 \text{ C}}{1.60 \times 10^{-19} \text{ C}} = 5.06 \times 10^{19}$$

There are several kinds of electric current:

1. When particular types of charge, such as negatively charged electrons, always move in the same direction through a conductor, the flow of charge is called **direct current** (dc). If a direct current exists, electric charge is actually displaced from one point to another in the conductor.

2. A *steady current* exists when the time rate of flow of electric charge is a constant.

3. If a current only exists for a short period of time, it is known as a *transient current*. For example, if two conductors of different charges are joined by a conductor (Fig. 20-2), the attraction between unlike charges and the repulsion between like charges produces an electric current from the higher or more positive electric potential toward the lower potential. This current tends to equalize the electric potentials of the two regions and then it ceases.

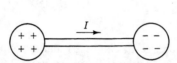

Figure 20-2 Two charged conductors joined by a conductor. The conventional current I is *from* the higher (more positive) *towards* the lower (more negative) potential.

4. When a conductor carries an *alternating current* (ac), all of the moving charges continually change their directions of motion, and they vibrate about their average positions in the conductor.

PROBLEMS

1. Determine the average current in a conductor if a net charge of 2.5 C flows past a fixed point in 5.0 s.

2. Find the average current in a copper wire if 5.6×10^{16} electrons flow past a fixed point in 2.8 ms.

3. If the current in a conductor is steady at 350 mA, **(a)** how much charge flows past a point in that conductor in 6.0 s; **(b)** how many electrons flow past that point in 6.0 s?

4. How long does it take for 1.3×10^{15} electrons to flow past a point in a copper wire when that wire carries 650 μA?

5. Determine the average electric current in a conductor if a total charge of 5.0 C flows past a fixed point in 2.5 s.

6. If the electric current in a copper wire is 3.2 A, how many electrons flow past a fixed point in 2.8 ms?

7. If the current in a wire is 1.75 A, how many electrons flow past a fixed point in 18.0 ns?

8. Determine the average current in a wire if 650 μC of charge flow past a given point every 180 ms.

9. What is the total electric charge that flows past a point in a wire in 5.00 min if the wire carries 350 mA?

10. What is the average current in a copper wire if 3.50×10^{18} electrons flow past a point in 850 ms?

11. Determine **(a)** the net charge that flows past a given point and **(b)** the average current in a wire if 1.80×10^{20} electrons flow past that point in 8.50 min.

20-2 CURRENT, VOLTAGE, AND RESISTANCE

Consider a simple circuit consisting of a battery, a light bulb, and a switch connected by some metal wires (Fig. 20-3). Obviously, this type of diagram is very awkward to draw and it is therefore usually simplified by using standard symbols to represent

Sec. 20-2 / Current, Voltage, and Resistance

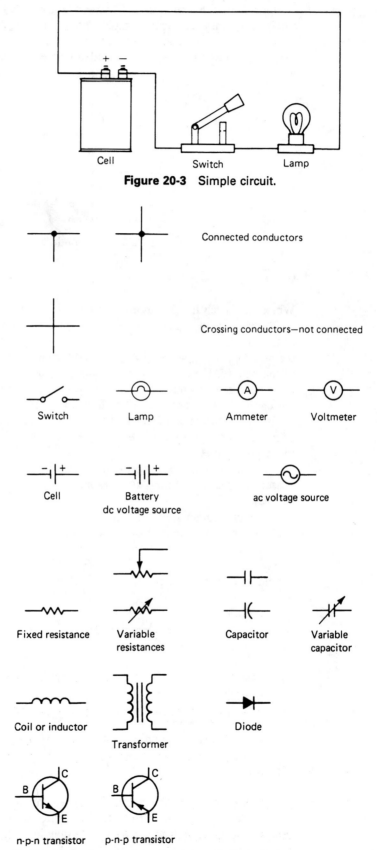

Figure 20-3 Simple circuit.

Figure 20-4 Some of the basic symbols used for electronic components.

circuit elements. Some of these basic symbols are illustrated in Fig. 20-4. Using these symbols, we may redraw the simple circuit as in Fig. 20-5. The interconnecting wires between the various components are represented by horizontal and vertical lines. If these wires are joined, a dot is included where they are connected. No dot is used if they merely cross without being connected.

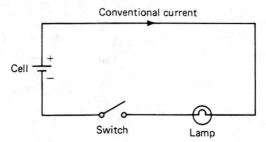

Figure 20-5 Schematic of the simple circuit.

Current

When the switch in this circuit is closed, the lamp glows indicating the presence of an electric current as the charges (electrons in this case) move around the closed path of the circuit. These electrons flow from the negative terminal of the battery, through the interconnecting wires, the lamp, and the switch and back to the positive terminal of the battery. However, by convention, we take the direction of the **conventional current** as that in which positive charges would move. Therefore, the conventional current is clockwise in Fig. 20-5, from the positive terminal of the battery, through the lamp and switch, and back to the negative battery terminal. Remember, the motion of negative charges in one direction is equivalent to the motion of positive charges in the opposite direction.

If the switch is now opened the lamp goes out, indicating that the current has ceased. *To maintain a current, there must be a closed path around which the charges can flow.*

An instrument known as an **ammeter** is used to measure the electric current in a circuit. Charges must flow through the ammeter before it registers the electric current; therefore, we must break the circuit to insert it. The ammeter may be installed at any location in the circuit (Fig. 20-6). Usually, the red terminal (R) of the ammeter should be closer to the positive terminal of the battery than the black terminal (B).

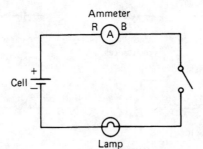

Figure 20-6 Simple circuit with an ammeter.

Voltage

The **battery** delivers the electric energy which makes charges (free electrons in this case) move around the circuit. These electrons are able to move relatively freely through the metal connecting wires, but they interact with the atoms in the thin wire filament of the lamp, causing them to vibrate with greater amplitudes, thereby heating the filament.

As the temperature of the filament rises, the lamp glows as the kinetic energy of the electrons is converted into thermal energy, light, and other energy forms. To maintain an electric current so that the lamp will continue to glow, this electric energy must be replaced. This is the function of the battery or some other source, such as a generator or power supply.

When the switch is closed, the battery acts like a pump, doing work on the electrons to drive them from its negative terminal through the wires, the lamp, and the switch and back to the positive terminal (Fig. 20-7). Electrons have a higher potential energy at the negative terminal and return with lower energy at the positive terminal. As the electrons arrive back at the positive terminal, the source does work to move them internally back to the negative terminal, where they then flow around the circuit once more.

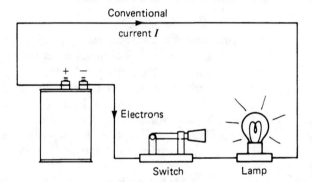

Figure 20-7 Path of electrons when the switch is closed.

It should be noted that while electrons orbit atomic nuclei at high speed (about 10^6 m/s), they drift very slowly, at speeds less than 1 cm/s, around the circuit. However, when the switch is closed the current appears very rapidly because of the interaction of the electrons. These interactions can be described in terms of the interactions between a touching row of marbles (Fig. 20-8). A push on a marble at one end A of the chain is transmitted very quickly, from marble to marble, to the other end B of the chain, even though the marble at A did not travel to B.

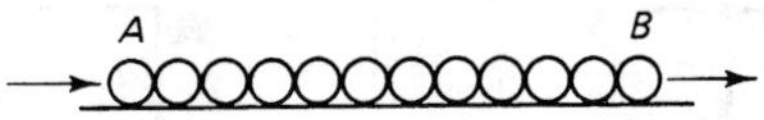

Figure 20-8 If the marble at A is pushed, the force is transmitted to B.

When the source does work (W) in moving a charge (Q) internally from one terminal to the other, the ratio of the work to the charge is called the **electromotive force**, emf ($\mathscr{E}$), of the source.

$$\mathscr{E} = \frac{W}{Q} \tag{20-3}$$

The SI unit of emf is therefore a joule per coulomb; this derived unit is given the special name of **volt** (symbol V).

$$1 \text{ volt} = 1 \frac{\text{joule}}{\text{coulomb}} \quad \text{or} \quad 1 \text{ V} = 1 \frac{\text{J}}{\text{C}}$$

EXAMPLE 20-3

Determine the emf of a source that does 2.70 J of work to transfer 6.80×10^{17} electrons.

Solution

Data: $W = 2.70$ J, $N = 6.80 \times 10^{17}$, charge on an electron $e = 1.60 \times 10^{-19}$ C, $\mathscr{E} = ?$

Equations: $\mathscr{E} = \dfrac{W}{Q}$, total charge transferred $Q = Ne$

Rearrange: $\mathscr{E} = \dfrac{W}{Ne}$

Substitute: $\mathscr{E} = \dfrac{2.70 \text{ J}}{(6.80 \times 10^{17})(1.60 \times 10^{-19} \text{ C})} = 24.8$ V

It should be noted that the emf of a source is not the same as the voltage between its terminals because some energy is lost internally.

While the battery or some other source is used to increase the energy of the moving charges (e.g., electrons), other circuit elements, such as the lamp filament, act as **loads** dissipating that energy by converting it into other energy forms. We describe this "loss" as a reduction in potential difference or **voltage drop** rather than energy because voltages are independent of charge. Since energy is conserved, in our simple circuit, the voltage of the source must equal the voltage drop around the circuit.

To summarize, to maintain a current, we require:

1. A **closed path** around which the charges can flow
2. Some **source,** such as a battery or power supply, to deliver energy continually to the circuit, to balance the energy that is dissipated in the circuit components
3. A **load** to use the energy

An instrument known as a **voltmeter** is used to measure potential differences or voltages. To measure a voltage across a device, the voltmeter is connected directly to its terminals, with the positive terminal of the voltmeter (usually red R) always closer to the positive terminal of the source (Fig. 20-9).

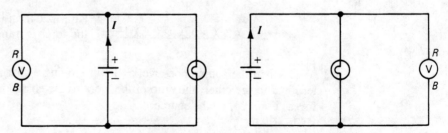

Figure 20-9 Connection of a voltmeter.

Resistance

With the exception of materials called superconductors, all known materials offer some opposition to the flow of electric charges. This opposition is called the **electrical resistance** R; it results in a loss of electrical energy.

The electrical resistance of any component is defined as the ratio of the potential difference (voltage V) across that object to the current I in it.

Sec. 20-2 / Current, Voltage, and Resistance

$$R = \frac{V}{I} \qquad (20\text{-}4)$$

This definition is valid for all circuit components. The SI unit for resistance is therefore a volt per ampere and it has the special name **ohm** (symbol Ω, the capital Greek letter omega).

$$1\ \Omega = 1\ \frac{\text{V}}{\text{A}}$$

EXAMPLE 20-4

Determine the resistance of a circuit component if a 30.0 V potential difference produces a 2.00 A current in it.

Solution

Data: $V = 30.0$ V, $I = 2.00$ A, $R = ?$

Equation: $R = \dfrac{V}{I}$

Substitute: $R = \dfrac{30.0\ \text{V}}{2.00\ \text{A}} = 15.0\ \Omega$

In general, the electrical resistances of uniform materials depend on their size, shape, the nature of the material, and the external conditions, such as temperature. Many objects have resistances that also vary with the potential difference, so that the ratio of the potential difference to the electric current varies. The resistance of some circuit elements, such as diodes, even depends on the direction of the current.

In semiconductors and insulators the resistance is mainly due to the absence of free charges. Their resistance decreases as their temperature increases because the thermal energy frees charges, which can then contribute to the current. However, the resistance of a metallic conductor increases with temperature because the thermal vibrations of the atoms tend to restrict the flow of charges.

Resistors are devices that conduct some electricity but also dissipate some electric energy as heat (Fig. 20-10). They are represented schematically by the symbols

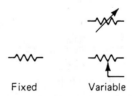

Fixed   Variable

Heater elements and light bulb filaments are resistors that dissipate electric energy as heat and light.

When a potential difference V exists between the ends of a conductor that has a resistance R, the conductor must carry an electric current:

$$I = \frac{V}{R}$$

The potential difference can exist only when the conductor carries the current. In circuits, the potential difference between the ends of a conductor is often referred to as the **voltage, potential drop** or *IR* **drop**. That is, from Eq. 20-4,

$$V = IR \qquad (20\text{-}5)$$

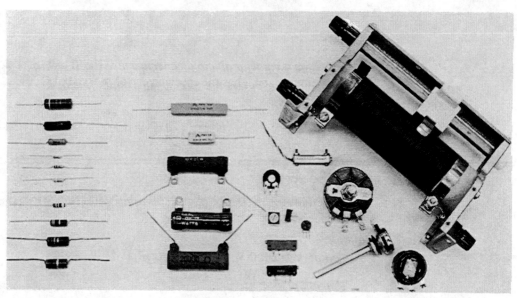

Wire wound and carbon composition Higher power Variable

Figure 20-10 Resistors.

EXAMPLE 20-5

Determine the potential drop across a 220 Ω resistor when it carries 450 mA.

Solution

Data: $R = 220$ Ω, $I = 0.45$ A, $V = ?$
Equation: $V = IR$
Substitute: $V = (0.45$ A$)(220$ Ω$) = 99$ V

PROBLEMS

12. Determine the emf of a source that does 4.50 J of work to transmit a charge of 3.25 C.
13. If the emf of a source is 12.0 V, how much work does it do to move 3.2×10^{18} electrons from the lower potential to the higher potential terminal?
14. If a source has an emf of 9.00 V and it does 8.60 J of work, how much charge does it transmit?
15. What is the resistance of a wire if a 2.00 V potential difference produces a 750 mA current in it?
16. What is the potential difference across the terminals of a 220 Ω resistor when it carries 120 mA?
17. Determine the resistance of a device if a 36.0 V potential difference produces a 120 mA current in it.
18. What is the current in a 560 kΩ resistor when the potential difference across it is 25.0 V?
19. Determine the resistance of a heater when it draws 12.0 A from a 110 V source.
20. Calculate the potential drop across a 220 Ω resistor when it carries 520 mA.
21. What is the potential drop across a 1.35 Ω busbar in a circuit box when it carries 95.0 A?

20-3 OHM'S LAW

While the resistance of many conductors and devices varies with the applied potential difference, Georg Simon Ohm (1787–1854) observed that if the temperature is constant, the electrical resistance of metallic conductors remains constant over wide ranges of potential difference. Thus in metallic conductors the electric current I is directly proportional to the applied potential difference V.

$$I \propto V \quad \text{or} \quad \frac{V}{I} = R = \text{constant} \tag{20-6}$$

where R is the resistance of the metallic conductor. This statement is known as **Ohm's law**; it should not be confused with the definition of electrical resistance. The resistance of all objects is defined by Eq. 20-4. The ratio of potential difference to the electric current is only a constant for metallic conductors and some special resistors.

Ohm's law can be demonstrated by a simple experiment. Consider the circuit in Fig. 20-11. As we increase the voltage (by adjusting the variable power supply), the current I through the load R increases. If we plot a graph of voltage V versus the current I, we obtain a straight line through the origin indicating that the voltage was proportional to the current (Fig. 20-12).

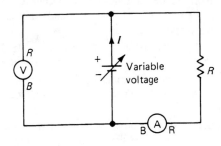

Figure 20-11 Circuit to demonstrate Ohm's law.

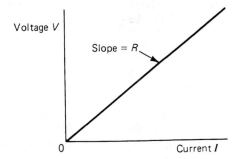

Figure 20-12 Current–voltage characteristics for a resistance obeying Ohm's law.

$$V \propto I$$

The constant of proportionality (the slope of the graph) is the resistance R. Thus

$$V = IR \quad \text{or} \quad R = \frac{V}{I}$$

EXAMPLE 20-6

A 110 V potential difference maintains a 2.00 A current in the metallic filament of a light bulb. Determine (a) the resistance of the filament; (b) the potential difference that would produce a 2.20 A current if the temperature remains constant.

Solution

Data: $V_1 = 110$ V, $I_1 = 2.00$ A, $I_2 = 2.20$ A, $R = ?$, $V_2 = ?$

Equations: $R = \dfrac{V_1}{I_1}$, $V_2 = I_2 R$

Substitute:

(a) $R = \dfrac{110 \text{ V}}{2.00 \text{ A}} = 55.0 \ \Omega$ 　　　　(b) $V_2 = (2.20 \text{ A})(55.0 \ \Omega)$
　　　　　　　　　　　　　　　　　　　　　　　$= 121$ V

An **ohmmeter** is a convenient instrument that allows us to measure resistance values. To measure the resistance, we merely apply the ohmmeter contacts to each end of the component. However, the ohmmeter should not be connected in a circuit or used on a component that is carrying an electric current. This instrument is also used to check circuits for continuity, that is, to see if there is a break in the circuit.

PROBLEMS

22 What resistance is required to limit the current to 350 mA when the applied potential difference is 15.0 V?

23 A 26.8 V potential difference maintains a 58.0 mA current in a metallic resistor. (a) What is the resistance of the resistor? (b) What potential difference would maintain a 120 mA current at the same temperature?

24 Determine the resistance that is required to produce a 25.0 V potential drop where the current is 12.5 mA.

25 A 15.0 V potential difference maintains a 350 mA current in a resistor. Determine (a) the resistance of the resistor; (b) the current that would produce a 35.0 V potential drop across the same resistance.

26 If a 25.0 V potential difference maintains a current in a resistor, what potential difference is required to triple the current at the same temperature?

27 If a 5.00 V potential difference is applied to a length of copper wire that has a resistance of 4.20 Ω, find (a) the current in the wire; (b) the number of electrons that pass a fixed point in 2.50 s.

20-4 RESISTIVITY

Ohm also investigated variations of the electric resistance in metallic conductors of different dimensions. He found that for conductors constructed of the same material, if the temperature is constant, their electric resistance R is directly proportional to their length L and inversely proportional to their cross-sectional area A:

$$R \propto \frac{L}{A} \quad \text{or} \quad R = \rho \frac{L}{A} \tag{20-7}$$

The proportionality constant ρ is called the **resistivity**. Its value depends on the temperature and the nature of the material present and it is independent of the size and shape of the material (Table 20-1). In SI resistivities are expressed in units of ohm meters ($\Omega \cdot$ m).

EXAMPLE 20-7

Determine the resistance of a 10.0 m long copper wire if its resistivity is $1.72 \times 10^{-8} \ \Omega \cdot$ m and its diameter is 0.800 mm.

TABLE 20-1
Resistivities at 20°C

Material	Resistivity ρ $\Omega \cdot m \times 10^{-8}$
Aluminum	2.6
Carbon	3500
Constantan	50
Copper	1.7
Nichrome	115
Platinum	11
Silver	1.6
Tin	11
Tungsten	5.5

Solution

Data: $L = 10.0$ m, $\rho = 1.72 \times 10^{-8}\,\Omega \cdot m$, $d = 8.00 \times 10^{-4}$ m, $R = ?$

Equations: The cross section is circular; therefore,

$$A = \frac{\pi d^2}{4} \quad \text{and} \quad R = \frac{\rho L}{A}$$

Rearrange: $R = \dfrac{\rho L}{\pi d^2/4} = \dfrac{4\rho L}{\pi d^2}$

Substitute: $R = \dfrac{4(1.72 \times 10^{-8}\,\Omega \cdot m)(10.0\,m)}{\pi(8.00 \times 10^{-4}\,m)^2} = 0.342\,\Omega$

Note that this resistance is quite small. The resistances of the interconnecting wires in a circuit are usually negligible compared with the other components.

PROBLEMS

28. Determine the resistivity of a wire 15.0 m long which has a resistance of 0.150 Ω and a cross-sectional area of $5.25 \times 10^{-5}\,m^2$.

29. Determine the resistivity of a wire that is 45.0 m long if it has a diameter of 0.750 mm and its resistance is 2.50 Ω.

30. Determine the resistance of a copper wire 125 m long that has a cross-sectional area of 0.750 mm².

31. What is the cross-sectional area of a copper wire 75.0 m long that has a resistance of 1.25 Ω at 20°C?

32. What length of 0.75 mm diameter copper wire with a resistivity of $1.7 \times 10^{-8}\,\Omega \cdot m$ has a resistance of 5.0 Ω at 20°C?

33. A 4.60 m long nichrome wire with a diameter of 0.500 mm and resistivity of $1.15 \times 10^{-6}\,\Omega \cdot m$ is made into a resistance coil for a small heater. What is the resistance of the coil?

34. A 7.50 V potential difference is dropped in a 325 m long aluminum wire when it carries 4.50 A. Determine **(a)** the cross-sectional area; **(b)** the diameter of the wire.

20-5 COMBINATIONS OF RESISTANCES

Two or more resistors are often connected by metallic wires of negligible resistance. The total resistance of the combination depends on their arrangement. There are two basic configurations.

Resistors in Series

Two or more resistors are said to be in **series** when they are connected end to end so that there is only one path and they all carry the same electric currents. Suppose that resistors with resistances R_1, R_2, and R_3 are connected in series with a source of electrical energy, such as a power supply or battery, that develops a total potential difference V (Fig. 20-13). Using a voltmeter to measure the potential differences, we find that the voltage across the source equals the sum of the voltages across the resistances.

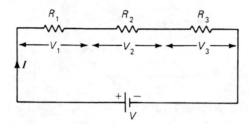

Figure 20-13 Three resistors in series.

$$V = V_1 + V_2 + V_3$$
$$= IR_1 + IR_2 + IR_3$$

or

$$V = I(R_1 + R_2 + R_3)$$

These resistors could be replaced by a single resistor with an **equivalent resistance** R without changing the current if

$$V = IR$$

Thus

$$R = R_1 + R_2 + R_3$$

In general, **series circuits** have the following **characteristics**:

1. The electric current I is the same at each point and in each resistor in the circuit.

$$I = I_1 = I_2 = I_3 \tag{20-8}$$

This is due to the conservation of charge.

2. Each resistor opposes the current and the total equivalent resistance is the sum of the individual resistances.

$$R = R_1 + R_2 + R_3 \tag{20-9}$$

3. The total applied potential difference is equal to the sum of the potential drops across all the resistors.

$$V = V_1 + V_2 + V_3 \tag{20-10}$$

This is due to the conservation of energy.

4. The ratio of the total voltage to the total resistance is equal to the current. This is merely an extension of Eq. 20-4.

$$\frac{V}{R} = I = \frac{V_1 + V_2 + V_3}{R_1 + R_2 + R_3} \tag{20-11}$$

Sec. 20-5 / Combinations of Resistances

5. Any break in the circuit stops the current because there is only one closed path.

EXAMPLE 20-8

Determine (a) the total resistance, (b) the current, and (c) the potential drops across each resistor of the circuit in Fig. 20-14.

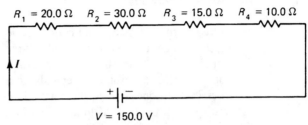

Figure 20-14

Solution

Data: $R_1 = 20.0 \ \Omega$, $R_2 = 30.0 \ \Omega$, $R_3 = 15.0 \ \Omega$, $R_4 = 10.0 \ \Omega$,
$V = 150.0$ V, $R = ?$, $I = ?$, $V_1 = ?$, $V_2 = ?$, $V_3 = ?$

(a) Equation: $R = R_1 + R_2 + R_3 + R_4$
Substitute: $R = 20.0 \ \Omega + 30.0 \ \Omega + 15.0 \ \Omega + 10.0 \ \Omega$
$= 75.0 \ \Omega$

(b) Equation: $I = \dfrac{V}{R}$

Substitute: $I = \dfrac{150.0 \text{ V}}{75.0 \ \Omega} = 2.00$ A

(c) Equation: $V = IR$
Substitute: $V_1 = (2.00 \text{ A})(20.0 \ \Omega) = 40.0$ V, $V_2 = (2.00 \text{ A})(30.0 \ \Omega) = 60.0$ V, $V_3 = (2.00 \text{ A})(15.0 \ \Omega) = 30.0$ V, $V_4 = (2.00 \text{ A})(10.0 \ \Omega) = 20.0$ V
Check: $V = V_1 + V_2 + V_3 + V_4$ or

$150.0 \text{ V} = 40.0 \text{ V} + 60.0 \text{ V} + 30.0 \text{ V} + 20.0 \text{ V}$

EXAMPLE 20-9

Calculate the value of the resistance R_2 in the series circuit illustrated in Fig. 20-15.

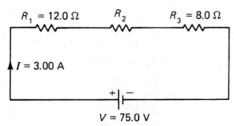

Figure 20-15

Solution

Data: $R_1 = 12.0 \ \Omega$, $R_3 = 8.0 \ \Omega$, $V = 75.0$ V, $I = 3.00$ A, $R_2 = ?$

Equations: $I = \dfrac{V}{R}$, $R = R_1 + R_2 + R_3$

Rearrange: $R = \dfrac{V}{I}$, $R_2 = R - R_1 - R_3$

Substitute: $R = \dfrac{75.0 \text{ V}}{3.00 \text{ A}} = 25.0 \text{ }\Omega$

Therefore, $R_2 = 25.0 \text{ }\Omega - 12.0 \text{ }\Omega - 8.0 \text{ }\Omega = 5.0 \text{ }\Omega$.

Resistors in Parallel

When two or more resistors are connected in **parallel**, they form different paths for the electric current. A typical parallel circuit with three resistors and a source is illustrated in Fig. 20-16. In this circuit, the current is divided between the branches and the current in each branch depends on the resistance values. If there is a break (**open circuit**) in one branch, there are still currents in the other branches. This type of circuit is commonly used in house wiring, so that when one branch, containing a light for example, is turned off, the others continue to operate.

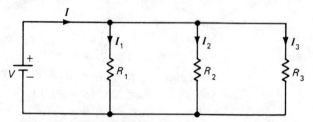

Figure 20-16 Three resistors in parallel.

In general, **parallel circuits** have the following characteristics:

1. The total electric current is equal to the sum of the currents in each branch.
$$I = I_1 + I_2 + I_3 \quad (20\text{-}12)$$
This is due to the conservation of charge.

2. The reciprocal of the total equivalent resistance is equal to the sum of the reciprocals of the individual resistances.
$$\frac{1}{R} = \frac{1}{R_1} + \frac{1}{R_2} + \frac{1}{R_3} \quad (20\text{-}13)$$
These resistors could be replaced by a single resistor of resistance R without changing the total current if
$$I = \frac{V}{R}$$
R is called the **equivalent resistance**.

3. The same potential drop occurs across each branch.
$$V = V_1 = V_2 = V_3 \quad (20\text{-}14)$$
This is a consequence of the conservation of energy.

4. If there is a break in any of the parallel resistance branches, the others continue to operate.

5. The current is the ratio of the voltage to the resistance in each component.
$$I = \frac{V}{R}, \quad I_1 = \frac{V}{R_1}, \quad I_2 = \frac{V}{R_2} \quad I_3 = \frac{V}{R_3} \quad (20\text{-}15)$$

EXAMPLE 20-10

Determine (a) the equivalent resistance, (b) the total current, and (c) the current in each resistance in the circuit illustrated in Fig. 20-17.

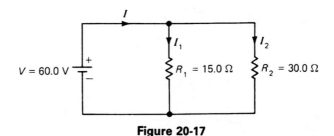

Figure 20-17

Solution

Data: $V = 60.0$ V, $R_1 = 15.0$ Ω, $R_2 = 30.0$ Ω, $I = ?$, $R = ?$, $I_1 = ?$, $I_2 = ?$

(a) Equation: $\dfrac{1}{R} = \dfrac{1}{R_1} + \dfrac{1}{R_2}$

Substitute: $\dfrac{1}{R} = \dfrac{1}{15.0\ \Omega} + \dfrac{1}{30.0\ \Omega} = \dfrac{2+1}{30.0\ \Omega} = \dfrac{3}{30.0\ \Omega}$

Therefore,

$$R = \dfrac{30.0\ \Omega}{3} = 10.0\ \Omega$$

This may be done directly with a calculator.

(b) Equation: $I = \dfrac{V}{R}$

Substitute: $I = \dfrac{60.0\ \text{V}}{10.0\ \Omega} = 6.0$ A

(c) Equations: $I_1 = \dfrac{V}{R_1}$, $I_2 = \dfrac{V}{R_2}$

Substitute: $I_1 = \dfrac{60.0\ \text{V}}{15.0\ \Omega} = 4.00$ A

$I_2 = \dfrac{60.0\ \text{V}}{30.0\ \Omega} = 2.00$ A

Check: $I = I_1 + I_2$, or 6.00 A $= 2.00$ A $+ 4.00$ A.

EXAMPLE 20-11

Determine (a) the equivalent resistance and (b) the resistance R_2 of the circuit in Fig. 20-18.

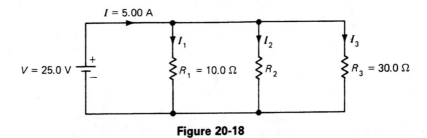

Figure 20-18

Solution

Data: $V = 25$ V, $I = 5.00$ A, $R_1 = 10.0$ Ω, $R_3 = 30.0$ Ω, $R = ?$, $R_2 = ?$

(a) Equation: $I = \dfrac{V}{R}$

Rearrange: $R = \dfrac{V}{I}$

Substitute: $R = \dfrac{25.0 \text{ V}}{5.00 \text{ A}} = 5.00$ Ω

(b) Equation: $\dfrac{1}{R} = \dfrac{1}{R_1} + \dfrac{1}{R_2} + \dfrac{1}{R_3}$

Rearrange: $\dfrac{1}{R_2} = \dfrac{1}{R} - \dfrac{1}{R_1} - \dfrac{1}{R_3}$

Substitute: $\dfrac{1}{R_2} = \dfrac{1}{5.00 \text{ Ω}} - \dfrac{1}{10.0 \text{ Ω}} - \dfrac{1}{30.0 \text{ Ω}} = \dfrac{6 - 3 - 1}{30.0 \text{ Ω}} = \dfrac{2}{30.0 \text{ Ω}}$

or

$$R_2 = \dfrac{30.0 \text{ Ω}}{2} = 15.0 \text{ Ω}$$

Series–Parallel Resistors

To determine the equivalent resistance of a complex combination of resistances, compute the resistance of each group in parallel and each group in series, and replace these groups by their equivalent resistance. Repeat the process until the entire combination is reduced to a single equivalent resistance, that is, equivalent from the point of view of the total current.

EXAMPLE 20-12

Find **(a)** the total equivalent resistance, **(b)** the total current, and **(c)** the current in each resistance of the circuit in Fig. 20-19.

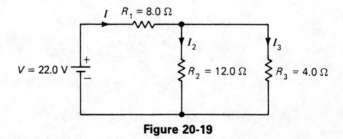

Figure 20-19

Solution

Data: $V = 22.0$ V, $R_1 = 8.0$ Ω, $R_2 = 12.0$ Ω, $R_3 = 4.0$ Ω, $R = ?$, $I = I_1 = ?$, $I_2 = ?$, $I_3 = ?$

We must first solve for the equivalent resistance R_{23} of the parallel resistances R_2 and R_3.

Equation: $\dfrac{1}{R_{23}} = \dfrac{1}{R_2} + \dfrac{1}{R_3}$

Substitute: $\dfrac{1}{R_{23}} = \dfrac{1}{12.0 \text{ Ω}} + \dfrac{1}{4.0 \text{ Ω}} = \dfrac{1 + 3}{12.0 \text{ Ω}} = \dfrac{4}{12.0 \text{ Ω}}$

and

Sec. 20-5 / Combinations of Resistances

$$R_{23} = \frac{12.0\ \Omega}{4} = 3.00\ \Omega$$

We may now replace R_2 and R_3 with the equivalent resistance R_{23} (Fig. 20-20).

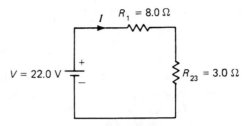

Figure 20-20

To find the total resistance, we now consider the resistances R_1 and R_{23} in series.

(a) *Equation:* $R = R_1 + R_{23}$
Substitute: $R = 8.00\ \Omega + 3.00\ \Omega = 11.00\ \Omega$

(b) *Equation:* $I = \dfrac{V}{R}$

Substitute: $I = \dfrac{22.0\ V}{11.00\ \Omega} = 2.00\ A$

(c) The total current $I = 2.00$ A occurs in R_1 and R_{23} because they are in series. Therefore, the potential drops across R_1 and R_{23} are given by

$$V_1 = IR_1 \quad \text{and} \quad V_{23} = IR_{23}$$

$$V_1 = (2.00\ A)(8.00\ \Omega) = 16.0\ V$$

$$V_{23} = (2.00\ A)(3.00\ \Omega) = 6.00\ V$$

Check: $V = V_1 + V_2$, or $22.0\ V = 16.0\ V + 6.00\ V$ for series resistances. Since R_2 and R_3 are in parallel, they have the same potential drop $V_{23} = 6.00$ V. Therefore,

$$I_2 = \frac{V_{23}}{R_2} \quad \text{and} \quad I_3 = \frac{V_{23}}{R_3}$$

and so

$$I_2 = \frac{6.00\ V}{12.0\ \Omega} = 0.500\ A \qquad I_3 = \frac{6.00\ V}{4.00\ \Omega} = 1.50\ A$$

Check: $I = I_2 + I_3$ or $2.00\ A = 0.500\ A + 1.50\ A$.

PROBLEMS

35 Determine the total resistance of resistors $R_1 = 5.00\ \Omega$, $R_2 = 15.0\ \Omega$, and $R_3 = 20.0\ \Omega$ when they are connected **(a)** in series; **(b)** in parallel; **(c)** with a parallel combination of R_1 and R_2 in series with R_3.

36 In Fig. 20-21, the resistances have values $R_1 = 7.00\ \Omega$, $R_2 = 8.00\ \Omega$, and $R_3 = 12.0\ \Omega$, and $V = 54.0$ V. Determine **(a)** the total equivalent resistance; **(b)** the total current; **(c)** the voltage drop across each resistance.

37 In Fig. 20-21, find **(a)** the total equivalent resistance, **(b)** the resistance R_1, and **(c)** the voltage drop across each resistance if $R_2 = 20.0\ \Omega$, $R_3 = 30.0\ \Omega$, $I = 1.50$ A, and $V = 90.0$ V.

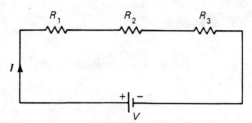

Figure 20-21

38. If $I = 125$ mA, $R_1 = 220$ Ω, $R_2 = 100$ Ω, and $R_3 = 150$ Ω in Fig. 20-21, find V.
39. When four identical lamps are connected in series to a 110 V source, they draw 1.50 A. Determine (a) the total resistance; (b) the resistance of each lamp; (c) the voltage drop across each lamp.
40. In Fig. 20-22, if $R_1 = 30.0$ Ω, $R_2 = 50.0$ Ω, $R_3 = 75.0$ Ω, and $V = 60.0$ V, find the total current and the currents in each resistance.

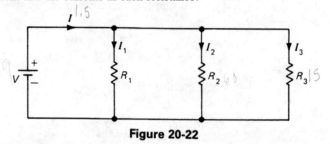

Figure 20-22

41. If $R_2 = 60.0$ Ω, $R_3 = 15.0$ Ω, $V = 9.00$ V, and the total current $I = 1.50$ A in Fig. 20-22, find (a) the total equivalent resistance; (b) the resistance R_1; (c) the currents in each resistance.
42. How many 60.0 Ω resistors must be connected in parallel to draw 750 mA from a 9.00 V source?
43. In Fig. 20-23, if $R_1 = 12.0$ Ω, $R_2 = 48.0$ Ω, $R_3 = 7.4$ Ω, and $V = 42.5$ V, find (a) the total resistance; (b) the total current; (c) the current and voltage drop across each resistance.

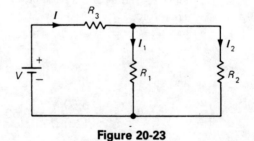

Figure 20-23

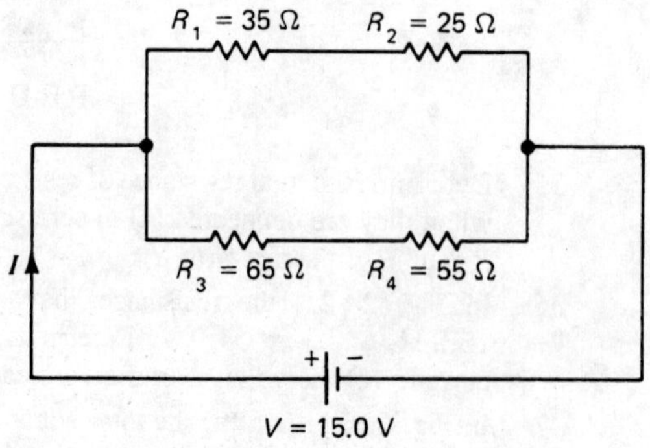

Figure 20-24

44 If $I = 600$ mA, $V = 30.0$ V, $R_1 = 60.0$ Ω, and $R_2 = 90.0$ Ω in Fig. 20-23, find (a) the total equivalent resistance; (b) the resistance R_3; (c) the currents in each resistance.

45 Determine the total equivalent resistance and the current in each resistor of the circuit illustrated in Fig. 20-24.

SUMMARY

An *electric current* is a flow of electric charge. SI unit is the base unit called an *ampere* (A). $I = \dfrac{Q}{t}$, and the charge $Q = It$. 1 C = 1 A·s.

Conventional current is taken as the direction of positive charge flow. Negative charges flow in the opposite direction to conventional current.

Emf of a source $\mathscr{E} = \dfrac{W}{Q}$. The SI unit is the *volt*

To maintain a current, we require:

1. A *closed path* for the charge flow
2. A *source* of energy to balance the energy losses
3. A *load* to use the energy

The *resistance* of any circuit element is defined as the ratio of the potential difference V across it to the current I in it. $R = \dfrac{V}{I}$.

The SI unit is the *ohm* (symbol Ω). 1 Ω = 1 V/A.
Potential drop: $V = IR$.
Ohm's law: For some conductors the resistance R is constant.

Resistivity: $\rho = \dfrac{RA}{L}$ and $R = \dfrac{\rho L}{A} = \dfrac{4\rho L}{\pi d^2}$ (circular conductor)

Resistors in series: Current I is the same in each device.

$$V = V_1 + V_2 + V_3 + \ldots \qquad R = R_1 + R_2 + R_3 + \ldots$$

Resistors in parallel: Potential drop is the same across each.

$$I = I_1 + I_2 + I_3 + \ldots \qquad \dfrac{1}{R} = \dfrac{1}{R_1} + \dfrac{1}{R_2} + \dfrac{1}{R_3} + \ldots$$

QUESTIONS

1 Define the following terms: (a) electric current; (b) coulomb of charge; (c) resistance; (d) ohm; (e) emf; (f) resistivity; (g) series circuit; (h) parallel circuit; (i) load resistance; (j) potential (IR) drop.

2 Describe the various types of electric currents.

3 What is the difference between electron flow and conventional current?

4 How would you connect (a) an ammeter and (b) a voltmeter in a circuit?

5 The electric circuitry in a house is normally in parallel. What are the reasons for this?

6 What is the difference between the definition of resistance and Ohm's law?

7 What are the characteristics of (a) series circuits; (b) parallel circuits?

8 (a) If a number of light bulbs are connected in series and one bulb burns out, what happens to the others? (b) What happens if the bulbs are in parallel?

9 Resistors are used to control currents and voltages in circuits. Describe how they accomplish these tasks.

10 Show that the equivalent resistance of a number of resistors in parallel is less than the individual resistances.

REVIEW PROBLEMS

1 Determine the current in a wire if (a) 2.50 C of charge flows past a fixed point in 30.0 s; (b) 4.9×10^{18} electrons flow past that point in 15.0 s.

2 If the current in a conductor is steady at 750 mA, (a) how much charge flows past a point in 18.0 s; (b) how many electrons flow past a point in 24.0 s?

3 What is the emf of a source that does 8.60 J of work to transfer a charge of 4.80 C?

4 If the emf of a source is 6.00 V, how much work does it do to transmit 9.80×10^{19} electrons?

5 What is the resistance of a diode when a 5.00 mV potential difference produces a 750 mA current in it?

6 What is the current in a 220 Ω resistor when a 2.50 V potential difference is applied across it?

7 Determine the voltage drop in a 470 Ω resistor when it carries 180 mA.

8 What is the resistance of a heater when it draws 9.60 A from a 110 V source?

9 A copper busbar is 2.50 mm thick, 2.00 cm wide, 75.0 cm long, and has a resistivity of 1.72×10^{-8} $\Omega \cdot$m. Determine (a) the resistance at 20°C; (b) the potential difference between its ends at 20°C if it carries 105 A.

10 Determine the resistivity of a wire that is 28.0 m long with a diameter of 1.10 mm if its resistance is 0.495 Ω.

11 A wire has a resistivity of 1.2×10^{-5} $\Omega \cdot$m and a diameter of 0.85 mm. What length of wire would be required to make the heating element for a toaster that draws 6.0 A from a 110 V source?

12 Determine the equivalent resistance when the following resistors are connected in series: (a) 220 Ω, 150 Ω, 470 Ω, and 560 Ω; (b) 100 Ω, 220 Ω, and 150 Ω; (c) 560 kΩ, 120 kΩ, and 380 kΩ.

13 Determine the equivalent resistance when the following resistors are connected in parallel: (a) 100 Ω, 150 Ω, and 75 Ω; (b) 220 Ω, 660 Ω, and 110 Ω; (c) 560 Ω, 470 Ω, 150 Ω, and 220 Ω.

14 Determine the equivalent resistance of the resistor combination in Fig. 20-25, if (a) $R_1 = 24$ Ω, $R_2 = 36$ Ω, $R_3 = 15$ Ω, $R_4 = 12$ Ω, and $R_5 = 18$ Ω; (b) $R_1 = 150$ Ω, $R_2 = 300$ Ω, $R_3 = 300$ Ω, $R_4 = 180$ Ω, and $R_5 = 50$ Ω.

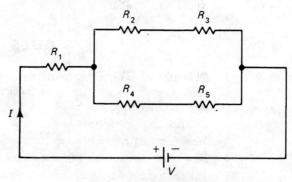

Figure 20-25

21
ELECTRIC ENERGY AND POWER

Most circuit elements consume electric energy when they carry an electric current. For example, electric motors are used to convert electric energy into mechanical work, whereas resistors and heating elements transform electric into thermal energy. Therefore, to maintain a current in a circuit, some source of electric energy is required.

21-1 ELECTRIC ENERGY AND POWER

When a potential difference is maintained across a conductor, electrons tend to drift from points of lower to points of higher (more positive) electric potential. As they drift, these electrons interact with the atoms in the conductor and they transfer some of their energy to those atoms. As a result, the vibratory energy of the atoms increases, some of the kinetic energy of the electrons is transformed into thermal energy, and the temperature of the conductor increases. If the conductor does not melt, it rapidly reaches an equilibrum state where it is able to dissipate the heat to its surroundings at the same rate that the heat is generated.

In an *electric welding process,* the conductor melts and unites with some other material before it reaches an equilibrium state. Fuses are devices containing a piece of wire that is able to carry some maximum current. If the current exceeds this value, the heat generated melts the wire and this opens the circuit (Fig. 21-1).

When some charge Q flows through a circuit component, the work done W by the electric field is equal to the energy E that is lost by the charge when it moves through a potential difference V. But the potential difference $V = W/Q$, therefore,

$$E = W = QV$$

This energy is transformed into thermal energy, mechanical work, chemical energy, light, and so on, in the circuit element. But from the definition of charge, in some elapsed time t an electric charge $Q = It$ flows past a fixed point in the conductor.

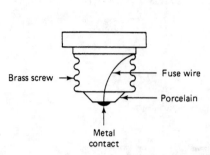
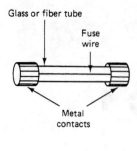

Figure 21-1 Fuses.

Consequently, the work done by the electric field is

$$W = QV = ItV$$

Therefore the time rate of energy dissipation, or the power dissipated (in units of watts) by any circuit element is

$$P = \frac{W}{t} = \frac{ItV}{t} = IV \tag{21-1}$$

EXAMPLE 21-1

Determine the power rating of a lamp that draws 1.25 A from a 120 V source.

Solution

Data: $I = 1.25$ A, $V = 120$ V, $P = ?$
Equation: $P = IV$
Substitute: $P = (1.25 \text{ A})(120 \text{ V}) = 150$ W

The cost of electricity is usually determined in terms of the number of kilowatthours of electric energy that are consumed. A kilowatthour (kW · h) is the electric energy used when 1 kW of electric power is dissipated for one hour. If a resistor dissipates electric energy at a rate of 1 kW, in 1 h it dissipates a total energy of

$$1 \text{ kW} \cdot \text{h} = (1000 \text{ W})(3600 \text{ s}) = 3.6 \times 10^6 \text{ J}$$

Also, since $V = IR$, the power dissipated in a resistance R is

$$P = IV = I^2R = \frac{V^2}{R} \tag{21-2}$$

EXAMPLE 21-2

What minimum power rating must a 2.2 kΩ resistor have if the current through it is 15 mA?

Solution

Data: $R = 2.2 \times 10^3$ Ω, $I = 15 \times 10^{-3}$ A $= 1.5 \times 10^{-2}$ A, $P = ?$
Equation: $P = I^2R$
Substitute: $P = (1.5 \times 10^{-2} \text{ A})^2(2.2 \times 10^3 \text{ Ω}) = 0.50$ W

If E is the total thermal energy produced in a resistor during an elapsed time t,

Sec. 21-1 / Electric Energy and Power

the average power that the resistor dissipates is

$$P = \frac{E}{t} = I^2 R \tag{21-3}$$

EXAMPLE 21-3

A 220 Ω electric heater draws an average current of 2.0 A for 3.0 h. **(a)** Determine the total heat produced. **(b)** If electric energy costs 7.3 cents/kW · h, calculate the cost.

Solution

Data: $R = 220\ \Omega$, $I = 2.0$ A, $t = 3.0$ h, $E = ?$, cost = ?

Equation: $P = \dfrac{E}{t} = I^2 R$

Rearrange: $E = Pt$

For convenience, we shall find the power first.

(a) Substitute: $P = I^2 R$
$= (2.0\ \text{A})^2 (220\ \Omega) = 880\ \text{W} = 0.88\ \text{kW}$

then the energy:

$E = Pt = (0.88\ \text{kW})(3.0\ \text{h}) = 2.64\ \text{kW} \cdot \text{h}$

$= (2.64)(3.6 \times 10^6\ \text{J})$

$= 9.5 \times 10^6\ \text{J}$

(b) Thus the cost = $(2.64\ \text{kW} \cdot \text{h})(7.3\ \text{cents/kW} \cdot \text{h}) = 19$ cents

---

Most electric circuits consist of a source of electric energy and a load composed of one or more devices that transform the electric energy into other energy forms. Normally, many components of an electric circuit dissipate electric energy. Consequently they contribute to the total load. The load of a circuit is often represented schematically by a single resistor that dissipates the equivalent amount of electric power.

PROBLEMS

1. Determine the electric power dissipated in a small motor that draws 650 mA from a 110 V power supply.
2. What minimum power rating must a 470 Ω resistor have if it is located in a circuit **(a)** where the current is 135 mA; **(b)** where it must withstand a 28.0 V potential difference?
3. What maximum voltage can be safely applied to a 560 Ω 0.50 W resistor?
4. Determine the power consumption of a portable radio that draws 2.80 mA from a 9.0 V source.
5. What current will a 12 kW load draw from a 115 V source?
6. A fuse has a 0.018 Ω resistance and a maximum power dissipation of 2.8 W. What maximum current will the fuse be able to carry?
7. Find the time required for a 120 Ω resistor to supply 250 J of heat if it carries 60.0 mA.
8. Determine the resistance of a heater if it dissipates energy at a rate of 1200 W when it carries 12.0 A.
9. **(a)** Calculate the resistance of a 60 W 110 V light bulb. **(b)** How much current does it draw from a 110 V source?
10. What is the total cost of operating a 375 W pump for 24 h if electricity costs 6.50 cents/kW · h?

11 (a) Determine the current that a 60.0 W lamp draws from a 115 V source. (b) If electricity costs 5.70 cents/kW · h, how much does it cost to operate the lamp for 5.00 h?

12 Eight 60.0 W lamps are operated in parallel from a 115 V source. (a) Determine the current drawn by each lamp. (b) How much does it cost to operate the lamps for 8.00 h per day for one week if electricity costs 6.20 cents/kW · h?

13 A 750 W heater, four 60.0 W lamps, and a 1.00 kW hair dryer are used together on a 115 V line with a 15.0 A fuse. Will the fuse blow?

14 How much does it cost to operate a circular saw at 10.0 A for 2.50 h on a 115 V line if electricity costs 5.15 cents/kW · h?

15 Determine the maximum number of 150 W lamps that can be operated together on a 110 V line with a 15 A fuse.

21-2 SOURCES OF ELECTRIC ENERGY

There are many different sources of emf.

Chemical Cell

In chemical cells, electricity is produced by chemical reactions between the components. These cells consist of two different conductors called **electrodes** that are located in a fluid conductor called an **electrolyte,** containing ions. Cells have emfs that are characteristics of the materials that are used. There are two classes of chemical cells.

1. **Primary cells** cannot be easily recharged and must eventually be discarded because their components are consumed during the chemical reactions. A common type of primary cell is the **dry cell** (Fig. 21-2). These cells are portable and each produces an emf of approximately 1.5 V. They consist of a carbon electrode (the positive electrode or **anode**) that is completely surrounded by powdered manganese dioxide and graphite. A zinc electrode (the negative electrode or **cathode**) is separated from the anode by an electrolytic paste (ammonium chloride and zinc chloride) and a porous barrier.

In recent years, cells with longer lifetimes have been developed. **Alkaline-manganese cells** are similar to dry cells but have more stored energy for a given size. These cells have emfs of about 1.5 V and they can deliver relatively large currents. Each cell consists of a zinc anode and a manganese dioxide cathode in a potassium hydroxide electrolyte.

A **mercury cell** consists of a mercuric oxide and graphite anode, a zinc cathode, and a potassium hydroxide electrolyte. They have emfs of approximately 1.3 V.

2. **Secondary cells** may be recharged by connecting them to larger sources of emf that reverse the direction of the current and also reverse the chemical reactions.

Lead-acid storage cells are secondary cells that produce emfs of approximately 2.2 V. Each cell consists of a lead cathode and a lead dioxide anode in an electrolytic solution of dilute sulphuric acid (Fig. 21-3). As this cell discharges, both electrodes become coated with lead sulfate and the concentration of the sulfuric acid in the electrolyte is reduced. Groups of these cells are often connected to form storage batteries.

Nickel-cadmium cells are rugged secondary cells that develop a relatively constant emf of approximately 1.2 V. The anode of each cell consists of nickel hydroxide with graphite, and the cathode is a mixture of cadmium oxide and iron oxide.

To recharge a secondary cell, a source, such as a generator, is used to reverse

Sec. 21-2 / Sources of Electric Energy

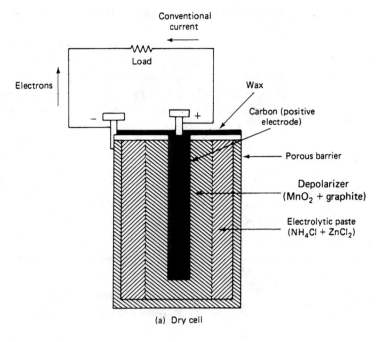

(a) Dry cell

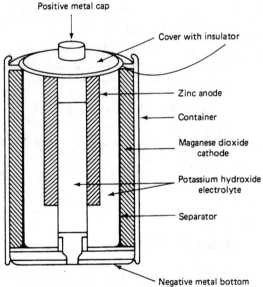

(b) Alkaline-manganese cell

Figure 21-2 Primary cells.

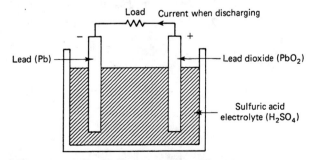

Figure 21-3 Storage cell.

the direction of the current. This provides the energy needed to reverse the chemical reactions.

The potential difference produced by most chemical cells varies with their use and they are sensitive to variations in temperature. All cells have some internal resistance r and they dissipate some of the electric energy that they generate internally. As a result, the **terminal potential difference** V that they deliver to a load is less than the emf $\mathscr{E}$. If the load draws a current I, the terminal potential difference of the source

$$V = \mathscr{E} - Ir \qquad (21\text{-}4)$$

Note that the terminal potential difference of the source therefore depends on the current that it delivers. As the load resistance decreases, the current that it draws increases and the terminal potential difference decreases. Cells and batteries should always be checked while they are connected to a load to see if the internal energy losses are too high.

EXAMPLE 21-4

A cell with a terminal potential difference of 8.90 V and an internal resistance of 0.520 Ω delivers 235 mA to a load. What is the emf of the cell?

Solution

Data: $V = 8.90$ V, $r = 0.520$ Ω, $I = 0.235$ A, $\mathscr{E} = ?$
Equation: $V = \mathscr{E} - Ir$
Rearrange: $\mathscr{E} = V + Ir$
Substitute: $\mathscr{E} = 8.90$ V $+ (0.235$ A$)(0.520$ Ω$) = 9.02$ V

Combination of Cells

Any combination of cells is called a **battery** and it is represented by the symbol

—|⊢ Cell

—||||⊢ Battery

Cells in Series

Two or more cells are connected in series when they are connected so that they provide only one path for an electric current. If the positive terminal of one cell is connected to the negative of the next in sequence they are said to be **series aiding** (Fig. 21-4). This is the most common series connection. For cells in series:

1. The current I is the same in each cell. $I = I_1 = I_2 = I_3 = \ldots$.
2. The total emf is the algebraic sum of the emfs.

$$\mathscr{E} = \mathscr{E}_1 + \mathscr{E}_2 + \mathscr{E}_3 + \ldots \text{ (aiding)}$$

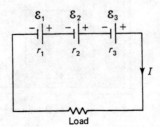

Figure 21-4 Series-aiding cells. $I = I_1 = I_2 = I_3$, $\mathscr{E} = \mathscr{E}_1 + \mathscr{E}_2 + \mathscr{E}_3$, $r = r_1 + r_2 + r_3$.

Sec. 21-2 / Sources of Electric Energy

3. The total internal resistance of the battery is the sum of the internal resistances of the cells.

$$r = r_1 + r_2 + r_3 + \ldots$$

EXAMPLE 21-5

Two cells with emfs of 9.00 V and internal resistances of 24.5 mΩ are connected in series to form a battery that delivers 825 mA to a load. Determine **(a)** the total emf; **(b)** the total internal resistance; **(c)** the terminal potential difference of the battery.

Solution

Sketch: See Fig. 21-5.

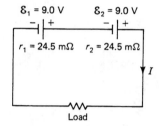

Figure 21-5

Data: $\mathcal{E}_1 = \mathcal{E}_2 = 9.00$ V, $r_1 = r_2 = 0.0245\ \Omega$, $I = 0.825$ A, $\mathcal{E} = ?$, $r = ?$, $V = ?$

(a) Equation: $\mathcal{E} = \mathcal{E}_1 + \mathcal{E}_2$
Substitute: $\mathcal{E} = 9.00\ \text{V} + 9.00\ \text{V} = 18.00\ \text{V}$

(b) Equation: $r = r_1 + r_2$
Substitute: $r = 0.0245\ \Omega + 0.0245\ \Omega = 0.0490\ \Omega$

(c) Equation: $V = \mathcal{E} - Ir$
Substitute: $V = 18.00\ \text{V} - (0.825\ \text{A})(0.0490\ \Omega) = 17.96\ \text{V}$

Identical Cells in Parallel

When identical cells are connected in parallel, their positive terminals are joined and their negative terminals are joined (Fig. 21-6). For this type of connection:

1. The emf is the same as the emf of an individual cell.

$$\mathcal{E} = \mathcal{E}_1 = \mathcal{E}_2 = \mathcal{E}_3 = \ldots$$

2. The total current is the sum of the individual currents.

$$I = I_1 + I_2 + I_3 + \ldots$$

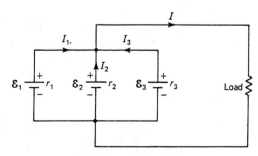

Figure 21-6

3. The total internal resistance is the resistance of one cell r_1 divided by the number of cells n.

$$r = \frac{r_1}{n}$$

EXAMPLE 21-6

Three identical cells with emfs of 1.50 V and internal resistances of 0.246 Ω are connected in parallel to form a battery. Determine **(a)** the total emf; **(b)** the total internal resistance of the battery; **(c)** the current that the battery delivers to a 2.25 Ω load.

Solution

Sketch: See Fig. 21-7.

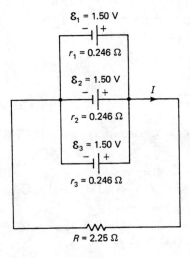

Figure 21-7

Data: $\mathscr{E}_1 = \mathscr{E}_2 = \mathscr{E}_3 = 1.50$ V, $r_1 = r_2 = r_3 = 0.246$ Ω, $R = 2.25$ Ω, $n = 3$, $\mathscr{E} = ?$, $r = ?$, $I = ?$

(a) *Equation:* $\mathscr{E} = \mathscr{E}_1$

Substitute: $\mathscr{E} = 1.50$ V

(b) *Equation:* $r = \dfrac{r_1}{n}$

Substitute: $r = \dfrac{0.246\ \Omega}{3} = 0.082\ \Omega$

(c) *Equations:* $V = IR$, $V = \mathscr{E} - Ir$

Rearrange: $IR = \mathscr{E} - Ir$; therefore, $I = \dfrac{\mathscr{E}}{R + r}$

Substitute: $I = \dfrac{1.50\ \text{V}}{2.25\ \Omega + 0.082\ \Omega} = 0.643$ A

Batteries and cells are often rated in terms of the total charge that they can transmit. This value is sometimes given in terms of ampere hours (A · h) rather than coulombs.

$$1\ \text{A} \cdot \text{h} = (1\ \text{A})(3600\ \text{s}) = 3600\ \text{C}$$

EXAMPLE 21-7

A 60.0 W lamp is connected to a fully charged 12 V 45 A·h battery. Determine **(a)** the total energy that the battery can deliver before it completely discharges; **(b)** the time that the bulb stays on if all the energy is transferred at a constant rate; **(c)** the total number of electrons that move through the bulb in this time.

Solution

Data: $Q = 45$ A·h $= 45 \times 3600$ C $= 1.62 \times 10^5$ C, $V = 12$ V, $P = 60.0$ W, $E = ?$, $t = ?$, $N = ?$

(a) Equation: $E = QV$

Substitute: $E = (1.62 \times 10^5 \text{ C})(12 \text{ V}) = 1.9 \times 10^6$ J

(b) Equation: $P = \dfrac{E}{t}$

Rearrange: $t = \dfrac{E}{P}$

Substitute: $t = \dfrac{1.9 \times 10^6 \text{ J}}{60.0 \text{ W}} = 3.2 \times 10^4$ s

or

$$t = \dfrac{3.2 \times 10^4 \text{ s}}{3600 \text{ s/h}} = 9.0 \text{ h}$$

This can be obtained another way:

Equation: $t = \dfrac{E}{P} = \dfrac{QV}{P}$

Substitute: $t = \dfrac{(45 \text{ A·h})(12 \text{ V})}{60.0 \text{ W}} = 9.0$ h

(c) Equation: $Q = Ne$, where N is the number of electrons of charge $e = 1.6 \times 10^{-19}$ C (Appendix E).

Rearrange: $N = \dfrac{Q}{e}$

Substitute: $N = \dfrac{1.6 \times 10^5 \text{ C}}{1.6 \times 10^{-19} \text{ C}} = 1.0 \times 10^{24}$

Fuel Cells

A **fuel cell** is a device in which chemical energy is converted directly into electrical energy. Although not yet commercially available, these devices are improving rapidly and will eventually become a common and important source of emf.

A simple fuel cell consists of two hollow porous tubes that are coated with a catalyst and immersed in an electrolyte, such as potassium hydroxide. Hydrogen gas is passed into one tube (the cathode) and oxygen gas is fed into the other (the anode) (Fig. 21-8). The chemical reaction between oxygen and hydrogen to form water releases electric energy.

Photoelectric Cells

The surfaces of some materials emit electrons when they are irradiated with light. These materials convert radiant energy directly into electrical energy, and they may be used to detect light or variations in light intensity.

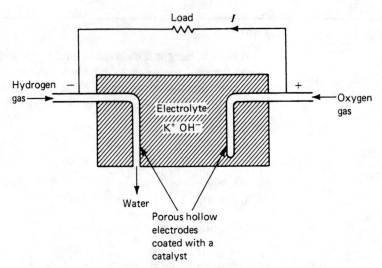

Figure 21-8 Fuel cell.

Generators

An emf is induced in any conductor that experiences a changing magnetic field. This variation in magnetic field may be produced by moving the conductor through the magnetic field. A **generator** or **dynamo** is a device in which mechanical energy is used to rotate conductors in a magnetic field so that an emf is produced; therefore, generators convert mechanical energy into electrical energy.

Thermoelectricity

A **thermocouple** is a device that is used to measure temperatures. It consists of two different metals that are connected so that they form two junctions. One junction is maintained at a constant temperature (e.g., inserted in ice water), whereas the other junction experiences the temperature to be measured (Fig. 21-9). The induced potential difference and the current in the device depend on the temperature difference between the junctions. Thermocouples can be calibrated so that the unknown temperature can be determined from the current provided by this thermoelectric potential difference.

Piezoelectricity

When crystals of some materials, such as quartz, are mechanically stressed, net charges of opposite polarity (sign) may be induced on opposing crystal faces. This

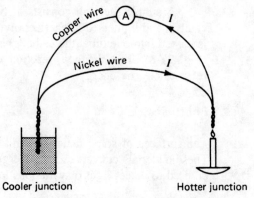

Figure 21-9 Thermocouple.

Sec. 21-2 / Sources of Electric Energy

produces an emf between the diametrically opposite faces in the crystal. This phenomenon is called the **piezoelectric effect.** The opposite also occurs. By applying a rapidly alternating electric field to a piezoelectric crystal, it is possible to produce high-frequency mechanical vibrations. Therefore, piezoelectric crystals can be used to produce high-frequency sound and ultrasonic waves. Piezoelectric materials are used in some microphones and phonograph pickups to convert mechanical pressures into small electrical impulses.

PROBLEMS

16 Determine the emf of a source that does 2.4×10^{-19} J of work as it transmits a single electron.

17 How much work is done by a 1.5 V cell to transmit a single electron?

18 What is the terminal potential difference of a battery that has an emf of 12.0 V and an internal resistance of 2.00 Ω when it delivers 350 mA to a load?

19 A cell has a terminal potential difference of 6.00 V, an internal resistance of 1.50 Ω, and it delivers a 135 mA current to a load. What is its emf?

20 Determine the internal resistance of a cell which has an emf of 9.00 V and it delivers 39.5 mA to a 220 Ω load.

21 Determine the current delivered by a battery with an emf of 12.0 V and internal resistance of 425 mΩ if its terminal potential difference is 11.7 V.

22 Five identical cells with emfs of 1.50 V and internal resistances of 152 mΩ are connected in series to form a battery that delivers 625 mA to a load. Determine (a) the emf; (b) the internal resistance of the battery. (c) What is the resistance of the load?

23 Three identical cells with emfs of 1.50 V and internal resistances of 126 mΩ are connected in parallel to form a battery. Determine (a) the emf; (b) the internal resistance of the battery. (c) What current would the battery deliver to a 7.50 Ω load?

24 Determine the current in the load resistance R of the circuit in Fig. 21-10 if the cells each have terminal potential differences V of 1.50 V and R is 220 Ω.

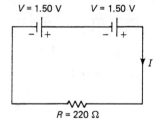

Figure 21-10

25 What are the terminal potential differences of the identical cells in Fig. 21-10 if $R = 1.20$ kΩ and $I = 7.50$ mA?

26 Three identical cells with emfs of 2.00 V and internal resistances of 450 mΩ are connected in series aiding to form a battery. Find (a) the emf; (b) the internal resistance of the battery. (c) What load will draw a 1.50 A current from the battery?

27 Three identical cells are connected in parallel to form a battery (see Fig. 21-11). If the terminal potential difference of each cell is 1.50 V, (a) what current will the battery deliver to a 150 Ω load; (b) what current is supplied by each cell?

28 What is the emf of each identical cell in Fig. 21-11 if $R = 470$ Ω, $I = 19.5$ mA and each cell has an internal resistance of 240 mΩ?

29 A 40 W lamp is connected to a fully charged 12 V 48 A·h battery. Determine (a) the total energy that the battery can deliver before it completely discharges; (b) the time that the lamp stays on if the energy is transferred at a constant rate; (c) the total number of electrons that move through the lamp in this time.

30 If a 9.0 V battery is connected to a 60 W lamp the battery lasts for 7.5 h. Find (a) the total energy that the battery delivers; (b) the rating of the battery.

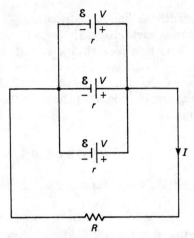

Figure 21-11

SUMMARY

The *power* dissipated by any circuit element $P = IV$.

Energy unit of $1.0 \text{ kW} \cdot \text{h} = 3.6 \times 10^6 \text{ J}$

Power dissipated in a resistance R: $P = IV = I^2R = \dfrac{V^2}{R}$

Primary cells cannot easily be recharged. *Secondary cells* can be recharged by reversing the current.

Terminal potential difference: $V = \mathscr{E} - Ir$

Cells in series: Current is the same in each.

$$I = I_1 = I_2 = I_3 = \ldots$$

$$\mathscr{E} = \mathscr{E}_1 + \mathscr{E}_2 + \mathscr{E}_3 + \ldots$$

$$r = r_1 + r_2 + r_3 + \ldots$$

Identical cells in parallel: Total emf is the same as the emf of one cell.

$$\mathscr{E} = \mathscr{E}_1 = \mathscr{E}_2 = \mathscr{E}_3$$

$$I = I_1 + I_2 + I_3 + \ldots$$

$$r = \dfrac{r_1}{n}$$

Charge unit: $1.0 \text{ A} \cdot \text{h} = 3600 \text{ C}$.

QUESTIONS

1. Define the following terms: **(a)** kilowatt hour; **(b)** primary cell; **(c)** secondary cell; **(d)** terminal potential difference; **(e)** battery; **(f)** series aiding cells.
2. What are the advantages of connecting cells in **(a)** series; **(b)** parallel?
3. A load should be used to draw a current from a cell when it is checked. Explain why.
4. List the common sources of electric energy.

REVIEW PROBLEMS

1. What minimum power rating must a 1.5 kΩ resistor have if it is located in a circuit (a) where the current is 25 mA; (b) where it is subjected to a voltage of 18.0 V?

2. Determine the current that a 0.024 Ω fuse can carry if its maximum power dissipation is 4.50 W.

3. Find the time required for a 220 Ω resistance heater to supply 450 J of heat if it carries 150 mA.

4. (a) What is the resistance of a 100 W 110 V light bulb? (b) How much current does it draw from a 110 V source?

5. Determine the emf of a cell that has an internal resistance of 2.00 Ω if it delivers 125 mA at 8.6 V to a load.

6. What is the terminal potential difference of a source that has an emf of 9.00 V and an internal resistance of 0.520 Ω when it delivers 1.25 A to a load.

7. Three cells with emfs of 9.00 V and internal resistances of 0.225 Ω are connected in series to form a battery. Determine (a) the emf; (b) the internal resistance of the battery. (c) What is the terminal potential difference when the battery delivers 775 mA to a load?

8. Four identical cells with emfs of 1.60 V and internal resistances of 0.500 Ω are connected in parallel to form a battery. What current would the battery deliver to a 1.20 kΩ load?

9. A 150 W lamp is connected to a fully charged 12 V 65 A·h battery. Determine (a) the total energy that the battery can deliver before it becomes completely discharged; (b) the time that the lamp will stay on if the energy is transferred at a constant rate.

10. A fully charged 12 V battery supplies energy at a constant rate to a 750 W motor for 4.5 h after which time the battery is completely discharged. Determine the rating of the battery in ampere hours.

11. Find the power developed by a 6.0 V source that transmits 6.5×10^{-5} C of charge in each second.

22
MAGNETISM

Electricity and magnetism are directly related because both types of phenomena are due to the interactions between electric charges. However, while electric forces are always exerted between charges, whether they are at rest or in motion, magnetic forces only occur between moving charges. In fact, most magnetic properties of materials originate from the motions of the orbital electrons in atoms.

The operation of many devices, such as ammeters, voltmeters, electric motors, particle accelerators, and television sets, depends on magnetic forces that are exerted on charges that move through a magnetic field. In addition, since magnetic fields originate from moving charges, electric currents are surrounded by magnetic fields. The operation of electromagnets, magnetic relays and switches, and many other devices depends on the magnetic fields of electric currents.

22-1 MAGNETIC PROPERTIES

Centuries ago it was noticed that pieces of a special kind of rock attracted each other. This type of rock is called magnetite or lodestone; it is a compound of iron and oxygen that is found in many parts of the world. This name originated from the discovery that it would align in a north-south direction; it was therefore called leading stone or lodestone. A sample of lodestone is a **natural magnet**.

Some materials, such as iron, cobalt, and nickel, are attracted by magnets even when they are not magnets themselves. This is known as **induced magnetism** (Fig. 22-1). By themselves, if they are not magnetized, these materials do not attract each other (except by the weaker force of gravity). However, when they are attracted to a magnet, they also act as magnets. They then have the ability to attract other pieces of similar materials. We can, for example, suspend many paper clips end to end from a single magnet. The paper clips may not be magnetized permanently themselves, but they act as magnets to attract others.

Pieces of iron, cobalt, and nickel can also become **magnetized** when they are

Sec. 22-1 / Magnetic Properties

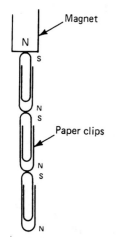

Figure 22-1 Induced magnetism.

either stroked many times in the same direction along their length with one end of a magnet, or if they are placed in the magnetic field of a current (Fig. 22-2; see Section 22-5). These materials are then **artificial magnets,** but they possess magnetic properties that are the same as those of natural magnets.

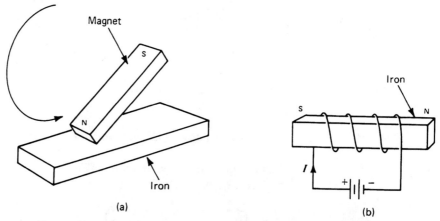

Figure 22-2 (a) Stroke a piece of iron with a magnet to make another magnet; (b) a piece of iron in the magnetic field of a current.

Most materials are **nonmagnetic.** They are not attracted by magnets, they cannot become magnetized, although they can have small influences on magnetic fields.

If iron filings are sprinkled on a piece of paper covering a magnet, they tend to cluster around two points, one at each end of the magnet. These points, where the magnetic attractions are the strongest, are called **magnetic poles** (Fig. 22-3).

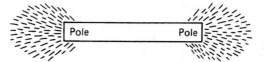

Figure 22-3 Iron filings show the location of the poles.

Magnets may be made in many different shapes, but they all have similar properties. A simple **magnetic compass** can be made by freely suspending a bar magnet on a pivot or from a string, so that the long axis of the magnet is horizontal (Fig. 22-4). If there are no other magnetic materials nearby, the suspended magnet eventually aligns with the magnetic field of the earth, in approximately a north-south direction. The same magnetic pole of the magnet always indicates a northerly direction; therefore, it is called a **north-seeking pole** or simply a **north pole.** The other magnetic pole is called a **south-seeking pole** or **south pole.**

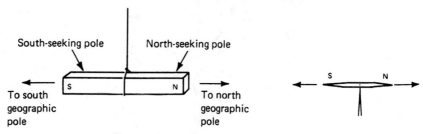

Figure 22-4 Magnetic compasses.

When the north pole of one magnet is brought close to the south pole of another, the two magnets attract each other. However, if a north pole of one magnet is moved close to the north pole of another, the magnets repel each other (Fig. 22-5). Therefore,

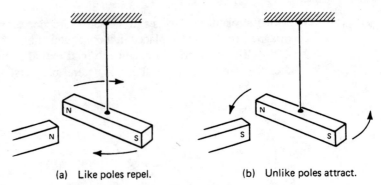

(a) Like poles repel.   (b) Unlike poles attract.

Figure 22-5

Like magnetic poles repel, and unlike magnetic poles attract each other.

Magnetic poles occur naturally in unlike pairs. Even if a magnet is cut into two or more pieces, each piece will contain two unlike poles (Fig. 22-6).

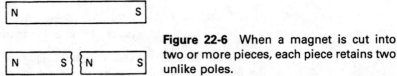

Figure 22-6 When a magnet is cut into two or more pieces, each piece retains two unlike poles.

A magnetic pole may attract other dissimilar magnetic poles and even pieces of some unmagnetized materials. However, magnetic **repulsion** can only occur between like magnetic poles. We can use this fact to determine whether a material is magnetized.

22-2 MAGNETIC FIELDS

We say that a **magnetic field** exists in any region where a magnetic pole experiences a magnetic force. Since magnetic forces occur between magnetic poles, magnetic fields are produced by magnetic poles. These magnetic fields are represented by **lines of magnetic force,** which indicate the direction of the force on a north magnetic pole. For example, exterior to the magnet, the lines of magnetic force are directed away

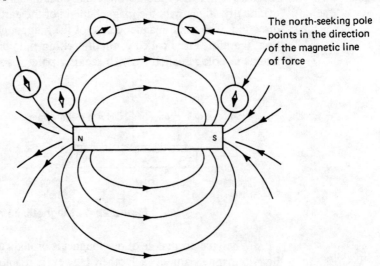

Figure 22-7 Mapping a magnetic field with the aid of a small compass.

Sec. 22-2 / Magnetic Fields

from a north pole because it would repel any other north pole. They are also directed toward a south pole because a south pole attracts north poles.

Magnetic fields are often mapped with the aid of a small magnetic compass. At each point, the direction indicated by the north pole of the compass is the direction of the magnetic field (Fig. 22-7). They can also be mapped by sprinkling iron filings on a sheet of paper that covers a magnet or system of magnets (Fig. 22-8). The filings become induced magnets and align themselves with the field.

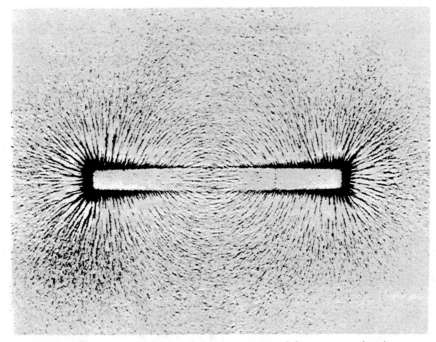

Figure 22-8 Iron filings indicate the lines of force around a bar magnet.

The earth itself behaves like a very large magnet with magnetic poles close to the geographical north and south poles. However, the earth's magnetic pole that is close to the geographic north pole is a south-seeking or south magnetic pole because it attracts the north poles of compasses and magnets. Similarly, the earth's north magnetic pole is located near the geographic south pole (Fig. 22-9).

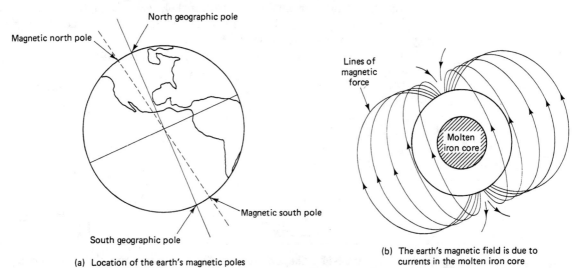

(a) Location of the earth's magnetic poles

(b) The earth's magnetic field is due to currents in the molten iron core

Figure 22-9

The magnetic poles of the earth are actually located below the earth's surface. As a magnetic compass approaches one of the earth's magnetic poles, it "points" into the earth. The angle that the compass makes with the horizontal is called the **angle of dip.** One end of a surveyor's compass may be made heavier than the other to compensate for dip.

22-3 FERROMAGNETISM

Magnetism originates because of the motions and interactions of moving charges, mainly the electrons in atoms. In most atoms, the electrons tend to "pair off" in their orbits so that the net magnetic effect is negligible. However, in solid crystals of materials such as iron, cobalt, and nickel, the electrons interact, and the magnetic fields of the atoms combine, to produce a small net magnetic field in regions called **magnetic domains.** These materials are called **ferromagnetic.**

Domains exist in ferromagnetic materials even when they are not magnetized. The domains are relatively small, and there is a random orientation of their magnetic fields, so that the net magnetic field of the entire crystal is approximately zero. When a ferromagnetic material is **magnetized,** the magnetic forces align the fields of the domains (Fig. 22-10). Similarly, a ferromagnetic material is **demagnetized** (by heating it, pounding it, or placing it in rapidly reversing fields) when the fields of the domains become randomly oriented.

(a) Unmagnetized material: — the magnetic fields of the domains have random orientation

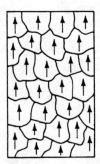
(b) Magnetized ferromagnetic material: — the magnetic fields of the domains are aligned

Figure 22-10 Magnetic domains.

Crystals of the elements iron, cobalt, nickel, gadolinium, and dysprosium, and many of their compounds, may have very large magnetic fields, and these materials may be made into **permanent magnets.**

Like poles of magnets repel each other, and this tends to push the magnetic

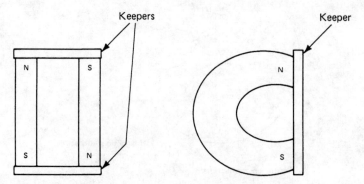

Figure 22-11 Keepers reduce the deterioration of the magnetic fields of permament magnets.

domains out of alignment so that the magnetization slowly deteriorates. To reduce this effect, bar magnets should be stored in pairs with opposite poles next to each other, and iron **keepers** should be placed across the ends. A single keeper is sufficient for horseshoe-shaped magnets (Fig. 22-11).

In many electric circuits, some components must be **shielded** from the stray magnetic fields of the electric currents and magnets. This is accomplished by placing the components within a hollow cavity of a ferromagnetic cylinder. Magnetic fields tend to concentrate in ferromagnetic materials; therefore, the lines of force are distorted so that they pass through the ferromagnetic shield. This leaves the cavity relatively free from external magnetic influences (Fig. 22-12).

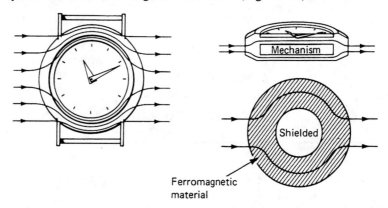

Figure 22-12 A ferromagnetic case shields the mechanism of a watch from magnetic fields. The lines of magnetic force concentrate in the ferromagnetic material.

22-4 MAGNETIC FORCES

A magnetic field at any location is described by a vector quantity called the **magnetic induction** or **magnetic flux density B.** The direction of the magnetic flux density at any location is taken as the direction of the line of magnetic force (corresponding to the force on a north magnetic pole). Its magnitude is defined in terms of its effect on moving electric charges.

If an electric charge Q moves with a velocity $\mathbf{v}$ at some angle θ with respect to the magnetic lines of force, the magnitude of the magnetic force varies directly as the charge Q, the speed v, and the sine of the angle θ:

$$F \propto Qv \sin \theta$$

The magnitude of the flux density B is defined as the proportionality constant:

$$F = QvB \sin \theta \quad \text{or} \quad B = \frac{F}{Qv \sin \theta} \tag{22-1}$$

The magnetic force **F** on the charge is always perpendicular to *both* the velocity **v** and the lines of magnetic force, represented by the flux density **B** (Fig. 22-13).

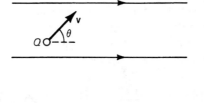

Figure 22-13 The force on a charge Q is directed into the page, in a direction that is perpendicular to the velocity **v** and the flux density **B**.

The SI unit for magnetic flux density is called the **tesla** (symbol T). It is defined as the magnetic flux density where a charge of 1 C moving at a speed of 1 m/s perpendicular to the magnetic field ($\theta = 90°$) experiences a magnetic force of 1 N.

$$1\text{ T} = \frac{1\text{ N}}{\text{C}\cdot\text{m/s}} = 1\frac{\text{N}}{\text{A}\cdot\text{m}} \quad \text{since } 1\text{ A} = 1\text{ C/s}$$

EXAMPLE 22-1

Determine the magnitude of the magnetic force on an electron in a television tube if it moves with a speed of 3.20×10^6 m/s at an angle of 90° with the field of the magnetic deflecting coils, where the flux density is 40.0 mT.

Solution

Data: $Q = 1.6 \times 10^{-19}$ C (Appendix E), $v = 3.20 \times 10^6$ m/s, $\theta = 90°$, $B = 0.0400$ T, $F = ?$
Equation: $F = QvB \sin\theta$
Substitute: $F = (1.6 \times 10^{-19}\text{ C})(3.20 \times 10^6\text{ m/s})(0.0400\text{ T})\sin 90°$
$\quad\quad\quad\quad = 2.05 \times 10^{-14}$ N

PROBLEMS

1. An electron moves horizontally at 6.25×10^6 m/s through a perpendicular magnetic field in the deflection coils of a television. If the flux density is 3.50 mT, determine **(a)** the magnitude of the magnetic force on the electron; **(b)** the acceleration of the electron.

2. When an electron moves at 4.80×10^5 m/s at an angle of 60° with the magnetic field, it experiences a force of 3.60×10^{-16} N. Calculate the magnitude of the flux density.

3. Determine the magnitude of the magnetic force on an electron that moves with a speed of 5.00×10^5 m/s through a perpendicular magnetic field with a 2.50 mT flux density.

4. An electron is moving at 6.25×10^6 m/s perpendicular to a magnetic field where the flux density is 25.0 mT. **(a)** Determine the magnitude of the magnetic force on the electron. **(b)** Compare the magnetic force with the weight of the electron at the earth's surface.

22-5 MAGNETIC FIELDS OF CURRENTS

Electric currents are moving charges and therefore they experience magnetic forces in magnetic fields. However, electric currents also produce magnetic fields. This fact was discovered by Hans Christian Oersted (1777–1851). He observed that the needle of a magnetic compass deflected from its normal north-south alignment when placed near a current-carrying conductor. We can observe the effect by passing a straight wire through a piece of paper which has iron filings scattered on it. If there is no current in the wire, the filings have no fixed pattern, but when the wire carries a current, the filings indicate the presence of a magnetic field and form circular patterns centered at the wire (Fig. 22-14). A magnetic compass can also be used to map the lines of force. The directions of the lines of force can be determined by a right-hand rule. "Hold" the conductor in the right hand with the straightened thumb pointing in the direction of the conventional current. The curled fingers indicate the direction of the magnetic flux density surrounding the conductor (Fig. 22-15).

At any point a distance r from a conductor that carries a current I, the magnitude of the magnetic flux density is given by

$$B = \frac{\mu I}{2\pi r} \quad \text{(long straight wire)} \quad\quad (22\text{-}2)$$

Sec. 22-5 / Magnetic Fields of Currents

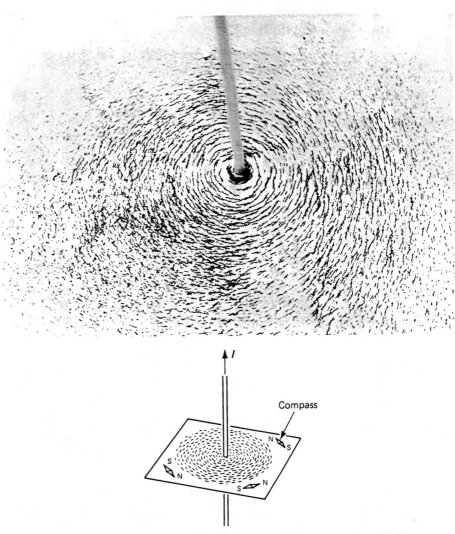

Figure 22-14 Iron filings indicate the lines of force around a straight current-carrying conductor.

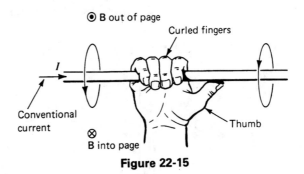

Figure 22-15

The term μ is called the **permeability of the medium;** its value depends on the nature of the material surrounding the conductor. For a vacuum, and approximately for air, the permeability has a value

$$\mu_0 = 4\pi \times 10^{-7} \text{ T} \cdot \text{m/A}$$

The subscript 0 is used to indicate the vacuum.

EXAMPLE 22-2

Determine the flux density in air at a point 4.80 cm from a long straight wire that carries 1.50 A.

Solution

Data: $r = 4.80$ cm $= 0.0480$ m, $I = 1.50$ A, $\mu_0 = 4\pi \times 10^{-7}$ T·m/A, $B = ?$

Equation: $B = \dfrac{\mu I}{2\pi r}$

Substitute: $B = \dfrac{(4\pi \times 10^{-7}\ \text{T·m/A})(1.50\ \text{A})}{2\pi(0.0480\ \text{m})} = 6.25 \times 10^{-6}$ T

Note that the earth's magnetic field is approximately 10^{-5} T; therefore, wires should either be shielded or kept away from any compasses that are used for navigation.

If the current-carrying conductor is bent into a loop, the lines of force due to each segment of the wire are concentrated in the enclosed region, and they are in the same direction, producing a stronger field in the center of the loop (Fig. 22-16).

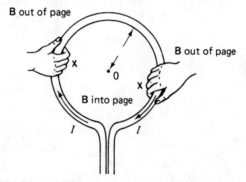

Figure 22-16 Lines of force concentrate in the interior of the loop.

The direction of the magnetic field due to current-carrying loops, coils, or solenoids may also be determined by a different **right-hand rule**: Align the curled fingers of the right hand in the direction of the conventional electric current; the straightened thumb points toward the north pole (Fig. 22-17).

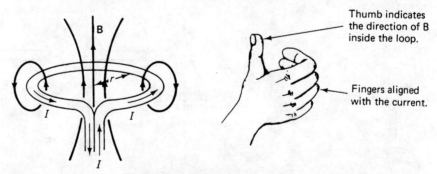

Figure 22-17 Right-hand rule.

The effect of concentration of the magnetic field can be even greater if the loop is made from several turns of the conductor forming a **flat coil** (Fig. 22-18).

A **solenoid** is a relatively common magnetic device that is used to store energy (in its magnetic field), in electrical circuits. It consists of a single wire that is wound

Sec. 22-5 / Magnetic Fields of Currents

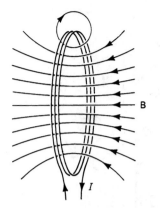

Figure 22-18 Concentration of the magnetic field in a flat coil.

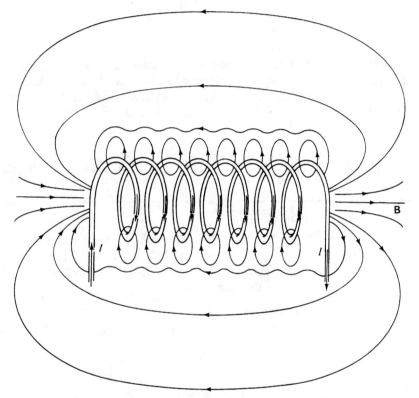

Figure 22-19 Magnetic field due to a current in a long solenoid.

into a helix (Fig. 22-19). Normally, the length L of the solenoid is much larger than its diameter, and the magnetic field in its interior (but not near its ends) is very uniform.

At a point on the interior axis of the solenoid, the magnetic flux density has a magnitude given by

$$B = \frac{N\mu I}{L} \quad \text{(interior axis of a solenoid)} \qquad (22\text{-}3)$$

where μ is the **permeability of the material** inside the solenoid, I the current in the wire, N the number of turns, and L the length of the solenoid. The direction of the flux density is given by the right-hand rule.

One type of solenoid, called a **toroid,** is frequently used in magnetic measurements and in electronic circuits. It consists of a tightly wound coil of wire around

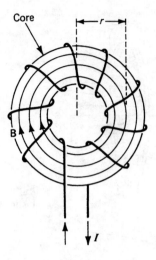

Figure 22-20 Toroid.

a doughnut-shaped material, such as an iron ring (Fig. 22-20). The magnetic field is very uniform inside the ring or **core** of material, and there is almost no magnetic field outside the toroidal core. In this case L is the mean circumference $2\pi r$ in Eq. 22-3, where r is the mean radius of the ring-shaped core. Therefore, the magnetic flux density

$$B = \frac{N\mu I}{L} = \frac{N\mu I}{2\pi r} \quad \text{(toroid)} \tag{22-4}$$

EXAMPLE 22-3

The core of a toroid has a mean radius of 12.5 cm and it is wound with 5000 turns of wire. When the current in the wire is 3.25 A, the magnetic flux density inside the core has a magnitude of 1.40 T. Determine the permeability of the core.

Solution

Data: $N = 5000$, $r = 0.125$ m, $I = 3.25$ A, $B = 1.40$ T, $\mu = ?$

Equation: $B = \dfrac{N\mu I}{2\pi r}$

Rearrange: $\mu = \dfrac{B(2\pi r)}{NI}$

Substitute: $\mu = \dfrac{(1.40 \text{ T})2\pi(0.125 \text{ m})}{(5000)(3.25 \text{ A})} = 6.77 \times 10^{-5}$ T·m/A

Electromagnets

An **electromagnet** consists of a solenoid (or coil) that is usually wound around a high-permeability ferromagnetic core. Any electric current in the solenoid magnetizes the core. Electromagnets are capable of lifting very heavy loads of materials, such as scrap iron and steel, and are used to separate ferromagnetic solids from other materials.

Electromagnets are used in **dynamic loudspeakers** and **microphones** to convert electric impulses into mechanical sound waves, and vice versa (Fig. 22-21). As the electromagnet is magnetized, it attracts a ferromagnetic material which is attached to a speaker cone. When the current is turned off, the electromagnet releases the ferromagnetic material, and the speaker cone returns to its original position.

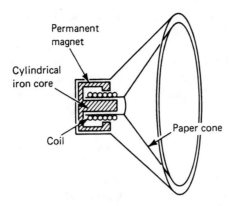

Figure 22-21 Moving-coil speaker. The magnetic field of the time-varying current in the coil interacts with the permanent magnet, causing the coil to move in and out. A paper cone attached to the coil also moves and causes the air to vibrate.

Relays and Switches

A simple **relay** is a type of switch in which an electromagnet is used to control the current in a circuit (Fig. 22-22). When the switch S of the control circuit is closed, the iron core of the electromagnet is magnetized by the current in the solenoid. As the iron armature is attracted to the core of the electromagnet, the armature pivots and closes the contacts, completing the main circuit. If the switch S is opened, the core of the electromagnet becomes demagnetized and the spring breaks the contacts in the circuit. Similar circuits are used for bells and circuit breakers.

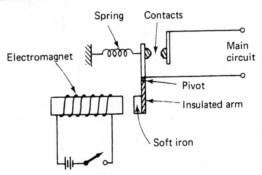

Figure 22-22 Simple relay.

PROBLEMS

5 The magnetic flux density has a magnitude of 6.50 μT in air at a point 6.80 cm from a long straight wire. Find the current in the wire.

6 Determine the magnitude of the magnetic flux density at a point 2.50 m from a long straight wire carrying 15.0 A in air.

7 At what distance from a long straight wire carrying a 12.0 A current will the magnetic flux density be 8.25 μT?

8 What is the magnetic flux density at the earth's surface due to a power line 8.50 m above the ground carrying 625 A?

9 If the magnetic flux density at a distance of 2.50 cm from a current-carrying wire is 2.50 mT, and if there are no other influences, what is the flux density at a distance of 4.20 cm from the same wire?

10 Determine the magnetic flux density at a point on the interior axis of a solenoid that has 85 turns, is 2.75 cm long, and carries 3.50 A if the permeability of the core is **(a)** $4\pi \times 10^{-7}$ T·m/A; **(b)** 3.200×10^{-4} T·m/A.

11 The core of a toroid has a mean diameter of 25.0 cm, and it is wound with 3500 turns of wire. When the current in the wire is 3.20 A, calculate the magnetic flux density in the core if the core is composed of a material that has a permeability of **(a)** $4\pi \times 10^{-7}$ T·m/A; **(b)** 1.50×10^{-5} T·m/A.

12 The core of a toroid has a mean diameter of 12.5 cm and is wound with 125 turns of wire.

If the current in the wire is 750 mA and the core has a permeability of 3.50×10^{-4} T·m/A, calculate the magnitude of the magnetic flux density in the core.

13 Determine the permeability of the core of a toroid that has a mean diameter of 2.85 cm if it has 25 turns and produces a flux density of 28.0 mT when it carries 2.38 A.

14 What current would produce a magnetic flux density of 2.90 mT in a toroid that has a mean diameter of 1.25 cm, 12 turns, and a core with a permeability of 2.50×10^{-4} T·m/A?

22-6 MAGNETIC FORCES ON CURRENTS: MOTOR PRINCIPLE

Moving charges experience magnetic forces, and electric currents are moving charges. Therefore, when a current-carrying conductor is located in a magnetic field it may experience a magnetic force. This can be demonstrated with a simple experiment. If we pass a current through a wire that is suspended in a magnetic field, the wire is deflected to one side (Fig. 22-23). Reversing the current produces a deflection in the opposite direction.

These magnetic forces arise because of the interactions between the applied magnetic field and the magnetic field due to the current in the wire. In Fig. 22-24, for

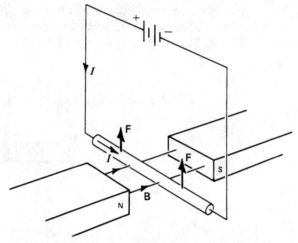

Figure 22-23 Current-carrying conductor in a magnetic field experiences a magnetic force.

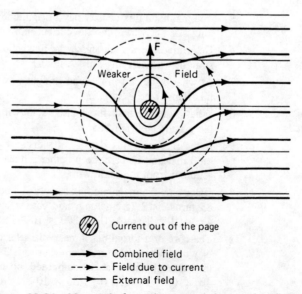

Figure 22-24 Magnetic force is toward the weaker field.

example, the external field is from left to right, and the field caused by the current is counterclockwise. Therefore, the total flux density is reduced in the region above the current (where the fields oppose), and it is increased in the region below the current (where the fields are in the same direction). The magnetic force experienced by the conductor is always directed away from the region where the magnetic flux density is largest toward the region where the flux density is smallest. We often assume that the lines of magnetic force act like stretched rubber bands that try to straighten out.

Consider a wire loop (or a flat coil) that is suspended freely between the poles of a permanent magnet (Fig. 22-25). A current in the loop sets up a magnetic field,

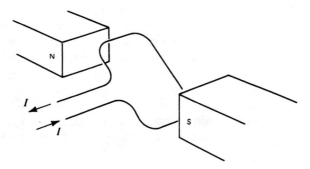

(a) Current-carrying coil in a magnetic field.

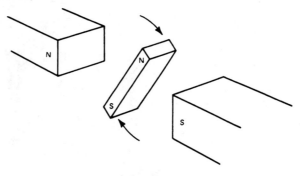

(b) Current in coil sets up a magnetic field; therefore, it behaves like a small magnet in the external field; like poles repel and unlike attract.

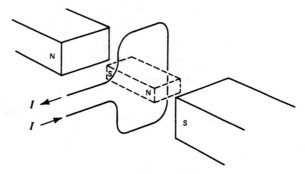

(c) Coil rotates until unlike poles face each other—the plane of the coil is then perpendicular to the lines of force.

Figure 22-25 Deflection of a current-carrying coil in a magnetic field.

and the loop itself then behaves like a small magnet. The forces between the poles of this magnet and the poles of the permanent magnet produce a torque that causes the loop to rotate until the unlike poles face each other and the plane of the loop is perpendicular to the lines of force. The operations of moving-coil meters and electric motors depend on the magnetic torques that are experienced by current-carrying coils in magnetic fields.

The D'Arsonval Galvanometer

A **D'Arsonval galvanometer** is a device that is sensitive to electric currents; its basic structure is illustrated in Fig. 22-26. A wire coil is wound lengthwise around a cylindrical iron core, and the system is mounted on bearings in a cylindrical cavity between the poles of a permanent magnet. Lines of magnetic force concentrate in the iron core, creating a more uniform magnetic field in the air gap between the permanent magnet and the iron core (Fig. 22-27). The magnetic torque depends on the current in the coil. Two spiral springs, one at each end of the iron core, control the deflection of the coil and the pointer. When the coil deflects, it distorts the spiral springs, and they develop a countertorque that is proportional to their distortion. Therefore, the magnitude of the deflection depends on the current in the coil.

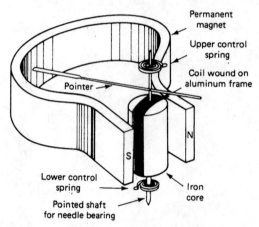

Figure 22-26 Moving-coil galvanometer.

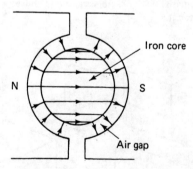

Figure 22-27 Lines of force take the "easiest path" through the soft iron core; as a result, the magnetic field is more uniform in the air gap.

DC Motors

Electric motors are used to convert electrical energy into mechanical energy. The structure of a simple **dc motor** is illustrated in Fig. 22-28. A coil of wire called the **armature** is suspended so that it is free to rotate (and therefore, it is called the **rotor**)

Sec. 22-6 / Magnetic Forces on Currents: Motor Principle

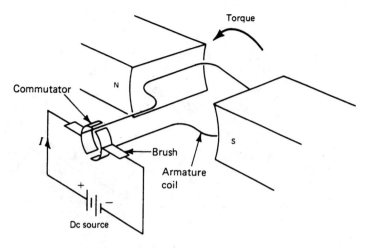

(a) Current-carrying coil experiences a torque

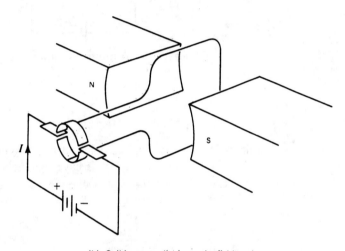

(b) Coil is perpendicular to the field, and the brushes change contacts to the other side of the commutator

Figure 22-28 Dc motor.

in the magnetic field of a permanent magnet located in a frame called a **stator.** The ends of the wire coil are attached to a split-ring conductor called a **commutator,** which makes electrical contact through graphite **brushes** to a dc source, such as a battery or power supply.

As discussed previously, a magnetic torque is exerted on the coil, and it rotates until its plane is perpendicular to the lines of force so that the unlike poles of the coil and permanent magnet face each other. However, when the coil rotates to that point, the brushes change contacts to the other half of the commutator, and this reverses the current and therefore the magnetic field of the coil. As a result, the poles of the coil's field are again repelled by the unlike poles of the permanent magnet, producing a torque that makes the coil continue to rotate. This process is repeated each time the coil aligns perpendicular to the lines of force, and the current in the coil reverses each half revolution of the coil.

The magnitude of the magnetic torque depends on the orientation of the coil. Most dc motors are constructed with armatures that are composed of a number of coils wound with different orientations on iron cores, which increase the magnetic fields of the coils and therefore the torques (Fig. 22-29). In addition, electromagnets, instead of permanent magnets, are frequently used to generate the fields around the armature.

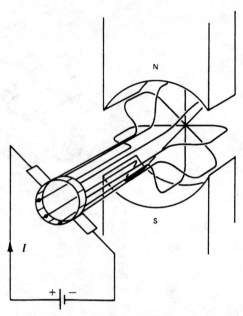

Figure 22-29 A multicoil motor has several coils with different orientations.

SUMMARY

Like poles repel and unlike poles attract. Magnetic lines of force represent magnetic fields. *Ferromagnetic materials* have relatively large magnetic fields.

The *magnetic force* on a moving charge Q: $F = QvB \sin \theta$.

The SI unit for *magnetic flux density (induction)* **B** is the *tesla* (symbol T).

$$B = \frac{\mu I}{2\pi r} \qquad \text{(long straight wire)}$$

$$B = \frac{N\mu I}{L} \qquad \text{(interior axis of solenoid)}$$

$$B = \frac{N\mu I}{L} = \frac{N\mu I}{2\pi r} \qquad \text{(toroid)}$$

Current-carrying conductors experience magnetic forces in magnetic fields.

A simple *dc electric motor* consists of a coil of wire (armature) suspended so that it is free to rotate in the magnetic field of a permanent magnet. Current is supplied to the coil through brushes and a split-ring commutator.

QUESTIONS

1. Define the following terms: **(a)** natural magnet; **(b)** induced magnet; **(c)** north magnetic pole; **(d)** magnetic domain; **(e)** ferromagnetic material; **(f)** magnetic induction or flux density; **(g)** a tesla; **(h)** relay.
2. Describe the structure of the following: **(a)** solenoid; **(b)** toroid; **(c)** electromagnet; **(d)** D'Arsonval galvanometer; **(e)** simple dc motor.
3. How can you tell whether a material is magnetized? Describe a simple experiment to investigate this.

4 What is the difference between induced magnetism and natural magnetism?
5 Gyrocompasses rather than magnetic compasses must be used in the Arctic and Antarctic regions. Explain why.
6 In the northern hemisphere, which end of a surveyor's compass should be the heavier? Explain why.
7 The cases of some watches are made of a ferromagnetic material. Explain why.
8 Describe what happens to the domains of a ferromagnetic material when it is stroked with one end of a bar magnet.
9 Large deposits of iron ore can be discovered with airborne measuring equipment using the principles of the deflections of magnetic compass needles. Explain the principles of this technique.
10 Describe how you would increase the strength of an electromagnet.
11 What is the main function of a circuit breaker?
12 What is the function of the commutator in a dc motor?
13 Explain why a D'Arsonval galvanometer can be used as an ammeter if the current is relatively small.
14 Why are the armature coils of the rotor usually wound on an iron core?
15 What is the main difference between the magnetic properties of copper and iron? Describe the origin of this difference.
16 Care must be taken when one is using a magnetic compass near a wire carrying electric current. Explain why.
17 Wire coils around the neck of a television tube are used to deflect electrons across the screen. Explain the principles involved.
18 In cyclotrons, electric charges are made to move in circles when they enter a region that has a perpendicular magnetic field. Describe in terms of the magnetic forces on the electrons how the effect arises.

REVIEW PROBLEMS

1 An electron moves at 4.75×10^6 m/s through a perpendicular magnetic field where the flux density is 450 mT. Calculate the magnetic force on the electron.
2 An electron moves horizontally due north with a speed of 3.50×10^6 m/s at a point where the earth's magnetic field has a flux density of 250 mT that is also due north but inclined at 30° below the horizontal. Determine the magnetic force on the electron.
3 What is the magnitude of the magnetic flux density if an electron experiences a magnetic force of 2.40×10^{-16} N when it moves at 3.50×10^6 m/s through a perpendicular magnetic field?
4 Determine the magnetic flux density at a distance of 32.0 cm from a straight wire in air that carries a current of 350 mA.
5 If the flux density at a distance of 25.0 cm from a straight current-carrying wire is 450 mT, what is the flux density at 15.0 cm from the same wire?
6 The core of a toroid has a diameter of 12.0 cm, and it is wound with 230 turns of wire. If the current in the wire is 2.50 A and the core has a permeability of 2.20×10^{-5} T·m/A, what is the flux density in the core?
7 The core of a toroid has a permeability of 3.80×10^{-4} T·m/A, a mean diameter of 2.75 cm, and it is wound with 16 turns of wire. What current in the wire will produce a flux density of 18.0 mT in the core?

23
ELECTROMAGNETIC INDUCTION

In 1819, Hans Christian Oersted observed that magnetic fields existed around any conductor that carried an electric current. The reverse effect, where magnetic fields produce currents, is called **electromagnetic induction.** It was first discovered by Joseph Henry in 1831, and then, independently, a few months later, by Michael Faraday.

The operation of several important devices, such as generators, transformers, and inductors, depends on the principles of electromagnetic induction.

23-1 ELECTROMAGNETIC INDUCTION

Faraday first conducted a series of experiments to determine the properties of electromagnetic induction. He found that a current (and emf) is induced in a conductor only if it experiences a magnetic field that varies with time. If the magnetic field is constant, there is no induced current. These time-varying magnetic fields can be produced by three different techniques:

1. Motion of a conductor through a magnetic field. If we connect a wire conductor to a sensitive galvanometer and move the conductor through the magnetic field between the poles of a horseshoe magnet, the galvanometer needle deflects, indicating the presence of an induced current (Fig. 23-1). If the conductor is moved more rapidly through the magnetic field, the induced current increases. The direction of the induced current depends on the direction of motion of the wire through the magnetic field; it can be determined by a **right-hand rule:**

Align the fingers of the right hand parallel to the velocity **v** of the wire so that the palm faces the south magnetic pole. The straightened thumb then gives the direction of the current in the conductor (Fig. 23-2).

No current is induced in the wire if it is held stationary or moved parallel to the lines of magnetic force (Fig. 23-1c). An emf is induced in the conductor only when it cuts through the lines of magnetic force.

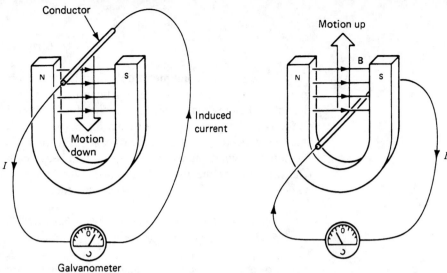

(a) Conductor moves downward; the induced current deflects the galvanometer needle to the right

(b) Conductor moves upward; the induced current deflects the galvanometer needle to the left

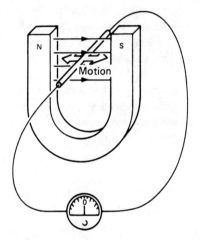

(c) No induced current when the wire is stationary or moves parallel to a line of force.

Figure 23-1

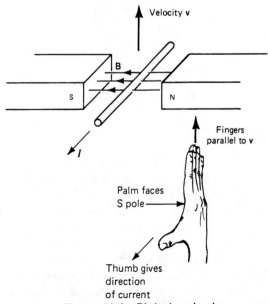

Figure 23-2 Right-hand rule.

2. Motion of a magnetic pole near a conductor. If we connect a coil of wire to a galvanometer and move a magnetic pole toward the coil, the galvanometer needle deflects indicating that a current is induced in the coil. If the magnetic pole is moved in the opposite direction (away from the coil), the galvanometer's deflection shows that the induced current is in the opposite direction. Electric currents are induced in the coil only when there is a relative motion between the magnetic pole and the coil. The direction of the induced current may be predicted by a simple rule that was suggested by Heinrich Lenz (1804–1865). **Lenz's law** states:

The induced electric current is always in a direction so that it sets up a magnetic field that opposes the original change that caused it.

For example, as a north magnetic pole moves toward the coil, the direction of the induced current is such that it sets up a magnetic field that repels the north magnetic pole (Fig. 23-3). The motion of the north pole toward the coil causes the induced current. The magnetic field of the induced current then opposes the motion of the magnetic pole. This is a direct consequence of the conservation of energy. Work must be done, by some external agent, to move the north magnetic pole toward the coil and to create the electric energy. The induced current could not be in the opposite direction, because it would then set up a magnetic field that attracted the north magnetic pole, and no external work would be required to move the pole and create electric energy. In other words, we would have perpetual motion. When the north magnetic pole is moved away, the induced current in the coil develops a south pole to oppose the movement, and we must do work against the magnetic forces to move the pole and create electric energy.

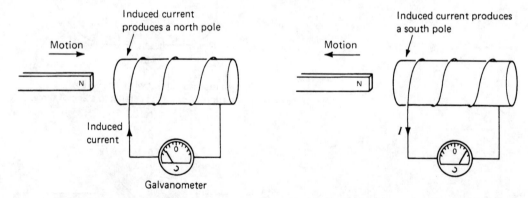

(a) North magnetic pole moves toward the coil, and the induced current sets up a magnetic field that repels the north pole.

(b) As the north pole moves away, the induced current sets up a magnetic field that attracts it.

Figure 23-3

3. Variation of a magnetic field near a conductor. Michael Faraday also investigated the induced current in one coil due to the magnetic field of a current in another adjacent coil (Fig. 23-4). When the switch is closed, a magnetic field builds up while the current in coil 1 (which is called the **primary coil**), increases from zero to some steady value. As the magnetic field of the current in coil 1 increases, the galvanometer indicates that an electric current is induced in coil 2 (the **secondary coil**). In accordance with Lenz's law, the direction of the induced current is such that its magnetic field opposes the changes of the magnetic field due to the changing current in coil 1. When the current in coil 1 is steady, there is no induced current in coil 2.

Sec. 23-1 / Electromagnetic Induction

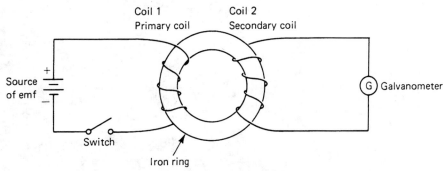

Figure 23-4 Induction.

If the switch is then opened, the current in coil 1 decreases rapidly to zero, and a transient current is induced in coil 2, which tries to maintain the magnetic field.

A current is induced in the secondary coil only when the current in the primary is changing, or when the two coils are moved relative to each other. This is called **mutual induction.** Faraday determined that the magnitude of the induced emf depends on the number of turns in the coils and the time rate of change of the magnetic field.

Self-Induction

While a changing electric current induces emfs in neighboring circuits, it also induces an emf in its own circuit. As the current varies in a single isolated circuit, its magnetic field changes, and there is a corresponding change in the magnetic field surrounding that circuit. This changing magnetic field induces an emf that opposes the original change of the current in the same circuit. It is called **self-induction.** The self-induced emf is known as the **counter** or **back emf.** The SI unit for inductance is called the **henry (H).**

In a circuit that has a self-inductance of 1 H, a time rate of change of current equal to 1 A/s induces a counter emf of 1 V.

Devices called **inductors** are often included in circuits because of their self-inductances. These devices are used to store energy for a period of time in their magnetic fields, and they return the energy to the circuit as the field collapses. Because the induced emf in an inductor opposes the changing magnetic field that causes it, inductors are used to filter or smooth variations in currents. In circuit schematics, inductors are usually represented by the symbols

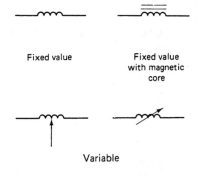

Most common inductors (**coils** or **chokes**) consist of coils of insulated wire. The type of core and the number of windings depend on the frequency at which the inductor is to operate. An inductor that operates at low frequencies usually has many turns of wire on a laminated iron core. Typical inductors are shown in Fig. 23-5.

In addition, the iron is a conductor of electricity, and any change in magnetic

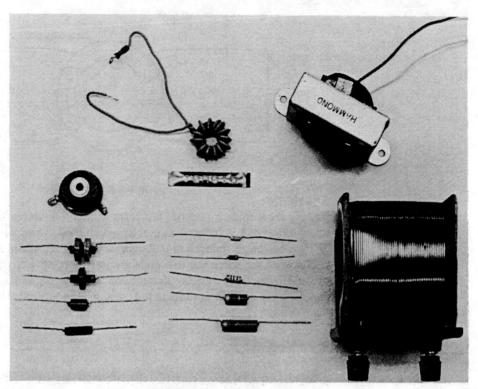

Figure 23-5 Typical inductors.

field normally induces small circulating currents in the iron core. These small currents are called **eddy currents.** They are undesirable, because they generate heat losses in the iron core and set up magnetic fields in opposition to the changing magnetic field that causes them. Eddy currents are normally reduced by using laminated iron cores; a number of sheets of iron are coated with an insulator such as varnish and stuck together with their flat surfaces parallel.

Low-frequency inductors normally have large inductances, and they are physically large. High-frequency inductors have fewer turns of wire and air, powdered iron, or ferrite cores.

23-2 TRANSFORMERS

A **transformer** is a device in which the principles of induction are normally employed to increase or decrease the voltages of alternating currents. It is represented schematically by the symbol

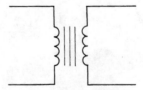

The device consists of two coils, the **primary** and the **secondary,** that are wound on a common laminated iron core (Fig. 23-6). In operation, power is applied to the primary coil and is removed at the secondary coil. As the current in the primary varies, the magnetic field in the iron core changes.

Sec. 23-2 / Transformers

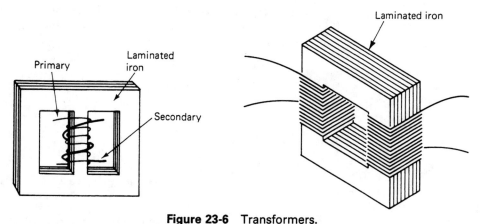

Figure 23-6 Transformers.

In an **ideal transformer** there would be no losses, and the same magnetic field would be present in both the primary and secondary coils. For these devices,

$$\frac{\text{Emf (voltage) at the secondary}}{\text{Emf (voltage) at the primary}} = \frac{\text{number of turns in the secondary}}{\text{number of turns in the primary}}$$

or

$$\frac{\mathscr{E}_s}{\mathscr{E}_p} = \frac{N_s}{N_p} \qquad (23\text{-}1)$$

In a **step-up transformer,** the secondary has more turns than the primary, and the output emf $\mathscr{E}_s$ at the secondary is greater than the input emf $\mathscr{E}_p$ at the primary. If the primary has more turns than the secondary, the output emf $\mathscr{E}_s$ is less than the input emf $\mathscr{E}_p$ and the device is called a **step-down transformer.**

Most transformers have relatively high efficiencies (in excess of 95%). Losses may arise because of the leakage in the magnetic field, the resistance of the windings, eddy currents, and hysteresis in the core. To minimize losses, the coils of high conductivity wire are wound together on a laminated iron core (Fig. 23-6). Hysteresis losses may be reduced by the use of materials with small hysteresis loops, but generally it is more convenient to use high permeability iron to ensure large magnetic fields. In an ideal transformer, which has a 100% efficiency, the output power P_out is equal to the input power P_in; thus

$$P_\text{out} = P_\text{in}$$

or

$$\mathscr{E}_s I_s = \mathscr{E}_p I_p$$

where I_p and I_s are the currents in the primary and secondary, respectively. Therefore,

$$\frac{\mathscr{E}_s}{\mathscr{E}_p} = \frac{I_p}{I_s} = \frac{N_s}{N_p} \qquad (23\text{-}2)$$

When a transformer steps up the voltage, it also reduces the current, and vice versa.

In an **autotransformer** the same coil is used as the primary and the secondary (Fig. 23-7). These devices are economical and efficient, but they have a limited transformation ratio.

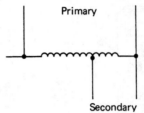

Figure 23-7 Autotransformer.

Transformers are used to step up or step down voltages in the transmission of ac power. Step-down transformers are used to obtain lower voltages and higher currents from a high-voltage line.

EXAMPLE 23-1

An ideal step-down transformer produces a 230 A current at the secondary for use in an arc welder. If the transformer draws 1.15 A at 240 V from the 60.0 Hz mains, determine (a) the ratio of turns in the transformer; (b) the voltage at the secondary.

Solution

Data: $I_p = 1.15$ A, $I_s = 230$ A, $\mathscr{E}_p = 240$ V, $N_s/N_p = ?$, $\mathscr{E} = ?$

(a) Equation: $\dfrac{N_s}{N_p} = \dfrac{I_p}{I_s}$

Substitute: $\dfrac{N_s}{N_p} = \dfrac{1.15 \text{ A}}{230 \text{ A}} = \dfrac{1}{200}$

(b) Equation: $\dfrac{\mathscr{E}_s}{\mathscr{E}_p} = \dfrac{I_p}{I_s}$

Rearrange: $\mathscr{E}_s = \mathscr{E}_p \dfrac{I_p}{I_s}$

Substitute: $\mathscr{E}_s = (240 \text{ V})\left(\dfrac{1.15 \text{ A}}{230 \text{ A}}\right) = 1.20$ V

The Induction Coil

An **induction coil** (Fig. 23-8) is a type of transformer that is used to produce high voltages. The primary coil (a few turns of heavy insulated wire) and the secondary coil (many turns of insulated fine wire) are both wound on a laminated iron core. Current enters the primary coil through the screw B and magnetizes the iron core, which attracts the iron hammer H. As the hammer moves toward the core, the platinum contact A breaks, and the primary current ceases. The iron core is then demagnetized, and the hammer H springs back, closing the primary circuit once more. The making and breaking of the circuit causes rapid changes of the magnetic field. This, combined with the large number of secondary turns, develops large voltages in the secondary at each "make" or "break" of the contact. The capacitor C is used to reduce arcing at contact A.

Sec. 23-3 / Generators

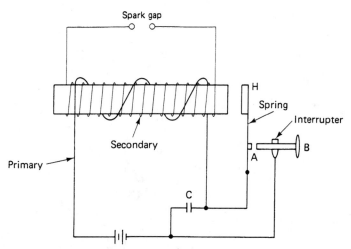

Figure 23-8 Induction coil.

PROBLEMS

1. An ideal transformer is used in an arc welder to produce a 250 A current at 2.0 V at the secondary. If the primary is connected to a 220 V, 60 Hz mains, determine **(a)** the ratio of turns in the transformer, **(b)** the current in the primary.

2. An ideal step-down transformer delivers 30.0 V to a 3.20 Ω load at the secondary. If the primary current is 4.80 A, determine **(a)** the primary voltage; **(b)** the secondary current.

3. A 12.5 MW ac generator delivers power at 220 kV to the 500 turn primary of an ideal step-up transformer. If the secondary coil has 6000 turns, find **(a)** the current; **(b)** the emf induced in the secondary.

4. An ideal transformer with a 540-turn primary and a 25-turn secondary is connected to a 115 V 60 Hz mains. What is the voltage delivered at the secondary?

23-3 GENERATORS

The principles of electromagnetic induction are used in **generators** (or **dynamos**) to convert mechanical energy into electricity.

AC Generators

The structure of a simple **ac generator** or **alternator** is illustrated in Fig. 23-9. A closely wound coil is mounted in a magnetic field of a permanent magnet or an electromagnet so that it is free to rotate about an axis that is perpendicular to the magnetic lines of force. The coil is usually wound on a laminated iron core to increase the magnetic field.

Circular metal conductors called **slip rings** are attached to the terminals of the coil, and graphite **brushes** make electrical contact as they slide over the slip rings. In operation the coil is rotated by the application of mechanical energy, and the induced electric current is delivered through the slip rings and brushes to the external load.

As the coil is rotated, its sides move in opposite directions through the magnetic field, but the induced currents combine in the coil itself. Let us consider what occurs as the coil is rotated through one revolution (360°). When the loop is vertical, the sides move parallel to the lines of force and there is no induced current. In the horizontal position (90° of rotation), the sides of the coil cut through the lines of force at a maximum rate, and the induced current and emf are a maximum in one direction. The

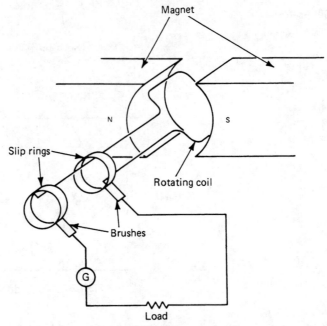

Figure 23-9 Ac generator.

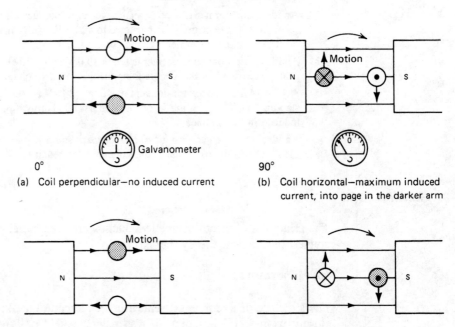

(a) Coil perpendicular—no induced current 0°

(b) Coil horizontal—maximum induced current, into page in the darker arm 90°

(c) Coil perpendicular—no induced current 180°

(d) Coil horizontal—maximum induced current in the opposite direction—out of page in the darker arm 270°

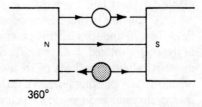

360°

(e) Coil perpendicular as in (a).

Figure 23-10 Induced currents as the coil is rotated.

induced current becomes zero again when the coil has rotated through 180° back to the vertical position since the sides move parallel to the magnetic field once more. As the coil becomes horizontal (after 270° of rotation) the induced current (and emf) is again a maximum, but *in the opposite direction* to that after 90° of rotation. Finally, after 360° the coil is vertical once more, and the induced current is zero.

The induced emf and current are both sinusoidal functions. They continually reverse direction or *alternate* as the coil rotates (Fig. 23-10). One complete rotation of the coil corresponds to a single cycle of the induced emf (Fig. 23-11).

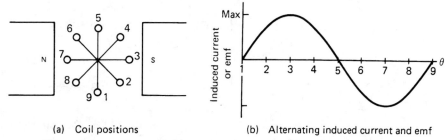

(a) Coil positions   (b) Alternating induced current and emf

Figure 23-11 Relationship between coil positions and induced current.

Many modern ac generators (alternators) are of the **revolving field** type in which the armature coils are located in fixed positions in a frame or **stator** around a rotating magnetic field. Direct current is supplied from a separate source through brushes and slip rings to the field coil windings on the **rotor,** so that the magnetic field rotates with the rotor.

Revolving field generators normally develop larger emfs than rotating coil generators, and they require less insulation because there are only two slip rings and brushes (for the field coils; Fig. 23-12).

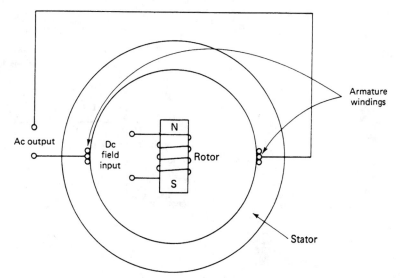

Figure 23-12 Rotating field generator.

DC Generators

A simple ac generator may be converted into a dc generator by replacing the two slip rings with a single split-ring **commutator** (Fig. 23-13). Each terminal of the coil is connected to a different half of the commutator. As the coil rotates, the brushes

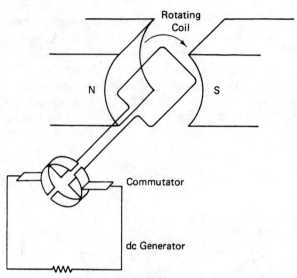

Figure 23-13 Dc generator.

alternately make contact with each of the halves. At the instant when the induced emf reverses polarity, the brushes change contacts on the commutator, so the pulsating potential difference is always in the same direction across the load (Fig. 23-14). To reduce the fluctuations (or **ripple**) in the emf, most dc generators are constructed with a number of different coils that are wound on the armature with their planes at different orientations. As the induced emf in one coil becomes zero, the emf induced in a perpendicular coil is a maximum; therefore, the net emf is much smoother.

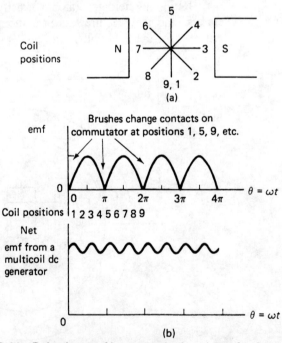

Figure 23-14 Pulsating emf between the brushes of a dc generator.

In dc generators the armature is always wound on the rotor, and the field coils are located in the stator. The stator consists of an annealed steel frame that supports the laminated steel pole pieces of the field coils and acts as a path for the lines of force. Armature windings are located in slots in the cylindrical steel core of the rotor. Brushes make electrical contact with a multisegment commutator.

SUMMARY

An emf and current are only induced in a conductor when the conductor experiences a time-varying magnetic field.

Lenz's law states that the induced electric current is in a direction so that it sets up a magnetic field that opposes the original change that caused it.

As the current in a conductor changes, the time-varying magnetic field that is produced induces an emf in the same conductor. This emf opposes the changes and it is called a *back* or *counter emf*. This is called *self-induction*.

The unit of inductance is called the *henry* (symbol H).

A *transformer* has two coils, the primary and the secondary, wound on the same laminated iron core.

$$\frac{\mathscr{E}_s}{\mathscr{E}_p} = \frac{N_s}{N_p}$$

For *step-up transformers*:

$$\mathscr{E}_s > \mathscr{E}_p \quad \text{and} \quad N_s > N_p$$

For *step-down transformers*:

$$\mathscr{E}_s < \mathscr{E}_p \quad \text{and} \quad N_s < N_p$$

For *ideal transformers*:

$$\frac{\mathscr{E}_s}{\mathscr{E}_p} = \frac{N_s}{N_p} = \frac{I_p}{I_s}$$

A simple *ac generator* consists of a coil mounted so that it is free to rotate in a magnetic field. Currents and emfs are induced in the coil when it is rotated in the field. These currents are delivered through slip rings and brushes to the load.

In a *dc generator* the slip rings are replaced by a single split ring called a *commutator*.

QUESTIONS

1. Define the following terms: **(a)** electromagnetic induction; **(b)** mutual induction; **(c)** self-induction; **(d)** a henry; **(e)** back emf; **(f)** eddy currents; **(g)** step-up transformer; **(h)** step-down transformer; **(i)** commutator; **(j)** rotor.
2. Describe the currents that are induced in a coil of wire **(a)** when a south pole is thrust into the core; **(b)** when the pole is stationary in the core; **(c)** when the south pole is removed from the core.
3. What current is induced in a secondary coil by mutual induction if the current in the primary remains steady? Explain.
4. Describe what is meant by eddy currents.
5. How does an inductor store energy in a circuit?
6. Describe the function of the following generator components: **(a)** brushes; **(b)** slip rings; **(c)** commutator.
7. Simple motors and generators have the same structures. Explain the differences in their operations.
8. How would you increase the output voltage from a generator?

9. Describe the output voltages from the following: (a) an alternator; (b) a simple dc generator; (c) a dc generator with a multisegment commutator.
10. What are the main differences between revolving coil and revolving field generators?
11. Describe the structures and operations of transformers.
12. Describe the structures of (a) a simple ac generator; (b) a dc generator; (c) a revolving field generator.

REVIEW PROBLEMS

1. An ideal step-down transformer produces a 240 A current at its secondary for use in an arc welder. If the transformer draws 1.5 A at 220 V from the 60 Hz mains, determine (a) the ratio of turns in the transformer; (b) the voltage at the secondary.
2. An ideal step-down transformer delivers 115 V to a 120 Ω load at the secondary. If the primary current is 180 A, determine (a) the primary voltage; (b) the secondary current.
3. A 15.0 MW ac generator delivers power at 230 kV to the 300-turn primary of an ideal step-up transformer. If the secondary coil has 7500 turns, determine (a) the current; (b) the emf induced in the secondary.

24
ALTERNATING CURRENT

An **alternating current** (ac) periodically reverses its direction. The free charges vibrate to and fro about fixed positions, so that there is no net transfer of charge around the circuit.

Most of our electricity is generated as alternating current because the time-varying voltage may be increased or decreased in transformers to reduce transmission losses. It is also more versatile and cheaper to produce than direct current.

A large number of devices operate directly with alternating current, but others require direct current. Therefore, special circuits are frequently used to **rectify** the alternating current (i.e., to convert it to direct current).

The normal operation of modern industries and households requires vast quantities of energy. Most of this energy is usually obtained from some distant source, and it is converted into electrical energy before it can be transmitted economically to the consumer. Some energy sources, such as waterfalls and geysers, produce mechanical energy directly. Many other sources, such as nuclear and fossil fuels, are used to produce mechanical energy in heat engines. Generators are then used to convert the mechanical energy into electric energy.

24-1 ALTERNATING CURRENT

Most alternating currents are produced in alternators. The instantaneous emf that is induced in an alternator coil as it rotates in a uniform magnetic field is given by

$$e = \mathcal{E}_p \sin \theta \qquad (24\text{-}1)$$

where $\mathcal{E}_p$ is the maximum or peak value of the induced emf and θ is called the **phase angle** (Fig. 24-1). It depends on the orientation of the coil in the magnetic field.

The time required for one complete cycle is called the **period** T, and the number of cycles per second is the **frequency** f of the alternating emf. In North America the mains frequency is usually 60 Hz.

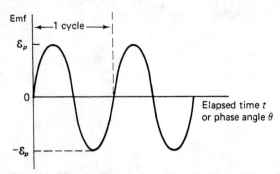

Figure 24-1 Alternating emf.

If an ac voltage is applied to a pure resistance, the current in the resistance and the voltage drop across it are *in phase*. That is, they have maximum values simultaneously, and they pass through zero together (see Fig. 24-2).

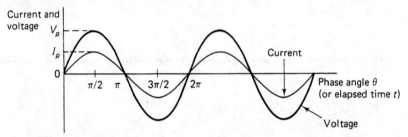

Figure 24-2 Pure resistive circuit. The current is in phase with the voltage.

Effective Current and Voltage

In an alternating current the charges vibrate. There is no *net* flow of charge, and the average current in one complete cycle is zero. However, electric power is still dissipated as heat when an alternating current is applied to a resistor.

The *effective value I* of the alternating current is equivalent to the steady direct current that produces the same average power dissipation in the resistor.

$$P = I^2 R \qquad (24\text{-}2)$$

In terms of the **peak current** I_p (the maximum value during a cycle), the effective current

$$I = \frac{I_p}{\sqrt{2}} = 0.707\, I_p \qquad (24\text{-}3)$$

Similarly, the *effective value V* of the alternating voltage is the steady voltage that produces the same power dissipation in the resistance:

$$P = \frac{V^2}{R} \qquad (24\text{-}4)$$

and in terms of the peak voltage V_p, the effective voltage:

$$V = \frac{V_p}{\sqrt{2}} = 0.707\, V_p \qquad (24\text{-}5)$$

Alternating currents and voltages are normally measured and stated in terms of their effective values.

EXAMPLE 24-1

When a metallic resistance heater is connected to a standard 120 V 60 Hz mains, it dissipates an average power of 180 W. Calculate (a) the peak voltage; (b) the effective current in the heater; (c) the peak current in the heater.

Solution

Data: $V = 120$ V, $P = 180$ W, $V_p = ?$, $I = ?$, $I_p = ?$

(a) Equation: $V = 0.707 V_p$

Rearrange: $V_p = \dfrac{V}{0.707}$

Substitute: $V_p = \dfrac{120 \text{ V}}{0.707} = 170$ V

(b) Equation: $P = IV$

Rearrange: $I = \dfrac{P}{V}$

Substitute: $I = \dfrac{180 \text{ W}}{120 \text{ V}} = 1.50$ A

(c) Equation: $I = 0.707 I_p$

Rearrange: $I_p = \dfrac{I}{0.707}$

Substitute: $I_p = \dfrac{1.50 \text{ A}}{0.707} = 2.12$ A

PROBLEMS

1. Calculate the peak value of an 18.0 A alternating current.
2. Determine the peak voltage of a European 220 V 50 Hz mains.
3. If an oscillator develops a peak voltage of 25.0 V, what effective current would it deliver to a 220 Ω load?
4. A circuit draws 5.00 A from a 120 V 60 Hz mains. Calculate (a) the peak current; (b) the peak voltage in the circuit.
5. If the peak voltage from a 500 Hz oscillator is 12.0 V, what is the effective voltage?
6. (a) Calculate the potential difference that would be indicated on an ac voltmeter when it is connected in parallel with a 120 Ω resistor that dissipates an average power of 60.0 W. (b) What is the effective current?
7. A 300 Ω heater filament is connected to a 120 V 60 Hz mains. How many kilowatt hours of electric energy does it dissipate in 10.0 min?

24-2 INDUCTIVE REACTANCE

Even if a circuit could have an inductance and no electric resistance, the induced emf would still oppose *changes* in the current; thus inductances oppose *alternating* currents. The opposition of an inductance to an alternating current is described by a term called the **inductive reactance** X_L. The inductive reactance of an inductance L is defined as the ratio of the effective voltage V across it to the effective current I through it, or as the ratio of the peak voltage V_p to the peak current I_p. If the frequency of the alternating current is f,

$$X_L = \frac{V}{I} = \frac{V_P}{I_P} = 2\pi f L$$

In SI, inductive reactances are expressed in units of ohms.

EXAMPLE 24-2

Determine (a) the inductive reactance and (b) the effective current when a 30.0 mH inductor with a negligible resistance is connected to a 12.0 V 1.30 kHz oscillator.

Solution

Data: $f = 1.30 \times 10^3$ Hz, $L = 30.0 \times 10^{-3}$ H, $V = 12.0$ V, $X_L = ?$, $I = ?$
(a) Equation: $X_L = 2\pi f L$
Substitute: $X_L = 2\pi(1.30 \times 10^3 \text{ Hz})(30.0 \times 10^{-3} \text{ H}) = 245 \, \Omega$

(b) Equation: $X_L = \dfrac{V}{I}$

Rearrange: $I = \dfrac{V}{X_L}$

Substitute: $I = \dfrac{12.0 \text{ V}}{245 \, \Omega} = 0.0490$ A or 49.0 mA

PROBLEMS

8 Calculate the inductive reactance when a 2.00 H coil is connected to a 220 V 50 Hz source.

9 A pure inductive circuit draws a 220 μA current from a 100 V 500 kHz source. What is the inductive reactance of the circuit?

10 An inductor has an inductive reactance of 500 Ω in a circuit when the ac frequency is 1.00 kHz. What is the inductive reactance when the ac frequency is 50.0 kHz?

11 At what frequency will a 120 μH inductor have an inductive reactance of 240 Ω?

12 What effective and peak currents will a 0.500 H choke draw from a 115 V 60 Hz mains if its resistance and capacitances are negligible?

13 Determine the inductive reactance of a 250 mH source at a frequency of (a) 60 Hz; (b) 7.5 kHz; (c) 1.5 MHz.

14 A 1.50 mH inductor is connected directly to a variable-frequency oscillator that produces a terminal potential difference of 25.0 V. Determine the current in the resistor when the oscillator frequency is (a) 600 Hz; (b) 2.50 kHz; (c) 12 kHz; (d) 4.5 MHz.

15 Determine the inductive reactances and the currents in the following pure inductors if they are connected directly to a 115 V 60 Hz source: (a) 2.00 H; (b) 750 mH; (c) 225 mH.

24-3 CAPACITIVE CIRCUITS

Any region that contains a net electric charge has an electric potential, and it can do work by attracting or repelling other charges. This property is used in devices called **capacitors** which store electric potential energy for periods of time in terms of a potential difference and an electric field.

There are many different types of capacitors (Fig. 24-3), and they have many applications. Most capacitors consist of two or more closely spaced conductors sepa-

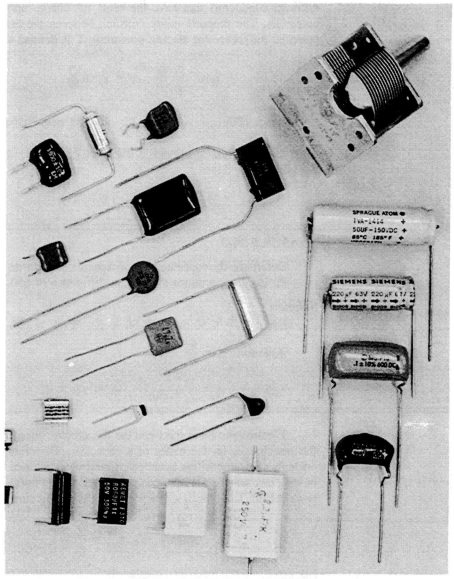

Figure 24-3 Typical capacitors.

rated by insulators. In operation the conductors normally have equal magnitude but opposite polarity (sign) charges. Capacitors are represented by the symbols

$$\text{---}|\text{---} \quad \text{or} \quad \text{---})|\text{---}$$

Fixed

$$\text{---})|^+\text{---}$$

Electrolytic

$$\text{---}\not|\text{---} \quad \text{or} \quad \text{---})\!\!/\text{---}$$

Variable

The electric potential of any conductor is directly proportional to its charge, but different conductors that have the same net charge may have completely different potentials. The proportionality constant between the net charge Q and the electric potential difference (or electric potential), V is defined as its capacitance C:

$$C = \frac{Q}{V} \qquad (24\text{-}7)$$

Capacitance is directly related to the physical size, shape, and nature of the conductors.

The SI unit of capacitance is the **farad** (symbol F). A farad is equal to a coulomb per volt:

$$1 \text{ F} = 1 \frac{\text{C}}{\text{V}}$$

EXAMPLE 24-3

Determine the capacitance of a capacitor if opposite polarity charges of 4.50 mC on its plates produce a potential difference of 650 V.

Solution

Data: $Q = 4.50 \text{ mC} = 4.50 \times 10^{-3}$ C, $V = 650$ V, $C = ?$

Equation: $C = \dfrac{Q}{V}$

Substitute: $C = \dfrac{4.50 \times 10^{-3} \text{ C}}{650 \text{ V}} = 6.9 \times 10^{-6}$ F $= 6.9$ μF

Electrostatic forces between the like electric charges tend to oppose increases in the net charges on the plates of a capacitor. Therefore, capacitors oppose electric currents. This opposition is called **capacitive reactance** X_C. It is defined as the ratio of the peak voltage V to the peak current I, or the ratio of the peak voltage V_p to the peak current I_p. If the frequency of the alternating current is f, then

$$X_C = \frac{V}{I} = \frac{V_p}{I_p} = \frac{1}{2\pi f C} \qquad (24\text{-}8)$$

The unit for capacitive reactance is the ohm.

EXAMPLE 24-4

Determine the capacitive reactance of a 215 pF capacitor in a circuit where the frequency is 1.50 MHz.

Solution

Data: $f = 1.50 \times 10^6$ Hz, $C = 2.15 \times 10^{-10}$ F, $X_C = ?$

Equation: $X_C = \dfrac{1}{2\pi f C}$

Substitute: $X_C = \dfrac{1}{2\pi(1.50 \times 10^6 \text{ Hz})(2.15 \times 10^{-10} \text{ F})} = 494 \ \Omega$

PROBLEMS

16 Determine the charge on a 25 μF capacitor when it is connected to a 120 V power supply.
17 Calculate the capacitance of a capacitor if a 250 μC charge on its plates produces a 35.0 V potential difference.

Sec. 24-4 / AC Series Circuits

18 Determine the capacitive reactance of 1.50 μF capacitor when it is connected to a 110 V 60 HZ source.
19 (a) What is the capacitive reactance of a pure capacitive circuit that draws 50.0 mA from a 110 V 60 Hz source? (b) Calculate the capacitance of the circuit.
20 At what frequency will a 20.0 μF capacitor have a reactance of 220 Ω?
21 What value capacitor will have a reactance of 60.0 Ω at 1.20 kHz?
22 What effective and peak currents are there in a 15.0 nF pure capacitive circuit that is connected to a 120 V 60 Hz mains?

24-4 AC SERIES CIRCUITS

In a **pure inductive circuit** (which has no resistance or capacitance), the induced back emf causes the current to *lag behind* the voltage by a quarter cycle, which is equal to a phase angle of 90°. This relationship is illustrated in Fig. 24-4. The peak current occurs a quarter cycle *after* the peak voltage. Therefore, the current is a maximum when the voltage is zero, and the voltage is a maximum when the current is zero. In practice, inductors have some electric resistance, and the current lags the voltage by some phase angle less then 90°.

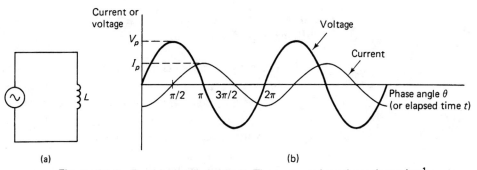

Figure 24-4 Pure inductive circuit. The current *lags* the voltage by $\frac{1}{4}$ cycle or a phase angle of $\pi/2$.

In a **pure capacitive circuit,** which has no resistance or inductance, the net charge and the potential difference is a result of the current. Therefore, the current *leads* the voltage by a quarter of a cycle, or a phase angle of 90°. This relationship is illustrated in Fig. 24-5. The peak current occurs a quarter cycle *before* the peak voltage.

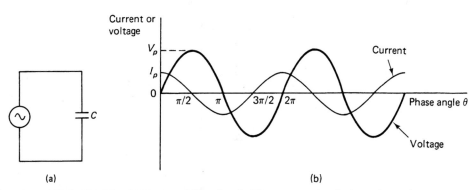

Figure 24-5 Pure capacitive circuit. The current *leads* the voltage by $\frac{1}{4}$ cycle or a phase angle of $\pi/2$.

Impedance

The collective opposition of a resistance R, an inductance L, and a capacitance C (Fig. 24-6) to an alternating current is called the **impedance** Z. It is defined as the ratio of the effective voltage V to the effective current I in the circuit.

$$Z = \frac{V}{I} \qquad (24\text{-}9)$$

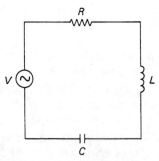

Figure 24-6 Series circuit with resistance R, inductance L, and capacitance C.

The SI unit is the ohm.

Remember, for ac circuits:

1. The current and voltage in a pure resistance are in phase.
2. In a pure inductor, the current lags the voltage by a 90° phase angle. This can be restated as: The voltage leads the current by 90°.
3. In a pure capacitor, the current leads the voltage by a 90° phase angle. Thus the voltage lags the current by 90°.

Using these phase relationships, we may represent the effective voltages by vectors. The potential drop V_R across the resistance R is drawn in the positive x-direction, and the phase angles are measured from this direction (Fig. 24-7). The potential drop V_L across the inductance L is drawn in the positive y-direction, and the potential drop V_C across the capacitance C is drawn in the negative y-direction. Therefore, if V is the total potential difference,

$$V = \sqrt{V_R^2 + (V_L - V_C)^2} \qquad (24\text{-}10)$$

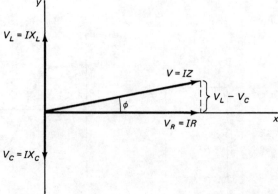

Figure 24-7 Vector diagram of effective voltages.

Sec. 24-4 / AC Series Circuits

The impedance Z can also be found from a vector diagram with the resistance in the x-direction, the inductive reactance X_L in the positive y-direction, and the capacitive reactance X_C in the negative y-direction (see Fig. 24-8). The impedance Z is then the vector sum

$$Z = \sqrt{R^2 + (X_L - X_C)^2} \qquad (24\text{-}11)$$

$$Z = \sqrt{R^2 + \left(2\pi f L - \frac{1}{2\pi f C}\right)^2} \qquad (24\text{-}12)$$

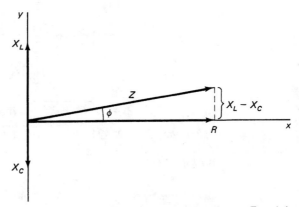

Figure 24-8 Vector diagram to find the impedance Z and the phase angle ϕ.

where f is the frequency. Also, the phase angle ϕ between the current and the voltage is found from:

$$\tan \phi = \frac{X_L - X_C}{R} \qquad (24\text{-}13)$$

In addition,

$$\cos \phi = \frac{R}{Z} \qquad (24\text{-}14)$$

EXAMPLE 24-5

An ac circuit consists of a 215 Ω resistance and a 255 mH inductance in series with a 115 V 60.0 Hz source. Determine **(a)** the inductive reactance; **(b)** the circuit impedance; **(c)** the effective current in the circuit; **(d)** the phase angle between the current and the voltage.

Solution

Sketch: See Fig. 24-9.

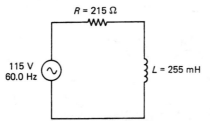

Figure 24-9

Data: $R = 215\ \Omega$, $L = 0.255$ H, $V = 115$ V, $f = 60.0$ Hz, $X_L = ?$, $Z = ?$, $I = ?$, $\phi = ?$

(a) *Equation:* $X_L = 2\pi f L$
Substitute: $X_L = 2\pi(60.0\ \text{Hz})(0.255\ \text{H}) = 96.1\ \Omega$
Sketch: See Fig. 24-10.

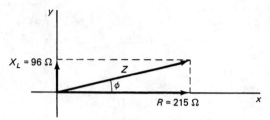

Figure 24-10

(b) *Equation:* $Z = \sqrt{R^2 + X_L^2}$ (no circuit capacitance)
Substitute: $Z = \sqrt{(215\ \Omega)^2 + (96.1\ \Omega)^2} = 236\ \Omega$

(c) *Equation:* $Z = \dfrac{V}{I}$

Rearrange: $I = \dfrac{V}{Z}$

Substitute: $I = \dfrac{115\ \text{V}}{236\ \Omega} = 0.488\ \text{A}$

(d) *Equation:* $\tan\phi = \dfrac{X_L}{R}$

Substitute: $\tan\phi = \dfrac{96.0\ \Omega}{215\ \Omega} = 0.447$

$\phi = \arctan 0.477 = 24.1°$

EXAMPLE 24-6

A circuit consists of a 56.0 Ω resistance and a 2.20 μF capacitance in series with a 35.0 V 1.20 kHz source. Determine **(a)** the capacitive reactance; **(b)** the impedance; **(c)** the effective current; **(d)** the phase angle.

Solution
Sketch: See Fig. 24-11.

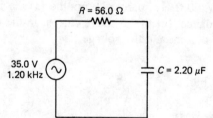

Figure 24-11

Data: $R = 56.0\ \Omega$, $C = 2.20 \times 10^{-6}$ F, $V = 35.0$ V, $f = 1.20 \times 10^3$ Hz, $X_C = ?$, $Z = ?$, $I = ?$, $\phi = ?$

(a) *Equation:* $X_C = \dfrac{1}{2\pi f C}$

Substitute: $X_C = \dfrac{1}{2\pi(1.20 \times 10^3\ \text{Hz})(2.20 \times 10^{-6}\ \text{F})} = 60.3\ \Omega$

Sketch: See Fig. 24-12.

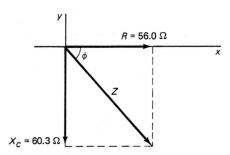

Figure 24-12

(b) *Equation:* $Z = \sqrt{R^2 + X_C^2}$ (no inductance)
Substitute: $Z = \sqrt{(56.0\ \Omega)^2 + (60.3\ \Omega)^2} = 82.3\ \Omega$

(c) *Equation:* $I = \dfrac{V}{Z}$

Substitute: $I = \dfrac{35.0\ \text{V}}{82.3\ \Omega} = 0.425\ \text{A}$

(d) *Equation:* $\tan \phi = \dfrac{X_C}{R}$

Substitute: $\tan \phi = \dfrac{-60.3\ \Omega}{56.0\ \Omega} = -1.08$

$\phi = \arctan(-1.08) = -47.1°$

Note that this phase angle is below the *x*-axis, and therefore it is negative valued.

EXAMPLE 24-7

Determine **(a)** the reactances, **(b)** the impedance, **(c)** the phase angle, and **(d)** the effective current, when a 115 V 60.0 Hz source is connected in series with a resistance of 1.20 kΩ, an inductance of 0.300 H, and a capacitance of 2.20 μF.

Solution

Sketch: See Fig. 24-13.

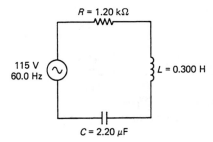

Figure 24-13

Data: $R = 1.20 \times 10^3\ \Omega$, $L = 0.300\ \text{H}$, $C = 2.20 \times 10^{-6}\ \text{F}$, $V = 115\ \text{V}$, $f = 60.0\ \text{Hz}$, $X_L = ?$, $X_C = ?$ $Z = ?$, $\phi = ?$, $I = ?$

(a) *Equations:* $X_L = 2\pi f L$, $X_C = \dfrac{1}{2\pi f C}$

Substitute: $X_L = 2\pi(60.0\ \text{Hz})(0.300\ \text{H}) = 113\ \Omega$

$X_C = \dfrac{1}{2\pi(60.0\ \text{Hz})(2.20 \times 10^{-6}\ \text{F})} = 1.21 \times 10^3\ \Omega$

Sketch: See Fig. 24-14.

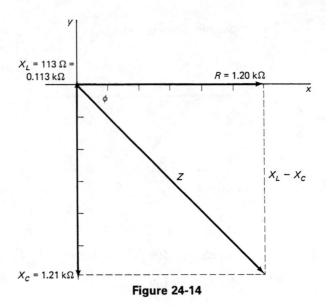

Figure 24-14

(b) *Equation:* $Z = \sqrt{R^2 + (X_L - X_C)^2}$

Substitute: $Z = \sqrt{(1.20 \times 10^3 \, \Omega)^2 + (113 \, \Omega - 1.21 \times 10^3 \, \Omega)^2}$

$\qquad = 1.63 \times 10^3 \, \Omega$

(c) *Equation:* $\tan \phi = \dfrac{X_L - X_C}{R}$

Substitute: $\tan \phi = \dfrac{113 \, \Omega - 1.21 \times 10^3 \, \Omega}{1.20 \times 10^3 \, \Omega} = -0.914$

$\qquad \phi = \arctan(-0.914) = -42.4°$

(d) *Equation:* $I = \dfrac{V}{Z}$

Substitute: $I = \dfrac{115 \, \text{V}}{1.63 \times 10^3 \, \Omega} = 7.06 \times 10^{-2} \, \text{A}$

PROBLEMS

23 A circuit consists of a resistor and a pure inductor in series with an ac source. If the effective potential drops across the resistor and inductor are 30.0 V and 40.0 V, determine **(a)** the voltage of the source; **(b)** the phase angle.

24 A 0.200 H coil and a 122 Ω resistance are connected in series with a 30.0 V, 500 Hz source. Calculate **(a)** the impedance; **(b)** the phase angle; **(c)** the effective current in the circuit.

25 Determine the inductive reactance, the impedance, the effective current, and the phase angle of the series circuits with the following components: **(a)** a 256 Ω resistance and a 135 mH inductance with a 115 V, 60.0 Hz source; **(b)** a 2.20 kΩ resistance and a 235 mH inductance with a 25.0 V 1.20 kHz source; **(c)** a 47.0 kΩ resistance and a 125 μH inductance with a 25.0 V 2.30 MHz source.

26 A 0.300 H inductor draws 120 mA from a 60.0 V 250 Hz source. Determine **(a)** the reactance; **(b)** the resistance of the inductor.

27 An circuit consists of a resistor and a capacitor in series with an ac source. If the effective potential drops across the resistor and capacitor are 120 V and 50.0 V, determine **(a)** the effective voltage of the source; **(b)** the phase angle.

Sec. 24-5 / Resonance

28. Determine (a) the impedance, (b) the phase angle, and (c) the effective current when a 220 pF capacitor and a 470 kΩ resistor are connected in series with a 250 V 1.50 kHz source.

29. Determine the capacitive reactances, impedances, phase angles, and currents in the ac series circuits with the following components: (a) a 465 Ω resistance and a 4.70 μF capacitor with a 115 V 60.0 Hz source; (b) a 56.0 Ω resistance and a 1.20 μF capacitor with a 115 V 3.20 kHz source; (c) a 15.0 Ω resistance and a 25.0 μF capacitor with a 115 V 60.0 Hz source.

30. Determine (a) the impedance, (b) the phase angle, and (c) the effective current in an ac series circuit containing a 220 V 50.0 Hz source with a resistance of 220 Ω, an inductance of 2.00 H, and a capacitance of 50.0 μF.

31. A circuit has a 105 resistance, a 12.0 μF capacitance, and a 9.50 mH inductance in series with a 45.0 V 235 Hz source. Determine (a) the impedance; (b) the phase angle; (c) the effective current.

32. A circuit contains a 56.0 Ω resistance, a 35.0 μF capacitance, and a 535 mH inductance in series with a 25.0 V 50.0 Hz source. Determine (a) the impedance; (b) the phase angle; (c) the effective current.

33. A 515 nF capacitor, a 345 Ω resistance, and a 145 mH inductance are connected in series with a 500 Hz ac source. If the circuit draws an effective current of 250 mA, determine (a) the impedance; (b) the voltage of the source; (c) the phase angle.

24-5 RESONANCE

In a circuit containing inductors and capacitors, the capacitance and inductance of the circuit both store energy, but during different half-cycles of the alternating current. As the energy stored in the electric field of the capacitance increases, the energy stored in the magnetic field of the inductance decreases, and vice versa. In a circuit containing inductors, capacitors, and resistors in series with an ac source, if the inductive reactance is equal to the capacitive reactance, there is a maximum effective current. That is,

$$I = \frac{V}{Z} = \frac{V}{\sqrt{R^2 + (X_L - X_C)^2}}$$

is a maximum value when $X_L = X_C$. In this case

$$Z = R \quad \text{and} \quad I = \frac{V}{R}$$

This condition is called **electric resonance**. When resonance occurs,

$$2\pi f_0 L = \frac{1}{2\pi f_0 C}$$

and the **resonant frequency** of the applied voltage is

$$f_0 = \frac{1}{2\pi\sqrt{LC}} \tag{24-15}$$

At this frequency the inductance and the capacitance merely exchange energies, they do not draw energy from the source.

Circuits that have specific ac frequencies are called **oscillators**.

Resonant circuits have important applications in communications systems. They are used to produce electric oscillations, and to detect modulated carrier waves in antenna circuits. Different frequencies may be selected by varying the inductance or the capacitance.

EXAMPLE 24-8

Determine the resonant frequency of an ac series antenna circuit that has a resistance of 20.0 Ω, an inductance of 280 μH, and a capacitance of 70.0 pF.

Solution

Data: $R = 20.0 \ \Omega$, $L = 280 \ \mu H$, $C = 70.0 \ pF$, $f_o = ?$

Equation: $f_0 = \dfrac{1}{2\pi\sqrt{LC}}$

Substitute: $f_0 = \dfrac{1}{2\pi\sqrt{(2.80 \times 10^{-4} \ H)(7.00 \times 10^{-11} \ F)}}$

$= 1.14 \times 10^6 \ Hz = 1.14 \ MHz$

PROBLEMS

34. Determine the resonant frequency of an ac series circuit that has a 7.5 Ω resistance, an inductance of 27 μH, and a capacitance of 300 pF.
35. What inductance must be placed in series with a 20.0 μF capacitor to produce a circuit with a resonant frequency of 3.00 kHz?
36. What capacitance must be connected in series with a 500 μH coil to produce a circuit with a resonant frequency of 1.50 MHz?

24-6 AC POWER TRANSMISSION

At present there are four methods that are used to produce most commercial electric power;

1. **Internal combustion engines**, such as diesels (Fig. 25-15), in which fuel is burned and the rotary output is used to turn the rotor of a generator. This type of power plant is used at isolated locations or in regions that have vast quantities of natural gas or fuel oil.

Figure 24-15 Diesel-powered generators. (Courtesy of Siemens.)

2. **Fossil fuels**, such as coal, oil, or natural gas, are used to heat water, producing high-pressure steam in a boiler. The steam then flows past the blades of a turbine into a low-pressure area, and the rotation of the turbine is transferred to a generator.
3. In **hydroelectric power plants**, a dam is used to create a large reservoir of water (Fig. 24-16). As the water is allowed to fall down a **penstock**, its potential energy is converted into kinetic energy, and at the bottom the fast-moving water turns the blades of a turbine. This rotary motion is then transferred to a generator (Fig. 24-17).

Figure 24-16 Typical hydroelectric dam. (Courtesy B.C. Hydro).

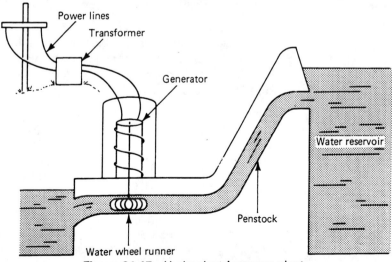

Figure 24-17 Hydroelectric power plant.

4. A **nuclear reactor** can be used to generate a considerable amount of heat, which superheats steam to high pressures. The steam then passes through the blades of a turbine, the shaft of which is connected directly to the armature of a generator.

Wind generators are used experimentally (Fig. 24-18).

Figure 24-18 Experimental wind generator (Courtesy B.C. Hydro).

Once the electric energy has been produced, it is normally transmitted from one place to another in a conductor called a **transmission (power) line** (Figs. 24-19 and 24-20). However, there are a number of ways in which energy is lost during the transmission process:

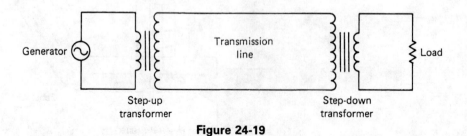

Figure 24-19

Sec. 24-6 / AC Power Transmission

Figure 24-20 Typical power transmission lines. (Courtesy B.C. Hydro).

1. **Line resistance losses.** With the exception of superconductors, all materials have a measurable electric resistance. Consequently, some electric energy is lost during the transmission process. If R is the resistance of the transmission line, and I is the current that it carries, the rate at which electric energy is dissipated as heat is

$$P_{loss} = I^2 R \qquad (24\text{-}16)$$

These losses are usually reduced by using a step-up transformer to decrease the current *before* the energy is transferred to the transmission line. At the receiver the current may be increased once more in step-down transformers (Fig. 24-21).

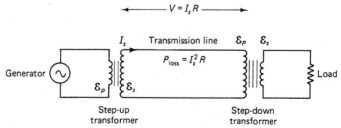

Figure 24-21 Transmission of electric energy from the source to the load.

EXAMPLE 24-9

Determine the transmission losses **(a)** if a 6.00 MW generator delivers power at 20.0 kV directly to a 60.0 Ω transmission line; **(b)** if an ideal (100% efficient) transformer is used to step up the voltage to 230 kV before transmission.

Solution

(a) *Data:* $P_{gen} = 6.00$ MW, $V = 20.0$ kV, $R = 60.0$ Ω, $P_{loss} = ?$

Equations: $P_{gen} = IV$, $P_{loss} = I^2R$

Rearrange: The current delivered to the transmission lines $I = \dfrac{P}{V}$

Substitute: $I = \dfrac{6.00 \times 10^6 \text{ W}}{20.0 \times 10^3 \text{ V}} = 300$ A

Therefore, the power lost in the line

$$P_{loss} = (300 \text{ A})^2(60.0 \text{ Ω}) = 5.40 \times 10^6 \text{ W} = 5.40 \text{ MW}$$

Note that only 6.00 MW is generated; therefore, only 0.60 MW is delivered to the load, 90% of the energy is lost in transmission.

(b) *Sketch:* See Fig. 24-22.

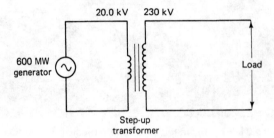

Figure 24-22

Data: $I_P = 300$ A [part (a)], $\mathscr{E}_p = 20.0$ kV, $\mathscr{E}_s = 230$ kV, $P_{loss} = ?$, $I_s = ?$

Equations: $\dfrac{\mathscr{E}_s}{\mathscr{E}_p} = \dfrac{I_p}{I_s}$ and $P_{loss} = I_s^2 R$

Rearrange: $I_s = \dfrac{I_p \mathscr{E}_p}{\mathscr{E}_s}$

Substitution: $I_s = \dfrac{(300 \text{ A})(20.0 \text{ kV})}{230 \text{ kV}} = 26.1$ A

Thus the power loss in the transmission line

$$P_{loss} = I_s^2 R = (26.1 \text{ A})^2(60.0 \text{ Ω}) = 4.08 \times 10^4 \text{ W} = 0.040 \text{ MW}$$

In this case only 0.68% of the generated power is lost in the transmission. Obviously, it is beneficial to step up the voltage before the transmission!

2. **Transformer losses.** While transformers are normally quite efficient, there are losses due to hysteresis and eddy currents in the core, magnetic field leakage, and the resistance of the windings.

3. **Leakage losses.** There is an upper limit to the transmission voltage, because high voltage ionizes the air surrounding the transmission lines, and there is some leakage of current. These are greater in damp climates.

4. **Power factor.** The inductance and capacitance of the load may cause more power to be drawn through the transmission line than is used. As a result, power is sent back to the transmitter, with the usual losses in the line. This effect can be compensated for by ensuring that the inductive reactance is equal to the capacitive reactance: $X_L = X_C$.

A typical ac power transmission network is illustrated in Fig. 24-23. In North America, electric power is produced at a frequency of 60 Hz and a voltage of about 20 kV at the generating station. The voltage is stepped up to 230 kV or 500 kV in a

Sec. 24-6 / AC Power Transmission

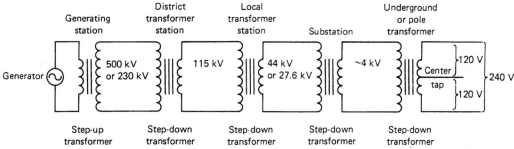

Figure 24-23 Typical ac power transmission network.

transformer before transmission through the familiar transmission lines. **District transformer stations** act as receivers for geographic zones. These stations step down the voltage to 115 kV and retransmit through lines to **local stations** that serve each community. The local stations step down the voltage to 44 kV or 27.6 kV, and transmit the electric energy to **substations** in each neighborhood, where the voltage is again stepped down to about 4 kV. These substations send the electricity through wires to underground transformers or hydro-pole transformers (Fig. 24-24), where it is stepped down to 240 V and delivered to the homes. A **center tap transformer** is used to produce both 240 V (for ovens, electric heaters, etc.) and 120 V (for lights and wall plugs) in the home.

Figure 24-24 Pole transformers.

PROBLEMS

37 A 50.0 kW generator delivers power at 2.00 kV to a 40.0 Ω transmission line. Calculate the power lost in transmission **(a)** if the generator delivers the power directly to the line; **(b)** if an ideal transformer (100% efficient) steps up the potential difference to 25.0 kV before the power is supplied to the line. **(c)** Determine the ratio of the number of turns in the coils of the step-up transformer.

38   A power transmission system consists of a 300 kW generator that delivers power at 1.50 kV to a 300-turn primary coil of an ideal step-up transformer. The 6000-turn secondary coil of the transformer delivers power to a 60.0 Ω transmission line. At the receiver, the energy is passed into an ideal step-down transformer, and the secondary coil produces 15.0 kV potential difference across the load. If there are no leakage losses, calculate (a) the voltage drop in the transmission line; (b) the power loss in the line; (c) the power delivered to the receiver; (d) the current in the load; (e) the ratio of the turns in the step-down transformer.

SUMMARY

Alternating emf: $e = \mathcal{E}_p \sin \theta$
   Effective current: $I = 0.707\, I_P$ *effective voltage* $V = 0.707\, V_P$
   Inductive reactance: $X_L = 2\pi f L$
   Capacitance $= \dfrac{\text{charge}}{\text{potential difference}}$ $C = \dfrac{Q}{V}$.
   The *fared* (F) is the SI unit for capacitance.
   Capacitive reactance: $X_C = \dfrac{1}{2\pi f C}$

In a resistance the current is in phase with the voltage. The current lags the voltage by a 90° phase angle in a pure inductor, and it leads the voltage by a 90° phase angle in a pure capacitor.

   Impedance: $Z = \dfrac{V}{I} = \sqrt{R^2 + (X_L - X_C)^2}$ $\tan \phi = \dfrac{X_L - X_C}{R}$

For *resonance*, $X_L = X_C$ and the current is a maximum. The resonant frequency,

$$f_0 = \dfrac{1}{2\pi \sqrt{LC}}$$

In *transmission lines*, $P_{\text{loss}} = I_{\text{line}}^2 R$.

QUESTIONS

1   Define the following terms: (a) alternating current; (b) period; (c) frequency; (d) effective current; (e) effective voltage; (f) inductive reactance; (g) capacitance; (h) capacitive reactance; (i) a farad; (j) impedance; (k) phase angle; (l) resonant frequency; (m) oscillator; (n) leakage losses.
2   Describe the effects of changing the frequency on the inductive and capacitive reactances.
3   An oscillator is a resonant circuit in which the inductance and the capacitance exchange energies. What oscillates? Describe how this circuit functions.
4   What are the functions of transformers in the transmission of electric power?
5   List and describe the common methods that are used to generate electric power.
6   Describe the various mechanisms that result in a loss in electric energy during the transmission from the generator to the home.
7   Describe how inductors and capacitors oppose alternating electric currents.

REVIEW PROBLEMS

1   A 100 W bulb is connected to a 120 V 60 Hz mains. Calculate (a) the peak voltage; (b) the effective current; (c) the peak current in the filament.
2   A motor of a power saw draws 8.00 A from a 120 V 60 Hz mains. Determine (a) the peak current in the circuit; (b) the average power dissipated by the motor.

Review Problems

3 If the peak voltage of a 1.50 kHz audio oscillator is 18.0 V, calculate the effective voltage.

4 A 15.0 Ω heater filament is connected to a 120 V 60 Hz mains. (a) How many kilowatt hours of energy does it dissipate in 3.00 h? (b) If electricity costs 6.00 cents per kilowatt hour, how much does it cost to run the heater for 3.00 h?

5 Calculate the inductive reactance when an 80.0 mH choke is connected to a 30.0 V 60 Hz source.

6 A pure inductive circuit draws 30.0 mA from a 5.00 V 250 Hz source. What is the inductance of the circuit?

7 An inductor has an inductive reactance of 392 Ω at 1.50 kHz. What is the inductive reactance at 700 Hz?

8 Determine the capacitance of a capacitor if a charge of 32.0 μC produces a potential difference of 2.50 V.

9 Determine the charge on a 2.20 μF capacitor when there is a 25.0 V potential difference between its plates.

10 What is the capacitive reactance of a pure capacitive circuit that has a capacitance of 200 μF when it is connected to a 220 V 50 Hz mains?

11 Determine (a) the capacitive reactance and (b) the capacitance of a pure capacitive circuit that draws 300 mA from a 120 V 60 Hz mains.

12 If a capacitor has a capacitive reactance of 620 Ω at 300 Hz, what is its capacitive reactance at 1.20 kHz?

13 A circuit consists of a resistor and a pure inductor in series with an ac source. If the potential drops across the resistor and inductor are 15.0 V and 26.0 V, determine (a) the voltage of the source; (b) the phase angle.

14 An 80.0 mH choke and a resistance of 59.0 Ω are connected in series with a 15.0 V 60.0 Hz source. Determine (a) the impedance; (b) the phase angle; (c) the effective current in the circuit.

15 A 30.0 mH inductor draws 20.0 mA when it is connected in series with a resistance and a 12.0 V 120 Hz source. Determine (a) the impedance; (b) the inductive reactance; (c) the resistance of the curcuit.

16 Determine (a) the impedance, (b) the phase angle, and (c) the effective current when a 315 nF capacitor and a 560 Ω resistance are connected in series with a 30.0 V 1.20 kHz source.

17 A 585 nF capacitor, a 415 Ω resistance, and a 182 mH inductance are connected in series with a 115 V 705 Hz source. Determine (a) the impedance; (b) the phase angle; (c) the effective current.

18 A 625 nF capacitor, a 375 Ω resistance, and a 125 mH inductance are connected in series with a 725 Hz source. If the circuit draws an effective current of 185 mA, determine (a) the impedance; (b) the phase angle; (c) the effective voltage of the source.

19 Determine the resonant frequency of an ac series circuit that has a 120 Ω resistance, an inductance of 30.0 μH, and a capacitance of 200 pF.

20 What inductance must be connected in series with a 200 pF capacitor to produce a circuit with a resonant frequency of 25.0 kHz?

21 A 750 kW ac generator delivers power to a 500 Ω transmission line. Calculate the power loss in the line (a) if the generator delivers power at 50.0 kV directly to the line; (b) if an ideal transformer steps-up the voltage to 250 kV before the power is delivered to the line. (c) What is the ratio of the turns in the transformer?

Appendix A
TECHNICAL MATHEMATICS

This section is intended for use as required. Check with your instructor to determine the topics that you should study.

A-1 SIGNED NUMBERS

We must be very familiar with the mathematical operations of signed numbers. There are a number of simple rules:

Rule 1: Addition of a negative number is equivalent to a subtraction.

EXAMPLE A-1

$$8 + (-6) = 8 - 6 = 2$$
$$24 + (-20) + (-8) = 24 - 20 - 8 = -4$$

Rule 2: Subtraction of a negative number is equivalent to addition.

EXAMPLE A-2

$$12 - (-8) = 12 + 8 = 20$$
$$-15 - (-7) = -15 + 7 = -8$$
$$-8 - (5) = -8 - 5 = -13$$

Rule 3: When *two* numbers are multiplied the result is positive valued if their signs are alike and negative valued if the signs are different.

Sec. A-1 / Signed Numbers

EXAMPLE A-3

$12 \times 3 = 36$	Both signs are positive.
$(-8) \times (-3) = 24$	Both signs are negative.*
$(-9) \times (8) = -72$	Unlike signs.

Rule 4: To multiply more than two signed numbers, the answer is negative valued if there are an *odd* number of negative numbers; otherwise, the result is positive valued.

EXAMPLE A-4

$(-8)(-2)(-7)(4) = -448$	The result is negative because there is an odd number (three) of minus signs.
$(-2)(-4)(7)(-3)(-8) = 1344$	This answer is positive because there is an even number (four) of negative values.

Rule 5: For the division by signed numbers, the result is positive valued if there is an even number of negative signs and negative valued if there is an odd number of negative signs.

EXAMPLE A-5

$\dfrac{-8}{-2} = 4$	An even number (two) of minus signs.
$\dfrac{-27}{3} = -9$	Odd number (one) of minus signs.
$\dfrac{(-8)(-5)}{-4} = -10$	Odd number (three) of minus signs.
$\dfrac{(4)(-8)(-3)}{(-6)(-2)} = 8$	Even number (four) of minus signs.

PROBLEMS

1 Solve.
  (a) $(-12) + 16$
  (b) $(-8) + (-4)$
  (c) $(-9) + (-7) + (-4)$
  (d) $16 + (-9)$
  (e) $(-8) + (-12)$
  (f) $-14 - (-12)$
  (g) $21 - (-24)$
  (h) $(-12)(-8)$
  (i) $(-3)(4)(8)$
  (j) $(-6)(-7)(-2)$
  (k) $\dfrac{-18}{3}$
  (l) $\dfrac{16}{-2}$
  (m) $\dfrac{-36}{-4}$
  (n) $\dfrac{(-27)(2)(-12)}{(-3)(-8)}$
  (o) $\dfrac{(-16)(-9)(-7)}{(-3)(-2)}$

*Multiplication can also be written without the $\times$ sign by using parentheses, as in $(-8)(-3) = 24$.

2   A tank of water has an initial temperature of 23°C. Determine the final temperature if its temperature (a) increases by 15°C; (b) decreases by 8°C; (c) decreases by 75°C.
3   A steel girder expands by 0.18 cm for each 1°C increase in its temperature. By how much does it expand when its temperature changes from 22°C to 45°C?
4   A steel beam is attached so that it has room for a 7.20 mm expansion and a 5.40 mm contraction at 23°C. If the beam expands when heated and contracts when cooled by 0.120 mm for each 1°C temperature change. Determine the maximum and minimum allowable temperatures.

A-2 EXPONENTS

Exponents are frequently used to express numbers in an abbreviated form without changing the value.

Rule: If a number is written as x^n, it is defined to be equivalent to x multiplied by itself n times. The number x is called the **base** and the number n is called the **exponent**.

EXAMPLE A-6

$2^3 = 2 \times 2 \times 2 = 8$ Base 2, exponent 3; 2 multiplied by itself 3 times.

$3^5 = 3 \times 3 \times 3 \times 3 \times 3 = 243$ Base 3, exponent 5

$(-2)^4 = (-2)(-2)(-2)(-2) = 16$ Base -2, exponent 4

$(-3)^3 = (-3)(-3)(-3) = -27$ Base -3, exponent 3

Multiplication

Rule: To multiply two numbers *with the same base*, the result is the *same base* raised to an exponent which is the *algebraic sum* (using the rules for signed values) of the original exponents.

$$(x^n)(x^m) = x^{n+m}$$

EXAMPLE A-7

$$2^3 \times 2^5 = 2^{3+5} = 2^8$$

That is,

$$(2)(2)(2) \times (2)(2)(2)(2)(2) = 2^8 = 256$$

Division

Rule: To divide two numbers with the same base, the result is the same base with an exponent equal to the exponent of the numerator minus that of the denominator.

$$\frac{x^m}{x^n} = x^{m-n}$$

EXAMPLE A-8

$$\frac{2^5}{2^3} = 2^{5-3} = 2^2 = 4$$

Sec. A-2 / Exponents

That is, $$\frac{(2)(2)(2)(2)(2)}{(2)(2)(2)} = (2)(2) = 2^2 = 4$$

Negative Exponents

Rule: Comparing the rules for multiplication and division, we see that to move a number from the numerator to the denominator, or vice versa, we merely change the sign of its exponent.

$$\frac{1}{x^n} = x^{-n}$$

since

$$\frac{x^m}{x^n} = x^{m-n} \quad \text{and} \quad (x^m)(x^{-n}) = x^{m+(-n)} = x^{m-n}$$

are equivalent using the rules for algebraic addition.

EXAMPLE A-9

$$\frac{1}{2^3} = 2^{-3} \quad \text{and} \quad \frac{1}{8^{-2}} = 8^2$$

Zero Exponents

Rule: Any number raised to a zero exponent is equal to 1.

$$x^0 = 1 \quad \text{since} \frac{x^n}{x^n} = x^{n-n} = x^0 = 1$$

EXAMPLE A-10

$$\frac{2^3}{2^3} = \frac{(2)(2)(2)}{(2)(2)(2)} = 1 \text{ or } 2^{3-3} = 2^0 = 1$$

Roots

Rule: From the rules of multiplication and division it can also be seen that we can write the nth root of a number as that number with an exponent of $1/n$.

$$\sqrt[n]{x} = x^{1/n}$$

since the nth root implies that n times the root yields the specified number.
For example,

$$(x^{1/3})(x^{1/3})(x^{1/3}) = x^{1/3+1/3+1/3} = x$$

EXAMPLE A-11

$$\sqrt[4]{16} = 16^{1/4} = 2$$

and

$$2^4 = (2)(2)(2)(2) = 16$$

Four times the root gives the original number, 16.

Powers of Numbers in Exponential Form

Rule: When a number in exponential form is raised to some power it is equivalent to the original base raised to the product of the two powers.

$$(x^m)^n = x^{mn}$$

EXAMPLE A-12

$$(2^3)^2 = 2^{3 \times 2} = 2^6 = 64$$

since

$$(2^3)^2 = [(2)(2)(2)]^2 = 8^2 = 64$$

Also,

$$(16^{1/4})^2 = 16^{(1/4)(2)} = 16^{1/2} = \sqrt{16} = 4$$

$$(2^{-2})^3 = 2^{(-2)(3)} = 2^{-6} = \frac{1}{2^6} = \frac{1}{64}$$

PROBLEM

5 Simplify.
(a) 3^2
(b) $27^{1/3}$
(c) $(-2)^3$
(d) $81^{1/4}$
(e) $3^3 \times 3^4$
(f) $5^8 \times 5^{-3}$
(g) $6^{-3} \times 6^{-5}$
(h) $8^{1/3} \times 8^4$
(i) $\dfrac{6^3}{6^5}$
(j) $\dfrac{a^7}{a^3}$
(k) $\dfrac{12^4}{12^3}$
(l) $\dfrac{4^8 \times 4^3}{4^{11}}$
(m) $\dfrac{8^3}{8^{-5}}$
(n) $\dfrac{4^3}{2^6}$

A-3 ROUNDING OFF AND SIGNIFICANT FIGURES

There are actually several different techniques for rounding off a number to the correct number of significant figures, but for simplicity we shall adopt the following conventions:

Rule: The last digit to be retained is left unchanged if the following digit is less than 5.

Rule: The last digit to be retained is increased by one if the following digit is 5 or greater.

EXAMPLE A-13

Round off the following numbers to three significant figures: (a) 3.682; (b) 0.000 059 68; (c) 4.6925×10^8; (d) 4.999×10^5.

Solution: In each case the third significant digit is indicated by a caret $_\wedge$.
   (a) 3.682 becomes 3.68 since the digit 2 is less than 5.
   (b) 0.000 059 68 becomes 0.000 059 7, we increase the 6 by one because the following digit 8 is greater than 5. Note that the zeros are not significant because they can be replaced by a power of 10, (i.e., $0.000\,059\,68 = 5.968 \times 10^{-5}$).
   (c) 4.6925×10^8 becomes 4.69×10^8, the power of 10 is not considered and 2 is less than 5.
   (d) 4.999×10^5 becomes 5.00×10^5. In this case we must retain the two zeros

Sec. A-3 / Rounding Off and Significant Figures

in order to specify three significant figures. Thus the 9 was converted into 10, which then changed the previous digit and so on.

Operations Involving Significant Figures

To avoid extravagant claims for precision after a calculation, we must follow certain rules for operations with significant figures. This is especially important when we are using electronic calculators, which often retain all the digits.

Addition and Subtraction

Rule: We can only add or subtract quantities directly if they have the same units. For example, we cannot add lengths in meters to times in seconds, because the units do not represent the same quantities.

Rule: The answer must be rounded off to the least precise quantity used (i.e., the same precision as the least significant digit of the quantities used). If scientific notation is involved, all quantities should be written in terms of the same power of 10 in order to find the least significant digit.

EXAMPLE A-14

Add the following lengths: (a) 72.684 m + 5.91 m; (b) 8.049 73 cm − 1.6 cm; (c) 29 000 km + 8650 km; (d) 3.2×10^8 m + 1.5×10^9 m.

Solution: In each case the least significant digit of all the numbers involved will be indicated by a caret and the answer must be rounded off to the same precision (i.e., the same decimal place or column of 10).

(a) 72.684 m + 5.91 m = 78.594 m which must then be rounded off to 78.59 m, since the digit 9 is uncertain.

(b) 8.049 73 cm − 1.6 cm = 6.449 73 cm, which must be rounded off to 6.4 cm because the first decimal place is the least significant digit of the original numbers.

(c) 29 000 km + 8650 km = 37 650 km, which is rounded off to 38 000 km since the thousands are the least significant digits. Remember that the zeros are significant only when they cannot be written as a power of 10. In this case 29 000 = 2.9×10^4.

(d) 3.2×10^8 m + $1.5 + 10^9$ m. In this case we must write the numbers in terms of the same power of 10 to find the least significant digit. Thus we obtain

$$3.2 \times 10^8 \text{ m} + 15 \times 10^8 \text{ m} = 18.2 \times 10^8 \text{ m}$$

which must be rounded off to 18×10^8 m or 1.8×10^9 m in scientific notation. Note that the answer has the same power of 10 as the original numbers. This can be checked by writing out the numbers in full.

Multiplication and Division

Rule: Find by inspection which of the given numbers has the least number of significant digits. Your answer must be rounded off to the same number of digits.

EXAMPLE A-15

Perform the following operations and round off the answer to the correct number of significant figures.

(a) 76.15×2.4 (b) $\dfrac{8.73}{1.1}$;

(c) $\dfrac{7620 \times 4.93}{1.987}$ (d) $\dfrac{1.84 \times 10^4 \times 3.286 \times 10^{-3}}{4.6 \times 10^{-8}}$

Solution

(a)
$$\begin{array}{rl} 76.15 & \text{(four significant figures)} \\ \times\ 2.4 & \text{(two significant figures)} \\ \hline =\ 182.76 & \text{(five significant figures)} \end{array}$$

The answer is therefore 180, because in this case the answer must be rounded off to two significant figures. Note that once again we need the zero, which is not significant, to keep the "size" of the number.

(b)
$$\begin{array}{rl} 8.73 & \text{(three significant figures)} \\ \div\ 1.1 & \text{(two significant figures)} \\ \hline =\ 7.936\,363\,63 & \end{array}$$

which must be rounded off to two significant figures to give 7.9.

(c) $\dfrac{7620 \times 4.93}{1.987} = 18\,906.190\,24$, which must be rounded off to 18 900 (three significant figures) since 7.620 and 4.93 both have three significant figures. Again we must retain the zeros here; otherwise, we would change the value of the number.

(d) $\dfrac{1.84 \times 10^4 \times 3.286 \times 10^{-3}}{4.6 \times 10^{-8}} = 1.3144 \times 10^9$ which is rounded off to 1.3×10^9 (two significant figures) since 4.6×10^{-8} has only two significant figures.

PROBLEMS

6 Round off the following to three significant figures: (a) 7 688 000 m; (b) 0.000 832 6 s; (c) 1.9834×10^4 km; (d) 4.9973×10^{-6} m; (e) 17.159 s; (f) 1083 kg.

7 Solve the following to the correct number of significant figures: (a) $26.831 + 8.32$; (b) $108.053 - 26.794\,61$; (c) $0.003\,86 - 0.1932 + 4.2187$; (d) $63\,800 - 1820$; (e) $1.98 \times 10^7 + 4.21 \times 10^6$; (f) $9.18 + 10^{-6} - 4.236 \times 10^{-5}$ (g) 29.1×3.2; (h) $8.66 \times 10^{-8} \times 1.6 \times 10^{-4}$; (i) $\dfrac{9.1 \times 10^{-31}}{1.6 \times 10^{-19}}$; (j) $\dfrac{4.2 \times 10^{-8}}{3.1 \times 10^7}$; (k) $\dfrac{29.6 \times 3.7}{1.86}$; (l) $\dfrac{3.2 \times 10^{-7} \times 1.61 \times 10^6}{1.93 \times 10^{-9}}$; (m) $\dfrac{7.23 \times 10^{-8} \times 6.41 \times 10^7}{9.387 \times 10^9}$

A-4 FORMULA MANIPULATION

In practice, formulas and equations must often be rearranged in order to solve for some other variable. This process involves the manipulation of algebraic quantities. The basic rule for algebraic manipulation is:

Rule: *Any change (or alteration) may be made to one side of an equation (equal sign) provided that the same change is made to the other side.*

Some of the possible processes are:

1. A term may be added to (or subtracted from) each side.
2. Both sides may be multiplied (or divided) by the same factor.
3. Both sides may be raised to the same power (or the same root of both sides may be taken).

Sec. A-4 / Formula Manipulation

The basic steps for algebraic manipulation are as follows (some of these can often be omitted).*

1. *Clear the fractions.* Multiply by the lowest common denominator.
2. *Remove the parentheses.* Any factor must apply to each term or group of terms inside the parentheses.
3. *Transpose.* Move all the terms containing the required symbol to one side and all others to the other side.
4. *Factor.* Separate the required symbol from the other terms.
5. *Divide.* This leaves the required factor on one side.
6. *Determine the roots.* Take the root if necessary.

EXAMPLE A-16

Solve $V = \frac{4}{3}\pi r^3$ for the term r.

Solution

Step 1: Multiply both sides by 3 to clear the fraction.

$$3V = 4\pi r^3$$

Steps 2–4: Not required in this case.

Step 5: Divide by 4π:

$$\frac{3V}{4\pi} = r^3$$

Step 6: Take the cube root of both sides:

$$\sqrt[3]{\frac{3V}{4\pi}} = r$$

EXAMPLE A-17

Solve $A = \dfrac{(a + b)h}{2}$ for a.

Solution

Step 1: $2A = (a + b)h$	Clear fractions by multiplying by 2.
Step 2: $2A = ah + bh$	Remove parentheses by multiplying terms in the parentheses by h.
Step 3: $2A - bh = ah + bh - bh = ah$	Transpose by subtracting bh from each side.
Step 4: Not required.	
Step 5: $\dfrac{2A - bh}{h} = a$	Divide each term by h to isolate the a.
Step 6: Not required.	

---

*We shall not consider general quadratic equations.

EXAMPLE A-18

Solve $\dfrac{f(ax - b)}{3} = c - dx$ for x.

Solution

Step 1: $f(ax - b) = 3c - 3dx$	Clear fractions by multiplying by 3.
Step 2: $fax - fb = 3c - 3dx$	Clear parentheses by multiplying the terms in the parentheses by f.
Step 3: $fax + 3dx = 3c + fb$	Transpose by adding $3dx + fb$ to both sides.
Step 4: $(fa + 3d)x = 3c + fb$	Divide each term on the left side by x to find the factor.
Step 5: $x = \dfrac{3c + fb}{fa + 3d}$	Divide by the multiple of x to isolate the x.
Step 6: Not required	

PROBLEMS

8 Solve the following for x: **(a)** $2x - 3 = 7$; **(b)** $7x - 12 = 3x + 4$; **(c)** $12x = 108$; **(d)** $ax - 4 = 11$; **(e)** $7(x - 3) = \dfrac{x - 9}{5}$; **(f)** $\dfrac{1}{x} - \dfrac{1}{p} = \dfrac{3}{r}$.

9 Solve the following for t: **(a)** $at - b = 5d$; **(b)** $s = \tfrac{1}{2}at^2$; **(c)** $q = \dfrac{(a + b)t}{2}$; **(d)** $\dfrac{(a + b)}{t} = c + 2d$.

Appendix B
CONVERSION FACTORS

Length

1 meter (m) = 3.281 ft = 39.37 in.
1 centimeter (cm) = 0.3937 in.
1 kilometer (km) = 3281 ft = 0.6214 mi
1 foot (ft) = 12 in. = 0.3048 m
1 inch (in.) = 0.0833 ft = 2.54 cm
1 mile (mi) = 5280 ft = 1.609 km

Area

1 m^2 = 10^4 cm^2 = 10.76 ft^2 = 1550 in^2
1 cm^2 = 10^{-4} m^2 = 0.1550 in^2
1 km^2 = 10^6 m^2 = 0.3861 mi^2 = 1.076 × 10^7 ft^2
1 hectare (ha) = 10^4 m^2 = 2.47 acres
1 ft^2 = 144 in^2 = 0.092 90 m^2 = 929 cm^2
1 in^2 = 6.94 × 10^{-3} ft^2 = 6.452 cm^2
1 mi^2 = 650 acres = 2.79 × 10^7 ft^2 = 2.59 km^2
1 acre = 43 560 ft^2 = 0.405 ha = 4050 m^2

Volume

1 m^3 = 1000 L = 35.31 ft^3 = 6.102 × 10^4 in^3
1 L = 10^3 cm^3 = 10^{-3} m^3 = 0.0353 ft^3 = 61.0 in^3 = 0.264 U.S. gallon = 0.220 Imperial gallon
1 ft^3 = 1728 in^3 = 7.47 U.S. gallon = 0.028 32 m^3 = 28.3 L

1 in^3 = 5.79 × 10^{-4} ft^3 = 16.39 cm^3 = 0.016 39 L
1 U.S. gallon = 231 in^3 = 0.134 ft^3 = 3.785 L
1 Imperial gallon = 4.546 L

Time

1 year (a) = 8770 h = 5.26 × 10^5 min = 3.15 × 10^7 s
1 day (d) = 2.74 × 10^{-3} a = 1440 min = 86 400 s
1 hour (h) = 1.14 × 10^{-4} a = 0.0417 d = 60 min = 3600 s
1 min = 60 s

Mass

1 kg = 0.0685 slug
1 unified mass unit (u) = 1.660 53 × 10^{-27} kg
1 metric ton (tonne) (t) = 1 Mg
1 slug = 14.6 kg

Force

1 newton (N) = 0.2248 lb
1 pound (lb) = 4.445 N = 16 oz
At the earth's surface a mass of 1 kg has a weight of approximately 2.2 lb.

Speed or velocity

1 m/s = 3.60 km/h = 3.281 ft/s = 2.25 mi/h
1 ft/s = 0.3048 m/s
88 ft/s = 60 mi/h
1 km/h = 0.911 ft/s = 0.278 m/s = 0.621 mi/h
1 mi/h = 1.47 ft/s = 1.609 km/h = 0.447 m/s

Pressure

1 atm = 101.325 Pa = 760 mm mercury = 14.69 lb/in^2 = 2120 lb/ft^2

Work and Energy

1 joule (J) = 0.738 ft·lb = 9.48 × 10^{-4} Btu = 2.78 × 10^{-7} kW·h
= 0.239 cal
1 kW·h = 3.6 MJ = 1.34 hp·h = 8.6 × 10^5 cal = 2.66 × 10^6 ft·lb
= 3410 Btu
1 Btu = 778 ft·lb = 3.93 × 10^{-4} hp·h = 1.06 kJ = 252 cal
1 ft·lb = 5.05 × 10^{-7} hp·h = 1.29 × 10^{-3} Btu = 1.36 J = 0.324 cal
1 cal = 4.186 J = 3.97 × 10^{-3} Btu

Power

1 kW = 1.34 hp = 3410 Btu/h = 239 cal/s = 738 ft·lb/s
1 hp = 550 ft·lb/s = 2540 Btu/h = 746 W = 178 cal/s
1 Btu/h = 3.93 × 10^{-4} hp = 2.93 × 10^{-4} kW

Light

1 foot lambert = 3.43 cd/m^2
1 lx = 0.0929 fc
1 fc = 10.76 lx

Charge

1 electronic charge unit (e) = 1.60×10^{-19} C
1 ampere hour (A·h) = 3600 C

Angles

1° = 60′ = 3600″ = 0.0175 rad
1 rad = 57.3° = 0.159 rev
1 rev = 360° = 2π rad
1 rev/min = $\dfrac{2\pi}{60}$ rad/s

Appendix C
DEFINITIONS OF SI BASE UNITS

Length: A **meter** (symbol m) is $\frac{1}{299\ 792\ 458}$ th of the distance traveled by light in 1 second in a vacuum.

Mass: A standard **kilogram** (symbol kg) mass is a cylindrical platinum–iridium alloy kept at the International Bureau of Standards.

Time: A **second** (symbol s) is the duration of 9 192 631 770 cycles of radiation from the transititon between the two ground-state energy levels in an atom of the isotope cesium-133.

Quantity: A **mole** (symbol mol) is the amount of a substance that contains as many elementary entities as there are atoms in 12 g of carbon-12, that is, Avogadro's number—approximately 6.023×10^{23} elementary entities of that substance.

Electric current: If two long, straight conductors are one meter apart in a vacuum and carry equal electric currents of one **ampere** (symbol A) there is a magnetic force per unit length of 2×10^{-7} N/m between them.

Temperature: The **kelvin** (symbol K) is the unit of thermodynamic temperature which is 1/273.16 of the temperature of the triple point of water.

Luminous intensity: A **candela** is the luminous intensity perpendicular to the surface of 1/60 of a square centimeter of a black body that is maintained at the temperature of freezing platinum (2046 K) and a pressure of 101 325 Pa.

Appendix D
SI UNITS AND PREFIXES

D-1 RULES FOR UNITS

The following rules have been adopted for the use of SI units.

1. All units are written in lowercase (i.e., without a capital letter). However, if the unit is derived from the name of a person, the unit symbol is capitalized. For example, meter is abbreviated as m, kilogram as kg, newton as N, and hertz as Hz. The liter (symbol L) is the exception to the rule.
2. A space is left between the numeric value and the symbol. For example, two meters is written as 2 m. However, if the unit is not SI (Table 1-5), no space is left when the first character of the symbol is not a letter. For example, we write 23°C, not 23 °C, and 49°12′18″, not 49° 12′ 18″.
3. Symbols are written without a period except at the end of a sentence, and they are not changed to indicate the plural. For example, twelve newtons is written as 12 N, not 12 N. or 12 Ns.
4. Symbols should be used with numerical values, but the full unit must be written if numbers are not involved. For example, we write 12 m^2, not 12 square meters, but the watt—not the W—is a unit of power.
5. Do not mix names and symbols. For example, we write kg · m or kilogram meter, not kg meter.
6. Central dots are used to indicate the multiplication of units, and like units may be squared, cubed, and so on. A space is left between multiplied units in written form. For example, we write (2 kg)(5 m) = 10 kg · m, (2 m)(3 m) = 6 m^2, and twelve newton meters, not 12 newton-meters.
7. A stroke or negative exponent is used to indicate division: for example,

$$\frac{m}{s^2} = m/s^2 \quad \text{or} \quad m \cdot s^{-2}$$

The negative power represents division in exactly the same way that

$$10^{-2} = \frac{1}{10^2}$$

The word "per" also means division (e.g., speed in meters per second). The central dot must be included when multiplication of different units occurs.

8. Long numbers are written with a space between each group of three digits starting at each side of the decimal point: for example,

$$93\ 042.135\ 679\ 2$$

9. A zero must be written before the decimal point if a number is less than 1: for example, 0.05, not .05.

D-2 RULES FOR PREFIXES

1. Prefixes are written without a space between the prefix symbol and the unit symbol, and the prefix name is combined with the unit name. For example, one megameter (1 Mm) is equal to one million meters (10^6 m), and one microsecond (1 μs) is equivalent to one one-millionth of a second (10^{-6} s).
2. Prefixes are considered to be combined directly with the unit for all algebraic manipulations. This is somewhat different from the rules for algebra! For example, in algebraic form; $x \cdot y^2 = x \cdot y \cdot y$. But with SI prefixes and units, 1 cm^2 = (1 cm)(1 cm) = (10^{-2} m)^2 = 10^{-4} m^2. That is, the prefix "centi" is assumed to be a part of the total unit that is being squared.
3. Compound prefixes should not be used; this rule also applies to the base unit kilogram. For example, $\mu\mu$F should be written pF, and μkg should be written as mg.
4. When a prefix is used to replace scientific notation, the numerical value should be between 0.1 and 1000. For example, we could write 8.9 × 10^{-7} s as either 0.89 μs or 890 ns. However, 0.02 μs or 1200 ks must be rewritten as 20 ns and 1.2 Ms, respectively.
5. Avoid use of the prefixes deci, centi, deka,* and hecto. Exceptions are "centimeter" and "square hectometer," 1 hm^2 (abbreviated 1 ha = 10^4 m^2), which is used to measure areas and is called a **hectare**.
6. Only one prefix should be used in a compound unit, and it should appear in the numerator. The exception is the base unit kilogram. For example, we write Mg/m^3, not g/cm^3, and kJ/kg, not J/g.

*Also written as deca.

Appendix E
PHYSICAL CONSTANTS

Constant	Symbol	Value
Absolute zero	0 K or 0°R	−273.16°C or −459.7°F
Acceleration due to gravity at the earth's surface	g	9.81 m/s² or 32.2 ft/s²
Avogadro's number	N_0	6.0225×10^{23}/mol
Coulomb constant (vacuum)	k	9.00×10^9 N·m²/C²
Earth's mass	M_E	5.975×10^{24} kg or 4.093×10^{23} slug
Earth's mean radius	R_E	6.371×10^6 m or 2.09×10^7 ft = 3960 mi
Electron rest mass	m_e	9.109×10^{-31} kg
Elementary charge	e	1.6021×10^{-19} C
Faraday constant	F	$9.648\ 67 \times 10^4$ C
Gas (universal) constant	R	8.3143 J/(K·mol)
Gravitation constant	G	6.673×10^{-11} N·m²/kg² or 3.44×10^{-8} lb·ft²/slug²
Mechanical equivalent of heat	J	4186 J/kcal or 788 ft·lb/Btu
Moon mass	m_m	7.35×10^{22} kg or 5.02×10^{21} slug
Moon mean radius	R_m	1.739×10^6 m or 5.71×10^6 ft
Neutron rest mass	m_n	1.6748×10^{-27} kg
Permeability of free space	μ_0	$4\pi \times 10^{-7}$ T·m/A
Permittivity of free space	ϵ_0	8.85×10^{-12} C²/N·m²
Planck's constant	h	6.6262×10^{-34} J·s
Proton rest mass	m_p	1.6726×10^{-27} kg
Speed of light (vacuum)	c	$2.997\ 925 \times 10^8$ m/s or 9.82×10^8 ft/s = 186 000 mi/s
Standard atmospheric pressure	p_0	101.325 kPa or 14.7 lb/in² or 760 mm mercury
Sun mass	M_s	1.987×10^{30} kg, 1.36×10^{29} slug
Sun mean radius	R_s	6.965×10^8 m, 2.29×10^9 ft
Unified atomic mass unit	u	$1.660\ 53 \times 10^{-27}$ kg or 931 MeV

Appendix F
ANSWERS TO ODD-NUMBERED PROBLEMS

CHAPTER 1

1. (a) 9.3×10^7 (b) 8.2×10^{-8} (c) 5.26×10^5 (d) 4.79×10^6 (e) 4.2×10^{-3} (f) 5.80×10^{-5} (g) 3.000×10^{-3} (h) 1.2×10^{-8} (i) 3.4×10^5 (j) 7.36×10^{-3} (k) 5.1×10^{-10} (l) 1.2×10^9 (m) 1.2×10^{-3}
3. (a) 2 (b) 4 (c) 4 (d) 4 (e) 4 (f) 2
5. 2.4×10^4 m^2 7. (a) 35.6 m^3 (b) 3.56×10^4 kg
9. (a) 4.6 km (b) 2 μg (c) 83 GHz (d) 0.52 μs or 520 ns (e) 12 pF (f) 3.2 Mg (g) 42 MN/C (h) 0.86 MV/m or 860 kV/m (i) 7.8 Mg/m^3
11. (a) 8.0×10^{-4} cm (b) 0.5 cm (c) 4×10^5 cm 13. 6.8×10^3 kg/m^3
15. 72 km/h 17. 1609 m
19. (a) 1.609 km (b) 2.4 slug (c) 533 N (d) 9.1 m/s (e) 82 ft/s^2 (f) 2300 kg/m^3 (g) 11.1 m^2 (h) 42 ft^3 (i) 2.10 lb/ft^2 or 0.0146 lb/in^2 (j) 7450 W

Review Problems

1. (a) 4.6×10^4 m (b) 1.90×10^{-6} s (c) 5.2×10^3 m (d) $4.200\,00 \times 10^{-7}$ s (e) 1.86×10^5 m (f) 4.26×10^8 ft (g) 5.23×10^{-6} s (h) 5.2×10^{-8} s (i) 8.20×10^4 m
3. (a) 9.20×10^3 m (b) 4.94×10^{-4} s (c) 4.19×10^7 g (d) 3.92×10^{-3} s (e) 4.40×10^3 s (f) 2.50×10^{-5} s (g) 4.00×10^{-3} s (h) 1.00×10^6 m (i) 3.60×10^8 m (j) 1.86×10^5 mi 5. (a) 4500 m^3 = 4.5×10^6 L (b) 4.5×10^6 kg
7. (a) 3600 cm = 3.6×10^{10} nm = 0.036 km (b) 0.188 L = 1.88×10^{-4} m^3 = 188 cm^3 (c) 1.2×10^8 mg = 0.12 Mg = 1.2×10^5 g (d) 3.0×10^4 mm^3 = 3.0×10^{-5} m^3 (e) 90 cm^3 = 9×10^{-5} m^3
9. (a) 57 ns (b) 12 Mg (c) 0.34 mm or 340 μm (d) 72 mL (e) 90 GHz (f) 0.125 nF or 125 pF (g) 38.0 ps (h) 490 m or 0.49 km (i) 0.78 cm^2 (j) 0.39 km^3 (k) 0.700 mm^3 11. 16.0 m/s
13. (a) 2400 kg/m^3 (b) 3.6 W/m^2 (c) 5.0×10^9 V/m or 5.0 GV/m (d) 26 J/kg (e) 36 A/m^2 (f) 1.5×10^4 A/m^2 15. 648
17. (a) 1.609×10^5 cm (b) 175 kg (c) 86 ft^2 (d) 0.708 m^3 (e) 1.29×10^7 mi^2 (f) 15.3 slug/ft^3

Answers to Odd-Numbered Problems

CHAPTER 2

1. (a) 165 cm² (b) 37.4 in² 3. (a) 0.358 cm² (b) 0.110 in²
5. (a) 0.530 m³ (b) 14.3 ft³ 7. (a) 1.86×10^5 ft² (b) 4.26 acres
9. (a) 0.181 cm² (b) 0.0243 in²
11. (a) (1) 9.00 cm² (2) 30.0 cm² (3) 450 cm² (4) 54 cm²
(b) (1) 1.39 in² (2) 4.65 in² (3) 69.7 in² (4) 8.35 in²
13. (a) (1) 20 000 ft² (2) 30 000 ft² (3) 20 000 ft² (4) 30 000 ft²
(b) (1) 0.459 acre (2) 0.689 acre (3) 0.459 acre (4) 0.689 acre 15. 29.7 m³ = 2.97×10^4 L
17. 232 L 19. 9.00 21. (a) 0.348 m³ (b) 13.1 ft³
23. (a) 775 cm², 1250 cm³ (b) 121 in², 77.5 in³
25. (a) 75.4 cm², 905 cm³, 520 cm² (b) 11.6 in², 48.3 in³, 73.4 in²
27. (a) 236 cm³ (b) 14.1 in³ 29. (a) 11.3 (b) 10.0 (c) 11.2
31. (a) $w = (2.35 \text{ oz/in}^3)d^3$ (b) 0.0603 oz
33. (a) $I = (0.200 \text{ A/V})V$ (b) 6.20 A (c) 10.0 V
35. (a) $d = (55.0 \text{ mi/h})t$ (b) 413 mi (c) 2.73 h
37. (a) $d = (16.01 \text{ ft/s}^2)t^2$ (b) 400 ft (c) 8.66 s
39. (a) $D = (4.7 \times 10^{-3} \text{ cm/m}^3)L^3$ (b) 1.0 cm (c) 8.8 m
41. (a) $P = (165 \text{ W/A}^2)I^2$ (b) 20.2 W (c) 0.270 A
43. (a) $V = \dfrac{180 \text{ lb/ft}^3/\text{in}^2}{p}$ (b) 4.00 ft³ (c) 10.0 lb/in²
45. (a) $I = \dfrac{14.4 \ \mu\text{W}}{d^2}$ (b) 0.400 μW/m² (c) 3.46 m
47. (a) $s = (20 \text{ ft/s})t + 50 \text{ ft}$ (b) $s = -(58.3 \text{ km/h})t + 200 \text{ km}$ (c) $P = (7.5 \text{ W/A}^2)I^2$
(d) $F = (0.2 \text{ N} \cdot \text{s}^2/\text{m}^2)v^2$ (e) $D = (0.044/\text{cm}^2)L^3$ (f) $w = (20 \text{ oz/in}^3)d^3$
(g) $p = (3.64 \text{ Pa/K})T$ (h) $F = \dfrac{0.50 \text{ N} \cdot \text{m}^2}{d^2}$ (i) $m = (12.5 \text{ g/mm}^3)d^3$
(j) $I = \dfrac{175 \text{ cd} \cdot \text{m}^2}{d^2}$ (k) $D = (0.0667 \text{ in./ft}^3)L^3$ (l) $T = (1 \text{ s}/\sqrt{\text{m}})\sqrt{L}$
49. (b) $F = (5.0 \text{ N/mm})e$ 51. (c) $m = (3.9 \text{ g/cm}^3)d^3$ 53. (c) $F = \dfrac{10 \text{ lb} \cdot \text{ft}^2}{d^2}$
55. (a) 1.99×10^{30} kg (b) 1.37×10^{29} slug 57. 9.0×10^3 kg/m³
59. (a) 7.98 g (b) 1.10×10^{-4} kg (c) 1.72 g (d) 1.94×10^{-3} slug
61. (a) 3.22×10^5 N (b) 72 600 lb
63. (a) 9.19×10^{-5} m³ (b) 4.78×10^{-5} m³ (c) 16.3 in³
65. (a) 1.83 cm (b) 2.42 cm (c) 3.13 cm (d) 0.316 ft 67. 1880 kg
69. First, 11 300 kg/m³, not gold; second, 19 300 kg/m³, gold
71. (a) 10.7 kg (b) 0.774 slug 73. A: 0.556, 0.833, 0.667; B: 0.833, 0.556, 1.50
75. (a) 4.12 m (b) 38.2°, 51.8° 77. 58.1 m

Review Problems

1. (a) 24 m², 258 ft² (b) 19.0 cm², 2.95 in² (c) 75 cm², 11.6 in²
(d) 5400 m², 58 100 ft² 3. (a) 2710 cm² (b) 372 in² 5. 183 L 7. 692 cm², 107 in²
9. (a) $E = [0.170 \text{ ft} \cdot \text{lb}/(\text{ft/s}^2)]v^2$ (b) 2.71 ft·lb (c) 59.4 ft/s
11. (a) $a = \dfrac{15 \text{ kg} \cdot \text{m/s}^2}{m}$ (b) 1.0 m/s² (c) 10 kg
13. (a) $R = (4.0 \ \Omega/\text{mm}^2)d^2$ (b) $a = \dfrac{18.0 \text{ m}}{T^2}$ (c) $d = (14 \text{ ft/s}^2)t^2$
15. (a) 8290 kg/m³ (b) 1.12 slug/ft³ 17. 8330 kg/m³
19. (a) 4.41 cm (b) 3.81 cm (c) 1.48 in. 21. (a) 18.9 g
23. a = 16.0 m, 52.5 ft, b = 20.5 m, 67.2 ft

CHAPTER 3

1. (a) 8.0 cm (b) 3.0 cm (c) 18.0 cm (d) 13.0 cm (e) 12.8 cm (f) 1.4 cm
3. (a) 3.0 cm (b) 16.8 cm (c) 28.8 cm (d) 1.4 cm
7. (a) 567 km/h (b) 564 mi/h 9. (a) 6.21 m/s (b) 22.4 km/h
11. (a) 4.85 h (b) 228 km 13. (a) 220 mi (b) 4.45 h
15. (a) 59 500 mi (b) 8.16×10^5 mi (c) 5670 mi (d) 70.8 mi 17. 368 km/h
19. 51.9 mi/h 21. 568 km/h north
23. (a) 1950 km W (b) 542 km W (c) 8.13 km W 25. (a) 2.64 km (b) 8640 ft
27. (a) 22.8 m/s E (b) 64.0 ft/s E 29. (a) 2.13 m/s² (b) 4.40 ft/s²
31. (a) 20 km E (b) 210 mi N
33. (a) 130 km (b) 40.0 km E (c) 52.0 km/h (d) 16.0 km/h E
35. (a) 47.5 m (b) 22.5 m down (c) 11.1 m/s (d) 5.27 m/s down

Review Problems

3. 62.2 km/h   **5. (a)** 6.69 km   **(b)** 6.06 mi   **7.** 7.5 km/h E   **9.** 1.60×10^5 m/s
11. (a) 1.2 m/s² S   **(b)** 3.0 m/s² S   **(c)** 0.833 m/s² S   **(d)** 1.50 m/s² S
(e) 11.7 ft/s² S   **(f)** 5.50 ft/s² S
13. (a) 8.00 km   **(b)** 2.00 km E   **(c)** 3.20 km/h   **(d)** 0.800 km/h E
15. (a) 90.0 mi   **(b)** 40.0 mi S   **(c)** 18.0 mi/h   **(d)** 8.00 mi/h S

CHAPTER 4

1. (a) 9.80 m/s   **(b)** 29.4 ft/s   **3. (a)** 1.04 m/s²   **(b)** 486 m
5. (a) 0.838 ft/s²   **(b)** 513 ft   **7. (a)** 70.0 m/s   **(b)** 175 m
9. (a) 0.617 ft/s²   **(b)** 10 800 ft   **11. (a)** 6.55 s   **(b)** 600 ft
13. (a) 2.77×10^{11} m/s²   **(b)** 31.1 cm
15. (a) 44.4 m, 5.55 m/s²   **(b)** 183 ft, 22.9 ft/s²
17. (a) 81.6 m/s   **(b)** 933 ms   **19. (a)** 29.0 m/s   **(b)** 122 m
21. (a) 19.9 m/s   **(b)** 84.5 m
23. (a) 452 m   **(b)** 968 m   **(c)** 51.2 m/s², −44.8 m/s²
25. (a) 0.455 m/s²   **(b)** 24.4 s   **27. (a)** 45.7 ft/s   **(b)** 20.7 s
29. (a) −1.04 m/s², 0, 1.11 m/s²   **(b)** 529 m   **31. (a)** 2.67 s   **(b)** 26.2 m/s
33. (a) 157 m/s   **(b)** 14.8 s   **35. (a)** 50.7 m, 32.5 m/s   **(b)** 163 ft, 106 ft/s
37. (a) 237 m/s   **(b)** 2670 m   **39. (a)** 221 m/s   **(b)** 22.6 s
41. (a) 3.54 s   **(b)** 55.5 m/s

Review Problems

1. 44 min 37 s   **3.** 54 min 24 s   **5. (a)** 3.60×10^{14} m/s²   **(b)** 16.7 ns
7. (a) 12.0 m/s   **(b)** 48.0 m   **9. (a)** 286 m/s²   **(b)** 21.0 s
11. (a) 77.5 ft/s   **(b)** 506 ft   **13. (a)** 4.69 m/s²   **(b)** 26.8 s   **(c)** 854 m
15. 189 m or 619 ft   **17. (a)** 1120 m   **(b)** 151 m/s
19. (a) −181 ft/s   **(b)** 507 ft   **(c)** 206 ft/s
21. (i) (a) Yes   **(b)** 2.0 m/s²   **(c)** 50 m   **(ii) (a)** Yes   **(b)** 6.6 ft/s²   **(c)** 164 ft

CHAPTER 5

1. (a) 60.0 N   **(b)** 9.34 lb
3. (a) 20.0 m/s²   **(b)** 2.67 cm/s²   **(c)** 34.0 cm/s²   **(d)** 18.1 ft/s²   **(e)** 0.872 ft/s²
(f) 70.0 ft/s²   **(g)** 2.23×10^4 m/s²   **(h)** 3.75×10^9 m/s²   **5.** 90.2 ft/s²   **7.** 60.0 lb
9. 1.00 m/s²   **11.** 70.0 m/s²
13. (a) 122 N   **(b)** 6.93 N   **(c)** 4080 N   **(d)** 4080 N   **(e)** 20.3 lb   **(f)** 134 lb
(g) 936 lb   **(h)** 1.48×10^{-30} N   **(i)** 4.16×10^{-6} N   **(j)** 5710 N   **(k)** 4.48×10^{-6} N
(l) 1960 N   **(m)** 7.99×10^{-9} N   **15.** 5.91 kg   **17. (a)** 30.0 lb   **(b)** 32.8 lb
19. 5.01 ft/s² up   **21. (a)** 2220 N   **(b)** 1030 lb   **23.** 2.02×10^9 N
25. (a) 159 N   **(b)** 116 N   **27.** 0.588   **29. (a)** 214 kg   **(b)** 18.6 slug
31. (a) 17 500 kg·m/s E   **(b)** 13 200 slug·ft/s S   **33.** 3.99×10^{10} kg·m/s
35. 5.46×10^{-25} kg·m/s E   **37.** 670 slug·ft/s down

Review Problems

1. (a) 0.80 m/s²   **(b)** 5.26 ft/s²   **3. (a)** 81.9 kg   **(b)** 135 N
5. (a) 736 N   **(b)** 647 N   **(c)** 287 N   **(d)** 1950 N   **(e)** 280 N   **(f)** 863 N
7. 110 ft/s²   **9. (a)** 569 m   **(b)** 21.3 s   **(c)** 45 500 N   **11. (a)** 67.5 lb   **(b)** 44.6 lb
13. (a) 178 kg   **(b)** 74.8 slug   **15.** 10.1 kg·m/s   **17.** 2.7 N·s   **19.** −2.36 m/s

CHAPTER 6

1. (a) 79 km 55° E of N   **(b)** 445 mi 38° W of S   **3.** 75 N at 27°
5. (a) 13 km/h at 18.5°   **(b)** 20 mi/h at 24°   **7.** 114 N at 52°
9. (a) 61.0 m at 65°   **(b)** 48 km at 280°   **(c)** 510 ft at 1.5°   **(d)** 166 N at 210°
(e) 9.8 ft/s at 14°   **(f)** 5.3 m/s² at 216°   **(g)** 99 lb at 3.5°
11. (a) 170 km N   **(b)** 255 km   **(c)** 51 km/h   **(d)** 34 km/h N

Answers to Odd-Numbered Problems

13. (a) 738 km, 548 km 63° W of N, 295 km/h, 219 km/h 63° W of N **(b)** 1200 mi, 875 mi at 59° S of W, 218 mi/h, 159 mi/h 59° S of W **15.** 30.3 N, 17.5 N
17. (a) 21.2 km, 21.2 km **(b)** 7.5 km/h, 19 km/h **(c)** -5.13 m/s^2, 14.1 m/s^2
(d) -10.6 ft/s, -22.7 ft/s **(e)** 6.16 mi/h, -16.9 mi/h **(f)** 17 m, -4.7 m
(g) -306 ft, -170 ft
19. (a) 37.7 N, 26.4 N **(b)** 3.97 lb, 37.8 lb **(c)** -45.1 N, 31.5 N
(d) 21.5 lb, -59.2 lb **21.** 225 km/h E, 125 km/h S **23.** -47.5 lb, -21.0 lb
25. (a) 17.2 N at 64.5° **(b)** 347 m at 207° **(c)** 6.8 m/s^2 at 249° **(d)** 73.0 lb at 299°
(e) 72.1 ft at 190° **(f)** 41.7 mi at 301° **(g)** 20.9 N at 255° **27.** 425 N at 32° W of N

Review Problems

1. 19 lb 39° **3.** 112 lb 23° **5. (a)** 109 mi 28° W of S **(b)** 14.5 mi/h 28° W of S
7. (a) A = $(-43.0$ km, -61.4 km), **B** = $(23.8$ km, -38.2 km), **C** $(10.4$ km, 59.1 km)
(b) A = $(109$ mi, 50.7 mi), **B** = $(127$ mi, -79.5 mi), **C** = $(-70.7$ mi, -70.7 mi)
9. (a) 600 km **(b)** 572 km 16° W of S **(c)** 240 km/h **(d)** 229 km/h 16° W of S
11. (a) 40.6 m/s, 52° W of N **(b)** 25 mi/h 53° E of S

CHAPTER 7

1. 135 lb **3.** 275 N **5. (a)** 444 N, 314 N **(b)** 48.1 lb, 34.0 lb
7. (a) 5000 N, 4330 N **(b)** 3500 lb, 3030 lb **(c)** 29 000 lb, 25 100 lb
(d) 34 300 N, 29 700 N **9. (a)** 30.3 N, 17.5 N **(b)** 7.14 lb, 4.13 lb
11. (a) 42 500 N, 36 000 N **(b)** 5580 lb, 4570 lb **(c)** 54 000 N, 46 700 N
13. 1600 N **15.** 125 N, 42.9 N
17. (a) 20.0 N 53° above the horizontal **(b)** 40.6 lb 38° below the horizontal
19. (a) 75.0 N·m **(b)** 52.5 lb·ft **21. (a)** 650 N·m **(b)** 46.0 cm
23. (a) 2660 N·m, 253 N·m **(b)** 880 lb, 770 lb·ft **25.** 325 lb **27.** 175 lb, 225 lb
29. 545 N, 554 N **31.** 131 kN, 121 kN **33.** 429 N **35.** 1670 N

Review Problems

1. 509 500 lb **3.** 54.7 lb 50.2°
5. (a) 13 600 N, 11 100 N **(b)** 5380 lb, 4560 lb **(c)** 17 200 N, 14 900 N
7. (a) 2210 N, 2630 N **(b)** 112 lb, 134 lb **(c)** 3700 N, 4400 N **(d)** 1760 lb, 2100 lb
9. (a) 184 lb·ft **(b)** 8.80 in **11.** 117 lb, 148 lb **13. (a)** 2880 N **(b)** 1680 lb
15. 60 lb **17.** 517 kN, 485 kN **19.** 5.2 ft from the end

CHAPTER 8

1. (a) 338 J **(b)** 595 ft·lb **3. (a)** 1.25×10^6 J **(b)** 4.75×10^5 ft·lb **5.** 38.0 ft
7. (a) 3.10 m **(b)** 2450 J **9. (a)** 3.37×10^4 J **(b)** 2.28×10^4 ft·lb
11. (a) 1.65×10^3 ft·lb **(b)** 75.0 lb **(c)** 1.65×10^3 ft·lb **(d)** 175 lb **(e)** 0
13. 685 ft·lb **15.** 919 N **17. (a)** 3.99×10^5 J **(b)** -8.86×10^4 ft·lb
19. (a) 1130 J **(b)** 1.0×10^5 J **21.** 3.28 m
23. (a) 28.9 J **(b)** 371 N **25. (a)** 5000 J **(b)** 2980 J **(c)** 0 J **(d)** 0 J
(e) 2020 J **(f)** 2020 J **27.** 1.55×10^4 J
29. (a) 4.78×10^3 ft·lb **(b)** -2.67×10^3 ft·lb **31.** 1.92×10^{-17} J
33. (a) 56.0 m/s **(b)** 78.6 m/s **(c)** 89.7 ft/s **(d)** 170 ft/s
35. (a) 9810 J **(b)** 9810 J **37.** 99.0 cm/s
39. (a) 62.4 m **(b)** 199 m **(c)** 140 ft **(d)** 283 ft **41. (a)** 57.5% **(b)** 59.1%
43. (a) 29.4 W **(b)** 0.0353 hp **45.** 64.9 kW **47.** 55.4 hp
49. (a) 53.1 hp **(b)** 70.8 hp

Review Problems

1. 4.80 kJ **3.** 1.07×10^5 ft·lb **5.** 75.7 kg **7.** 3920 J
9. (a) 613 J **(b)** 5150 J **(c)** 8.66×10^4 ft·lb **(d)** 1.12×10^4 ft·lb
11. 1.53×10^{-17} J **13.** 434 J **15. (a)** 1.45×10^5 J **(b)** -3.97×10^4 J **17.** 491 J
19. (a) 59.4 m/s **(b)** 135 ft/s **21. (a)** 5.15×10^3 J **(b)** 71.5 W **(c)** 74.5%
23. 1.31 MW **25.** 0.193 hp **27.** 53.1 hp

CHAPTER 9

1. (a) 3.25 m/s (b) 34.0 m/s² 3. 92.4 s 5. 2.99 × 10⁴ m/s 7. 12.5 N
9. 2.18 mi 11. (a) 26.5 m/s (b) 89.2 ft/s 13. 0.499 15. 21.8°
17. (a) 17.5 m/s (b) 57.3 ft/s
19. (a) 12.4 rev, 4470° (b) 1200 rev, 431 000° (c) 0.404 rev, 146°
(d) 5570 rev, 2 010 000° 21. 5.00 rad 23. 96.8 m
25. (a) 270 rpm (b) 56.3 rpm (c) 239 rpm (d) 0.497 rpm
27. (a) 1.47 rad/s 14.0 rpm (b) 323 rad/s, 3090 rpm (c) 845 rad/s, 8070 rpm
29. (a) 36.7 m/s (b) 19.1 cm 31. 2.04 m/s 33. (a) 87.3 rad/s (b) 116 ft/s
35. (a) 8.73 rad/s² (b) 225 rev 37. 4.14 ft/s 39. 212 rpm
41. (a) 157 rad/s or 1500 rpm (b) 851 rad or 135 rev 43. (a) 7.08 rad/s (b) 392 rev
45. 113 lb·ft 47. 24.6 N·m or 18.1 ft·lb 49. 52.7 W 51. 2.43 rad/s
53. 14.8 N·m 55. 23.6 rad/s

Review Problems

1. (a) 2.77 m/s² (b) 4150 N 3. (a) 1.51 h (b) 8.93 m/s² 5. 63.3 ft/s
7. 16.1 m/s 9. (a) 738 m (b) 2.68 × 10³ ft 11. (a) 0.455 rev (b) 164°
13. (a) 347 rev (b) −6.44 rad/s² 15. (a) 67.4 rad/s² (b) 510 rev
17. (a) 61.8 rad/s (b) 500 ms 19. 3.26 hp
21. (a) 12.9 hp (b) 25.8 hp (c) 45.2 hp 23. 8.55 N·m or 6.31 lb·ft

CHAPTER 10

1. 43% 3. 87.3%
5. (a) 60%, 14.0, 23.3 (b) 51%, 14.1, 27.8 (c) 338 N, 3.61 m, 62.5%
(d) 71.3 lb, 10.5 ft, 59% (e) 2360 N, 300 cm, 66.1% (f) 609 lb, 59.2 ft, 62%
(g) 1960 N, 11.0 m, 8.82 (h) 275 lb, 37.5 ft, 9.38 (i) 59.1 N, 7.69 cm, 6.34
(j) 13.0 lb, 1.84 ft, 6.53 (k) 5850 N, 13.0, 15.9 (l) 3380 lb, 135, 163
7. 4.49 gal 9. 138 hp 11. (a) 234 N (b) 5.25 13. 4450 N 15. 1230 lb
17. (a) 185 N (b) 3.45 19. (a) 38.8 cm (b) 45.9 cm (c) no
21. (a) 9.6 (b) 14
23. (1) (a) 1 (b) 8.00 m (c) 12.5 ft (d) 1290 N (e) 124 lb
(2) (a) 2 (b) 16.0 m (c) 25.0 ft (d) 645 N (e) 248 lb
(3) (a) 2 (b) 16.0 m (c) 25.0 ft (d) 645 N (e) 248 lb
(4) (a) 3 (b) 24.0 m (c) 37.5 ft (d) 430 N (e) 373 lb
(5) (a) 3 (b) 24.0 m (c) 37.5 ft (d) 430 N (e) 373 lb
(6) (a) 5 (b) 40.0 m (c) 62.5 ft (d) 258 N (e) 621 lb
(7) (a) 4 (b) 32.0 m (c) 50.0 ft (d) 323 N (e) 497 lb
(8) (a) 4 (b) 32.0 m (c) 50.0 ft (d) 323 N (e) 497 lb
(9) (a) 5 (b) 40.0 m (c) 62.5 ft (d) 258 N (e) 621 lb
(10) (a) 6 (b) 48.0 m (c) 75.0 ft (d) 215 N (e) 745 lb
(11) (a) 6 (b) 48.0 m (c) 75.0 ft (d) 215 N (e) 745 lb
25. (a) 1.05 m, 736 N, 775 N (b) 1.40 m, 552 N, 581 N or 1.75 m, 441 N, 465 N
(c) 2.10 m, 368 N, 387 N (d) 2.10 m, 368 N, 387 N or 2.45 m, 315 N, 332 N
(e) 2.45 m, 315 N, 332 N 27. (a) 2.17 (b) 325 rpm
29. (a) 250 rpm, 2250 N·m (b) 120 rpm, 4690 N·m (c) 320 rpm, 1760 N·m
31. (a) C: 114 rpm counterclockwise; D: 25.3 rpm clockwise (b) A: 1930 rpm counterclockwise;
E: 398 rpm clockwise 33. (a) 4.13 (b) 5.76 (c) 72% 35. 12.8 N 37. 23.0

Review Problems

1. 77% 3. 38.2% 5. (a) 160 N (b) 10.7 7. (a) 250 N (b) 3.33
9. 13.0 11. (a) 4.67 (b) 686 rpm
13. (a) 150 rpm counterclockwise (b) F: 900 rpm counterclockwise; A: 720 rpm
clockwise 15. 4.44 17. 117 N 19. 24%

CHAPTER 11

1. 15 600 Pa, 14 600 Pa, 29 200 Pa 3. 50 600 N 5. 157 lb 7. 665 kg/m³
9. 60.6 lb
11. (a) 16.2 kPa, 2.35 lb/in² (b) 220 kPa, 31.9 lb/in² (c) 38.9 kPa, 5.63 lb/in²

Answers to Odd-Numbered Problems

13. (a) 886 kPa (b) 100 N   **15.** 1020 kg/m³
17. (a) 10.3 m, 33.9 ft   (b) 75.7 cm, 2.49 ft   (c) 15.1 m, 49.9 ft   **19.** 471 kPa
21. 417 kPa   **23.** 113 kPa   **25.** 111 kPa   **27.** 30.2   **29.** (a) 52.6 N   (b) 26.8 lb
31. 25.2   **33.** 37.8 cm   **35.** 193 N   **37.** (a) 206 N   (b) 2000 times   (c) 10.3 s
39. (a) 116 lb   (b) 38.9 times   (c) 5.45 hp
41. (a) 53.3 N   (b) 2.30   (c) 5.43×10^{-3} m³
43. (a) 5.15 N   (b) 3.50 N   (c) 4.53 N   **45.** (a) 9.0 cm   (b) 212 N   **47.** 8.33 ft²
49. 220 s
51. 168 ft/s   **53.** 3.57 cm

Review Problems

1. 700 lb   **3.** 1.97×10^4 N   **5.** 566 kPa   **7.** 15.4 m or 50.9 ft
9. 30.7 lb/ft², 36.5 lb/ft², 19.2 lb/ft²   **11.** (a) 24.5 kPa   (b) 3.25 lb/in²
13. 111 kPa   **15.** (a) 860 kPa   (b) 106 N   **17.** 1.36×10^4 N
19. (a) 109 N   (b) 4800 times   **21.** 4.18
23. (a) 18.6 L/s, 1120 L/min   (b) 59.1 m/s

CHAPTER 12

1. (a) 82°F   (b) −36°F   (c) 316°F   (d) 2462°F   (e) −292°F   (f) 72°F
3. (a) 308 K   (b) 225 K   (c) 653 K   (d) 148 K   (e) 63 K   (f) 0 K
5. (a) 580°R   (b) 492°R   (c) 386°R   (d) 274°R   (e) 1000°R   (f) 420°R
7. −40°   **9.** (a) 1900 J   (b) 10.5 J   (c) 14 900 J   (d) 942 J
11. (a) 4.05×10^6 ft·lb   (b) 4.08×10^5 ft·lb   (c) 9.34×10^6 ft·lb
(d) 3.31×10^7 ft·lb   **13.** −6550 J   **15.** 1.08×10^7 J   **17.** 1.04×10^9 J
19. (a) 647 W   (b) 5.57 s   **21.** 380 Btu/h   **23.** (a) 0.185 m²·°C/W
(b) 1.07 h·ft²·°F/Btu   **25.** (a) 4.88 cm   (b) 2.23 in.
27. 1.92 m²·°C/W   **29.** 1.64×10^5 J   **31.** 50.0 kJ   **33.** 76°F   **35.** 3.64×10^8 J
37. 371 J/(kg·°C)   **39.** 143°F   **41.** 1.22 kg   **43.** 45.1°C   **45.** 2.71×10^7 J
47. 4.0×10^6 J   **49.** 9.1×10^6 J   **51.** 12.6°C
53. (a) 3.91×10^8 J   (b) 2.10×10^5 Btu   **55.** (a) 4.92×10^5 J   (b) 5.6×10^{-3} m³
57. 18°C

Review Problems

1. (a) 17°C   (b) 91°C   (c) −44°C   **3.** (a) 245 K   (b) 362 K   (c) 1843 K
5. (a) 4.02×10^7 J   (b) 920 kcal   (c) 1.23×10^6 ft·lb   (d) 1.25×10^6 Btu
7. −1650 J   **9.** 1110 Btu   **11.** (a) 0.118 m²·°C/W   (b) 0.641 h·ft²·°F/Btu
13. 2.29 m²·°C/W   **15.** 7.04×10^7 J   **17.** 295 g   **19.** 80.1 g   **21.** 100°C
23. (a) 5.22×10^8 J   (b) 1.49×10^6 Btu

CHAPTER 13

1. 3.13×10^{-5}/°C   **3.** (a) 6.742 24 m   (b) 6.752 48 m   **5.** 0.002 84 m
7. (a) 25.011 38 ft   (b) 24.972 38 ft   **9.** 6.060×10^{-2} cm
11. (a) 352°F   (b) −320°F   **13.** 12.0094 in.   **15.** (a) 78.3°C   (b) −24.2°C
17. (a) 0.002 60 ft   (b) −0.003 12 ft   **19.** (a) 1.200 50 in²   (b) 1.198 63 in²
21. (a) 78°F   (b) 63.3°F   **23.** 88.3 L   **25.** 35.4°C   **27.** 4.51×10^{-4}/°C
29. 133 mL   **31.** (a) 45.742 50 L   (b) 42.277 50 L
33. (a) 40.3°C   (b) −16.5°C   **35.** (a) 843.50 lb/ft³   (b) 855.94 lb/ft³
37. (a) 5254.16 cm³   (b) 5298.20 cm³   (c) 5239.04 cm²
39. (a) 7.38×10^{-5} m³   (b) 7.43×10^{-5} m³   (c) 10 431 kg/m³   **41.** 6.76 ft³
43. 10.5 ft³   **45.** (a) 22.0 lb/in² gauge   (b) 32.1 lb/in² gauge   **47.** 556°C
49. 0.91 kg/m³   **51.** 215 lb/in² gauge   **53.** (a) 40.4 L   (b) 9.09 L   (c) 4.0 L
55. 195 lb/in² gauge   **57.** (a) 3.77 ft³   (b) 0.401 ft³   (c) −229°F

Review Problems

1. 2.02×10^{-5}/°C   **3.** 299.737 m   **5.** 0.245 m   **7.** 38.1 L   **9.** 203°F
11. 6.30 ft³   **13.** 1201°C   **15.** 13.3 L   **17.** (a) 7.20 L   (b) 1365°C

CHAPTER 14

1. 256 Hz   3. (a) 10.5 Hz   (b) 95.2 ms   5. 1.75 Hz, 571 ms
7. (a) 1.29 m   (b) 4.24 ft   9. (a) 1.67×10^{-15} s   (b) 1.64×10^{-6} ft
11. (a) 255 Hz   (b) 3.92 ms   13. 1500 Hz 650 μs
15. (a) 2000 m or 1.24 mi   (b) 4410 m or 2.74 mi
17. 4.74×10^{14} Hz   19. 11.7 ft   21. 965 m/s
23. (a) 5.50×10^{-7} m or 1.80×10^{-6} ft   (b) 3.56×10^{-7} m or 1.17×10^{-6} ft
25. 485 Hz or 415 Hz   27. (a) 2.90 cm, 10.3 GHz   (b) 0.760 in., 15.5 GHz

Review Problems

1. 1.74×10^{-15} s, 523 nm   3. 11 200   5. (a) 27.7 cm   (b) 0.917 ft, 833 μs
7. (a) 1.33×10^{-15} s   (b) 4.00×10^{-7} m or 1.31×10^{-6} ft   9. 0.905 ft, 13.6 ft
11. (a) 164 m, 1.83 MHz   (b) 490 ft, 2.00 MHz

CHAPTER 15

1. (a) 1.60 m   (b) 53.3 cm
3. (a) 5.88 Hz   (b) 60.7 cm, 17.6 Hz   (c) 36.4 cm, 29.4 Hz
5. (a) 0.625 ft   (b) 208 Hz   7. (a) 130 Hz   (b) 390 Hz
9. (a) 110 Hz, 10.0 ft   (b) 330 Hz, 3.33 ft   (c) 550 Hz, 2.00 ft
11. (a) 64.9 cm   (b) 127 Hz   (c) 381 Hz
13. (a) 302 m/s   (b) 319 m/s   (c) 352 m/s   (d) 1000 ft/s   (e) 1050 ft/s
(f) 1130 ft/s   15. (a) 24°C   (b) −10°C   (c) 9°F   (d) 45°F
17. 746 Hz or 754 Hz

Review Problems

1. (a) 73.5 Hz   (b) 221 Hz   (c) 28.3 cm   3. (a) 3.21 ft   (b) 2.14 ft
5. (a) 2.56 ft   (b) 84.4 Hz   7. (a) 18°C   (b) −18°C   (c) 36°F   (d) 22°F
9. (a) 3.33 ft   (b) 550 ft/s   11. (a) 106 cm   (b) 1310 m/s

CHAPTER 16

1. (a) 3.85×10^{-7} m   (b) 5.77×10^{-7} m   (c) 7.69×10^{-7} m
3. (a) 2.61×10^{19} Hz, gamma   (b) 1.00×10^{10} Hz, microwaves
(c) 1.20×10^{6} Hz, shortwave   (d) 5.41×10^{14} Hz, light
5. (a) 5.23×10^{-19} J   (b) 4.68×10^{-19} J   (c) 3.45×10^{-19} J   (d) 2.64×10^{-19} J
7. 12 000 cd   9. 76.2 lm   11. (a) 1160 lm   (b) 6940 cd   13. 23 500 cd
15. 17.0 lm/W   17. (a) 63.8 W   (b) 66.0 cd   19. 62.2 lx   21. 157 lx   23. 496 lm

Review Problems

1. (a) 6.58×10^{-7} m   (b) 4.21×10^{-7} m   3. (a) 3.10×10^{-19} J   (b) 4.63×10^{-19} J
5. (a) 2.79×10^{14} Hz, 1.08×10^{-6} m   (b) 5.43×10^{16} Hz, 5.53×10^{-9} m
(c) 6.97×10^{18} Hz, 4.31×10^{-11} m   7. (a) 315 cd   (b) 953 cd
9. (a) 452 lm   (b) 182 cd   11. 4310 cd   13. 20.0 lx   15. 150 lm

CHAPTER 17

1. 0.875 m or 2.87 ft   3. (a) −45.0 in., 12.0 in.   (b) enlarged, upright, virtual
5. (a) −3.33 in., 1.11 in.   7. (a) 2.00   (b) 1.50   9. (a) 10.3 in.   (b) −35.8 in.
11. (a) 67.9 cm, −1.00 cm   (b) 88.0 cm, −1.85 cm   (c) 165 cm, −5.09 cm
(d) ∞, ∞   (e) −31.9 cm, 3.19 cm   13. −3.75 cm, 2.50
15. (a) 22.2 cm   (b) 3.13 cm   17. (a) 10.3 cm   (b) 3.33 cm
19. −5.76 in., 0.248 in.

Answers to Odd-Numbered Problems

Review Problems

3. (a) 7.50 cm, −0.750 cm
5. (b) −7.78 cm, 1.87 cm   (c) real, inverted, same size as object   7. (a) 7.71 in.
9. −5.30 cm, 14.0 mm   11. (a) 15.6 cm   (b) 6.60 cm

CHAPTER 18

1. (a) 24.9°   (b) 2.26×10^8 m/s, 140 000 mi/s   3. (a) 1.40   (b) 2.14×10^8 m/s
5. 24.2°   7. (a) 24.4°   (b) 62.5°   9. 6.67 diopters   11. (a) −46.9 mm, 9.38 mm
13. (a) 2.77 m   (b) −79   (c) 8.0 diopters   15. (a) 30.0 in.   (b) 15.0 in.
17. (a) −15.0 cm, 3.75 cm   (b) 1.88

Review Problems

1. 2.26×10^8 m/s   3. (a) 24.2°   (b) 2.00×10^8 m/s   5. 48.9°
7. (a) 30.0 cm, −5.00 cm   (b) real
9. (a) Diverging   (c) −7.14 cm, 0.571 cm   (d) virtual
11. (a) Diverging   (c) −6.00 in., 2.00 in.   (d) virtual   13. −21.6 cm

CHAPTER 19

1. −2.03 μC   3. 1.12×10^{-18} C   5. 3.28×10^{11} 7. 5.76×10^{10} N   9. 0.622 N
11. 2.56×10^{-7} N   13. 2.38×10^{-10} C   15. 4.48 cm   17. 5.12×10^{-17} J
19. (a) 1.53×10^{-17} J   (b) 95.7 V   21. (a) 6.43×10^3 V/m   (b) 1.03×10^{-15} N

Review Problems

1. −60.8 mC   3. 1.49×10^{16} 5. 0.309 N
7. (a) 5.12×10^{-16} N   (b) 7.65×10^6 m/s   (c) 1.36×10^{-8} s
9. (a) 1.92×10^{-18} J   (b) 2.05×10^6 m/s
11. (a) 5630 V/m   (b) 9.00×10^{-16} N   (c) 2.16×10^{-17} J

CHAPTER 20

1. 0.50 A   3. (a) 2.1 C   (b) 1.3×10^{19} 5. 2.0 A   7. 1.97×10^{11} 9. 105 C
11. (a) −28.8 C   (b) 56.5 mA   13. 6.14 J   15. 2.67 Ω   17. 300 Ω   19. 9.17 Ω
21. 128 V   23. (a) 462 Ω   (b) 55.4 V   25. (a) 42.9 Ω   (b) 817 mA
27. (a) 1.19 A   (b) 1.86×10^{19} 29. 2.45×10^{-8} Ω·m   31. 1.02 mm^2 33. 26.9 Ω
35. (a) 40.0 Ω   (b) 3.16 Ω   (c) 23.8 Ω
37. (a) 60.0 Ω   (b) 10.0 Ω   (c) 15.0 V, 30.0 V, 45.0 V
39. (a) 73.3 Ω   (b) 18.3 Ω   (c) 27.5 V
41. (a) 6.00 Ω   (b) 12.0 Ω   (c) 750 mA, 150 mA, 600 mA
43. (a) 17.0 Ω   (b) 2.50 A   (c) $V_3 = 18.5$ V, $I_3 = 2.50$ A, $V_1 = V_2 = 24.0$ V, $I_1 = 2.00$ A, $I_2 = 0.50$ A   45. $R = 40$ Ω, $I_1 = I_2 = 0.25$ A, $I_3 = I_4 = 0.125$ A

Review Problems

1. (a) 83.3 mA   (b) 52.3 mA   3. 1.79 V   5. 6.67 mΩ   7. 84.6 V
9. (a) 2.58×10^{-4} Ω   (b) 27.1 mV   11. 87 cm
13. (a) 33.3 Ω   (b) 66.0 Ω   (c) 66.1 Ω

CHAPTER 21

1. 71.5 W   3. 16.7 V   5. 104 A   7. 579 s   9. (a) 202 Ω   (b) 0.55 A
11. (a) 0.522 A   (b) 1.71 cents   13. 17.3 V, yes   15. 11   17. 2.4×10^{-19} J
19. 6.20 V   21. 706 mA   23. (a) 1.50 V   (b) 42.0 mΩ   (c) 199 mA   25. 9.00 V
27. (a) 10.0 mA   (b) 3.33 mA   29. (a) 2.07 MJ   (b) 14.4 h   (c) 1.08×10^{24}

Review Problems

1. (a) 0.94 W (b) 0.22 W **3.** 90.9 s **5.** 8.85 V
7. (a) 27.0 V (b) 675 mΩ (c) 26.5 V **9.** (a) 2.81 MJ (b) 5.20 h
11. 3.9×10^{-4} A

CHAPTER 22

1. (a) 3.50×10^{-15} N (b) 3.85×10^{15} m/s² **3.** 2.00×10^{-16} N **5.** 2.21 A
7. 29.1 cm **9.** 1.49 mT **11.** (a) 17.9 mT (b) 214 mT **13.** 4.21×10^{-5} T·m/A

Review Problems

1. 3.42×10^{-13} N **3.** 4.29×10^{-4} T **5.** 750 mT **7.** 256 mA

CHAPTER 23

1. (a) 1 : 110 (b) 2.27 A **3.** 4.74 A, 2.64 MV

Review Problems

1. (a) 1 : 160 (b) 1.38 V **3.** (a) 2.61 A (b) 5.75 MV

CHAPTER 24

1. 25.5 A **3.** 80.3 mA **5.** 8.48 V **7.** 8.00 W·h **9.** 455 kΩ **11.** 318 kHz
13. (a) 94.2 Ω (b) 11.8 kΩ (c) 2.36 MΩ
15. (a) 754 Ω, 153 mA (b) 283 Ω, 407 mA (c) 84.8 Ω, 1.36 A **17.** 7.14×10^{-6} F
19. (a) 2.20 kΩ (b) 1.21 μF **21.** 2.21 μF **23.** (a) 50.0 V (b) 53.1°
25. (a) 50.9 Ω, 261 Ω, 441 mA, 11.2° (b) 1.77 kΩ, 2.82 kΩ, 8.85 mA, 38.8°
(c) 1.81 kΩ, 47.0 kΩ, 532 μA, 2.21° **27.** (a) 130 V (b) −22.6°
29. (a) 564 Ω, 731 Ω, −50.5°, 157 mA (b) 41.4 Ω, 69.7 Ω, −36.5°, 1.65 A
(c) 106 Ω, 107 Ω, -81.9°, 1.07 A **31.** (a) 113 Ω (b) −22.0° (c) 398 mA
33. (a) 381 Ω (b) 95.3 V (c) −25.2° **35.** 1.41×10^{-4} H
37. (a) 25.0 kW (b) 160 W (c) 12.5 : 1

Review Problems

1. (a) 170 V (b) 833 mA (c) 1.18 A **3.** 12.7 V **5.** 30.2 Ω **7.** 183 Ω
9. 55.0 μC **11.** (a) 400 Ω (b) 6.63 μF **13.** (a) 30.0 V (b) 60.0°
15. (a) 600 Ω (b) 22.6 Ω (c) 600 Ω **17.** (a) 591 Ω (b) 45.3° (c) 195 mA
19. 2.05 MHz **21.** (a) 113 kW (b) 4.50 kW (c) 5 : 1

APPENDIX A

1. (a) 4 (b) −12 (c) −20 (d) 7 (e) −20 (f) −2 (g) 45 (h) 96
(i) −96 (j) −84 (k) −6 (l) −8 (m) 9 (n) 27 (o) −168 **3.** 4.14 cm
5. (a) 9 (b) 3 (c) −8 (d) 3 (e) 2187 (f) 3125 (g) 5.95×10^{-7}
(h) 8192 (i) 0.028 (j) a^4 (k) 12 (l) 1 (m) 1.68×10^7 (n) 1
7. (a) 35.15 m (b) 81.258 s (c) 4.0294 cm (d) 6.20×10^4 m (e) 2.40×10^7
(f) -3.318×10^{-5} (g) 93 (h) 1.4×10^{-11} (i) 5.7×10^{-12} (j) 1.4×10^{-15}
(k) 59 (l) 2.7×10^8 (m) 4.94×10^{-10}
9. (a) $t = \dfrac{5d + b}{a}$ (b) $t = \sqrt{\dfrac{2s}{a}}$ (c) $t = \dfrac{2q}{a + b}$ (d) $t = \dfrac{a + b}{c + 2d}$

INDEX

Tables are in *italics;* for topics with multiple page references, the most important reference is in **boldface**.

A

Aberration:
 chromatic, **327,** 328
 spherical, **304,** 306, **326,** 328
Abscissa 64
Absolute pressure, **205,** 213
Absolute temperature,**216,** 244
Absolute zero, 244, **447**
AC (*see* alternating current)
Acceleration, 10, **58,** 61, 77
 angular, **163,** 169
 average, 65
 centripetal, **155,** 163, 169
 due to gravity, 70, 73, **79,** 447
 and mass, 76
 tangential, **163,** 169
 uniform, **64, 67,** 73
Achromatic lens, 327
Acoustics, 267
Acre, 441
Action, line of, 78
Action and reaction, 81, 91
Actual mechanical advantage, **173,** 177
Addition of vectors, 59, 95, **98, 105,** 109, 115
Adhesion, 197
Air columns, 267, **269,** 278
Air pressure, 199

Air wedge, 262
Algebraic manipulation, 432
Alkaline-manganese cell, 370
Alternating current, 348, **407,** 409
 effective current and voltage, **407,** 409
 frequency, 411
 period, 411
 phase, 411, 417, 430
 power, 407
 resonance, 423, 430
Alternating current generators, **405,** 409, **411,** 430
Alternator, 405
Altimeter, 204
AMA, **173,** 177
 hydraulic jack, 207
 hydraulic press, 207, 213
Ammeter, 350
Ampere, 9, **346,** 365, **444**
Ampere hour, **347,** 378, *442*
Amplitude modulation, 263
Amplitude of vibration, **251,** 271
Amplitude of a wave, 255
Aneroid barometer, 203
Angle:
 of bank, **158,** 169
 critical, **313,** 328
 of dip, 384
 of elevation, 311

 of incidence, 259, 309
 phase, 417, 419, 430
 of reflection, 259, 309
 of refraction, 309
 solid, 9
Angular acceleration, **163,** 169
Angular displacement, **162,** 169
Angular measurement, 9, 11, **159,** *442*
Angular motion, 162
Angular velocity, *10,* **162,** 169
Angular velocity, gears, **186,** 194
Anode, 370
Antinode, 264, 268
Aperture:
 camera, 324
 lens, 316
 mirror, 297
Archimedes' principle, **209,** 213
Area, *10, 11, 441*
Area, surface. 24
Area expansion, **239,** 247
Armature, 394, 396
Artificial magnets, 381
Astronomical telescope, 325
Atmosphere, 199, *442*
Atmospheric pressure, 199, *447*
Atmospheric refraction, 311
Atom, 1, **332,** 344
Atomic clock, 9

458

Atomic mass unit, 447
Atomic nucleus, 1, **332,** 244
Atomic structure, **332,** 344
Attenuation, 275
Atto, *12*
Autotransformer, 404
Average acceleration, 65
Average speed, **55,** 61
Average velocity, 64
Avogadro's number, *444,* 447
Axis of rotation, 124, 154

B

Back emf, **401,** 409
Balance, chemical, 81
Banked curves, **158,** 169
Barometer, 203
Base units, **8,** 9
Battery, 350, **372**
Bearing, 88, *89*
Beats, **263,** 274
Belt driven pulleys, 186, 194
Bevel gear, 188, *189*
Bimetallic strip, 238
Binoculars, 326
Block and tackle, 182
Boiling liquid, 229, **230**
Boiling point, 216, *225*
Boom, 118
Bourdon gauge, 205
Boyle's law, **242,** 247
Brinell hardness test, 3
British thermal unit, **218,** 233, *442*
Brushes, 395, 405
Buoyancy, 209

C

Calipers, 6
Calorie, **218,** 233
Calorimeter, 226
Cam, 191, **192**
Camera, 323
Candela, 9, **287,** 291, **444**
Candlepower, 287
Capacitance, *10,* **414,** 430
Capacitive reactance, 416
Capacitor, **414,** *415*
Carrier wave, 263
Cartesian coordinate system, 31
Cathode, 370
Caustic curve, 304
Cells, 370
Celsius temperature scale, *11,* **215**
Center of curvature:
　lens, 316
　mirror, 297
Center of gravity, **128,** 132
Center-tap transformer, 429
Centi, *12*
Centigrade temperature (*see* Celsius temperature)

Centrifugal force, 158
Centripetal acceleration, **155,** 163, 169
Centripetal force, 155, **156,** 169
Chain hoist, **183,** 194
Change in gravitational potential energy, 142
Charge, electric, *10,* **330,** *443*
　conservation, **332,** 344
　distribution, 340
　electron, 333
　induced, 334
　polarity, 331, **336,** 344
　proton, 333
　separation, 336
Charging by conduction, 335
Charging by induction, 335, **336**
Chemical balance, 81
Chemical cells, 370
Choke, 401
Chromatic aberrration, 327
Circuits, electric, 346
　alternating current, 348, **407,** 409
　direct current, 346
　parallel, **360,** 365
　resonant, **423,** 430
　series, **358,** 365, **372,** 378
Circuit symbols, *349*
Circular motion, 154
Circular waves, 258
Circulation, fluid, 223
Classes of levers, 172
Coefficient of friction, 86, *87,* 92, 157, 169
Coefficient of linear expansion, **236,** *237*
Coefficient of volume expansion, **240,** 247
Cohesion, 197
Coil, 393, 401
Collinear vectors, **59,** 61, 70
Color, light, 282
Combinations of cells, **372,** 378
Combinations of resistors, **357,** 365
Combustion, heat, 231, **232,** 234
Combustion engines, 424
Commutator, 395, 407
Compass, magnetic, 381
Components of vectors, **101,** 105, 109, 117
Compound, 2
Compound machine, 194
Compound microscope, 325
Compression ratio, 30
Compression wave, **254,** 273
Concave mirror, 297
Concurrent forces, 114
Condensation, wave, **254,** 272
Condensing lens, 324
Conductance, *10*
Conduction, **220,** 233
Conductivity, thermal, *10,* 220
Conductor:
　electricity, 333
　heat, 219

Index

Conservation of:
　charge, **332,** 344
　energy, **144,** 149, 151, 400
　mechanical energy, 144
　momentum, 91
　thermal energy, 226, 233
Constants, physical, *447*
Constant of proportion, **28,** 34, 45
Constructive interference, **261,** 274
Convection, 222
Conventional electric current, **346,** 350, 365
Converging lens, **315,** *319, 320,* 328
Conversion factors, 13, **17,** 57, *441*
Conversion of prefixes, 12
Conversion of units, 13, 16
Convex mirror, 297
Corona discharge, 343
Cosine, 42, 46
Cosmic ray, 281
Cost, electricity, 368
Coulomb constant, 339, *447*
Coulomb's law, **338,** 344
Coulomb unit, *10,* **331,** 344, **347**
Counter emf, **401,** 409
Crest, wave, 253, 258
Critical angle, **313,** 328
Cross-sectional area, 22
Current, 9, **332, 346,** 348, 350, 365
　alternating, 348, **407,** 409, **411,** 430
　conventional, **346,** 350, 365
　direct, 348
　eddy, 402
　effective, **412,** 430
　and magnetic fields, 385, 388, 396
　peak, 412
　steady state, 348
　transient, 348
Current loop, **388,** 396
Curvature:
　center of, **297,** 316
　radius of, 297, 316
Cycle:
　alternating current, 412
　vibration, 154, 250

D

Dam, 425
Damping, 251
D'Arsonval galvanometer, 394
Data extrapolation, 33
Data point, 31
Day unit, *11, 442*
Deci, *12*
Decibel, 275
Deformations, 2, 3
Degree, angle, 160, *443*
Degree, temperature, 215
Deka, *12*
Demagnetization, 384

Index

Demodulation, 263
Density, *10*, 37, **38**, 46
Density, relative, *38*, **40**, 46, 209, 213
Dependent variable, 19
Depth of field, 324
Derived unit, 8, **9**, *10*, **17**
Destructive interference, **261**, 274
Diaphragm, camera, 324
Dielectric, 333
Differential pulley, 183
Diffraction, 260
Diffuse reflection, 294
Diopter, 321
Dip, angle, 384
Direct current, 346, **348**
Direct current generators, **407**, 409
Direct current motors, **394**, 396
Directions, 54
Discharge tube, 285
Displacement, **53**, 61, 71, 99
Displacement, angular, **162**, 169
Distance, **53**, 61, 99
Distinct vision, 324
Distortion, 327, 328
Distribution of charge, 340
District transformer station, 429
Diverging lens, **315**, *319*, *320*, 328
Domain, magnetic, 384
Doppler effect, sound, **274**, 278
Double refraction, 315
Drag, 114
Driven gear, 188
Driving gear, 188
Dry cell, 370
Duality concept, 283
Ductility, 3
Duct size, 211
Dynamic equilibrium, 114
Dynamic pressure, 211
Dynamics, 76
Dynamic speaker, 390
Dynamo, 376

E

Ear, **267**, 275, 278
Earth, electric, 335
Earth dimensions, 447
Eclipse, 284
Eddy currents, 402
Effective current, **412**, 430
Effective voltage, **412**, 430
Efficacy, luminous, *288*, 291
Efficiency, **149**, 151, 173, 174, 194, 207
Einstein quantum theory, **283**, 291
Elastic material, 81
Elasticity, 3
Elastic potential energy, 142
Electrical ground, **335**, 344
Electric capacitance, *10*, **414**, 430
Electric charge, *10*, **330**, *443*
Electric circuits, 346

Electric current, 9, **332**, **346**, 348, 350, 365, 444
Electric discharge, **337**, 343
Electric energy, 367
Electric energy sources, 370
Electric field, 341
Electric field strength, *10*
Electric forces, **339**, 343, 344
Electric ground, **335**, 344
Electric insulator, 333
Electric load, 352, 369
Electric motor, **394**, 396
Electric potential difference, *10*, **341**, 344
Electric power, 367
Electric power transmission, **424**, 428
Electric relays, 391
Electric resistance, *10*, 348, **352**, 365
Electric resistivity, **356**, *357*, 365
Electric resonance, **423**, 430
Electric welding, 367
Electrification, 330
Electrode, 370
Electrolyte, **333**, 370
Electrolytic cell, 370
Electromagnet, 390
Electromagnetic induction, **398**, 409
Electromagnetic radiation, 253, **281**, 291
Electromagnetic spectrum, *282*
Electromagnetic waves, 250
Electromotive force, **351**, 365
Electron, 1, 332, 344
  charge, 333
  mass, 333, 447
Electronics, 346
Electroscope, 334
Electrostatic generator, 337
Electrostatic precipitator, 343
Electrostatics, **330**, 344
Element, 1, 332
Elementary charge, 333, *442*, 447
Elevation, angle, 311
Emf, **351**, 365
Energy, *10*, *15*, 135, **139**, 156, *442*
  conservation, **144**, 149, 151
  elastic, 142
  electric, 352, 367
  friction, 149, 151
  gravitational, **142**, 151
  heat, 218
  internal, **218**, 233
  kinetic, 139, **140**, 151
  light, 281, **283**, 291
  photon, **283**, 291
  potential, **142**, 151
  rotational, 166
  sound, 275
  thermal, 2, **218**, 226, 233
  work, 140
Engineering units, **15**, 17
Equation, 19
Equilibrant, 120

Equilibrium, 78, **112**, 114, 117, **127**, 132
  concurrent forces, **114**, 117
  dynamic, 114
  nonconcurrent forces, 127
  static, 114
Equipotential surface, 343
Equivalent resistance, 358, 360
Evaporation, 2, 230
Exa, *12*
Expansion coefficient:
  area, **239**, 247
  linear, **236**, *237*, 241, 247
  volume, **240**, 247
Expansion of a gas, **242**, **244**, 247
Exponents, 4, **434**
Exposure time, 324
Eye, 280, 323
Eyepiece, 325
Eye sensitivity curve, 286

F

f-number, 324
Fahrenheit temperature, 215
Falling objects, 70
Farad, unit, *10*, **416**, 430
Faraday constant, 447
Far principal focus, 316
Fatigue, 3
Femto, *12*
Ferromagnetism, **384**, 396
Fibers, optical, 314
Fields:
  electric, *10*, 341
  magnetic, 10, 380, **382**
First equilibrium condition, 112
First law of motion, 76, 92
First law of thermodynamics, **218**, 233
First principal focus, 316
Floatation, 209, 213
Flow rate, fluid, 211, 213
Fluid, 2, **197**, 212
  buoyancy, 209, 213
  pressure, 199
Fluorescence, 286
Fluorescent lamp, 286
Flux:
  luminous, *10*, **286**, *288*, 291
  radiant, **286**, 291
Flux density, *10*, **385**, 396
Focal length:
  lens, 318
  mirror, 298
Focus:
  principal, 298, 316
  virtual, 316
Footcandle, **290**, 292, *442*
Footlambert, *443*
Footpound, *15*, **135**, 139, 151
Foot unit, **15**, 441
Force, 2, *10*, 16, 76, 92, *442*
  buoyant, 209
  centrifugal, 158

Force (*cont.*)
  centripetal, 155, **156**, 169
  concurrent, coplanar, 114
  electrostatic (Coulomb), **339**, 343, 344
  friction, 86, 92, 119
  gravitational, 2, **79**, 80, 92
  line of action, 78
  magnetic, 381, **385**, 396
  moment of, *10*, **124**, 132
  nonconcurrent, **124**, 127
  normal, **82**, 86, 119
  reaction, 82, 129, 198
  tangential, 125
Forced convection, 223
Forced vibration, 271
Formula, **19**, 28, 31, 35
Formula manipulation, 438
Free body diagram, 82, 113, 117
Free fall, **70**, 114
Freezing point, 216, *225,* 228
Frequency, *10*
  alternating current, 411
  angular motion, 162
  beat, 263
  fundamental, **269**, 277
  harmonic, **269**, 277
  light, 282
  natural, 252
  resonant, 267, 423, 430
  rotation, 162
  vibration, 155, **250,** 265, 267
  wave, 255
Frequency modulation, 264
Frequency response, ear, *275,* 278
Friction, **86**, *87,* 92, 119, 149, 151, 156, 173
Fringe patterns, 262
Fuel cell, 375
Fulcrum, lever, 176
Fundamental frequency, **269**, 277
Fuse, 367
Fusion, latent heat, *225,* **228**, 233

G

Galilean telescope, 326
Gallon, 441
Galvanometer, 394
Gamma ray, 281
Gas, 2, 197
Gas constant, universal, 447
Gas discharge, 285
Gas laws, 242, 244, 246, 247
Gauge, Bourdon, 205
Gauge pressure, **205**, 243
Gears, 186, *189,* 194
Gear ratio, 187, 188
Gear train, 188, *189,* 194
General gas equation, 246, 248
Generators, 337, 376, **405, 407,** 409, 424
Geometrical optics, 283
Giga, *12*
Graph, 31, 45, 64

axes, 31
data point, 31, 32
displacement-elapsed time, 64
intercept, 32
slope, 32
velocity-elapsed time, 65
Gravitation constant, 447
Gravitation, force, 2, **79**, 80, 92
Gravitational potential energy, **142**, 151
Gravity:
  acceleration due to, 70, 73, **79**
  center of **128**, 132
  force of, *10*, 12, **79**, 92
  specific, *38,* **40**
Ground, electric, 335, 344

H

Hardness, 3
Harmonic, **269**, 277
Head, pressure, 201
Heat, 215, **218**
  combustion, **231**, *232,* 234
  conduction, **220**, 233
  convection, 222
  energy, 218
  flow, 220, 222, 223
  latent, *225,* **228**, 229, 233
  radiation, 223
Heat capacity, *10,* **224**, *225,* 233
Heat sink, 219
Heat transfer, 219, 226, 233
Hectare, *10,* 441
Hecto, *10*
Helical gear, 188, *189*
Henry unit, *10,* **401,** 409
Hertz unit, *10,* **251,** 265
Hooke's law, 81
Horsepower, *15,* **146**, 149, 442
Hour unit, *11,* 442
Human ear, 267, 275
Human eye, 280, 323
Hydraulic jack, 207
Hydraulic press, **172, 197, 206,** 213
Hydroelectricity, 425
Hydrometer, 210
Hydrostatic pressure, 211
Hypotenuse, 41

I

Ideal gas law, 246, 248
Ideal mechanical advantage (IMA), **174,** 179
  cam, **192,** 194
  chain hoist, 184
  gears, **186,** 194
  hydraulic press, **207,** 213
  inclined plane, 191
  jack, **193,** 194
  lever, **177,** 194
  screw, 192
  screwjack, **193,** 194
  sheave, 182

  wheel and axle, **180,** 194
Ideal transformer, **403,** 409
Illuminance, *10, 280,* **289,** 292
Illuminating system, 324
Illumination, *280,* **289,** 292
IMA, (*see* ideal mechanical advantage)
Image, 295, **296,** *300,* 306, 319, 320
Image distance, 296, 316, 318
Impedance, **418,** 430
Impulse, 89, **90,** 92
Incandescence, 285
Inch, *15,* 441
Incident angle, 255, 309
Inclined plane, **172,** 191
Independent variable, 19
Index of refraction, **310,** 328
Induced charge, 334
Induced emf, **398,** 409
Induced magnetism, 380
Inductance:
  mutual, 401
  self, **401,** 409
Induction, charge by, 335
Induction, electromagnetic, **389,** 409
Induction, magnetic, **385,** 396
Induction, mutual, 401
Induction, self, **401,** 409
Induction coil, 404
Inductive reactance, **413,** 430
Inductor, 401
Inertia, 2, **16,** 76, 92
Insulator:
  electricity, 333
  heat, 219
Intensity, electric, *10*
Intensity, luminous, 9, **287,** 291
Intensity, magnetic, *10*
Intensity, radiant, *10*
Intensity, sound, **275,** 278
Interference, sound, 274
Interference, waves, 261
Internal combustion engine, 424
Internal energy, 217, **218,** 233
Internal reflection, 312
Internal resistance, 372
International system of units (SI), 7, **8, 17**
Inverse proportion, 29
Inverse square law, 290, 292
Ion, 333
IR drop, 353

J

Jack, 207
Joule unit, *10,* **135,** 139, 151, *442*
Journal, bearing, 88

K

Keeper, magnet, 385
Kelvin temperature, 9, **216,** 244, **444**
Kilo, *12*

Kilogram, 8, 9, **444**
Kilowatt hour, **368**, 378
Kinematics, 64
Kinetic energy, 139, **140**, 151, 156, 250
Kinetic energy-work theorem, 140
Kinetic friction, *87*, 92

L

Lamps, 285
Latent heat, *225*, **228**, **229**, 233
Laws of motion, **76**, 83, 92, 112, 139
Laws of reflection, 259, 265, 295
Laws of refraction, 310
Lead acid storage cell, 370
Leaf electroscope, 334
Leakage loss, 428
Length, 9, *441*, **444**
Lens, 308, **315**
 achromatic, 327
 aperture, 316
 converging, **315**, 328
 diverging, **315**, 328
 focal length, 318
 linear magnification, **321**, 328
 objective, 325
 ocular (eyepiece), 325
 power, **321**, 328
 ray diagram, 316
 sign convention, 318, **320**
Lens aberrations, 326, 327, 328
Lens defects, 326
Lens equation, **318**, 328
Lenz's law, **400**, 409
Lever, **172**, 177, 194
Lift, 114
Light, 280, *443*
 characteristics, 286
 color, 282
 energy, 281, **283**, 291
 frequency, 282
 illumination, *280*, **289**, 292
 interference, 262
 nature, 281
 photon, **283**, 291
 quantum theory, **283**, 291
 reflection, 294
 refraction, 259, **308**, 328
 shadows, 283
 sources, 285
 speed, 281, 308, 447
 wavelength, 282, 291
 wave theory, 281
Line of action of a force, 78
Linear expansion, **236**, *237*, 241, 247
Linear magnification:
 lens, **321**, 328
 mirror, **236**, 237
Linear momentum, **89**, 92
Line resistance losses, 427
Lines of force, **382**, 396
Liquid, 2, 197

Liquid-in-glass thermometer, 241
Liquid pressure, 200, 206, 213
Liter unit, *11*, 441
Load, circuit, 352, 369
Local transformer station, 429
Longitudinal wave, 254
Losses, power transmission, 427
Loudness, sound, 275
Lumen unit, *10*, **286**, 291
Luminance, *10*
Luminous body, 280, 291
Luminous efficacy, **288**, 291
Luminous flux, *10*, **286**, **288**, 291
Luminous intensity, 9, **287**, 291, **444**
Lunar eclipse, 284
Lux unit, *10*, **290**, 292, *442*

M

Machine, 149, 151, **172**, 194
Magnet, 380, 384
Magnetic compass, 381
Magnetic domain, 384
Magnetic field of the earth, 383
Magnetic fields, 380, **382**, 386
 center of a current loop, **388**, 396
 around a solenoid, **388**, 396
 around a straight current, **386**, 396
 around a toroid, **389**, 396
Magnetic flux density, *10*, **385**, 396
Magnetic forces, 381, **385**, **392**, 396
Magnetic induction (induction), *10*, **385**, 396
Magnetic keepers, 385
Magnetic line of force, **382**, 396
Magnetic permeability, 387, 389
Magnetic poles, 381
Magnetic shielding, 385
Magnetic torque, 394
Magnetism, 380
Magnetization, 380, **384**
Magnification, linear, **301**, 306, **321**, 328
Magnifier, 324
Magnitude, 50, 61, 95
Malleable, 3
Manometer, 204
Mass, 2, 9, *11*, *15*, 77, **79**, 80, 92, *442*, **444**
Mass density, *10*, 37, *38*, 46
Mass of earth, 447
Mass of moon, 447
Mass of sun, 447
Matter, 1
Mean spherical luminous intensity, **287**, 291
Measurement, 1, **5**, 7, **17**
Mechanical advantage, 173, 177
Mechanical equivalent of heat, 218, 447
Mechanical work, 135
Mega, *12*
Melting point, *225*

Mercury barometer, 203
Mercury cell, 370
Meter unit, 9, 441, **444**
Micro, *12*
Micrometer caliper, 6
Microphone, 390
Microscope, 325
Microwave, 281, **282**
Mile, *15*, 441
Milli, *12*
Millibar, 198
Minute, angle, 160, *443*
Minute, time, *11*, 442
Mirror, 294
 parabolic, 304
 plane, **295**, 306
 spherical, **296**, 306
Mirror equation, **301**, 306
Miter gear, 188, *189*
Mixture, 2
Mode of vibration, **267**, 270
Modulation, 263
Mole, 9, **444**
Molecular motion, 198
Molecule, 2
Moment arm, 124
Moment of a force, **124**, 132
Momentum, 77, **89**
 conservation, 91, 92
 linear, 89, 92
Moon dimensions, 447
Motion:
 accelerated, **64**, **67**, 73
 angular, 162
 circular, 154
 dynamics, 76, 92
 energy, 139, **140**, 151
 free fall, 70, 144
 graphical analysis, 64
 kinematics, 64, **67**, 73
 laws, 76, **77**, 81, 90, 92, 198
 molecular, 198
 periodic, 154, 250, 265
 rotary, 154
 uniform, 67
Motor, electric, **394**, 396
Movable pulley, 180
Multiplication of units, 11
Mutual induction, 401

N

Nano, *12*
Natural convection, 223
Natural frequency, 252, 267, 268
Natural magnet, 380
Nature of light, 281
Near principal focus, 316
Negative charge, 331, 332, 344
Negative exponents, 435
Negative vector, **60**, 101, 108
Negative work, 137
Net charge, 331
Neutron, 332, 344, 447

Newton unit, *10*, **77**, 92, 112, 139, *442*
Newton's laws of motion:
  first law, 75, 92
  second law, **77**, 90, 92, 112, 139
  third law, 81, **91**, 92, 198
Nickel-cadmium cell, 370
Node, **264**, 268
Noise, 267
Nonconcurrent forces, **124**, 127
Normal force, **82**, 86, 119
Normal to surface, 259
North magnetic pole, 381
Nuclear reactor, 426
Nucleus, atomic, 1, **332**, 344
Number line, 13

O

Object distance, 296, 316
Objective lens, 325
Ocular, 325
Ohm unit, *10*, **353**, 365, 414, 416, 418
Ohm's law, **355**, 365
Ohmmeter, 356
Opaque, 283
Open circuit, 360
Opera glass, 326
Optical center, 316
Optical density (*see* index of refraction)
Optical fibers, 314
Optical instruments, 323
Optics, geometrical, 280, 283
Ordinate, graph, 64
Origin, vector, 50
Oscillations, 154
Oscillators, 423
Ounce, 15

P

Parabolic reflectors, 304
Parallel circuits, 360, 365
Parallelogram of vector addition, **98**, 109
Pascal unit, *10*, 198
Pascal's principle, **206**, 213
Peak current, 412
Peak voltage, 412
Pendulum, simple, 144
Penstock, 425
Penumbra, 284
Period:
  alternating current, 411
  circular motion, 162
  rotation, 162
  vibration, 255
  wave, 154, 169, 250, 265
Periodic motion, 154, 250, 265
Permanent magnet, 384
Permeability, 387, 389, *447*

Permittivity, *447*
Peta, *12*
Phase:
  alternating current, 411, 412, 417, 419, 430
  changes of, 228
  vibration, 255
Photoelectric cell, 375
Photometry, 280
Photon, 283, 291
Physical constants, *447*
Physical quantity, 7
Pico, *12*
Piezoelectricity, 376
Pitch, screw, 192
Pitch, sound, 276
Pivot point, 124
Planck constant, 283, *447*
Plane mirror, 295, 306
Plane wave, 258
Polar form, **51**, 103, 105, 109
Pole, magnetic, 381
Pole transformer, 429
Polygon method of vector addition, **95**, 109, 115
Position, 53
Positive charge, 331, 332, 344
Potential difference, *10*, **341**, 344
Potential drop, **342**, 353, 365
Potential energy, **142**, 151
Potential gradient, **343**, 344
Pound unit, 15, *442*
Power, *10*, 15, 135, **146**, 148, 151, *442*
  dissipation, 412
  electric, 367, **368**, 378
  radiant, 286
  rotational, 166, **167**
  transmission, 424, 428
Power factor, 428
Power of a lens, 321, 328
Power line, 424
Precision, 5
Prefix, SI, **11**, *12*, **17**
Prefix conversions, 12
Pressure, *10*, 15, **197**, 212, *442*
  absolute, 205, 213
  atmospheric, 199, *447*
  depth variation, 201
  dynamic, 211
  fluid, 199
  gauge, **205**, 243
  hydrostatic, 211
  liquid, 200, 213
  measurement, 203
  sound, 275
  transfer, 206
  vapor, 230
Pressure altimeter, 205, 213
Pressure head, 201
Primary cell, **370**, 378
Primary coil, 400, 402
Principal axis, 297, 316
Principal focus, 298, 316
Principle of superposition, 261, 265

Prism, 313
Prism binoculars, 326
Problem solving, 20
Projection lens, 324
Projector, 324
Properties of materials, 2
Proportion, **28**, 34, 45, 64
Proportion constant, 28, 34, 45
Proton, 332, 344, *447*
Pulley, **180**, 186, 194
Pulse, wave, 253
Pythagorean theorem, **41**, 46, 102, 105, 106, 109, 121

Q

Quantity of matter, 444
Quantum, **283**, 291
Quantum theory, 282

R

Radiant flux (power), **286**, 291
Radiant intensity, 10
Radian unit, 9, **160**, *442*
Radiation, electromagnetic, 223, 233, 253, **281**, 291
Radiative heat, 223
Radius of curvature:
  lens, 316
  mirror, 297
Radius of the earth, 447
Radius of the moon, 447
Radius of the sun, 447
Rankine temperature, 217, 244
Rarefaction, **255**, 272
Rating of a cell, 374
Ratio, **28**, 42, 45
Ray, 259, 283
Ray diagram:
  lens, 316
  mirror, 297
Reactance:
  capacitive, 416
  Inductive, **413**, 430
Reaction force, **82**, 129, 198
Reactor, nuclear, 426
Real image and object, 296, 306, 318
Real principal focus, 316
Recharging a cell, 370
Recording, sound, 277
Reduction of friction, 88
Reflection, 294
  diffuse or specular, 294
  internal, 312
  laws, 259, 265
  regular, 294
  sound, 276
  waves, 259, 265
Reflectors, parabolic, 325
Refracting telescope, 325
Refraction, 259, **308**, 328

Index

Refraction (cont.)
  angle, 309
  atmospheric, 311
  double, 315
  index, **310**, 328
  laws, 310
  sound, 273
Regular reflection, 294
Relative density, 38, **40**, 46, 209, 213
Relay, 391
Reproduction of sound, 277
Resistance:
  electric, *10*, 348, **352**, 365
  internal, 372
  thermal, 221
Resistivity, **356**, *357*, 365
Resistors, 353, *354*
  parallel, **360**, 365
  series, **358**, 365
  series-parallel, 362
  types, 353, *354*
Resolution of vectors, 101
Resonance, **252**, 271
Resonant circuits, **423**, 430
Resonant frequency, 267, 268, 423, 430
Resultant vectors, 59, 61, **95**, **101**, 106, 109, 114, 120
Reverberation time, 276
Revolution, 159, *443*
Revolving field generator, 407
Right hand rule, 386, 388, 398
Rigid body, 2
Ripple, 407
Ripple tank, 257
Rockwell hardness, 3
Rotation, axis, **124**, 127
Rotational energy, 166
Rotational motion, 154, **164**
Rotational power, 166, **167**
Rotor, 394, 407
Rounding off, 7, **436**
R-value, *221*, 233

S

Saturated vapor, 230
Scalar quantity, **50**, 61
Scale diagram, 50, 95, 109, 114, 121
Scale factor, 51
Scientific notation, 3, 17
Screw, 192
Screwjack, 192
Secondary cell, **370**, 378
Secondary coil, 400, 402
Second equilibrium condition, 127
Second law of motion, **77**, 90, 92, 112, 139
Second principal focus, 316
Second unit, angle, 160, *443*
Second unit, time, 9, *444*
Self inductance, **401**, 409

Semiconductor, 334
Sensitivity curve, eye, *286*
Series circuits, **358**, 365, **372**, 378
Shadows, 283
Shaft, bearing, 88
Sheave, pulley, 182
Shielding, magnetic, 385
Shutter, camera, 324
Shock wave, 273
Signed numbers, 432
Siemen unit, *10*
Sign convention:
  lens, 318, **320**
  mirrors, 301, 306
Significant figures, 5, **436**
Simple machines, 172
Sine, 42, 46
SI prefixes, **11**, *12*, 17, **446**
SI rules, 445
SI units, 7, **8**, **17**, 445
Sliding friction, *87*, 92
Slip ring, 405
Slope, graph, 32
Slug unit, 15, 77, *442*
Snell's law, **310**, 328
Solar eclipse, 284
Solenoid, 389
Solid, 2, 197
Solid angle, 9
Sonar, 276
Sound, **267**, 277
  energy, 275
  intensity, **275**, 278
  interference, 274
  loudness, 275
  pitch, 276
  pressure, 275
  reflection, 276
  refraction, 273
  reproduction, 277
  reverberation, 276
  sonic boom, 273
  sources, 267
  spped, **272**, *273*, 278
Sound level meter, 275
Sources:
  of emf, 370
  light, 285
  sound, 267
Specific gravity, *38*, **40**, 209, 213
Specific heat capacity, *10*, **224**, *225*, 233
Specific heat of combustion, **231**, *232*, 234
Specific latent heat, *225*, **228**, 233
Specular reflection, 294
Speed, *10*, **55**, 61, *442*
  electromagnetic radiation, light, 281, 308, *447*
  sound, **272**, *273*, 278
Spherical aberration, 304, 306, 326, 328
Spherical lens, 316
Spherical mirror, 296
Spring constant, 81

Spur gear, 188, *189*
Standard atmosphere, 199, *447*
Standard position, 51, 61, 103, 105, 109
Standards, **5**, 8
Standing waves, 264, 265, 268, 269
Static electricity, 330
Static equilibrium, 114
Static friction, 86, *87*, 92
Stationary waves, **264**, 265
Stator, 395, 407
Steady current, 348
Steady flow, 211
Step-up and step-down transformers, **403**, 409
Steradian, 9
Storage cell, 370
Straight line kinematics, 64
Streamlines, 211
Stress, thermal, 236, 237
Sublimation, 230
Substation, transformer, 429
Subtraction of vectors, 108, 109
Sun dimensions, *447*
Superposition theorem, 264, 265
Supersonic speed, 273
Supplementary units, 9
Surface area, 24
Symbols, electric circuit, *349*

T

Tangent, 42, 46
Tangential acceleration, **163**, 169
Tangential force, 125
Technical mathematics, 432
Telescope, 325, 326
Television wave, 281
Temperature 9, *11*, 215, **218**, 233
  absolute, 9, 216, 244, **444**
  conversions, 216
Temperature scales, 215, 216
  Celsius (Centigrade), 215
  Fahrenheit, 215
  kelvin, 9, **216**, 244
  Rankine, 216
Tension, force, 82
Tera, *12*
Terminal point, vector, 50
Terminal potential difference, **372**, 378
Terminal velocity, 70
Terrestrial telescope, 326
Tesla unit, *10*, **386**, 396
Theodolite, 326
Thermal conductivity, *10*, 220
Thermal conductor, 219, **220**, 233
Thermal energy, 2, 217, **218**, 233
Thermal expansion, 236
Thermal insulator, 219
Thermal resistance, 221
Thermal stress, 236, 237
Thermocouple, 376
Thermodynamics, first law, **218**, 233

Thermodynamic temperature, 218
Thermoelectricity, 376
Thermometers, 241
Thermostat, 238
Third law of motion, 81, **91,** 92, 198
Thrust, 114
Time, 9, *442,* 444
Ton, metric (tonne) *11, 442*
Toroid, 389
Torque, *10,* **124,** 132, 166, 394
Torque, gears, 187, *189*
Total internal reflection, 312
Total luminous flux, **287,** 291
Transducer, 277
Transformer, **402,** 409
Transformer losses, 428
Transformer stations, 429
Transient current, 348
Translational motion, 164
Translucent, 283
Transmission line, **426,** 430
Transparent, 283
Transverse wave, **253,** 281
Trigonometry, **41,** 102, 117
Trough, wave, 253, 258
Truss, 45
Turret nose, 325

U

Ultrasonic wave, 277
Umbra, 283
Uncertainty, 5
Unified mass unit, *442,* 447
Uniform circular motion, 154
Uniformly accelerated motion, **64,** 67, 73
Unit, 1, **5,** *8, 9, 10, 11,* **17,**
  base, **8,** *9*
  conversion factors, 13, 16
  derived, 8, **9,** *10,* **17**
  division, 11
  Engineering 8, **15,** 17
  International (SI), 7, **8, 17**
  multiplication, 11
  prefixes, *11, 12,* 17
  supplementary, 9
  systems, 8
  United States Customary (UCSC), 8, 15, 17
Universal gas constant, 447

V

Van de Graaf generator, 337
Vapor, 2
Vapor pressure, 230
Vaporization, 229
Vaporization, specific latent heat, *225,* **229,** 233
Variable, **19,** 31
Vectors, **50,** 61, 70, 73
  addition, 59, 95, 98, **105,** 106,115
  components, **101,** 105, 109, 117
  direction, 54
  magnitude, 50, 61, 95
  negative, **60,** 101, 108
  parallelogram, **98,** 109
  polygon, **95,** 109, 115
  resolution, 101
  resultant, 59, 61, **95, 101,** 106, 109, 114, 120
  scale diagram, 50
  subtraction, 108, 109
Velocity, *10,* 55, **56,** 61, *442*
  angular, *10,* **162,** 169
  average, 64
  elapsed time graph, 65
Velocity ratio, 174
Vernier calipers, 6
Vertex, mirror, 297
Vibrations, **154,** 250
  air columns, 267, **269,** 278
  forced, 271
  mode, **267,** 270
  phase, 255
  rods, 267
  strings, 267, **268,** 278
Virtual image, 296, 306, 318
Virtual principal focus, 316
Voltage, effective, **412,** 430
Voltage, peak, 412
Voltage drop, 352
Voltmeter, 352
Volt unit, *10,* **342,** 344, 348, **350, 351,** 365
Volume, *10, 11,* 441
Volume expansion coefficient, **240,** 241, 247

W

Watt unit, *10,* **146**
Wave, 250, **252**
  characteristics, 255
  compression, 254, 273
  diffraction, 260
  electromagnetic, 250
  energy, 250
  frequency, 255
  interference, 261
  light, 281
  longitudinal, 254
  period, 255
  phase, 255
  reflection, **259,** 265
  refraction, **256,** 265, 281, 291
  relationship, **256,** 265, 281, 291
  sound, **267,** 277
  speed, 260
  standing, **264,** 265
  superposition, **261,** 265
  transverse, 253
  ultrasonic, 277
Wavefront, 257
Wavelength, **255,** 265, 282, 291
Weber unit, *10*
Wedge, simple machine, 191, **192**
Weight, 2, *10,* 15, **79, 80,** 82, 92, 128
Weight density, *10,* **38,** 46
Welding, electric, 367
Weston differential pulley, 183
Wheel and axle, 180, 194
Wide angle mirror, 303
Wind generator, 426
Wimshurst static machine, 337
Work, *10, 15,* **135,** 137, 149, 151, 218, *442*
Worm gear, 188, *189*

X

X-ray, 281

Y

Year unit, *11, 442*

Z

Zero exponents, 434